1st International Conference on Mechanical Engineering Research

(ICMER2011)

AA002777

IOP Conference Series: Materials Science and Engineering Volume 36

Pahang, Malaysia
5 – 7 December 2011

ISBN: 978-1-63439-847-3
ISSN: 1757-8981

1st International Conference on Mechanical Engineering Research

(ICMER2011)

IOP Conference Series: Materials Science and Engineering Volume 36

Pahang, Malaysia
5 – 7 December 2011

ISBN: 978-1-63439-847-3
ISSN: 1757-8981

IOP Institute of Physics

Printed from e-media with permission by:

Curran Associates, Inc.
57 Morehouse Lane
Red Hook, NY 12571

Some format issues inherent in the e-media version may also appear in this print version.

Copyright© by the Institute of Physics
All rights reserved.

Printed by Curran Associates, Inc. (2015)

For permission requests, please contact the Institute of Physics
at the address below.

Institute of Physics
Dirac House, Temple Back
Bristol BS1 6BE UK

Phone: 44 1 17 929 7481
Fax: 44 1 17 920 0979

techtracking@iop.org

Additional copies of this publication are available from:

Curran Associates, Inc.
57 Morehouse Lane
Red Hook, NY 12571 USA
Phone: 845-758-0400
Fax: 845-758-2634
Email: curran@proceedings.com
Web: www.proceedings.com

TABLE OF CONTENTS

PREFACE

The year 2010 represented a significant milestone in the history of the Mechanical Engineering community with the organization of the first and second national level conferences (National Conference in Mechanical Engineering for Research, 1st and 2nd NCMER) at Universiti Malaysia Pahang on 26–27 May and 3–4 December 2010. The conferences attracted a large number of delegates from different premier academic and research institutions in the country to participate and share their research experiences at the conference. The International Conference on Mechanical Engineering Research (ICMER 2011) followed on from the first and second conferences due to good support from researchers. The ICMER 2011 is a good platform for researchers and postgraduate students to present their latest finding in research. The conference covers a wide range of topics including the internal combustion engine, machining processes, heat and mass transfer, fuel, biomechanical analysis, aerodynamic analysis, thermal comfort, computational techniques, design and simulation, automotive transmission, optimization techniques, hybrid electric vehicles, engine vibration, heat exchangers, finite element analysis, computational fluid dynamics, green energy, vehicle dynamics renewable energy, combustion, design, product development, advanced experimentation techniques, to name but a few.

The international conference has helped to bridge the gap between researchers working at different institutions and in different countries to share their knowledge and has helped to motivate young scientists with their research. This has also given some clear direction for further research from the deliberations of the conference. Several people have contributed in different ways to the success of the conference. We thank the keynote speakers and all authors of the contributed papers, for the cooperation rendered to us in the publication of the CD conference proceedings. In particular, we would like to place on record our thanks to the expert reviewers who have spared their time reviewing the papers. We also highly appreciate the assistance offered by many volunteers in the preparation of the conference proceedings. All papers in ICMER 2011 have the opportunity to be published in *IOP Conference Series: Materials Science and Engineering*, (indexed by Scopus, Ei Compendex, Inspec), *International Journal of Automotive and Mechanical Engineering* (IJAME) and *Journal of Mechanical Engineering and Sciences* (JMES).

Professor Dr Hj Rosli Abu Bakar
Chairman
ICMER 2011

Welcoming speech from Dean Faculty of Mechanical Engineering, UMP

In the Name of Allah, the Most Beneficent, the Most Merciful.

It is with great pleasure that I welcome the participants of the International Conference of Mechanical Engineering Research 2011. The Prophet Muhammad (peace be upon him) said 'Acquire knowledge and impart it to the people.' (Al Tirmidhi). The quest for knowledge has been from the beginning of time but knowledge only becomes valuable when it is disseminated and applied to benefit humankind. It is hoped that ICMER 2011 will be a platform to gather and disseminate the latest knowledge in mechanical engineering. Academicians, Scientist, Researchers and practitioners of mechanical engineering will be able to share and discuss new findings and applications of mechanical engineering. It is envisaged that the intellectual discourse will result in future collaborations between universities, research institutions and industry both locally and internationally. In particular it is expected that focus will be given to issues on environmental and energy sustainability.

Researchers in the mechanical engineering faculty at UMP have a keen interest in technology to harness energy from the ocean. Lowering vehicle emissions has been a primary goal of researchers in the mechanical engineering faculty and the automotive engineering centre as well including developing vehicles using alternative fuels such as biodiesel and renewable sources such as solar driven electric vehicles.

Finally I would like to congratulate the organizing committee for their tremendous efforts in organizing the conference. As I wrote this in the Holy Land of Makkah, I pray to Allah swt that the conference will be a success.

Prof. Dr. Zahari Taha CEng, MIED, FASc
Dean, Faculty of Mechanical Engineering
Universiti Malaysia Pahang

Message from Vice Chancellor, UMP

Assalamualaikumwarahmatullahiwabarakatuh and Salam i Malaysia

First and foremost, I want to thank the International Conference Mechanical Engineering Research (ICMER) organisers for inviting me to address and officiate at this conference. It is a privilege and an honour for me on this momentous occasion to grace the ceremony. The ICMER provides a platform to bring together not only researchers but also postgraduate students in Mechanical Engineering, Automotive Engineering, Manufacturing Engineering, Biomechanical Engineering, Material Engineering and Industrial Engineering. With this platform, ICMER will embark on a whole process of making new discoveries and then translating them into products and services for the marketplace; this is only made possible by people like all of you. It might be only a starting point but with hard work and perseverance I am sure you will succeed with flying colours. As one of Malaysia's Public Universities, UMP's main challenge is to remain competitive and relevant by offering high quality technical academic programmes and research activities, focusing on its niche areas. New knowledge and findings cannot be generated without research and development (R&D); therefore, Malaysia has had substantial investment in research and development facilities. These efforts will undoubtedly generate lots of interesting results and new knowledge as either further investigation or commercial activities. Therefore, researchers like you must see this as the generator of new knowledge to extend your research outcomes from laboratory experiments to the marketplace and towards commercialisation. Naybe this doesn't appear significant in the short term but it may make a tremendous impact in the future.

The Malaysian government has invested a huge sum of Ringgits in R&D over the years. Therefore, public universities such as UMP must produce more quality researchers and graduates to ensure Malaysia reaps the returns from these investments and consequently progressive economic growth for the country. Commensurate with these efforts, the results of research must be commercialised effectively. That is why our 2nd KRA UMP's 2011–2015 Strategic Plan determines that we strengthen and sustain its financial support by allocating research grants and industry collaboration and consultations through its business and commercialisation unit. Commercialisation of research outcomes doesn't happen naturally. It may need many man-hours and effort to be invested rather than just purchasing new tools and equipment for laboratories. Moreover, it would be unfortunate if many research results remain in the form of technical publications and scientific journals rather than actual products. These may of course be valuable in their own right but the impact on society and economies must be the main agenda and one of the objectives of the research. The process of taking technology from laboratories to the marketplace will face challenges and pitfalls. In that case we as researchers must be responsible and willing to create a supportive environment for those who undertake this risky business of commercialising R&D especially for innovative entrepreneurship that carries out a new technological approach out from the laboratories to high-tech ventures.

We have achieved many accolades from numerous competitors since we first embarked on our journey to make R&D activities play a more significant role in the economy in

Malaysia. Therefore, we must keep up the pace of this journey and maintain our investment in R&D despite facing the unfavourable global financial situation because ultimately we must to remain a top and competitive player in the global economy. This can be done by continuing to produce high quality research and technology which will set a milestone in transforming our economy. On behalf of UMP, I would like to express my appreciation to the all the committee members from the Faculty of Mechanical Engineering and Automotive Engineering Centre for their hard work and relentless effort. Without their commitment and contributions, this event would not have been possible at this time. And with that, thank you again and wassalam.

YH Profesor Dato' Dr Daing Nasir Ibrahim
Vice Chancellor
Universiti Malaysia Pahang

Committee for International Conference on Mechanical Engineering Research (ICMER 2011)

Scientific Advisory Committee:
1) Prof. Dr. Ahmad Kamal Ariffin (UKM)
2) Prof. Dr. Hj. Rosli Abu Bakar (UMP)
3) Prof. Dr. Hanafi Ismail (USM)
4) Prof. Ir. Dr. Mohd Jailani Mohd Nor (MoHE)
5) Prof. Dr. Zahari Taha (UMP)
6) Prof. Dr. Masjuki Haji Hassan
7) Prof. Ir. Dr. Ramesh Singh (UNITEN)
8) Prof. Dr. Razali Ayob (UTEM)
9) Prof. Dr. Wan Khairuddin (UTM)
10) Prof. Dr. Sulaiman Hj. Hasan (UTHM)
11) Prof. Dr. Zuraidah Mohd. Zain (UniMAP)
12) Prof. Dr. Horizon Gitano (USM)
13) Prof. Dr. K.V Sharma (UMP)
14) Prof. Dr. Shahrani Anuar (UMP)
15) Assoc. Prof. Dr. Abd Rashid Abd. Aziz (UTP)
16) Assoc. Prof. Dr. Aidy Ali (UPM)
17) Assoc. Prof. Dr. Saidur Rahman (UM)
18) Assoc. Prof. Dr. Md Abdul Maleque (UIA)

Organizing Committee
Chairman: Prof. Dr. Hj. Rosli Abu Bakar
Co-Chair: Prof. Dr. Zahari Taha
Co-Chair: Prof. Ir. Dr. Jailani Salihon
Secretary: Dr. Rizalman Mamat

Committee on Keynote Speaker
1) Kumaran Kadirgama (Chair)
2) Prof. Dr. K.V. Sharma
3) Haji Amirruddin Abdul Kadir
4) Miminorazeansuhaila Loman
5) Mohd Akramin Mohd Romlay

Technical Committee (Peer Review & Proceedings)
1) Dr. Abdul Adam Abdullah (Chair)
2) Dr. Ahmad Razlan Yusoff
3) Mohd Yusof Taib
4) Dr. Md. Mustafizur Rahman
5) Dr. Hjh. Yusnita Rahayu
6) Dr. Gigih Priyandoko
7) Dr. Agung Sudrajad
8) Muhammad Hatifi Mansor
9) Mohd Fadzil Abdul Rahim

Technical Committee (Panels & Session Chairs)
1) Dr. Mahadzir Ishak (Chair)
2) Prof. Dr. Shahrani Anuar
3) Dr. Maisara Mohyeldin Gasim Mohamed
4) Muhammad Ammar Nik Mu'tasim
5) Ahmad Basirul Subha bin Alias

Technical Committee (Journal Publication)
1) Dr. Ahmad Razlan bin Yusoff (Chair)
2) Mohd Yusof Taib
3) Dr. Mahadzir Ishak
4) Dr. Abdul Adam Abdullah
5) Hj. Amirruddin Abdul Kadir
6) Hadi Abdul Salaam

Bureau of Publicity & Website
1) Dr. Muhamad Arifpin Mansor (Chair)
2) Amir Abdul Razak
3) Idris Mat Sahat
4) Prof. Dr. Hj. Rosli Abu Bakar
5) Muhamad Zuhairi Sulaiman
6) Dr. Sugeng Ariyono
7) Asnul Hadi Ahmad
8) Mohd Tarmizy Che Kar
9) Mohd Padzly Radzi

Bureau of Special Task & Poster
1) Lee Giok Chui (Chair)
2) Mohd Fadzil Abdul Rahim
3) Che Ku Eddy Nizwan
4) Hazami Che Hussain
5) Mohd Fazli Ismail
6) Mahdhir Mohd Yusof
7) Mohd Padzly Radzi
8) Rahimah Che Ramli

Secretariats
1) Ir. Ahmad Rasdan Ismail (Chair)
2) Mohd Shahri Mohd Akhir
3) Luqman Hakim Ahmad Shah
4) Juliawati Alias
5) Nurazima Ismail
6) Mohamad Faizal Mohamed Zahri
7) Raja Allen Jordan Izzuddin Shah
8) Rosidah Mohd Norsat
9) Norshalawati Mat Yusof

10) Zainab Daud
11) Nur Sufiah Jamaludin
12) Azslinda Ibrahim
13) Nurul Azreen Zainal Abidin
14) Nurul Ashikin Mohd Khalil
15) Mohd Zaki Mohd Ali

Investigation of geometrical defect in cold embossing of a pin head

A B Abdullah[1,2], S M Sapuan[2], Z Samad[2], H M T Khaleed[2], and N A Aziz[2]

[1] School of Mechanical Engineering, Universiti Sains Malaysia, Engineering Campus 14300 Nibong Tebal, Penang, MALAYSIA
[2]Department of Mechanical and Manufacturing Engineering, University Putra Malaysia 43400 Serdang, Selangor, MALAYSIA.

E-mail[1]: mebaha@eng.usm.my

Abstract. Under-filled in one of the major geometrical defect can be found in cold forging part. For net-shape manufacturing, defect will influence the assembly accuracy and tend to affect the performance of the system. Therefore, it must be predicted and minimized as early soon as possible before the manufacturing begins. The paper intents to investigate the geometrical defect of the cold embossing pin head. The pin head accuracy is very important in the propeller blade assembly. In this work, the effect of punch and pin head geometries to the formation of defect was studied using commercially software namely DEFORM-2D. The result is validated through experimental works.

Keywords. Geometrical defect, cold embossing, filling-ability

1. Introduction

Net-shape manufacturing must be able to prevent or minimize secondary processes such as machining, therefore, it is essential to predict and prevent the defect that probably occur as early as possible. Cold forging governs by many factors such as friction, part geometry, die shape, and temperature setting of the die and workpiece, which accumulatively tends to cause defects. Arentoft and Wanheim classified forging part defect into six: fold, shear defect, surface defect, form defect, crack, and structural defect [1]. Lin and Lin studied the folding defect during the cold extrusion of aluminum [2]. The relationship between defect and geometrical shape were studied based on the radius ratio. A similar approach was employed to investigate the barreling formation during cylinder upsetting with hollow die [3]. Metal flow imperfection related defect has attracted many researchers. For example, Chan proposed a dynamic change of tooling geometry to control the most common flow-induced defect such as folding [4]. In other case, Yilmaz conducted an experimental investigation of conventional closed-die forging to study the effect of aspect ratios and indentation to the forging load and die filling [5]. Similar effort has been made by Khaleed et al., in cold forging of flashless connecting rod [6]. Lu and Zhang claimed that forging defects are mainly caused by improper die dimension [7]. Their study was carried out by taking into consideration the temperature dependency of the thermal and mechanical properties of materials. Abdullah et al., evaluated the geometrical error in cold forging of Autonomous Underwater Vehicle (AUV) using optical technique and found that large twist springback is become one of the major cause of inaccuracy [8]. Song and Im studied the effect of process design parameters to under-

Published under licence by IOP Publishing Ltd

filling and folding defect during closed-die forging of bevel gears [9]. Qamar investigated the effect of shape complexity and dead metal zone to the quality of cold extruded part [10]. In other case, Shen and Yu introduced a new type of die design to solve the folding defect during the orbital forging process [11]. The new design has influenced the flow of the preform, resulting in lower tool stresses and preventing the folding problem. Su et al. studied the material flow behavior of magnesium alloy in the press forging of notebook cases [12]. They studied the methods to prevent the material from flowing outwards during the formation of hook, sidewall and boss.

The paper intends to investigate the defect formation in ensuring accurate assembly of the blade to the hubs. The paper begins with the introduction, and then followed by methodology of the study. Afterward, the result is discussed, and the paper ends with conclusions.

2. Methodology

The 2D model of the simulation is as shown in the figure 1(a). The model consists of a workpiece, punch, top die and bottom die. Both dies are considered as rigid body. The workpiece was placed in between the dies, before the embossing take place. In this process, the workpiece is press by a punch at constant speed of 250 mm/min. The initial thickness of the AA6061 workpiece is 3.0 mm. Since the process does not use any lubricant and the material to be formed is aluminum, so the friction coefficient is assume to be 0.4, the initial temperature of the work-piece, punch and die is at the room temperature or 25°C. In this study, the rigid-plastic finite element method was applied in the analysis. The material properties were listed in the table 1 and the punch and die is made from D2 tool steel.

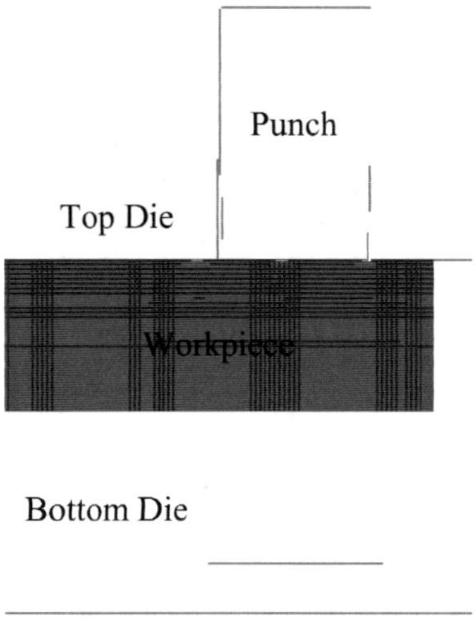

(a)

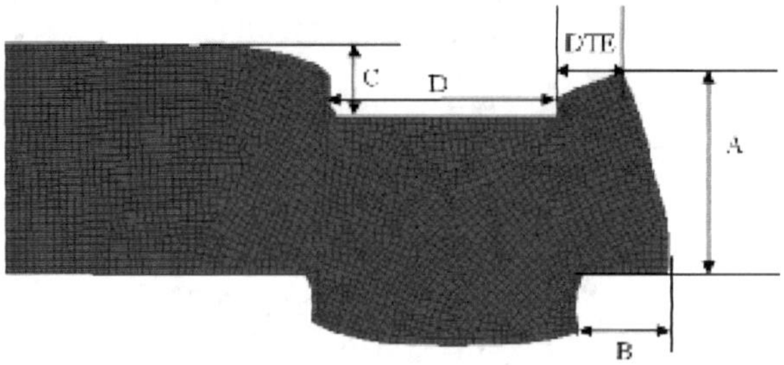

(b)

Figure 1. (a) the FE model of the embossing process at initial position and (b) the resulted embossed pin head and the geometries to be measured.

Table 1. Material properties of the workpiece

Parameter	Workpiece
Material type	AA 6061
Young's Modulus (GPa)	70
Poisson's ratio	0.35

In this study, two major defects are considered i.e. un-filling region and bulging. The un-filling region is quantifies based on filling ratio. The higher the ratio, the higher unfilled region depicts the larger defect. Filling ratio can be obtained from the ratio of unfilled region to the fully-filled region or $AR = \frac{\text{Unfilled Region}}{\text{Fully-Filled Region}}$. For consistency, a simple programm has been developed using Matlab. The image of the embossed pin head will be captured from FE simulation, and the geometry of the part will be determined using ordinary programming developed using Matlab. Similarly for the bulging defect, as shown in figure 1(b), which was measured from the image of the forged pin head obtained from the FE result. Here two features were measured i.e. A and B.

3. Results and Discussion

The pin head is one of the main features located at the head of AUV propeller (Figure 2). The pin head feature is intentionally designed to ensure the blade to be firmly fixed to the hubs. From the research, it is found that he geometry accuracy are depends of embossing depth (C), punch diameter (D), and the distance to edge (DTE), which can be referred in figure 1(b). To study the effect of these features on the embossed pin head, a parametric study was performed. The filling ratio of 1 depicts fully filled pin head or no defect, whereas lower ratio represent unfilled region i.e. defective part.

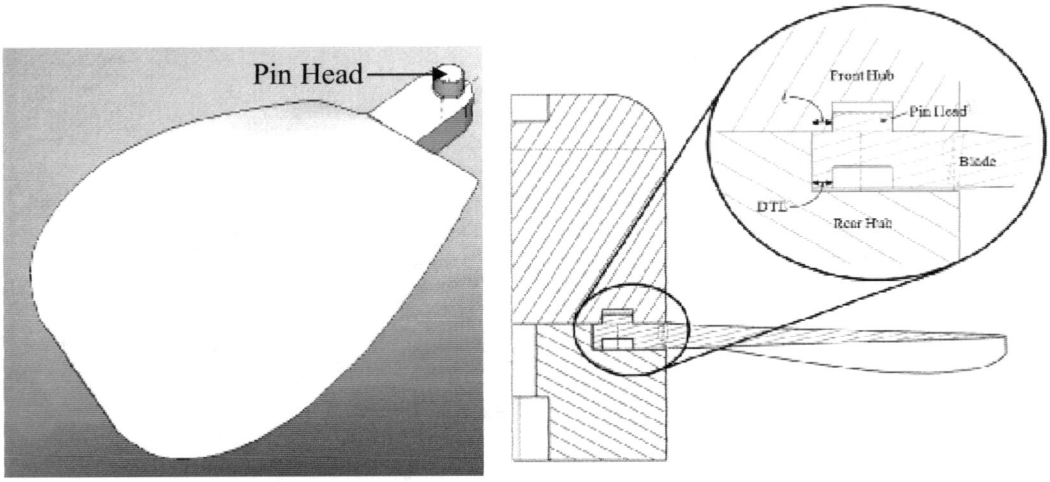

Figure 2. Propeller blade and final assembly of the AUV propeller

As a result, increasing the punch diameter will cause in less unfilled area as shown in figure 3(a) or the higher diameter, the higher the ratio of filled to unfillled area. In other case, the effect of embossing depth to the filling ratio is found to be less compared to the punch diameter. For example in figure 3(b), by increasing the 1 mm depth, i.e., from 1.5 to 2.5 mm, the increment is only about 0.04%. Significant effect can be seen for the DTE, as shown in figure 3(c). Where approximately 3% increment found by increasing the distance by 0.3 mm, i.e., from 0.7 to 1.0 mm. The effect of DTE was further investigated until distance of 2.0 mm, and the filling region for this case is found to be almost perfect at approximately 99.2% filled. Based on the result, it is clear that the DTE is the most significant parameter to the fillingability of the embossed pin head.

In other case, the width of the bulging was estimated. Figure 4 shows the result of DTE and effect to the bulging width. In this case, only DTE were studied and the effects of DTE up to 3.8 mm were observed. As the DTE increases, the lesser the size of bulging had been predicted. Based on the FE simulation results depicted a significant effect of DTE to the size of bulging or in other words by increasing the DTE, the effect of bulging can be clearly seen. It can be predicted that, there is no bulging occur as DTE is projected approximately at 3.8 mm.

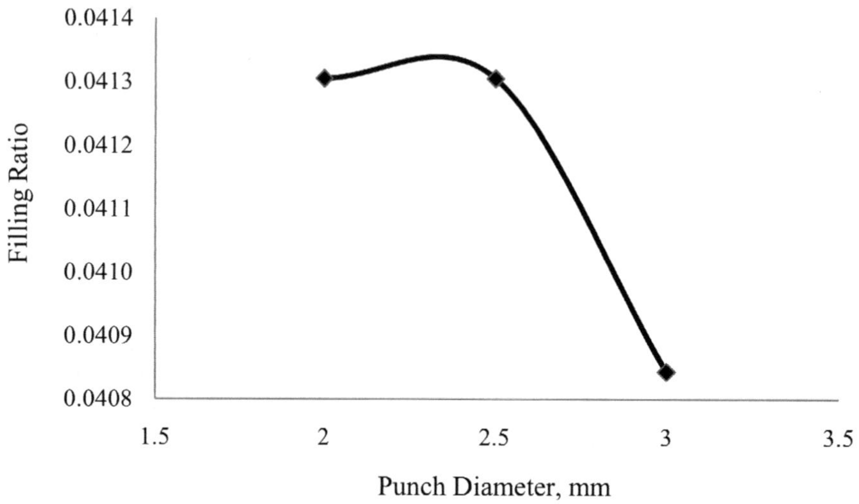

(a)

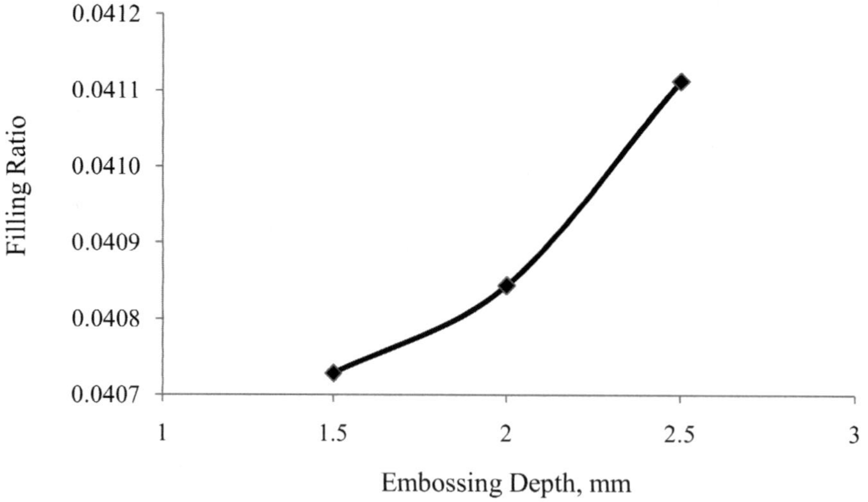

(b)

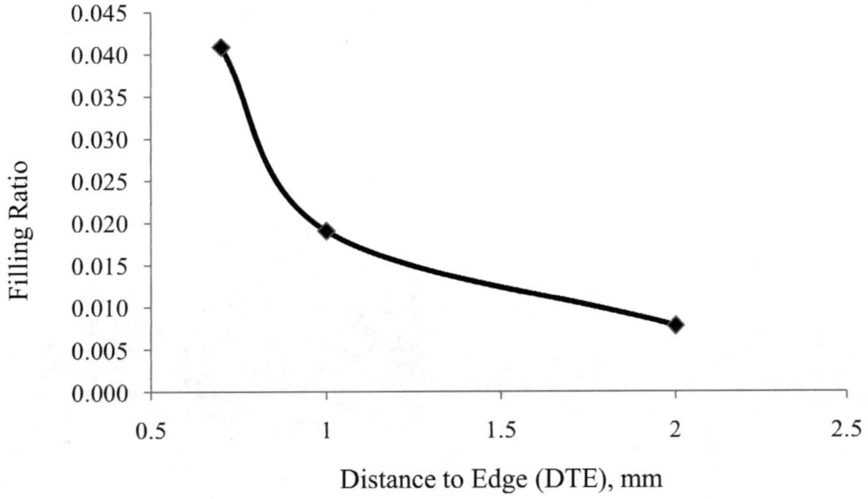

(c)

Figure 3. The effect of (a) punch diameter, (b) embossing depth and (c) distance to edge to the filling ratio

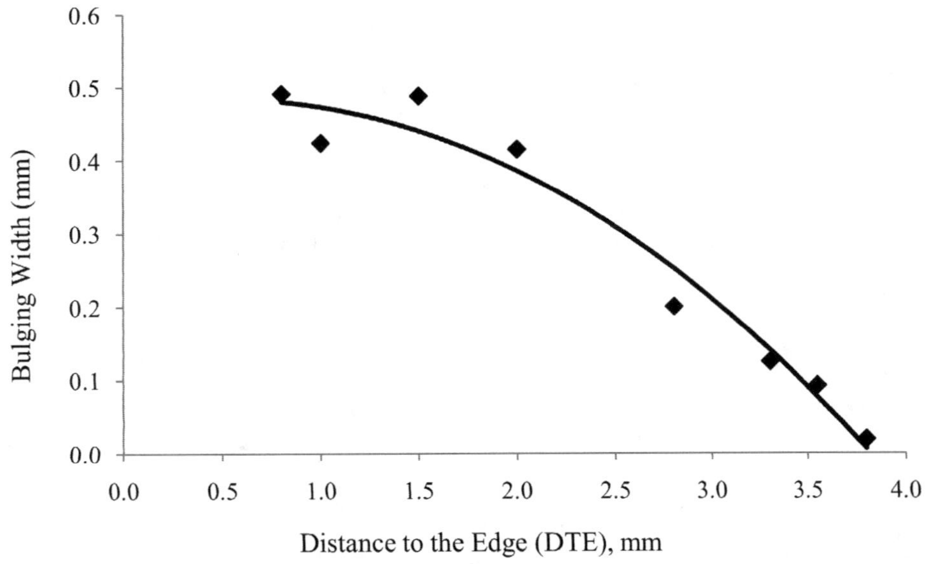

Figure 4. The effect of DTE to the width of the bulging

For verification, a comparison between the estimated and measured loads were performed, as shown in figure 5. The measured loads were obtained using Universal Tensile Machine. The maximum load in both approaches show good agreement. The load slightly different because of the condition of both approach. In the simulation, there is no gap between the punch and workpiece, while in the experiment, the gap cannot be controlled. Similar case occur at the stroke more than 1.6 mm, the material tend to fail due to shearing in the experiment, therefore the load suddenly drop. This is because, the materil in the experiment is more brittle compared than the material used in the simulation.

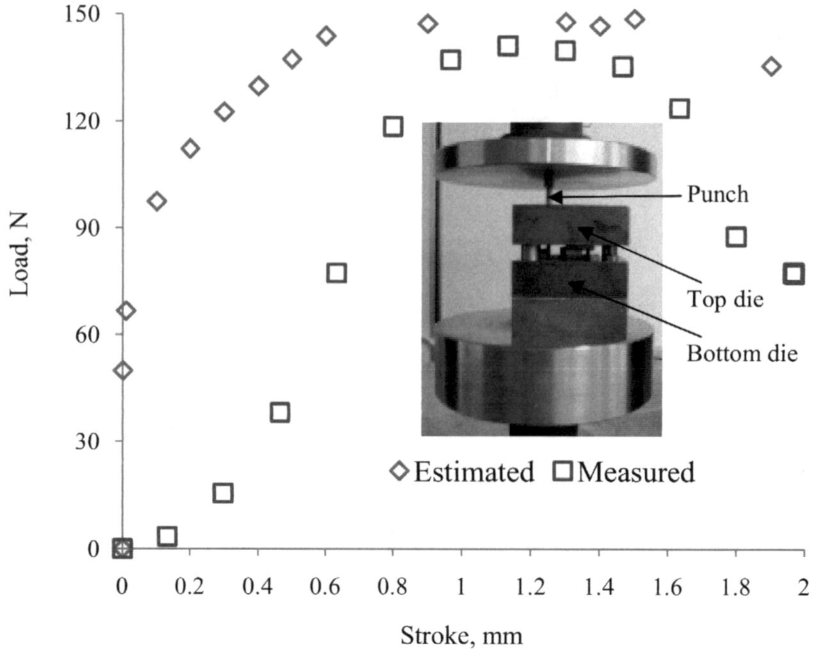

Figure 5. The comparison between the estimated and measured load

4. Conclusion

The main objective of the paper is to investigate the geometrical defect of the cold embossing part. The 2D FE analysis was performed to investigate the mechanism of the cold embossing process. In this paper, the effects of the few design parameters were studied including the distance to the edge (DTE) to the formation of the geometrical defect in terms of under-filling and bulging. From the simulation depict that, the most influencial parameter is the distance to edge. For validation of the FE result, comparison of the pin head geometries was performed and the result found an agreement between the FE result and the forged pin head.

Acknowledgement

The authors want to acknowledge the Universiti Sains Malaysia for providing the grant through Research University Research Grant Scheme.

References

[1] Arentoft M and Wanheim T 1997 *J. Mat. Process Technol.* **69** 1-3 227
[2] Lin F C and Lin S Y 2002 *Comp. & Struc.* **80** 24 1817
[3] Lin F C and Lin S Y 2003 *Finite Element in Anal. & Des.* **39** 4 325
[4] Chan W L, Fu M W, Lu J and Chan L C 2009 *J. Mater. Process Technol.* **209** 11 5077
[5] Yilmaz C, Tahir A M and Erol A H 2004 *Indian Eng. & Mater. Sci.* **11** 6 487
[6] Khaleed H M T, Samad Z, Othman A R, Mujeebu M A, Abdullah A B, Ab-Kadir A R, Badruddin I A and Salman Ahmed N J 2011 *Arabian J. for Sci & Eng. B: Eng.* **36** 5 855.
[7] Lu C and Zhang L W 2006 *Trans. Nonferrous Met. Soc. China* **16** 6 1386
[8] Abdullah A B, Sapuan S M, Samad Z, Khaleed H M T and Aziz N A 2012 *Adv. Mater. Res.* **383 – 390** 7117.
[9] Song J H and Im Y T 2007 *J. Mater. Process Technol.* **192-193** 1
[10] Qamar S Z 2010 *Mater. & Man. Processes* **25** 1454.
[11] Shen J J and Yu C H 2008 *J. Mater. Process Technol.* **201** 1-3 9.
[12] Sun C, Zhang S H, Tang W D and Wang Z T 2010 *Mater. & Man. Processes* **25** 534

IOP Conf. Series: Materials Science and Engineering **36** (2012) 012002 doi:10.1088/1757-899X/36/1/012002

An experimental study of gaseous exhaust emissions of diesel engine using blend of natural fatty acid methyl ester

Agung Sudrajad[1] , Ismail Ali[1] , Khalid Samo[2] , Danny Faturachman[3]

[1] Faculty of Mechanical Engineering University Malaysia Pahang, Campus Kuala Pahang 26600, Malaysia
[2] Maritime Studies University Malaysia Terengganu, Malaysia
[3] Post Graduate Program, University Malaysia Pahang, Malaysia

E-mail: sudrajad@ump.edu.my , goenkobe@yahoo.com

Abstract. Vegetable oil form in Natural Fatty Acid Methyl Ester (FAME) has their own advantages: first of all they are available everywhere in the world. Secondly, they are renewable as the vegetables which produce oil seeds can be planted year after year. Thirdly, they are friendly with our environment, as they seldom contain sulphur element in them. This makes vegetable fuel studies become current among the various popular investigations. This study is attempt to optimization of using blend FAME on diesel engine by experimental laboratory. The investigation experimental project is comparison between using blend FAME and base diesel fuel. The engine experiment is conducted with YANMAR TF120M single cylinder four stroke diesel engine set-up at variable engine speed with constant load. The data have been taken at each point of engine speed during the stabilized engine-operating regime. Measurement of emissions parameters at difference engine speed conditions have generally indicated lower in emission NO_x, but slightly higher on CO_2 emission. The result also shown that the blends FAME are good in fuel consumption and potentially good substitute fuels for diesel engine

1. Introduction

The environmental concern of the global warming and climate change has greatly increased the interest of study of renewable fuels to internal combustion engine. The sharply rising petroleum price on the markets has revived more and more interest in the use of natural fatty acid methyl ester (FAME) as substitute of fossil fuel. FAME have their own advantages: available everywhere in the world, renewable as vegetables which produce oil, "greener" to the environment as they seldom contain sulphur element them. Diesel engines are preferred prime movers due to their excellent drivability and high thermal efficiency [1]. Despite these advantages they produce higher levels of NOx and smoke which affect human health [2]. The original diesel engine that Rudolph Diesel designed ran with vegetable oil. He used peanut oil to fuel one his engines at the Paris Exposition in 1900 [3]. But soon afterward, the application of vegetable oils as fuel was dropped due to cheap supply of petroleum based fossil fuels available in the world during the last century.

Published under licence by IOP Publishing Ltd

Even many investigations of FAME to substitutes for diesel fuel [6-10], but there are still a lot work that needs to be done to make vegetable oil really apply to diesel engine instead of ordinary diesel fuel. The vegetable oil in this experiment study have a lower calorific value compare to pure diesel fuel. Vegetable oil in this experiment is produce from 100% palm oil and it's called natural fatty acid methyl ester. The same with others vegetable oils, it have disadvantages of high viscosity. Modern diesel engines have fuel injection systems that are sensitive to viscosity changes. High viscosity may lead to poor atomization of the fuel, to incomplete combustion, to coking of the fuel injector, to ring carbonization, and the accumulation of fuel in the lubricating fuel [4]. The objective of this study is to carry out an experimental study to investigate the performance and the exhaust emissions characters of diesel engine fuel with FAME and its blends, compared to those of ordinary diesel fuel.

2. Experimental Procedure

A direct-injection diesel engine (model TF120 YANMAR) one cylinder, four stroke, naturally aspirated, water-cooled, was employed to test whose specifications are shown in table 1. This engine experimental work was operated at 1200, 1400, 1600, 1800, 2000, 2200, and 2400 rpm with constant engine load. The experiment by six kinds of fuel (blends by volume) were tested, i.e. blends of 5% FAME with 95% diesel fuel (B5), 10% FAME with 90% diesel fuel (B10), 15% FAME with 85% diesel fuel (B15), 20% FAME with 80% diesel fuel (B20), 100% FAME (B100) and 100% diesel fuel (DO). The test was conducted and repeated five times for every kind of fuel, in order to increase reliability of the test results. The specification of vegetable oil and pure diesel is shown in table 2.

A hydraulic dynamometer type gear pump was used for engine load. Exhaust gas temperature was measured using a thermocouple located downstream of the exhaust valve. Figure 1 shows the diagram of experimental setup. It consist of a test-bed, a diesel engine, a dynamometer, a fuel tank, a data acquisition system, a computer, an operating panel, exhaust emissions analyzer, and various sensors to measure the exhaust temperature, water cooling temperature, etc.

The experimental works started with a preliminary investigation of engine running on pure diesel in order to determine the engine's operating characteristics and exhaust emissions levels, constituting the base-line that is compared with corresponding cases when using vegetable oil. The same procedure was repeated for each fuel blend by keeping the same operating condition. For every fuel change, the fuel line was cleaned and the engine was left to operate for about fifteen minutes to stabilize at its new condition. The differences in the measured performance and exhaust emission parameters from the baseline operation of the engine were determined and compared.

Table 1. Engine Specification

Engine Type	YANMAR TF120
Number of Cylinder	1
Bore x stroke (mm)	92 x 96
Displacement (L)	0.638
Continuous Output (HP)	10.5 HP at 2400 rpm
Rated output (HP)	12 HP at 2400 rpm
Cooling system	Hopper
Dry weight (kg)	102

Table 2. Fuel Characteristics

Fuel Type	DIESEL OIL	FAME
Density @15^{0}C, kg/L	0.87	0.874
Viscosity @40^{0}C,	4.0	4.4
Flash Point , ^{0}C	91.0	154
Sulphur , wt%	0.3	0.2
Carbon Residue , wt%	<0.1	<0.01
Cetane Index	55	67.5
Ash content, wt%	0.002	0.01
Copper corrosion	1a	1a

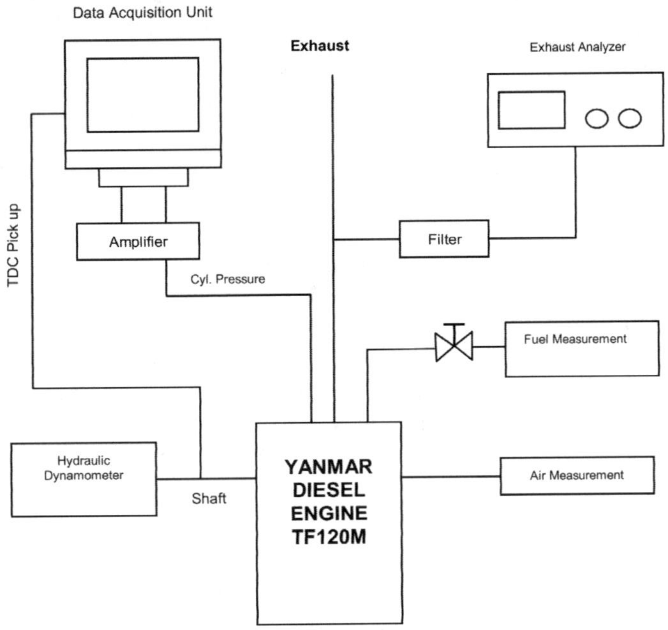

Figure 1. Experimental Diagram

3. Results and Discussions

The diesel engine ran well on all the fuels mentioned above. The performance of engine is greatly influenced by sources of bio-diesel, for example engine fuelled with palm oil bio-diesel is more efficient than bio-diesel produced from tallow and canola oil [5]. Bio-diesel is likely produce less power with high fuel consumption than diesel as the gross calorific value (energy content) of FAME is lower than pure diesel fuel. The calorific value of experimental fuels is shown in fig. 2.

Fig. 3 shows the brake torque results for experiment. The torque increases with the increase of engine speed until the limitation of speed at 1800, and decrease starting from 2000 rpm engine speed until maximum speed. From the figure also can explained that 100% FAME give result in lowest torque compare to other blends fuel and pure diesel fuel. Decrease in torque is due to their lower energy content of FAME. Lower energy content shows lower energy characteristic.

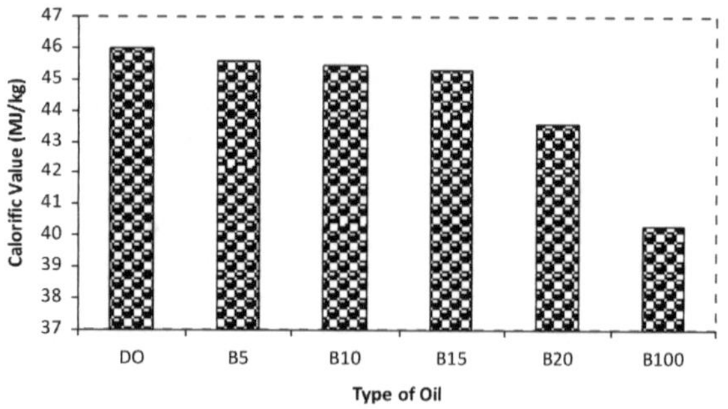

Figure 2. Calorific Value of Fuel Experiment

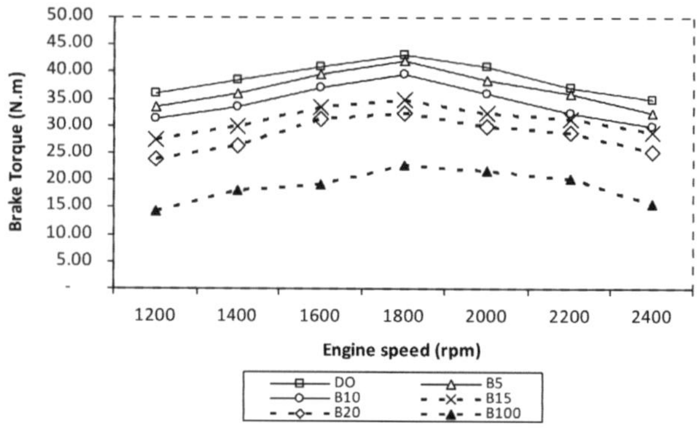

Figure 3. Brake Torque versus engine speed

The comparison of fuel consumption of experimental result is shown in fig. 4. It can be seen that fuel consumption generally increase with blend ratio compare to pure diesel fuel. A fuel consumption increases by 4%. This trend can be explained by lower calorific value of FAME compared with pure diesel fuel. Fig. 5 shows the result of NO_x emission for the all fuel test. NO_x emissions decreased with the increase of engine speed. Thermal NO_x refer to NO_x formed through high temperature oxidation of nitrogen (N_2) in combustion chamber. The formation rate of NO_x is primary function of combustion (flame) temperature, residence time of nitrogen at that temperature, and the contents of oxygen in the reaction regions in the combustion chamber. The NO_x emissions of blended fuels were slightly higher than those of the pure diesel fuel at all engine speed. The higher combustion temperatures and the presence of fuel oxygen in the blended fuels caused higher NO_x emissions. It is well known that the external oxygen supplied within air itself is less effective than the fuel borne oxygen in the production of NO_x. From the fact that when the speed increases, the friction horsepower decreases according to decrease in mechanical efficiency in order to maintain torque output, leading to an increase in the fuel consumption rate. This significant increase of consumption results in lesser evaporation rate which consequently lead to the reduction of NO_x emissions.

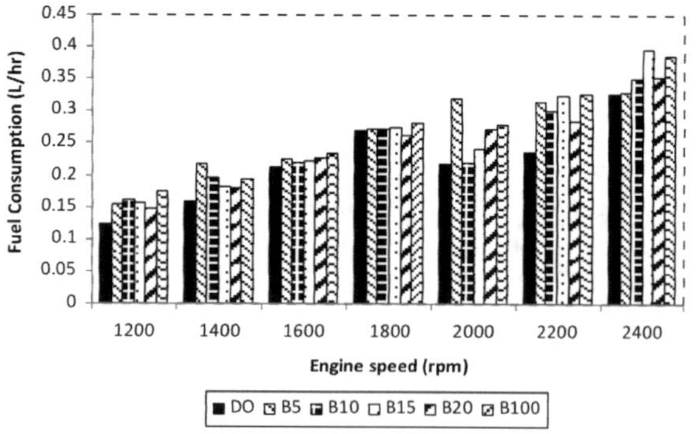

Figure 4. Fuel consumption versus engine speed

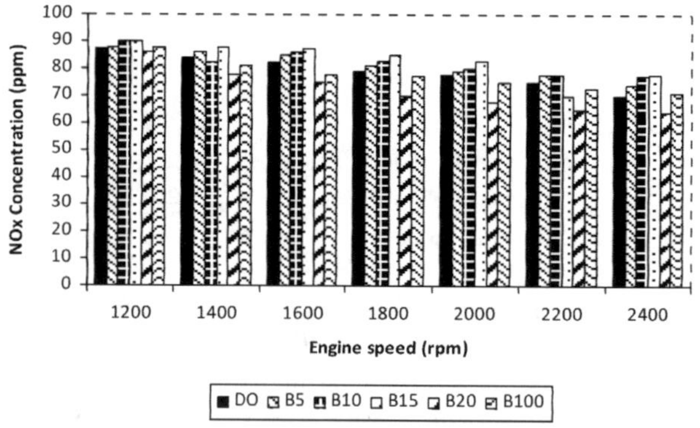

Figure 5. NO_x emission versus engine speed

Fig. 6 shows the CO emissions from the engine at the variable engine speed. The concentration of CO emissions was increase with increase of engine speed. From the figure show that the changed of CO emissions is not significant compare with blends fuel and pure diesel fuel. It can be explain that the in all condition the temperature in cylinder were high, which made the chemical reaction of fuel with oxygen be easier and the combustion become more complete.
Fig. 7 shows the exhaust gas temperatures versus engine speed. The exhaust gas temperatures rise with the increase of engine load for all of the fuel. From that figure can be seen that exhaust temperature at different fuels, are nearly the same at all engine speed condition. Only in the case of lower speed at 1200 rpm with 10% (B10) and 15% (B15) vegetable oil the exhaust temperature slightly higher than the others fuel. The differences are between 10% and 14% compare to the pure diesel fuel.

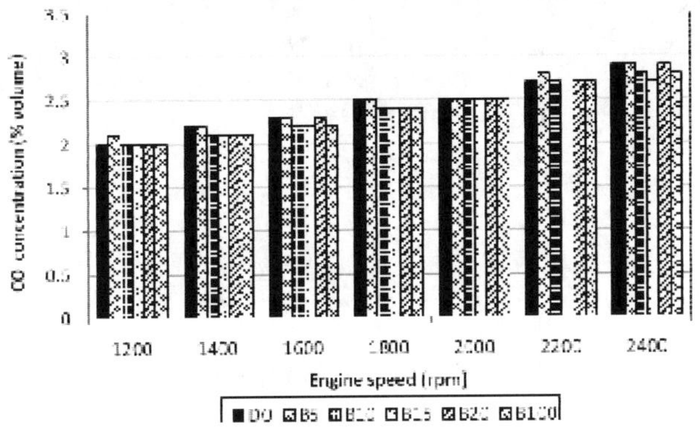

Figure 6. CO emission versus engine speed

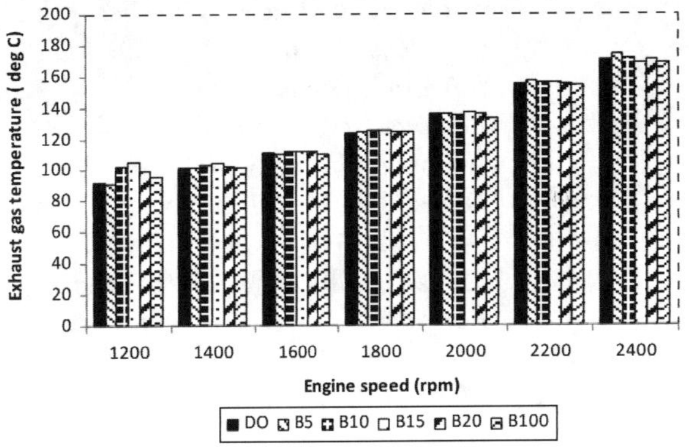

Figure 7. Exhaust temperature versus engine speed

3. Conclusions

This study investigated the use of fatty acid methyl ester (FAME) in unmodified diesel engine, with specific reference to performance and emissions with variable engine speed and blend ratio. The engine torque is lower when the engine fuelled with FAME and its blends compared with that of pure diesel fuel. The fuel consumption is higher when using FAME. The emissions of nitrogen oxide (NO_x) from FAME and its blends, at the range of tests, are generally lower than that pure diesel fuel. This is the most important gaseous emissions characteristic of FAME and its blends. FAME having higher oxygen content can lead less CO emissions with increasing blend ratio due to complete combustion in diesel engine. The results from the experiments prove that the FAME and its blends are potentially good substitute fuels for diesel engine in the near future when petroleum deposits become scarcer.

Acknowledgement

The support of the Faculty of Maritime University Malaysia Terengganu and the support of the University Malaysia Pahang through the Research Grant are all greatly appreciated.

REFERENCES

[1] Heywood J.B, 1988. *Internal Combustion Engine Fundamental*, NY-USA, Mc Graw Hill

[2] Y.D Wang et all, 2006. *An Experimental Investigation of the Performance and Gaseous Exhaust Emissions of Diesel Engine Using Blends of Vegetable Oil*, Journal of Applied Thermal Engineering 26 (2006) 1684-1691

[3] A. Demirbas, 2002. *Biodiesel from Vegetable Oils via Transesterification in Supercritical Methanol* , Energy Conversion and management 43 (2002) 2349-2356

[4] A.S Ramadhas, 2004. *Use of Vegetable Oils as IC Engine Fuels* – A Review, Renewable Energy 29 (5) (2004) 727-742.

[5] P. McCarthy, M.G Rasul, S.Moazzem, 2011. *Analysis and Comparison of Performance and Emissions of an Internal Combustion Engine Fuelled with Petroleum Diesel and Different Bio-diesel*, Fuel 90 (2011) 2147-2157

[6] Herchel T.C Machacon et all, 2001. *Performance and Emission Characteristic of Diesel Engine Fueled with Cocobut Oil-Diesel Fuel Blend*, Biomass and Bioenergy 20 (2001) 63-69

[7] C.D Rakopolus et all, 2006. *Comparative Performance and Emissions Study of a Direcvt Injection Diesel Engine Using Blends of Diesel Fuel with Vegetable Oil or Bio-Diesel of Various Origin*, Energy Conversion and Management 47 (2006) 3272-3287

[8] Jincheng Huang et all, 2009. *Experimental Investigation on the Performance and Emissions of Diesel Engine Fuelled with Ethanol-Diesel Blends*, Applied Thermal Engineering 29 (2009) 2484-2490

[9] Makame Mbarawa, 2010. *The Effect of Clove Oil and Diesel Fuel Blends on the Engine Perfiormance and Exhaust Emissions of a Compression-Ignition Engine*, Biomass and BIoenergy 34 (2010) 1555-1561

[10] Masjuki H, Zaki AM, Sapuan SM, 1993. *Rapid Test to Measure Performance Emissions and Wear of Diesel Engine Fuelled with Palm Oil Diesel*, Journal of American Oil Chemist Society, 1993:70 (10):1021

Design, development and performance of a disk plow combined with rotary blades

A Hashemi[1] , D Ahmad[1] , J Othman[1] and S Sulaiman[2]
[1]Department of Biological and Agricultural Engineering, University Putra Malaysia, Serdang, Selangor,43400, Malaysia
[2]Department of Mechanical and Manufacturing Engineering, University Putra Malaysia, Serdang, Selangor,43400, Malaysia

Email: Ali5_hashemi@yahoo.com,
desa @eng.upm.edu.my

Abstract. Disk plow combined with rotary blades, defined as comboplow, is used for soil preparation for planting. The comboplow includes four units: Chassis, concave disk, transmission system and rotary blades. A multiple tillage operation is reduced in a single pass resulting in a potential reduction of soil compaction, labor, fuel cost and saving in time. The comboplow was tested at University Putra Malaysia Research Park, Serdang, Selangor, Malaysia, on three different plots of 675 m^2 in the year 2010/2011.The treatments were three types of blade [(straight (S),curved (c) and L-shaped)] and three rotary speeds (130,147and 165 rpm). The parameters were Mean Weight Diameter Dry Basis (MWD_d), Mean weight Diameter Wet Basis (MWD_w), Aggregate Stability Index (SI) and Instability Index (II).

Keywords: tillage operation, combined implement, disk plow, Rotary blade

1.Introduction

Plow is as old as agriculture which originated 10–13 million years ago in the Fertile Crescent of the Near East, mostly along the Tigris, Euphrates, Nile, Indus and Yangtze River valleys, and also were introduced into Greece and Southeastern Europe 8000 years ago[1]. *Tillage* is one component in any system of soil management for crop production and is a process of applying energy to the soil to change its physical condition by disturbing it [2] . Tillage or soil preparation has been an integral part of traditional agricultural crop production practice. It is the mechanical manipulation of the soil and plant residue to prepare a seedbed where seeds are planted to produce grain for our consumption. Also tillage breaks soil, enhances the release of soil nutrients for crop growth, destroys weeds and enhances the circulation of water and air within the soil [3] .*Tillage operation* is also defined as a procedure for breaking up soil; the soil failure depends largely upon the soil properties, tool parameters, and cutting speed. Also it is one of the highest power-required processes of the agricultural production. In addition, the high cost of energy, makes the farmers to find alternative economic tillage methods [4]. The cost and the timeliness of operation assume critical importance while deciding the type of tillage tools. The large part of energy consumption in mechanized agriculture is related to tillage operation. Godwin [5] reported that the cost of tillage operation is a vital component to determine farm profitability and recent years have seen a significant move to reduce tillage operation. Operations simultaneously utilizing two or more different types of tillage tools or implements to simplify, control or reduce the number of operations over a field are called *combined tillage* [6], [7]. Combined tillage is the act of using two or more different types of tillage implements or tools at the same time to

Published under licence by IOP Publishing Ltd

manipulate control or reduce the number of operations in the field [8] . Machines for tillage operation usually pass the farm four times or more which causes soil compaction, increases cost of labour and energy. The compression of soil causes reduction of the moisture penetration, soil oxygen capacity, penetration of root in the soil, organic materials capacity in soil and increasing energy consumption [9].The past research work did not consider the combined effect of disk plow and rotary blades on farm. In this regard Ahmad and Amran [10], studied the energy prediction model for disk plow combined with a rotary blade in wet clay soil. In order to produce cheaper agricultural products, it is necessary to reduce expenditure. The concept of comboplow has the following advantages; higher degree of soil crumbling; better mixing of soil and mineral fertilizer; improve parameters of work on heavy soils; the guarantee of complete preparation of the field; reduction of draft and wheel slip; reduces soil compaction, energy consumption, fuel consumption; decreases labor and machinery cost. The Comboplow is simultaneously utilized to prepare an adequate seedbed and bury crop residues. It is particularly adapted for use in hard-dry soils, shrubby or bushy land. It is utilized in clay soil for reduction of energy consumption in tillage operation, soil compaction and structural degradation due to vehicular traffic; because current two-pass or three-pass tillage treatments could be replaced with single-pass. The outstanding benefit with current single-pass tillage treatment is to save fuel, labor and machinery cost.

2. Materials and Methods

2.1 Design of main working components of comboplow

A disk plow combined with rotary blades ‘Figure 1’ consists of the following components; a chassis, three point hitch, transmission system, disk, rear wheel, rotary blades and adjusting mechanism.

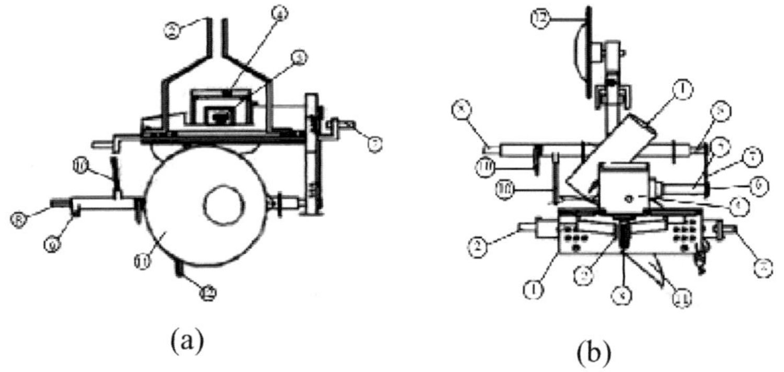

Part number	Part name
1	Chassis
2	Three point hitch
3	Input shaft
4	Gear box
5	Output shaft
6	Sprocket
7	Chain
8	Main shaft
9	Holders of blades
10	Blades
11	Disk
12	Rear wheel

Figure 1. Schematic of Comboplow (a) front view and (b) top view

2.1.1 Transmission system

The transmission system consists of PTO, universal joint shaft with safety clutch, gear box, input shaft, output shaft, chain, sprockets, main shaft, and holder of blades. The comboplow is mounted on a tractor three point linkage system and driven by the tractor PTO. Figure 2 presents the flow chart, schematic diagram and photograph of the transmission system of the comboplow implement. The sizes of the sprocket couples (13:23, 14:22 and 15:21) of the transmission system that give the speed of the rotary blades were selected based on the reported works from the literatures which reported that the speed of the rotary blades should be greater or equal to the travel speed of the tractor [6].

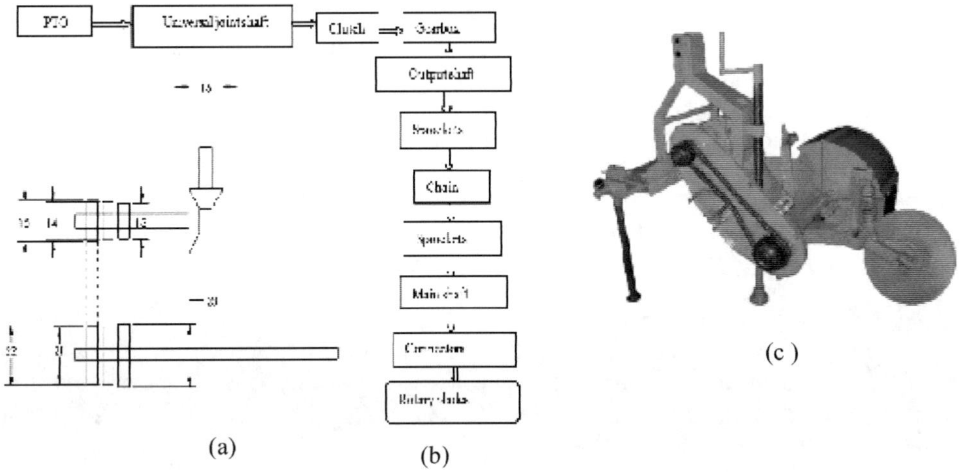

Figure 2. Transmission system of comboplow (a) schematic diagrams (b) flow chart (c) photograph

2.1.2Rotary blades

The most widely used rotary tillers are those having straight (S), curved (C) and L-shaped blades 'Figure 3'. The C type blades provide sliding cutting of the fibrous soil with the following advantages over the other type of blades; minimum resistance, have self-cleaning characteristics during working with damp soils, and also have sufficient resistance to breakage and abrasive wear.

Figure 3. Photograph of blades of comboplow (a) Straight blade, (b) L-shaped and (c) Curved (c)

2.2 Tillage site

The experiments were conducted at University Putra Malaysia Research Park, Serdang Selangor, Malaysia, on three different plots having 675 m^2 sizes each during the year 2010/2011. The site was located at longitude 101°, 42.916′E, latitude 2°, 58.812′N and an altitude of 46 to 48 m above sea level The experimental site has an average annual rainfall of 2548.5 mm with maximum and minimum temperatures of 33.1°C and 23°C respectively.

2.3 Soil properties

Soil samples were collected from the topsoil layer of 0-30 cm depth at nine different locations in the experimental site before tillage operation in three replications. A comprehensive understanding of the physical characteristics (bulk density, moisture content, porosity, particle size, and aggregate size distribution) and mechanical soil properties (Cohesiveness, internal friction angle, shear stress and penetration resistance) are important in the designed and fabricated of disk plow combined with rotary blades.

2.3.1 Soil penetration resistance,

A penetrograph was used to determine the soil penetration resistance in the field. The penetrograph has the ability to show the penetration reading in the field at different depths from 0-80 cm. In the Cone Penetration test, a cone penetrograph is pushed into the ground at a constant rate and direct and automatic measurements are made of the resistance to penetration of the cone.

2.3.2 Cohesiveness and internal friction angle

A Wykeham Farrance International (WFI) SLI 4Hw shear box laboratory apparatus was used to obtain cohesion, angle of internal friction and shear stress of the soil .

2.3.3 Aggregate size distribution

Experiments were conducted to determine the influence of various types of blade and rotational speeds on aggregate size distribution. In the experiment, soil physical characteristics investigated include MWD_d, MWD_w, to find out the 'SI' and 'II'. The method of wet-sieving was adapted from (W. D. Kemper & R. C. Rosenau, 1986) [11].

2.4 Test procedure

The test was performed based on 2×3 factorial treatment experimental design in RCBD with three replications. The analysis of variance (ANOVA) and Duncan's multiple range test were used to analyze the data using the statistical analysis systems (SAS) 2005 software.

2.5 Tractor characteristics

A 63.4 kW John Deere 6405 tractor (PTO power 51 kW) was used for the tillage experiments. The tractor has static weight distribution of 40% front and 60% rear with total mass of 3891 kg and 180 kg balancer (6×30 kg).The front tires were radial 12.4-24 single operated at 220 kPa inflation pressure and the rear tires were radial 18.4-34 single operated at 160 kPa inflation pressure.

3. Results and discussion

3.1. Soil texture and soil penetration resistance

Soil texture was determined by standard pipette method analysis achieved by USDA textural classification system . 'Table 1' shows result of the mean values of soil particle size with different texture content from clay, sand and silt as 23.91%, 53.54% and 22.55%, respectively indicating that under the classification of soil texture, the soil was found to be *sandy clay loam.* Figure 4 shows the mean values of nine cone penetration points randomly taken in the field using the penetrograph before comboplow tillage operation. The maximum and minimum soil penetration resistance values of 380 N/cm^2 and 180 N/cm^2 were found in plots "A" at 75 cm and "C" at 65 cm depths, respectively.

Table 1. Soil particle size from different textures in the study area

Texture	Min(%)	Max(%)	Mean(%)
Clay	23.87	23.95	23.91
Sand	37.33	69.75	53.54
Silt	6.38	38.75	22.55

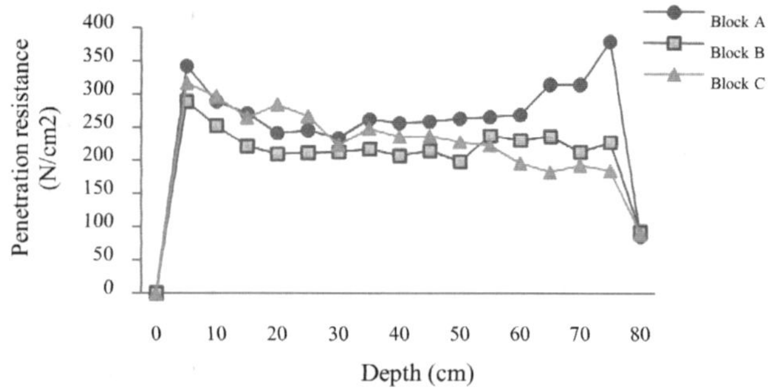

Figure 4. Soil penetration resistance at different sampling plots

3.2 Cohesiveness and internal friction angle

Table 2 presents the normal stress and shear stress in all three blocks with three replications. The normal stresses were 40.95 kN/m^2, 69.2 kN/m^2 and 97.45 kN/m^2. The typical graphs of maximum shearing strength versus normal stress load at the depth 0-30 cm are shown in 'Figure 5'. The results of the direct shear test show that the mean values of soil friction angle (φ) were 25°, 19.5° and 13.4 ° and mean values of cohesion (C) were 3.383 kN/m^2, 8.863 kN/m^2 and 12.29 kN/m^2.

Table 2. Mean values of shear stress and normal stress for block A, B and C

Normal Stress (kN/m^2)	Shear Stress (kN/m^2)		
	A	B	C
40.95	21.83	23.35	22.40
69.2	37.00	33.08	27.97
97.45	48.19	43.24	35.82

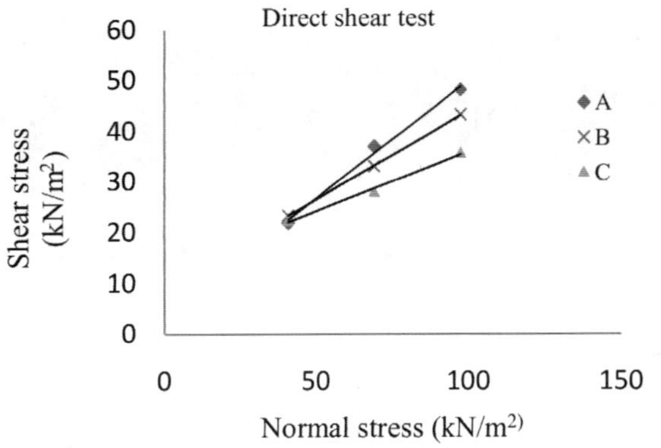

Figure 5. Maximum shear stress versus normal stress for block A, block B and block C

3.3 Effect of type of blade and rotary speed on MWD_w and MWD_d

Tables 3-4, 'Figure 6', 'table 5-6' and 'Figure 7'show the relationship between MWD_w and MWD_d with respect to type of rotary blades and rotational speeds of blades. The rotational speed of blade was found to have significant affect on the MWD_w at the lowest values of 1.39 mm for the speed of 130 rpm. The results for MWD_w proved that there was no significant difference among the kinds of blades. There was no significant interaction effects found between rotary speeds × type of blades 'table 3'. MWD_d was found to be significantly affected by the rotational speed of blade at the lowest value of 3.12mm for the rotational speed of 165 rpm. Result showed significant difference ($p<0.001\%$) at the 130 rpm rotational speed and was highest compared to the other speeds 'table 5'. The results for MWD_d showed that there was a significant difference among the types of blades. There was significant interaction effects found between rotary speeds × type of blades 'table 6'. However, it was noted that the L-type blade at 130 rpm and S- type blade at 165 rpm had maximum and minimum values of mean weight diameter dry basis and were 3.37mm and 3.12mm respectively. This was similar to the finding of Boyadas and Turgut [12] who reported that operating speeds of moldboard plow with rotary harrow significantly affected the mean weight diameter values. They pointed out that an increase in implement speed caused a decreased of mean weight diameter.

IOP Conf. Series: Materials Science and Engineering **36** (2012) 012003 doi:10.1088/1757-899X/36/1/012003

Table 3. Effect of type of blade and rotary speed on mean weight diameter wet basis

S.O.V	DF	MS	Fvalue	Pr>F
Block	2	5.36	72.98^{**}	0.0001<
Speed	2	0.28	3.87^{*}	0.0426
Blade	2	0.01	0.14^{ns}	0.8687
Blade×Speed	4	0.02	0.29^{ns}	0.8775
Error	16	0.07		

** Significant at 1% level,* significant at 5% level and ns not significant , C.V = 17.01% and R^2=90.65%

Table 4. Duncan's multiple range tests for effect of type of blades and speed of rotary on MWD_w

N	Blade Speed(rpm)	Mean (mm)	
9	S	1.62^{a}	
9	C	1.59^{a}	
9	L	1.56^{a}	
9	165		1.73^{a}
9	147		1.66^{ab}
9	130		1.39^{c}

Means with the same letter in the same column are not significantly different

Table 5. Effect of type of blade and rotary speed on MWD_d

S.O.V	DF	MS	Fvalue	Pr>F
Blade	2	0.075	3.75^{*}	0.0463
Speed	2	0.407	20.32^{**}	<.0001
Block	2	0.408	20.41^{**}	<.0001
Blade×Speed	4	0.085	4.25^{*}	0.0156
Error	16	0.020		

** Significant at 1% level,* significant at 5% level and ns not significant C.V = 4.27 and R^2=86.87%

Table 6. Duncan's multiple range tests for effect of type of blades on MWD_d

N	Blade Speed(rpm)	`Mean (mm)	
9	L	3.37^{a}	
9	C	3.35^{a}	
9	S	3.20^{b}	
9	130		3.54^{a}
9	147		3.28^{b}
9	165		3.12^{c}

Means with the same letter in the same column are not significantly different

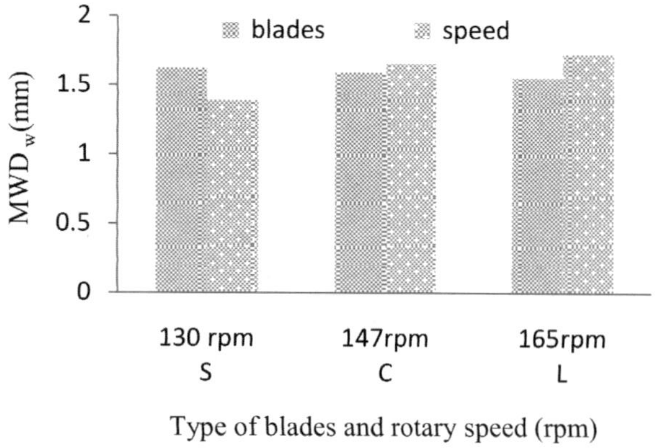

Figure 6. The effect of speed of rotary blades and rotary speed on MWD_w

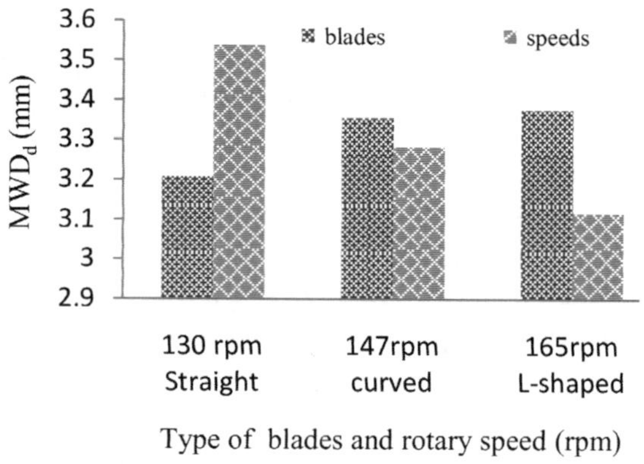

Figure 7. The effect of speed of rotary blades and rotary speed on MWD_d

3.4 Effect of type of blade and rotary speed on stability index and instability index

Tables 7-8, 'Figure 8', 'table 9-10' and 'Figure 9'show the relationship between stability index with respect to type of rotary blades and rotational speeds of blades. *SI* was found to be significantly affected by the rotational speed of blade at the lowest value of 0.52 for the rotational speed of 165 rpm. The results for *SI* showed that there was no significant difference among the types of blades 'table 8'. There was no significant interaction effects found between rotary speeds × type of blades 'table 7'. *II* was found to be significantly affected by the rotational speed of blade at the lowest value of 1.39 for the speed of 165 rpm. The results for *II* showed that there was no significant difference among the types of blades 'table 10'. There was no significant interaction effects found between rotary speeds × type of blades 'table 9'. However, it was noted that the L- type blade at 130 rpm and S- type blade at 165 rpm had maximum and minimum values of the instability index were 1.82 and 1.39 respectively.

Table 7. Effect of type of blade and rotary speed on Stability Index

S.O.V	DF	MS	F value	Pr>F
Blade	2	0.196	1.03^{ns}	0.3792
Speed	2	0.700	3.69^{*}	0.0482
Block	2	1.219	6.41^{**}	0.009
Blade×speed	4	0.231	1.21^{ns}	0.3433
Error	16	0.19		

** Significant at 1% level,* significant at 5% level and ns not significant C.V = 6.84 and R^2=82.8%

Table 8. Duncan's multiple range tests for effect of speed of rotary blades on Stability Index

N	Blade Speed(rpm)	Mean (mm)	
9	S	0.94^{a}	
9	C	0.69^{a}	
9	L	0.67^{a}	
9	130		1.07^{a}
9	147		0.72^{a}
9	165		0.52^{b}

Means with the same letter in the same column are not significantly different

Table 9. Effect of type of blade and rotary speed on Instability Index

S.o.V	DF	MS	Fvalue	Pr>F
Blade	2	0.1358	2.03^{ns}	0.1635
speed	2	0.1355	20.28^{**}	<0.0001
block	2	5.95	89.1^{**}	<0.0001
Blade×speed	4	0.231	1.57^{ns}	0.2294
Error	16	0.0669		

** Significant at 1% level,* significant at 5% level and ns not significant C.V = 15.03 and R^2=93.47%

Table 10. Duncan's multiple range tests for effect of speed of rotary blades on Instability Index

N	Blade	Mean (mm)	
9	L	1.82^{a}	
9	C	1.76^{a}	
9	S	1.58^{a}	
9	130		2.15^{a}
9	147		1.62^{a}
9	165		1.39^{b}

Means with the same letter in the same column are not significantly different

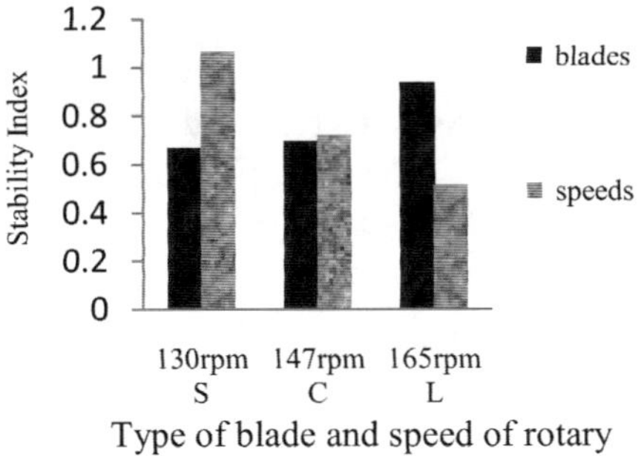

Figure 8. The effect of type of blades and speed of rotary blades on Stability Index

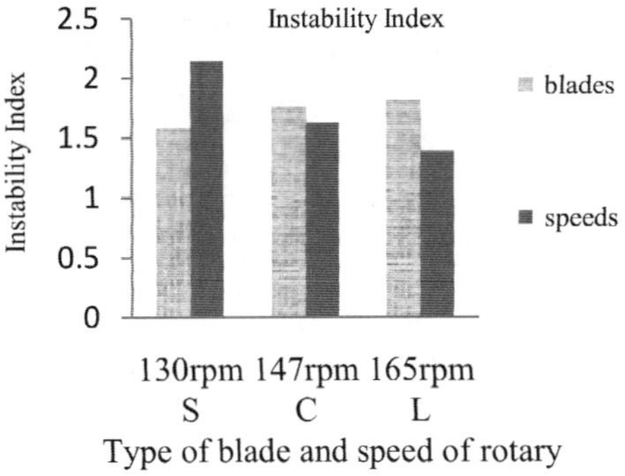

Figure 9. The effect of type of blades and speed of rotary blades on Instability Index

4. Conclusion

A Comboplow for land preparation has been successfully designed, fabricated and tested. The machine consists of a chassis, three point hitch, transmission system, disk, rear wheel, rotary blades and adjusting mechanism. The blades rotational velocity had high significant effect on selected parameters (MWD_d, MWD_w, II, SI). The results for MWD_d showed that there was a significant difference among the types of blades. However, it was noted that the L-type blade at 130 rpm and S-type blade at 165 rpm had maximum and minimum values of MWD_d .The effect of types of blades was similar on MWD_w, II, SI. The results of MWD_d , II and SI suggest that the rotary blade speed of 130 rpm was highly effective than 147 and 165 rpm.

5. References

[1] Lal R , Reicosky D C, Hanson J D, 2007, Evolution of the plow over 10,000 years and the rationale for no-till farming. *Soil and Tillage Research*. **93**(1) 1-12.

[2] Hill, R.L. and R. Cruse, Tillage effects on bulk density and soil strength of two Mollisols. Soil science .Soc.Am.J.(united States), 1985.49(5):p.1270-1273.

[3] Reicosky D C. and Allmaras R R 2003, Advances in tillage research in North American cropping systems. *Journal of Crop Production*. **8**(1) 75-125.

[4] Bayhan Y ,Kayisoglu B, Gonulol E, Yalcin H,Sungur N2006, Possibilities of direct drilling and reduced tillage in second crop silage corn. *Soil and tillage research*. **88** (1-2) 1-7.

[5] Godwin R J, 2007, Advances in labour and machinery management. *Agricultural Engineering International.The CIGR* Ejournal **4**, 1-11.

[6] Sahu,R. K. and H. Raheman, an approach for draft pridection of combination tillage implements in sandy clay loam soil.Soil and Tillage Research,2006.90(1-2)p:145-155.

[7] Manian,R., V. Nagaiyan, and K. Kayhirvel, Development and evaluation of combination tillage-bed furrow-former.Agric Mech Asia Afr Lat Am,1999.30(4) :p. 22-29

[8] Al-janobi,A. and S.Al-Suhaibani, Draft of primary tillage implements inh sandy loam soil. Applied engineering in agriculture,1998. 14(4): P.343-348

[9] Tebrügge F and Düring R A, 1999, Reducing tillage intensity a review of results from a long-term study in Germany. *Soil and tillage research*. **53**(1) 15-28.

[10] Ahmad D and Amran F A, 2004, Energy prediction model for disk plow combined with a rotary blade in wet clay soil. *International Journal of Engineering and Technology*. **1**(2) 102-114.

[11] Kemper, W. D., & Rosenau, R. C. (1986). Aggregate stability and size distribution.

[12] BoyadasM G and Turgut N, 2007,Effect of tillage implements and operating speeds on soil physical properties and wheat emergence. *Turk Journal Agriculture*, **31** 399-412

IOP Conf. Series: Materials Science and Engineering **36** (2012) 012004 doi:10.1088/1757-899X/36/1/012004

Investigation of propulsion system for large LNG ships

R P Sinha[1] and Wan Mohd Norsani Wan Nik[2]

[1]Department of Marine Engineering, Malaysian Maritime Academy, Batu Rakit, Kuala Terengganu
[2]Department of Maritime Technology, Faculty of Maritime Studies and Marine Science, Universiti Malaysia Terengganu

E-mail[1]: rajendra@alam.edu.my

Abstract. Requirements to move away from coal for power generation has made LNG as the most sought after fuel source, raising steep demands on its supply and production. Added to this scenario is the gradual depletion of the offshore oil and gas fields which is pushing future explorations and production activities far away into the hostile environment of deep sea. Production of gas in such environment has great technical and commercial impacts on gas business. For instance, laying gas pipes from deep sea to distant receiving terminals will be technically and economically challenging. Alternative to laying gas pipes will require installing re-liquefaction unit on board FPSOs to convert gas into liquid for transportation by sea. But, then because of increased distance between gas source and receiving terminals the current medium size LNG ships will no longer remain economical to operate. Recognizing this business scenario shipowners are making huge investments in the acquisition of large LNG ships. As power need of large LNG ships is very different from the current small ones, a variety of propulsion derivatives such as UST, DFDE, 2-Stroke DRL and Combined cycle GT have been proposed by leading engine manufacturers. Since, propulsion system constitutes major element of the ship's capital and life cycle cost, which of these options is most suited for large LNG ships is currently a major concern of the shipping industry and must be thoroughly assessed. In this paper the authors investigate relative merits of these propulsion options against the benchmark performance criteria of BOG disposal, fuel consumption, gas emissions, plant availability and overall life cycle cost.

1. Introduction

The trend in power generation of shifting away from coal to reduce the adverse effects of CO_2 and other gas emissions on the environment has made steep demand on the supply of natural gas. Because of these developments, natural gas is now the fastest growing energy source and in the last five years, the growth rate in LNG production has been about 60% with the current annual output now touching 260 million tons. Industry estimates also indicate that this trend may continue for another several years. The accelerated growth in LNG business is not difficult to visualize as there is crying demand for clean energy all over the world, and particularly in the developing countries. According to world energy review 2010 [1], the global energy use is growing by 36 % with major demands coming from developing economies such as China. This indicates, the natural gas is set to play key role in meeting world's energy needs and this has led to the exploration in remote and hostile environment of deep sea far away from shore. The search for gas in such remote areas is throwing intense technical and commercial challenges which need to be tackled for commercial success. According to trade estimates, the expected investment to enhance LNG production may be to the order of US $200 billion which is likely to increase World's LNG production by about 50% in next five years. Figure 1.

Published under licence by IOP Publishing Ltd

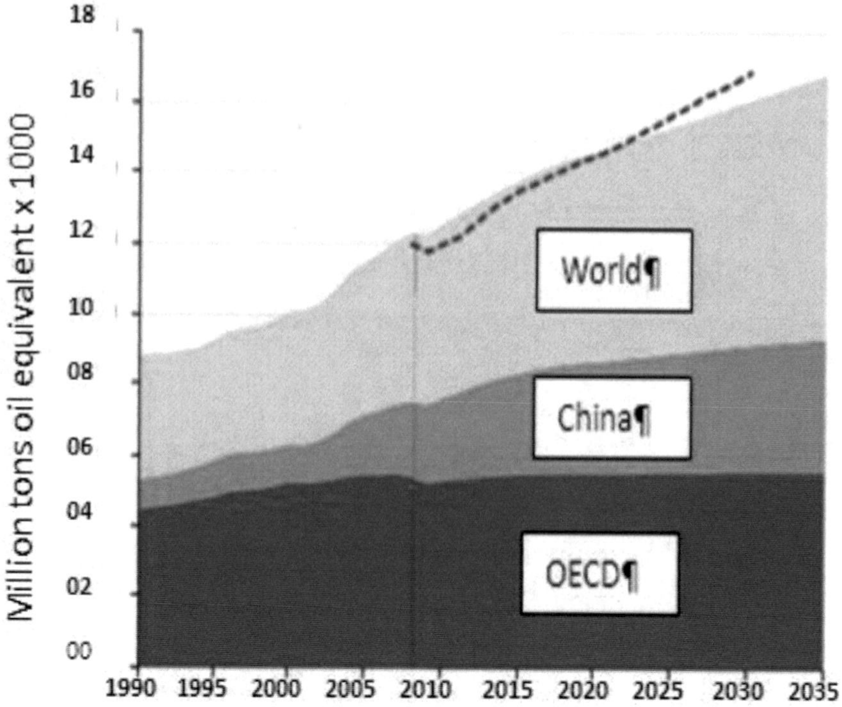

(a) World Energy demand

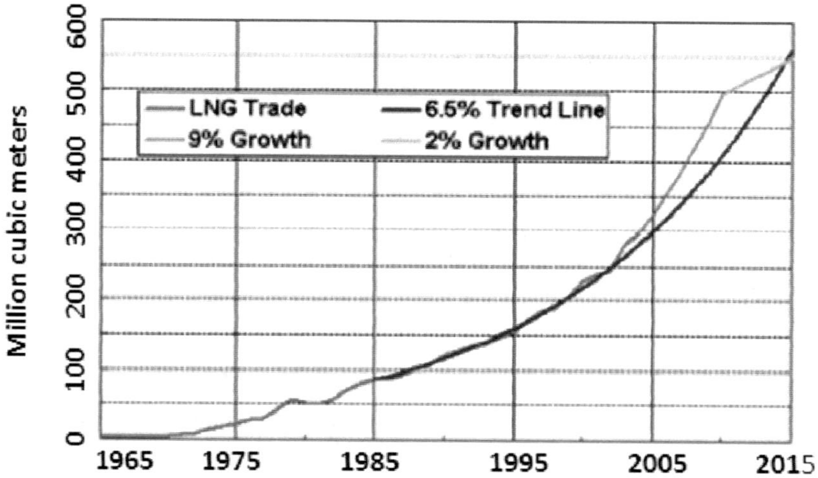

(b) LNG Trade Growth

Figure 1 World Primary Energy Demands, [1,2].

2. Trends in LNG ships Construction

The LNG trade in general has been characterized by long-term shipping contracts and dedicated fleet of ships sailing on fixed routes schedules between limited number of LNG terminals in the world. However, due to increasing demand for supply of liquefied gas this trend is now slowly changing with a steadily increase in the number of spot cargo deals being executed. This new form of LNG business is expected to grow even more in the future. Therefore, from the transportation point of view the operators will be looking for ships with more operational flexibility and efficiency in response to different contractual agreements. Because of these developments the current trend in LNG transportation is moving towards use of larger ships which can offer high utilization factor for economic viability of the project. A review of new orders [2,3] for LNG ships from 2007-2014 in Figure 2 confirms this trend and it is expected that most future LNG ships will be Q-flex or Q-max types with size varying from 200000-265000 m^3.

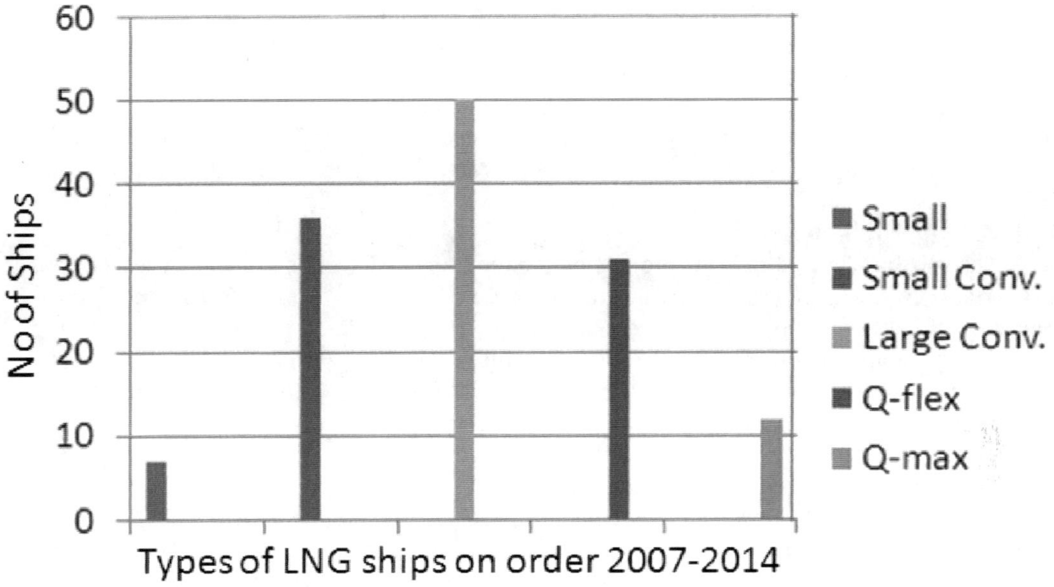

Figure 2 LNG Ships on order [3]

Propulsive power needs of Q-flex and Q-max type LNG ships operating at average speed of 20 knots have been estimated by MAN B&W [3] to about 35-45 MW and if we include about 30-40% electrical load of the auxiliary machinery then total power demand may come to about 50-60 MW. .

3. Propulsion Options and Issues

As the total power of Q-flex and Q-max class LNG ships falls in the range of about 50-60 MW any of the following alternative propulsion systems can be used to power these ships.

(i) Steam propulsion
(ii) Dual Fuel Diesel Electric
(iii) Gas Turbine in Combined Cycle Mode
(iv) 2-Stroke Slow Speed Direct coupled Diesel Engines with re-liquefaction.

However, which of these alternatives offers the most economical powering solution is the real issue to be investigated. In estimating total ship's power the share of auxiliary electrical load and its percentage utilization will vary considerably depending upon the nature and configuration of the power plants. For instance, unlike remaining three options the 2-stroke slow speed direct coupled diesel option will require a re-liquefaction plant for BOG disposal into the cargo tank which may consume 5-7 MW of additional power. On the other hand in DFDE system the total power demand can be considerably lower because of the integration of propulsion and auxiliary electrical powers systems. Therefore for meaningful comparison each of these alternative propulsion options are investigated against the following common performance benchmarks.

- Boil Off Gas Disposal
- Fuel Consumption
- Reliability, Availability and Maintainability
- Gas Emissions

- Overall life cycle cost

The very first issue that has the highest bearing on commercial operation as well as environmental emission is whether the BOG should be burnt or re-liquefied and sold as cargo. Although, the selling price of LNG varies from contract to contract the general pricing trend indicates that it costs about 30-40% less than HFO which is the common bunker for all merchant ships. If the cost of transporting and handling charges of the LNG at receiving terminal is also included then the cost of LNG vis a vis HFO may further fall to around 50% lower in comparison to its main competitor HFO. Additionally, because LNG is a clean fuel the exhaust emissions are greatly reduced which is becoming a critical factor in international shipping. Therefore, re-liquefaction of BOG, a relatively cheaper and cleaner fuel and that too after consuming 5-7 MW of electrical power does not appear a sound economic decision against using more expensive and environmentally less friendly HFO for propulsion.

3.1 Steam propulsion

Steam turbine has been the main propulsion system for LNG ships from very early days of gas transportation by sea primarily because of the ease with which the natural BOG can be burned in the boilers. A suitable marine steam power plant for Q-flex and Q-max class ships will employ two boilers each of around 80-90 tons per hour steaming capacity at 60-70 bar pressure and 520 ^{0}C superheat steam temperature [4]. Propulsion system will comprise of a high pressure, an intermediate pressure and a low pressure turbines with total power output of 35-45 MW. The low pressure turbine also carries an astern turbine on the same rotor shaft for speed reversal. The auxiliary electrical power of about 10 MW will be provided by two steam turbines generators and one medium speed diesel generator of 3 MW power ratings. Overall thermal efficiency of a typical 30 MW conventional marine steam power plant, Mitsubishi, used for propulsion of 157000 m^3 LNG tanker has been estimated to 35% [4]. Now the improved versions of these steam plants designed on UST technology are able to offer nearly 15% fuel saving [5] which amounts to an increase in the overall fuel efficiency to about 41% and is thus comparable to the fuel consumption of DFDE.

3.2 Dual Fuel four Stroke Diesel Engines (DFDE)

In the early stage of LNG transportation diesel engines lost to steam turbines mainly because they could not handle natural BOG from the cargo tanks. But now the new diesel engines particularly the 4-stroke medium speed design can alternatively burn both liquid as well as gas fuels and because of that they have been considered potential alternative to less fuel efficient steam turbines. A look at the new orders of LNG ships shows a good number of these contracts have been signed with 4 stroke medium speed diesel engines as prime mover in electrical propulsion modes. In gas mode when BOG is the fuel the engine operates with lean air/fuel ratio on the principle of Otto cycle with a small amount of diesel oil injection in the combustion chamber as pilot fuel for ignition [6]. However, when the BOG is insufficient then the engine is operated on liquid fuel such as DO or HFO. In this situation the BOG has to be disposed off by burning in the GCU with consequent penalty on energy loss. Therefore loss of BOG together with losses in associated electrical components of the propulsion system (6-8%) must be taken into account while comparing DFDE with its competitors.

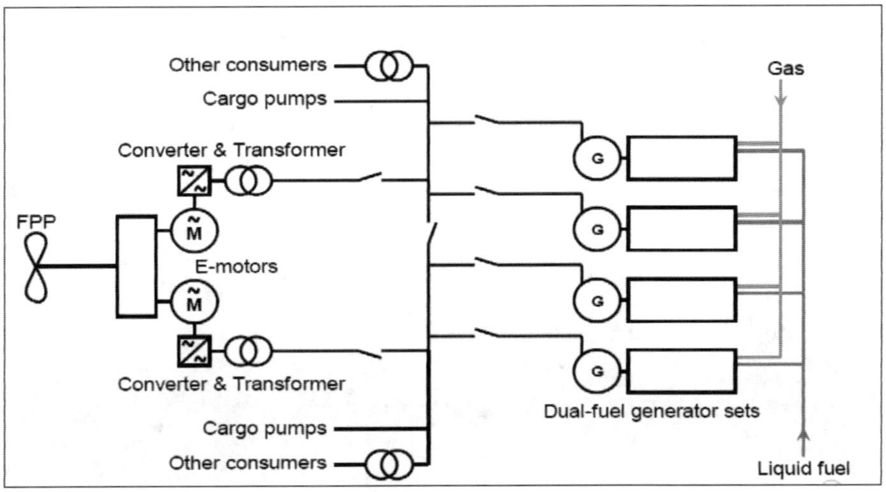

Figure 3 Schematic DFDE Propulsion [2].

In the DFDE design, figure 3, because electrical power for propulsion and cargo handling are in different operating time phase the net power requirement of the ship is considerably reduced unlike in other mechanical coupled propulsion systems which is a distinct advantage. However, the loss from BOG disposal during liquid fuel mode of operation will more than offset this advantage. Another serious shortcoming of DFDE is the high risk of detonation and misfiring as load increases. The DFDE have very narrow range of air/fuel ratio for detonation free operation which calls for a complex control system as each cylinder requires dedicated A/F ratio controller.

3.3 Slow Speed Two Stroke Diesel Engines

The two-stroke slow speed diesel engines are the most preferred propulsion choice for all large merchant ships primarily because of their improved efficiency and special feature that they can burn poor quality low cost fuel which the medium speed 4-stroke engines cannot consume. This special feature of 2-stroke slow speed diesel engines has drawn interest of the LNG ship owners of large capacity, over 200000 m^3 for use as substitute to steam propulsion. Because of large power needs of these ships, twin screw configuration is the preferred choice although even single screw layout can be successfully used. A typical layout of 2-stroke slow speed dual fuel diesel engine propulsion for Q-flex and Q-max class LNG ships is shown in Figure 4.

Though 2-stroke slow speed diesel engines have proven high fuel efficiency when used in LNG propulsion they require an onboard re-liquefaction plant to convert BOG into liquid gas and return to the cargo tanks. The re-liquefaction unit is a heavy duty equipment and depending upon the natural boil off rate it can consume from 5-7 MW of electrical power which substantially lowers the efficiency benefits of these engines. Because of that some estimates suggest that net efficiency of the 2-stroke slow speed diesel engines may be lower than DFDE propulsion. Further, as backup, the ship will require a GCU to dispose off the BOG in the event of a breakdown of the re-liquefaction plant. The net auxiliary electrical power due to re-liquefaction plant and increased size of cargo pumps may be of the order of 14-16 MW which is met by using four medium speed diesel engines

To retain high efficiency of 2-stroke designs, research is ongoing for high pressure gas injection (300 bar) into these engines to dispose of the natural BOG and eliminate the requirement of onboard re-liquefaction. But associated risk of high pressure gas pipe line in the ship has not gone down well with the ship owners and this proposal is still in the investigation stage. High vibration levels of 2 stroke slow speed diesel engines is another disadvantage as it may influence the integrity of the cargo containment system and this aspect needs to be addressed as well.

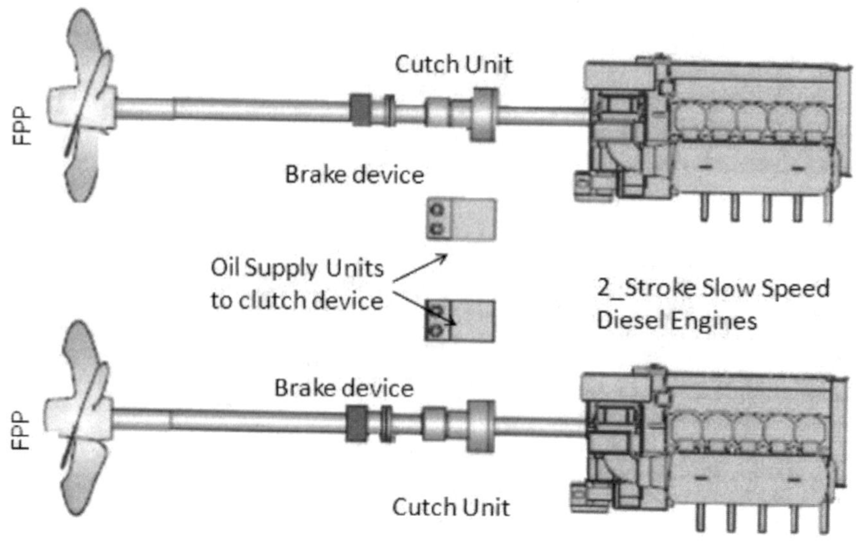

(a) MAN B&W ME with Liquid fuel

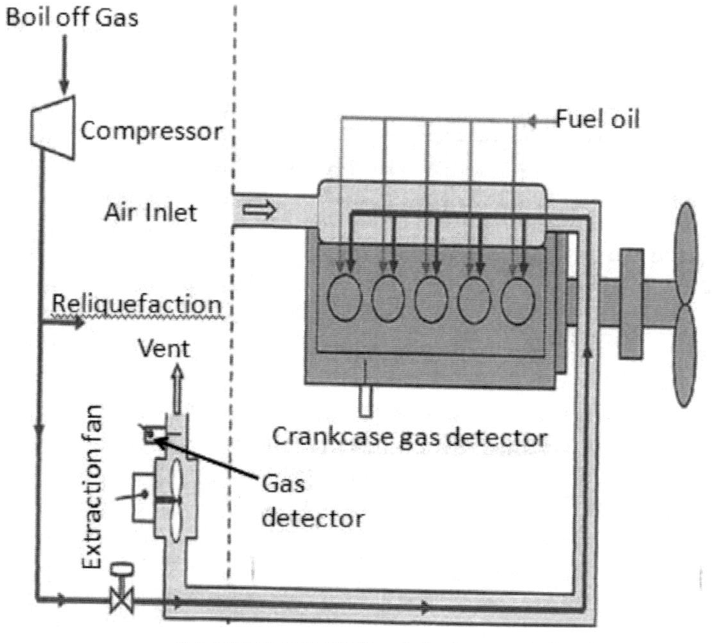

(b) MAN B&W Dual fuel - gas injection

Figure 4 2-Stroke Slow Speed Diesel Propulsion for LNG Ships [7,8] .

3.4 Gas Turbines in Combined Cycle

Gas turbines despite their many good features such as compact size, light weight, excellent reliability, high power to weight ratio and quick response to sudden power demand have not found favours with ship owners mainly because of low fuel efficiency. In marine field, gas turbines have found applications mostly in naval ships propulsion and to some extent in offshore industry. In naval applications the gas turbines are used in either CODOG or COGAG configurations in which the base cruising power is provided by a low powered diesel engine or gas turbine. The high power demand for the short sprint combat action comes from the large power gas turbine which makes overall power plant operation reasonably fuel efficient but still less economical. However, gas turbines in land based application have been used in combined cycle with steam turbines and achieved very favorable fuel efficiency often superior to all other prime movers. Figure 5 shows typical combined cycle gas turbine in combination with steam turbine (COGES) for electrical propulsion of LNG ships proposed by Rolls Royce and General Electric respectively. .

If individual efficiencies of gas turbine and steam plants are η_g and η_s then the combined plant's efficiency η_{comb} equals to

$$\eta_{comb} = \eta_g + \eta_s - \eta_g \eta_s \quad (1)$$

which explains improved efficiency of combined cycle gas turbine plants. One major drawback of gas turbine is the short blade life due to high temperature of flue gases which requires running hour based replacement of the turbine unit but this shortcoming has been addressed to by offering lease contract of the gas turbine to the ship owners. Under this arrangement, on completion of stipulated running hours the vendor will replace the turbine with a re-serviced unit at no extra cost.

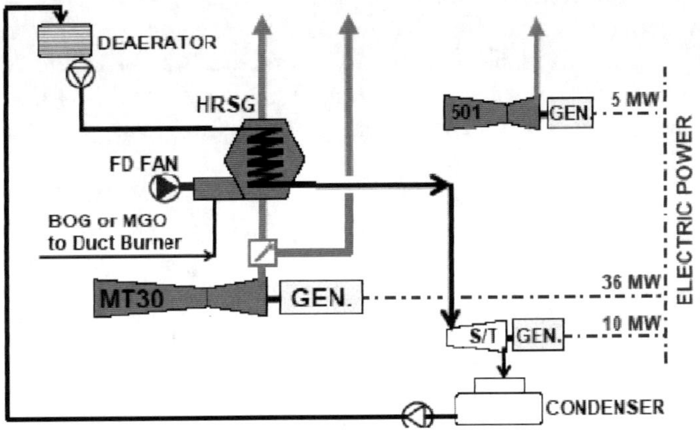

(a) Rolls Royce gas turbine

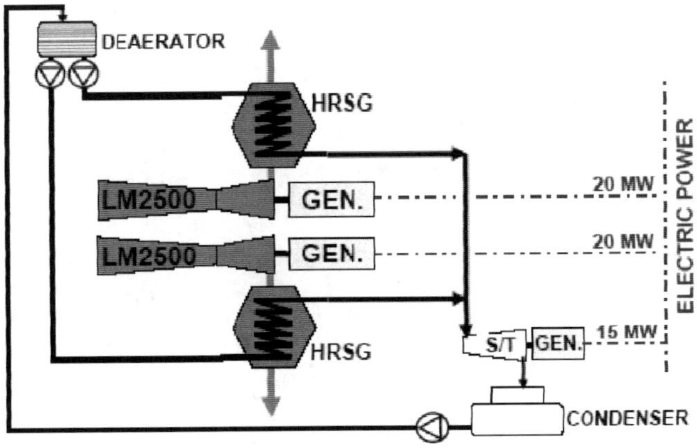

(b) General Electric Gas Turbine

Figure 5 Combined cycle gas turbine electric propulsion [9,10]

4. Comparative study

The performance of four available propulsion alternatives is compared against the stated criteria of BOG disposal, fuel consumption, plant reliability, gas emission and over all life cycle cost.

4.1 Boil Off Gas Disposal

Handling natural boil off gas is a major safety issue in LNG transportation which requires a very reliable and effective arrangement onboard for its safe disposal. However, there are only two available options for BOG disposal such as:

(i) Burn it as fuel in the engine to produce useful propulsive power
(ii) Re-liquefy and put back into the tank for sale as cargo.

Which of the above alternatives will be used in a particular ship is to be determined by the type of propulsion system and has major technical and economic impact on the cost of LNG transportation. Both options have been used in different ships but which one is the preferred choice still remains a matter of disagreement among experts. Burning the BOG in main engine or its re-liquefaction as cargo has technical as well as commercial aspects that needs to be considered for optimum benefits. For instance, burning the BOG results in bunker saving but at the same time a valuable cargo has been consumed which could have been sold as good quality fuel. Similarly, in re-liquefaction of BOG though cargo is saved but the liquefaction plant consumes about 6-7 MW of additional electrical power in converting the gas into liquid. Because of this the benefits of saving BOG as cargo is largely lost by excessive electrical power consumption. Thus both options are fraught with certain advantage/disadvantage which calls for more thorough evaluation to arrive at the final optimal choice.

The re-liquefaction is an option only if the BOG cannot be burned as fuel in the main engine, such as 2-stroke slow speed diesel engines which are otherwise excellent in fuel efficiency. Although in recent years MAN B&W have developed 2 stroke slow speed engines capable of burning natural gas but it utilizes high pressure gas injection technology which requires a high pressure gas compressor to raise the gas to injection pressure of 250-300 bar. Another equally important aspect of BOG re-liquefaction that needs to be considered is the price of natural gas versus bunker fuel. The current global price reference of oil and gas indicates that the natural gas is trading at nearly 50% of the oil price and perhaps this is one reason why LNG is being seriously considered for use as the bunker fuel for general shipping, particularly on European trading routes. The LNG being a clean fuel will also be more friendly to exhaust emissions another favorable point against re-liquefaction of the BOG.

The other alternative to burn the BOG as fuel can be achieved in boilers of the steam plant as well as in the DFDE and Gas turbine units. But which of these propulsion arrangements gives the best overall result in terms of economic and technical benefits is the factor to be considered for their comparison as propulsion options. In this respect steam and gas turbine propulsions have advantage because they can burn gas and liquid fuel simultaneously unlike the DFDE which can only operate either on gas or on liquid fuel at any one time. Because of this restriction if the amount of BOG is small or insufficient the DFDE will be using only liquid fuel leaving the BOG to be burned in the gas combustion unit as waste which lowers overall fuel efficiency.

4.2 Fuel oil consumption

Fuel efficiency is the most significant performance index of a propulsion system as it not only affects the operating cost of the ship but also hugely contributes to gas emissions. Steam power plants have lower fuel efficiency in comparison to internal combustion engines but recent developments in UST technology with increased pressure and temperature followed by steam reheating the fuel efficiency of marine steam plants has been raised almost to the level of marine diesel engines. Similarly operating gas turbines in combined cycle with a steam turbine gives much improved fuel efficiency often higher than conventional diesel engines. Fuel efficiency of a typical conventional steam turbine propulsion plant of a 157000 m^3 LNG tanker has been calculated from the heat balance data obtained during sea trials [4]. The estimated fuel efficiency of this plant works out to 35% and if additional 15% fuel saving from UST design is also taken into account then the overall fuel efficiency of steam plant comes to 42% which is very close to the DFDE. Fuel efficiency of different propulsion options is shown in Table 1[11].

Table 1 Fuel Oil Consumption Summary

S.No	Type of Plant	Thermal eff %	SFC g/kWh	FOC t/day
1	Steam	35	286	180.
2	Steam-UST	41.2	243	147
3	2-S Diesel+RL	40	230	143
4	DFDE	42	243	146
5	GT-MT 30-CHP	50	245	180

Table 2 Comparison of Electrical and Total Power

Propulsion Type	Steam (140k m^3)	DFDE (150k m^3)	2-Stroke (220 k m^3)	GT-CHP (Elec)
Elec. Power,MW	10-12	30-40	15-20	45-55
Voltage, KV	3.3	6.6	6.6	11
Utilization at Sea%	15-20	80	35-45	70-80
Utilization in port%	70-80	25	50-80	25
Total Power	45	38	56	50

Thus it appears that in terms of fuel consumption the UST, 2-S slow speed diesel and DFDE have almost similar performance and any one of these options can be equally suitable for the propulsion of LNG ships if fuel consumption is the only criteria of plant selection. The lower efficiency of 2-stroke slow speed engines is mainly due to extra electrical power (5-7 MW) consumed by the re-liquefaction plant. Table 2 shows auxiliary electrical load and total power requirement of the Q-max LNG tanker for four alternative propulsion options [12]. As expected, the DFDE needs overall minimum power mainly due to the feasibility of integrating auxiliary electrical and ship's propulsive loads together as both these powers are not utilized at the same time.

4.3 Emissions

Gas emission is now a major environmental concern because of its impact on global warming and to address the issue, strict time bound emission control levels for ships have been imposed by the IMO through appropriate MARPOL regulations. Because of MARPOL restrictions gas emission will constitute a major parameter of comparison for alternate propulsion systems. It is therefore important to understand the chemistry of gas emissions and their cost. .

Sulphur essentially enters fuel in the refinery with crude oil and converts into SO_2 or higher oxides when burns in combustion chamber. Since sulphur is present in fuel as impurity nothing can be really done in the combustion chamber to minimize its emission and the only option to prevent excess SO_2 release is to limit sulphur content in the fuel. The MARPOLE guideline [13] on permissible limits of sulphur in marine fuel is shown in Figure 6. The emission from sulphur is therefore fuel specific and will affect all prime movers equally, and therefore should not form basis of comparison for alternative propulsions options.

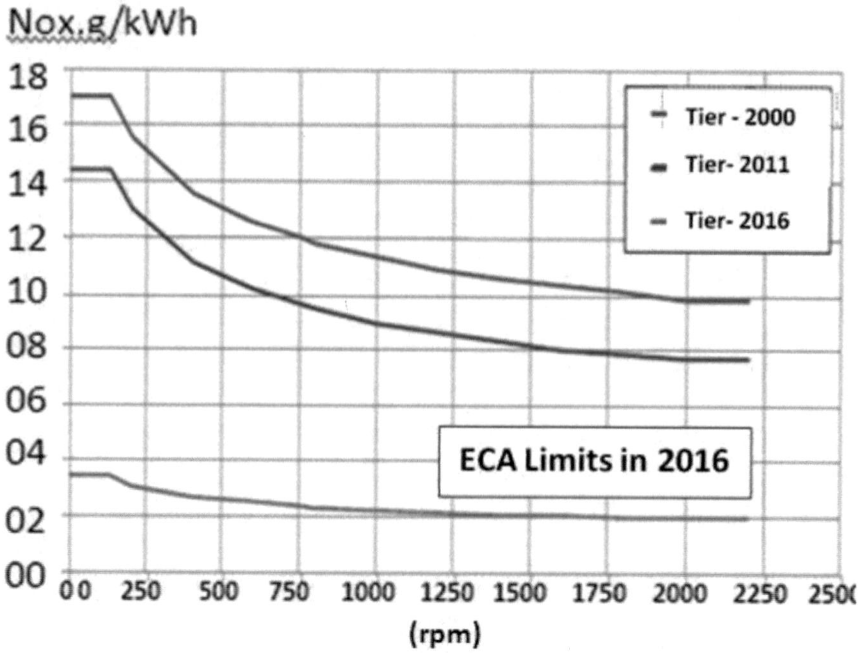

Figure 6 MARPOL limits on gas emission [13]

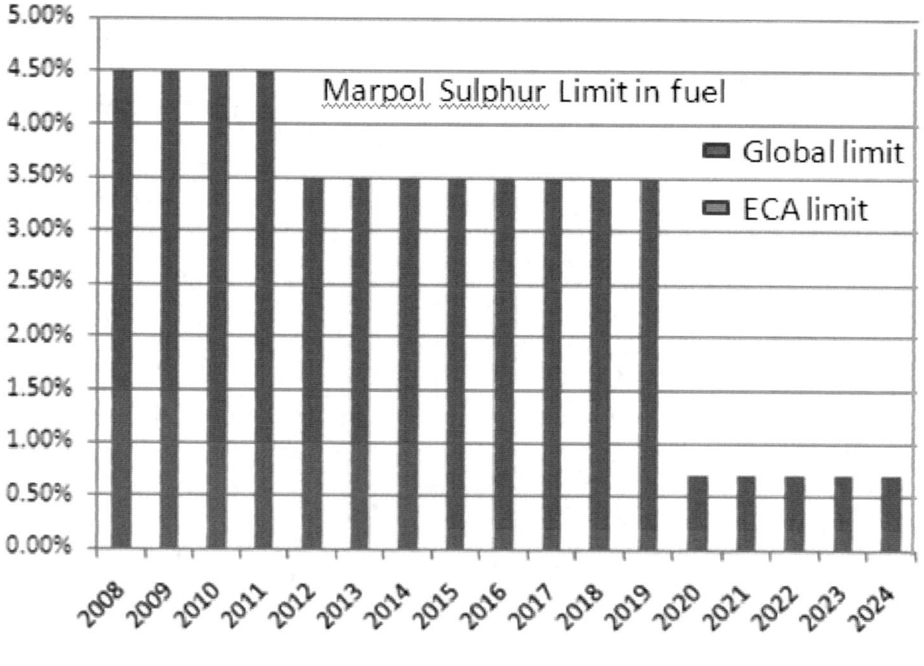

Figure 7 Sulphur limits on marine fuel [13]

The chemical kinetics of NOx on the other hand is quite different and depends upon the thermal conditions in the combustion chamber and also on the amount of nitrogen present in the fuel bond. However, it is the thermal NOx which contributes to over 80% of the emission from the engine and the fuel bounded nitrogen having very small overall impact. The thermal NOx is produced from reaction between N_2 and O_2 at high temperature. The formation of NOx may be explained as, at high temperature stable diatomic O_2 dissociates into hyper reactive atomic O which manages to break the strong triple bond of a very stable N_2 and thus forms nitric or nitrous oxides i.e NOx. And since presence of both O_2 and N_2 in the combustion chamber cannot be avoided because of the need of combustion air NOx control becomes quite tricky. The basic chemical equation of thermal NOx production may be expressed as- . . .

$$[NO] = Ae^{-\frac{b}{T}} N_2 \int \sqrt{O_2 \, dt} \tag{2}$$

where *A*, *b* are constants and *T* is temperature in the furnace [14].

As the quantity of N_2 is much large it may be assumed to remain approximately constant and then it can be deduced that NOx is a strong function of combustion temperature *T* and only weakly related to oxygen concentration in combustion chamber. This also suggests that thermodynamic conditions in the combustion chamber or furnace play significant role in the production of NOx. For instance, type of burners i.e diffusion or pre-mix will have effect on the NOx emission. Similarly the amount of excess air which depends upon the degree of difficulty in complete burning of the fuel in combustion chamber has two mutually opposing effects on NOx emission. First the excess air provides extra O_2 into the combustion chamber thus raising level of NOx emission but at the same time it also lowers the flame temperature which has opposite effect on the production of NOx. Which of these two mutually opposing kinetics dominate in the combustion process will ultimately determine the amount of NOx emissions from the prime mover of the plant.

Results of gas emission from different propulsion alternatives investigated by few selected researchers are presented below. Figure 8(a) shows results of investigation carried out by Mitsubishi [5] for set of alternative propulsions, all using gas burning and operating under full load conditions. The result suggests CO_2 emission which is an indicator of fuel efficiency is comparable for UST, DFDE and GT CC. This finding is in agreement with the result obtained by the authors in Table 1. The NOx emission is consistent as both DFDE and GTCC operate at high combustion temperature and more excess air in comparison to CST or UST. The result in Figure 8 (b) shows emissions with DFDE using only gas and other engines operating on liquid plus gas fuel in mixed fuel mode [15].

From these results it may be concluded that NOx emission from diesel engines, as expected is always relatively more mainly due to high temperature and excess air in the combustion chamber. On the other hand, steam and gas turbines because of better controlled combustion in the furnace release low NOx emissions. However, between the two turbines gas turbine will produce atleast 2-3 times more NOx in comparison to steam turbines, due to high temperature in the gas turbine combustion chamber. It is therefore, important to also include the cost of emission control for each option while assessing their comparative merits and demerits as alternative propulsion choice for LNG ships.

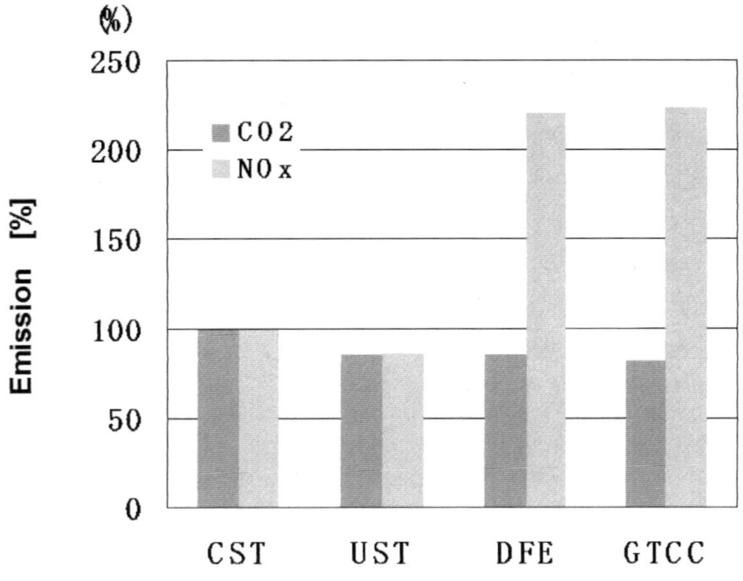

(a) HFO+BOG

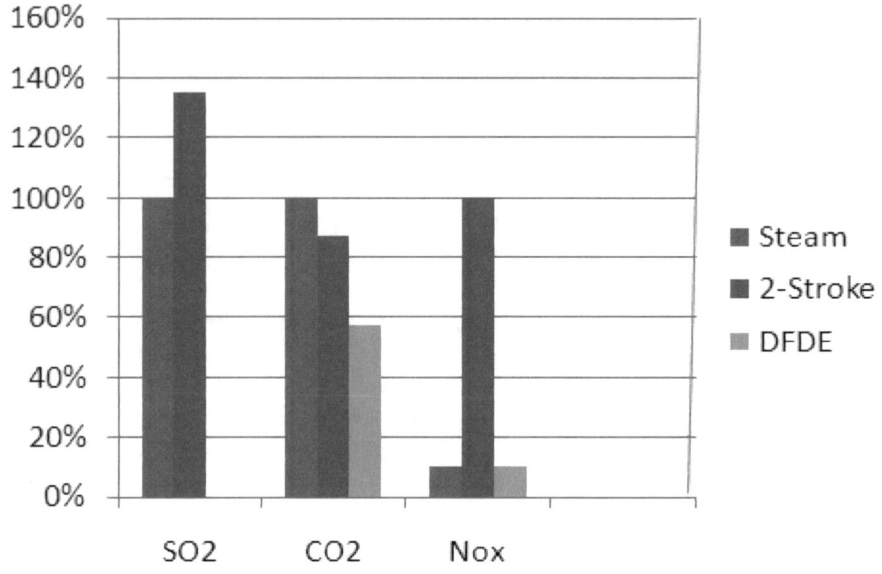

(b) DFDE (Gas) others liquid plus gas fuel [15]

Figure 8 Gas emissions-Comparison

4.4 Reliability

Reliability assessment of propulsion plants requires large operational data which is readily not available in most cases and therefore a comparative investigation of reliability for all alternative propulsion options is rarely possible. For the purpose of comparison in this paper, result of reliability investigation of typical LNG ship's UST propulsion and DFDE plant with three units of 12 cylinder Wartsila medium speed diesel engines from reference [11] is considered. The result shows, after initial settling down period the UST offers overall higher reliability which is consistent with the operational experience of both plants, figure 9. The MTTF for UST and DFDE, calculated from repairs and maintenance records of these two plants indicate that, the UST will have relatively higher availability.

Although such reliability results for 2-stroke slow speed diesel engine is not available but considering multi cylinder configuration and reciprocating nature of these engines their availability is expected to be similar to DFDE. Similarly since data on gas turbine propulsion is not available but based on the experience in naval ships, these engines may require internal inspection of a few days duration at 8000-10000 hours of operation and a major overhaul including blades renewal after 35000-40000 hours of continuous service. Therefore, it can be concluded that steam turbines surpass reliability criteria of all other alternative option.

$$MTTF(UST) = 45320\,hrs \qquad MTTF(DFDE) = 32161\,hrs$$

4.5 Life cycle cost

The global LNG business is very complex and to arrive at an accurate estimate of the life cycle operating cost of propulsion plant is rarely possible. Each LNG project is uniquely executed on long term charter contract under varying technical and commercial agreements. However, calculation of life cycle cost will need inputs of many financial and commercial data from the contract which are generally confidential information and not available to the general analyst. In this paper a very preliminary estimate of the plant life cycle cost has been made from the price information of repairs, maintenance/spares and services of three propulsion options obtained from few reputed vendors of LNG propulsion systems, at 2008 cost, [11] Table 3 .

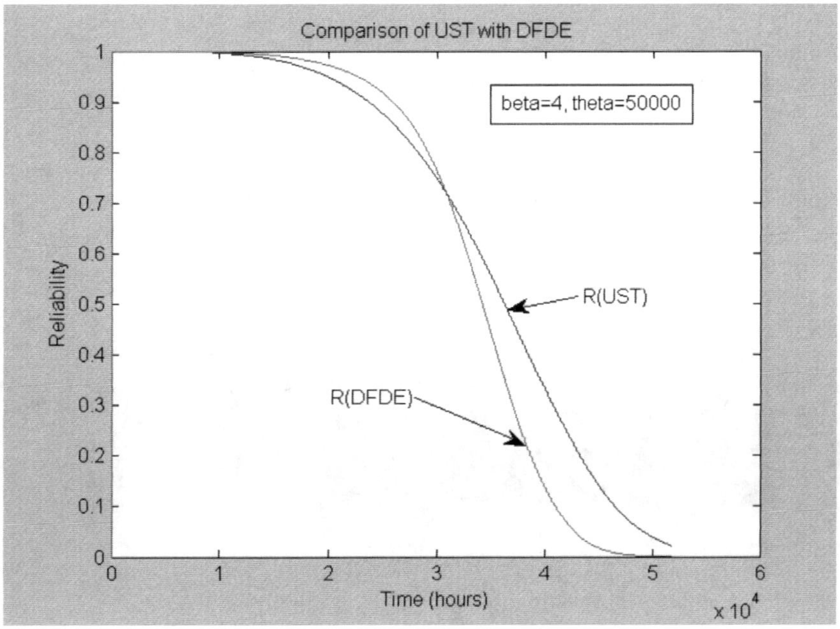

Figure 9 Reliability of UST versus DFDE [11]

Table 3 20 Years Life Cycle Maintenance and Spare Cost

Plant	Capex (Million USD)	Prime Mover Lub oil (Million USD)	Prime Mover Maintenance & Spares (Million USD)	Prime Mover Specific maint. cost (USD/MWh)
Steam Turbine	Base	Negligible	4.47	1.0
2-stroke + RL MAN B&W	+5	12.97#	18.03 *	7.0
DFDE Wartsila	+25	2.49 #	19.67	5.0

The specific and overall life cycle repair/maintenance cost for steam plant is low because it requires downtime for turbine, boiler and steam pipes at 30 months interval and call for a full inspection in dry docks only after every five years of operation. Unlike steam plant the DFDE engines need a pre-determined stoppage for routine maintenance at every 2000 operating hours.

The difference in cost of maintenance, spare parts and lube-oil replacement over a 20 year life cycle between steam and diesel plants is glaring. The total costs for the 2-stroke slow speed and DFDE medium speed engines are approximately **USD 31.0 million** and **USD 22.0 million,** respectively, whereas for a steam turbine it is only about **USD 4.5 million**. The pricing of these costs does not take into account any escalation over the 20 years period and therefore the actual figure could easily be a lot higher. This clearly indicates that the steam plant is far cheaper to maintain in the short as well as long terms. Result of similar investigation [15] on the overall economics of different LNG ship propulsions systems in terms of optimum viable freight rate is shown in Figure 10. The results indicate all propulsion options offer comparable economic viability which is in agreement with the findings in this paper.

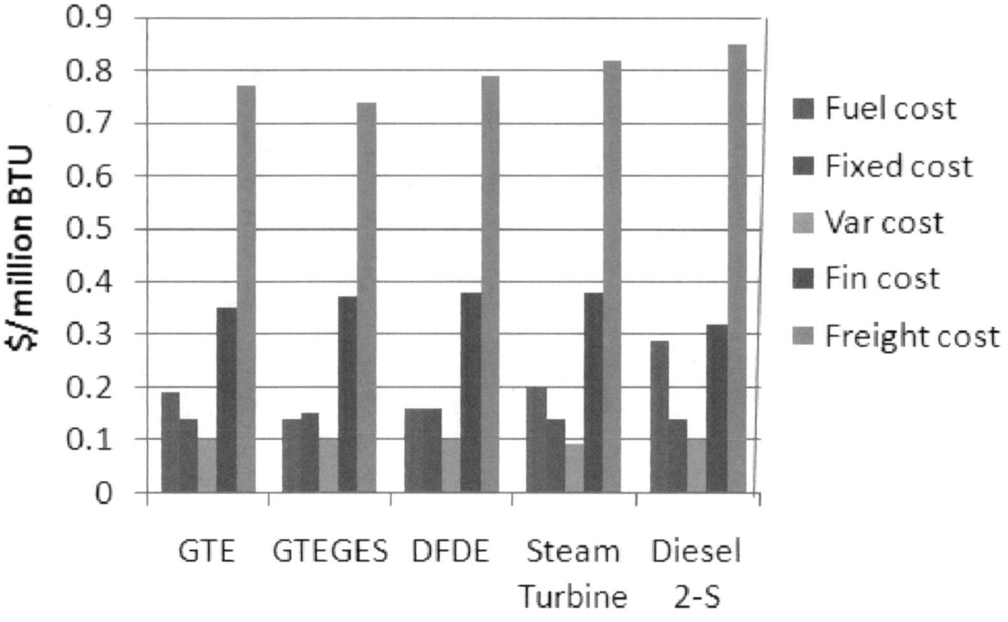

Figure 10 Freight rate comparison [10]

5. Conclusions

The benchmark features of four available propulsion options to be used for reliable comparison have been discussed in sufficient details to identify their relative merits and demerits. Results of investigations are summarized in Sections 5.1 and 5.2.

5.1 General

(i) Most investigation reported in the literature have not been conducted on common criteria hence interpretation of those results for and against specific choice of the propulsion options may not be fully valid.

(ii) For instance, is the reported favorable performance of DFDE really due to improved thermal efficiency or because it is operated in 100% gas burning mode for laden as well as ballast voyages may be still debatable.

(iii) Similarly for DFDE, the cost of high maintenance and low plant availability have not been considered explicitly.

(iv) In view of current low price of gas in comparison to HFO and also strict restriction on gas emission, is gas re-liquefaction still a valid choice from an economic view point ?.

(v) Cost of gas emission control has not been accounted for in arriving at the overall economic grading of different options.

5.2 Technical

(i) Except for the combined cycle gas turbine plans the fuel efficiency of other three propulsion options are comparable.

(ii) Results of reliability analysis, Figure 9, indicates that after initial running in period the UST plant offers improved reliability in comparison to the DFDE. This finding is in agreement with the operational experience of these plants.

(iii) The estimated MTTF of UST is 40 % higher in comparison to the DFDE plant which agrees well with operating experience.

(iv) Calculation of availability shows that even when based only on scheduled outage data the UST has 41% higher availability in comparison to the DFDE.

(v) In terms of NOx and SOx emissions, performance of the UST plant is far superior (Figure 8(a)). The result in Figure 8(b) for NOx emission of DFDE appears inconsistent.

(vi) The CO_2 emission being primarily a function of the amount of fuel burnt, thus for a specific power output it will depend on the SFC of the plant.

(vii) The overall life cycle maintenance and repair costs of the UST is lowest in comparison to other propulsion options.

In final conclusion it may be stated that although comprehensive investigations may constitute an elegant technical analysis of large LNG ships propulsion options but because of the complex financial nature of the global gas business that alone cannot be the actual basis to decide upon a specific choice of the propulsion unit. Many more factors, such as nature of agreement between the gas supplier and receiver, i.e long term supply contract or spot purchase for different clients with variable voyage routes, the terms of project finance and reliability of the gas supply etc will ultimately play key roles in the selection of propulsion system for large LNG ships.

Nomenclature

BOG	Boil Off Gas
CODAG	Combined Diesel Engine and Gas Turbine Propulsion
COGAG	Combined Cruising Gas Turbine with Booster Gas Turbine Propulsion
CST	Conventional Steam Turbine
UST	Ultra Steam Turbine
DFDE	Dual Fuel Diesel Electric
DE+RL	2 Stroke Slow Speed Diesel Engine with Re-liquefaction plant
GCU	Gas Combustion Unit
LNG	Liquefied Natural Gas
MARPOL	Marine Pollution
MTTF	Mean Time To Failure
NG	Natural Gas
UST	Ultra Steam Turbine

References

[1] US Department of Energy, Energy Information Administration, Energy Outlook 2010

[2] Kristen Schebach et-al, The next generation of large LNG carriers, Proc. of the 9th International Marine Design Conference, 15-19 May 2006, Ann Arbor,MI

[3] MAN Diesel and Turbo, Propulsion Trends in LNG Carriers, http//www.manb&w.com

[4] Sinha, R P , Marine Steam Turbine Theory and Practice, Lecture notes, Malaysian Maritime Academy

[5] Mitsubishi Heavy Industries, LNG-15 Ultra Steam Turbine 24-27 April 2007, Spain

[6] Barend Thijssen, 2007 Wartsila Finland Oy, Ship power solutions,: Dual Fuel Electric Electric LNG Carriers, LNG Shipping Operations, Hamburg, September 2006

[7] MAN B&W, LNG Carrier Propulsion by ME Engines and Reliquefaction

[8] Lloyd's Register ,2005, Liquefied Natural Gas Ship Training, Alternative propulsion for LNG ships, Document version 2.0 created Mar 2005

[9] Dong Yeo et-al, Propulsion Alternatives for Modern LNG Carriers, Paper PS6-S, Samsung Heavy Industries Co Ltd http/www.shi.samsung.co.kr

[10] W.S Waney, et-al, A Natural Evolution of the Modern LNG carriers, The application of Gas turbines for LNG carriers propulsion systems., GASTECH 2005

[11] M A Ganesen et-al 2008, Alternative Propulsion System for LNG Ship, International Gas Union Research Conference,(IGRC-2008) Paris 8 October 2008

[12] Jan Fredrik Hansen et-al, Comparison of Electric Power and Propulsion Plants for LNG Carriers with different Propulsion Systems, ABB-AS, Oslo, Norway, http//www.abb.com/marine

[13] Norwegian Marine Technology Research Institute, MARINETEK REPORT-D5-6 LNG supply chain feasibility study overall report,Nov 2008 MARINETEK Report No-222120.50.03.

[14] Joseph Collannino,Modelling of Combustion Systems-A Practical Approach, Taylor & Francis ISBN 0-8493-3365-2.

[15] Gupta B et-al, The future of LNG Transportation, Various Propulsion Alternatives, Martin's Marine Engineering, www.dieselduck.net

IOP Conf. Series: Materials Science and Engineering **36** (2012) 012005 doi:10.1088/1757-899X/36/1/012005

Eddy current braking experiment using brake disc from aluminium series of Al6061 and Al7075

M Z Baharom[1,2,a], **M Z Nuawi**[1,b], **G Priyandoko**[2,c] and **S M Harris**[1,d]
[1]Department of Mechanic and Material,
Universiti Kebangsaan Malaysia, 46300 Bangi, Selangor, Malaysia.
[2]Faculty of Mechanical Engineering,
Universiti Malaysia Pahang, 26600 UMP Pekan, Pahang, Malaysia.

E-mail: [a]mohamadzairi@ump.edu.my, [b]zaki@eng.ukm.my, [c]gigih@ump.edu.my, [d]salleh@eng.ukm.my

Abstract. The electromagnetic braking using eddy current was studied, focused on two series of aluminium as the brake disc which are Al6061 and Al7075. This paper presents the comparison for both series in a few varied parameters related to eddy current braking such as air-gap, number of turns and brake disc thickness. Optical tachometer has been used along with PULSE analyzer to capture the speed (rpm) and time (s). The findings shows that the smaller the air-gap, the larger of electromagnet turns and the thicker disc thickness is, will generate higher braking torque to stop the rotational motion of disc brake and give great performance for eddy current braking. Thos parameters that been evaluated also addressed a potential on expanding this knowledge to develop an electromagnetic braking system to replace the conventional braking system.

1. Introduction

Eddy current braking system has a lot of advantages compared to the conventional braking system. The advantages such as it can reduce the wear of brake pad, vibration and it is environmental friendly. Eddy current braking was said as environmental friendly because it can reduce the pollution of the wear debris from the brake pad itself [1].

In an early research, it was founded that aluminium is the best material to be use as the brake disc compared to copper and zink [13]. The reactions and eddy current effects on aluminium are more effective than copper and zink. As we know, the motion been slow down by a magnetic drag force that has been produced while a conducting material is moving through a stationery magnet or otherwise [2]. The changing magnetic field will induce eddy current in the conductor [2]. These currents will dissipate energy in the conductor and generate drag force [2].

With an experiment setup that has been done by Manuel I. Gonzalez in his laboratory [3], the research regarding eddy current braking behaviour using electromagnet has been expanded by us which focused on two series of aluminium which are Al6061 and Al7075. Both series are the most common alloys of aluminium for general purpose use. They have been used for aircraft structures, bicycle components, chassis plates and wheel spacers for car. Therefore, these two series of Aluminium have been chosen to be tested in this eddy current braking experiment as the brake disc.

Published under licence by IOP Publishing Ltd

2. Experimental Procedure

Figure 1 shows the experiment setup for this study while Figure 2 shows the 3D view of the experiment setup. Based on the eddy current braking experiment done before [3], both setups are quite similar with a few additional equipments plug-in. The use of digital tachometer been replaced by optical tachometer which connected to the PULSE analyzer to capture the speed (rpm) and time (s). Meaning that it also replaces the function of stopwatch in the previous study by Manuel. By doing this, the speed (rpm) can be recorded directly and more precise. Besides that, power supply that been used is higher power of 10A to induce the current for electromagnet coil. The experiment also been conducted in the semi-anechoic chamber to reduce the environmental error.

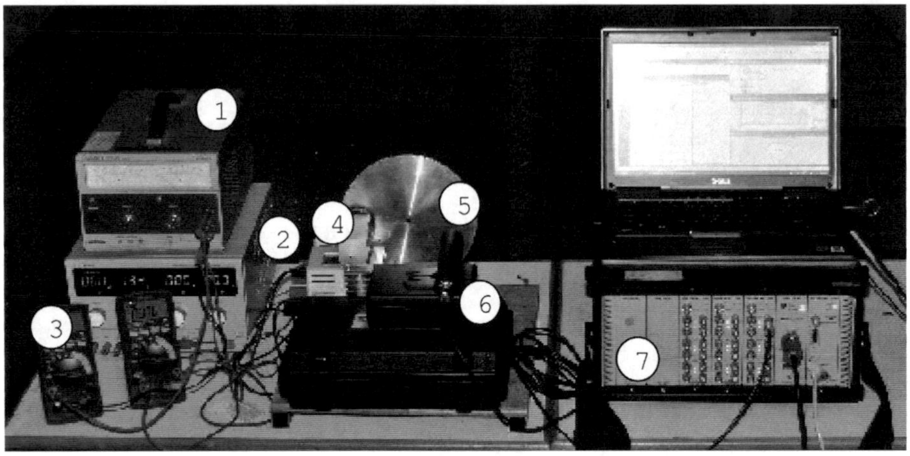

Figure 1. The experimental setup of eddy current braking experiment using electromagnet. 1:power supply for electromagnet, 2: power supply for motor, 3:multimeter, 4: electromagnet, 5: disc attached to motor, 6:optical

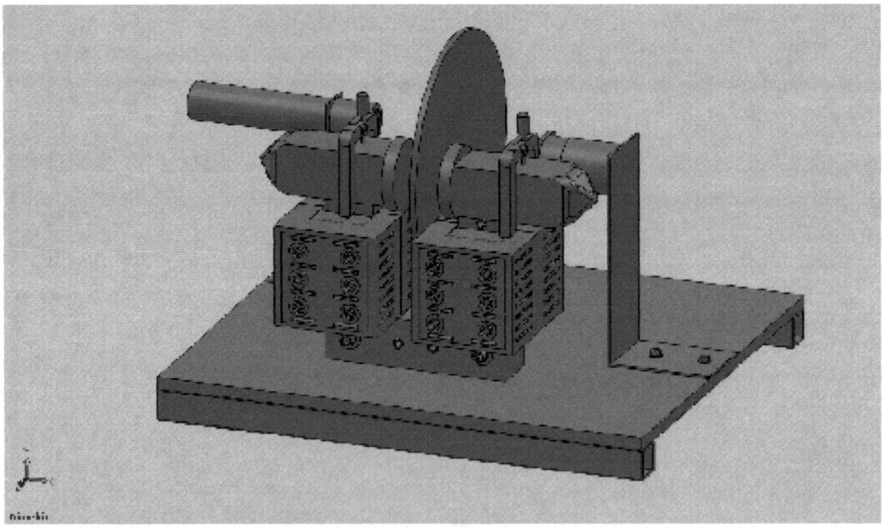

Figure 2. 3D view for the experimental setup of eddy current braking

The power supply for the motor is a 0-30V power supply. It will supply voltage about 11V to 15V for this experiment to achieve speed of 90 rpm to 130 rpm when unloaded. Then, the current will be induced to the electromagnet in increment of 0.5A each. The optical tachometer will record the speed (rpm) of the disc rotational motion. The experiment uses two different thicknesses of disc which are 4mm and 5mm. Therefore, this study will looks on the effect of different series of Al6061 and Al7075, the air-gap, thickness of the disc and the number of electromagnet turns that been used. Eddy current braking study, *a, V, t, d, I, N* will represent the air-gap, voltage supplied to the motor, disc thickness, disc diameter, current induced and number of electromagnet turns, respectively.

3. Results and discussions

This section will discussed three parts of this study which are the effect of different air-gap, number of electromagnet turns and material comparison of Al6061 and Al7075. Graph will shows the effect of parameters evaluated while increasing the current induced. It also shows the decelaration occured on the bake disc when current induced into the system.

3.1. Air-gap effect (a)

Figure 3(a) and 3(b) shows the effect regarding air-gap of 3mm and 5mm, respectively, for both series of aluminium. Constant voltage of 11V been supplied to the DC motor with the use of electromagnet with 250 turns. As we can see, the smaller the air-gap between the disc and the iron core of the electromagnet was, the better the eddy current braking effect is. Small air-gap will produce a stronger magnetic field, *B* as been mentioned by Kapjin Lee and friends in their research [4]. Higher magnetic flux been induced in the small air-gap of 1mm compared to 3mm & 5mm. It means that drag force has been generated in the smaller air-gap for both Al6061 & Al7075 disc will contribute to slow the motion faster.

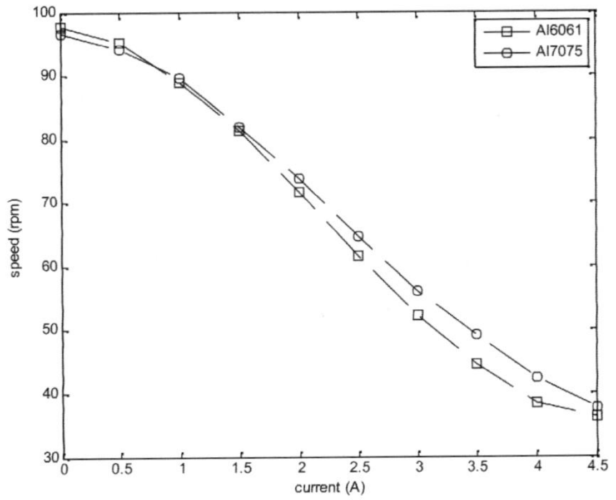

Figure 3(a). Speed vs current for Al6061 and Al7075 *a*(1mm), *N*(250), *d*(4mm) and *V*(11V).

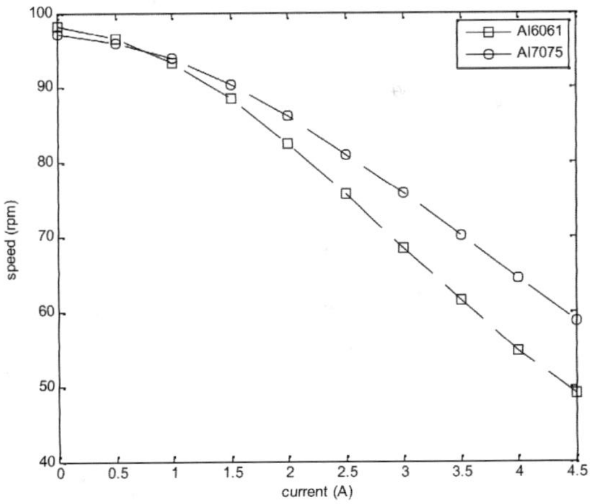

Figure 3(b). Speed vs current for Al6061 andAl7075 with *a*(3mm), *N*(250), *d*(4mm) and *V*(11V).

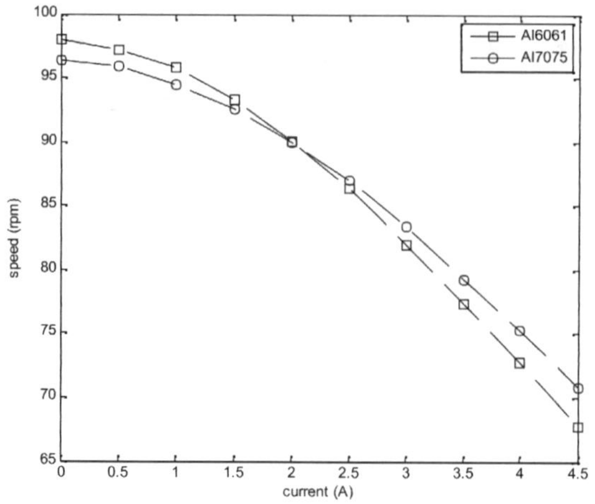

Figure 3(b). Speed vs current for Al6061 andAl7075 with *a*(5mm), *N*(250), *d*(4mm) and *V*(11V).

3.2. Number of electromagnet turns (N)

Figure 4 shows the graph of speed (rpm) vs current (A) for electromagnet with 50 and 250 turns with the air-gap of 1mm, 11V voltage supply and disc thickness of 4mm. During the test, due to safety issue, maximum permissible current that been induced for electromagnet with 50 and 250 turns are 8A and 4.5A, respectively. As we can see in the Figure 4, the larger the number of turns, the better the braking is. It is parallel to the Kapjin Lee and friends research as been mentioned that when we increase the number of turns, it will stronger the magnetic field [4]. Speed been dropped from about 98rpm to almost 38rpm with the use of electromagnet with 250 turns and maximum of 4.5A current induced. For electromagnet with 50 turns, the speed just drop to about 81rpm with the maximum of 8A current induced. Large number of electromagnet turns will require a small amount of current to produce higher braking torque compared to less number of turns which needed higher amount of current. It is really important to have an optimum number of electromagnet turns for eddy current braking to save amount of current that will be induced into the system.

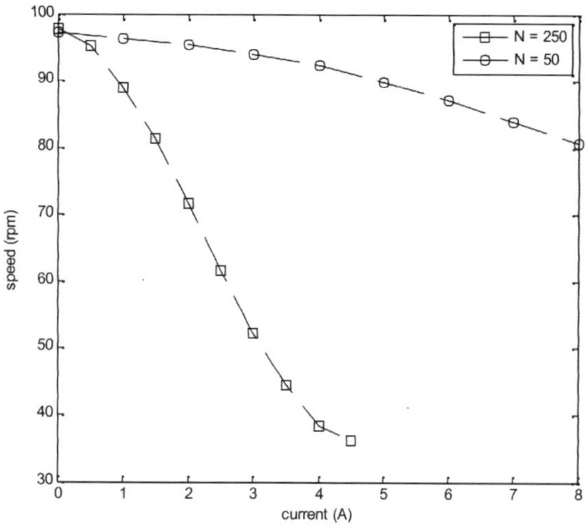

Figure 4. Speed vs current of Al6061 for different number of electromagnet turns with *a*(1mm), *d*(4mm) and *V*(11V).

3.3. Disc thickness (d)

In figure 5, we can see the effect of using difference thickness for the disc brake. For this experiment, disc of 4mm & 5mm in thickness has been tested, with the supplied of 11V to DC mototr, air-gap of 1mm and 250 turns electromagnet. Result shows that the larger the disc thickness is, the better the braking is. Torque produced on thicker disc is higher which will approaching the DC motor torque in order to stop the motion. Disc with thickness of 5mm shows better decelaration for both material of Al6061 and Al7075 compared to 4mm disc thickness.

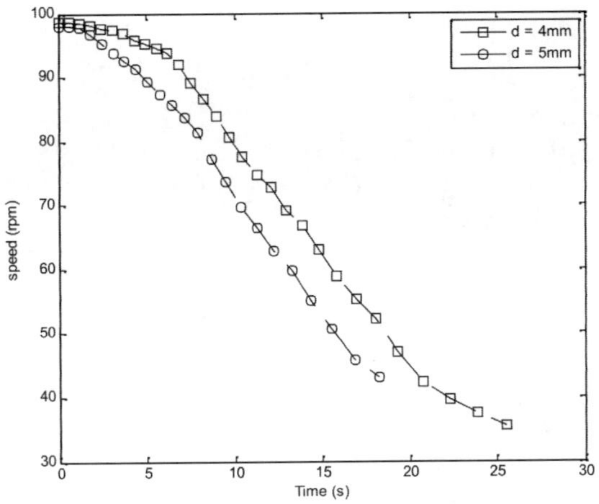

Figure 5(a). Speed vs time of Al6061 for different disc thickness with *a*(1mm), *N*(250) and *V*(11V).

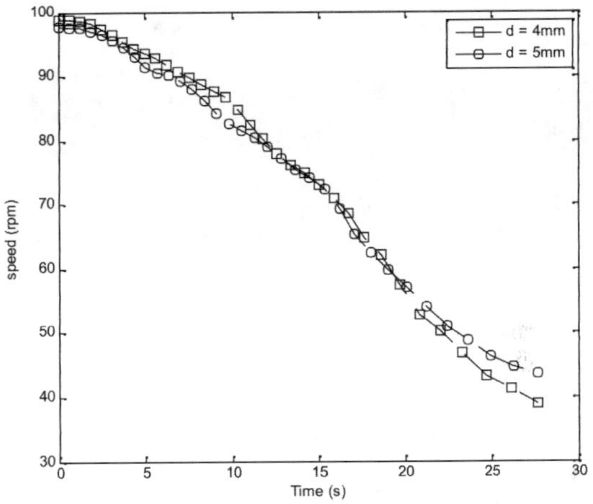

Figure 5(b). Speed vs time of Al7075 for different disc thickness with *a*(1mm), *N*(250) and *V*(11V).

3.4 Comparison of Al6061 and Al7075

From figure 6, we can see the comparison between two series of aluminium which are Al6061 and Al7075. For this test, the number of electromagnet turns fixed at 250, air-gap of 1mm, disc thickness of 4mm and motor voltage supply at 11V. For current induced of 4A, Al6061 has dropped its speed to about 38 rpm compared to Al7075 which achieved only 42 rpm. It means that Al6061 produced higher braking torque compared to Al7075. The electrical conductivity of Al6061 and Al7075 are 2.73 x 10^{7} Ωm^{-1} and 1.92 x $10^{7}\Omega m^{-1}$, respectively. Higher electrical conductivity influenced the generation of greater breaking torque.

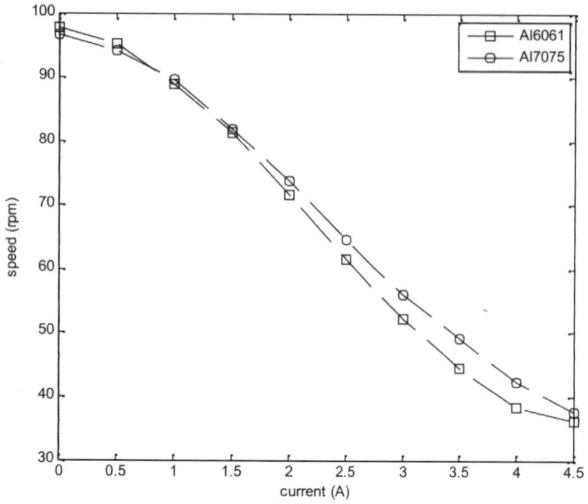

Figure 6. Speed vs current for Al6061 & Al7075 with *N*(250), *a*(1mm), *d*(4mm) and *V*(11V).

4. Conclusion

From the experiment, we may conclude that Al6061 is better than Al7075 to be used as the brake disc material for our electromagnetic braking system using eddy current project. Besides that, we may also conclude that the smaller the air-gap, the larger the number of turns is, and the larger the disc thickness is, will generate higher braking torque and performance for electromagnetic braking. Finding mentioned parameters from this experiment are parallel with the theory. Those parameters that have been evaluated show that there is a potential of developing the electromagnetic braking system to replace the conventional braking system.

5. Acknowledgement

The authors would like to thank Faculty of Engineering and Build Environment, Universiti Kebangsaan Malaysia for hosting the place and capital to carry out this study. Also to Mr.Jabbar, Mr.Kamal and other technicians that assist us during the fabrication and setup of the test rig until the completion on this first stage of this project.

6. References

[1] K. Kukutschovaa, V. Roubi˘ceka, K. Malachovab, Z. Pavli˘ckovab, R. Holu˘sab, J. Kuba˘ckovac, V. Mi˘ckac, D. MacCrimmond, P. Filip d. 2009. Wear Mechanism in Automotive Brake Materials, Wear Debris and its Potential Environmental Impact, *International Journal of Wear*.

[2] M. Jou, J.K. Shiau, C.C. Sun. 2006. Design of a Magnetic Braking System, *Journal of Magnetism and Magnetic Materials, 304(2006) c234-c236.*

[3] M.I. Gonzalez. 2004. Experiments With Eddy Currents: The Eddy Current Brake, *European Journal of Physics, 25(2004) 463-468.*

[4] K.Lee, K.Park. 1998. Environmental Optimal Robust Control of a Contactless Brake System Using an Eddy Current, *Journal of Mechatronics.*

[5] O. Uexkull, S. Skerfving, R. Doyle, M. Braungart. 2005. Carbide Antimony in brake pads—a carcinogenic component?, *J. Cleaner Prod. 13(2005) 19-31.*

[6] R.A. Peters. 2008. Environmental Effects of Copper in Brake Pad Wear Debris, *Brake Pad Partnership, Brake Colloquium & Exhibition.*

[7] M.T. Thomson. Permanent Magnet Electrodynamic Brakes Design Principles and Scaling Laws, *Online Symposium for Electronics Engineers.*

[8] H. Straky, M. Kochem, J. Schmitt, R. Isermann. 2002. Influences of Braking System Faults On Vehicle Dynamics, *Control Engineering Practice.*

[9] B.Ebrahimia, M.B Khameseea, M.F Golnaraghib. 2008. Design and Modelling of a Magnetic Shock Absorber on Eddy Current Damping Effect, *Journal of Sound and Vibration, 315 (2008) 875-889.*

[10] W.H Li, H. Du. 2003. Design and Experimental Evaluation of a Magnetorheological Brake, *Journal of Advanced ManufacturingTechnology, 21:508-515.*

[11] V.V Kickinyov. 2007. An Automatic Torque Control System for a Bicycle Ergometer Equipped with an Eddy Current Brake.*119-121.*

[12] B. Funieru, A. Binder. 2007. Losses and Forces due to Eddy Currents in a Magnetically Non-Linear Conductive Half Space. *90:43-54.*

[13] M.Z. Baharom, M.Z. Nuawi, G. Priyandoko, S.M. Harris, L.M. Siow. 2011. Eddy current braking study for brake disc of aluminium, copper and zinc, *Regional Engineering Postgraduate Conference.*

Productivity improvement using industrial engineering tools

H.A. Salaam[1], S. B. How, M. F. Faisae
Faculty of Mechanical Engineering, Universiti Malaysia Pahang (UMP),
Pekan, Pahang Darul Makmur, 26600, Malaysia

E-mail: hadisalaam@ump.edu.my

Abstract. Minimizing the number of defects is important to any company since it influence their outputs and profits. The aim of this paper is to study the implementation of industrial engineering tools in a manufacturing recycle paper box company. This study starts with reading the standard operation procedures and analyzing the process flow to get the whole idea on how to manufacture paper box. At the same time, observations at the production line were made to identify problem occurs in the production line. By using check sheet, the defect data from each station were collected and have been analyzed using Pareto Chart. From the chart, it is found that glue workstation shows the highest number of defects. Based on observation at the glue workstation, the existing method used to glue the box was inappropriate because the operator used a lot of glue. Then, by using cause and effect diagram, the root cause of the problem was identified and solutions to overcome the problem were proposed. There are three suggestions proposed to overcome this problem. Cost reduction for each solution was calculated and the best solution is using three hair drier to dry the sticky glue which produce only 6.4 defects in an hour with cost of RM 0.0224.

Keywords: Industrial engineering tools, Check sheet, Pareto chart and Cause and effect diagram.

1. Introduction

Survival of any industry in today's environment especially manufacturing industry not only depends on response time but also quality of the product produced. In this respect, seven quality control tools can be utilized as a method or process to reduce the manufacturing costs, production time and increasing the production line productivity.

Productivity can be defined as a ratio between output and input. Productivity improvement is a critical success factor and the foundation of profitability [1]. Productivity measurement is a long-term measurement. Any changes in dynamic potential show a growth or reduction of figures over a long period [2]. Industrial Engineering in the other hand concerned with the design, improvement, and installation of the integrated system of men, materials, and equipment. It draws upon specialized knowledge and skills in the mathematical, physical, and social sciences, together with the principles and methods of engineering analysis and design, to specify, predict and evaluate the result to be obtained from such a system [3].

Today, quality plays an important role as it leads to increasing number of product sold and also increment of company profit. Productivity, quality, and cost of operation relatively depended to each other. By improving the productivity, the quality also must be improved and hence lower the reject rates or defects [4].

[1]Corresponding author: H. A Salaam
Email address: hadisalaam@ump.edu.my

Published under licence by IOP Publishing Ltd

Engineer needs to know what the main problem is, the causes of the problem before they proposed any solutions. However, logical and systematic method should be applied to make the job easier by locating and eliminating the root cause of the problem. There are many techniques [5] available in problem solving, ranging from simple and easy to use methods to relatively complicate and advance statistical tools [6]. The Seven QC Tools are easier to apply and understand and yet proven scientific management tools. It can be used in all process phases, starting from the beginning of a product development up to management of a process, on day to day basis and in systematic manner [7]. They form the fundamental foundation for all problem solving and quality control activities.

Generally, the Seven QC tools are check sheet, Pareto chart, flowchart, cause and effect diagram, histogram, scatter diagram and control chart [8] and [9]. These tools are also known as Total Quality Management (TQM) tools [10].

Many researches have been done using seven quality control tools. Pierchala and Jyoti Surti [11] and Winco [6] used Statistical Process Control to solve problems their case study. Jafri and Chan [4] presents a case study in plastic injection molding company deployed check sheets, pareto chart, cause and effect diagram and control chart to improve the monthly defect quality from 13.49% to 7.4%. Naser [12] used histogram, scatter plot and pie chart to investigate their case study.

2. Methodology

This study starts with identifying suitable company to do this project. The main criteria used to select a company are small medium enterprise (SME). This is because SME companies did not have enough capital to hire expertise to help them to improve their production line productivity, quality of the product produced and at the same time reduce the number of defect products. The company that has been chooses is a paper box producer company located in Kuantan, Pahang.

Next, a few visits were organized to the company to study their production line layout, product process flows and the number of defects produced in each production line. Based on observation and discussions with the company representative, the selected line is the one which produce 1040 mm box. This is because it has the highest demand from customer, daily produced and at the same time it produces many defects. During the visits, observations at the production line were done to see how the operators perform their job at the production line.

Besides that, interview with operators were organized to get the feedback and also defect product data were collected and analyzed using Pareto Chart. Next, cause and effect diagram were used to identify the root cause of the defect problems and proposed four solutions to overcome the problem. Manufacturing cost calculation analysis was done before select the best solution to be implemented at the company.

3. Results and discussions

Based on the observation in the production line, interview with operators and supervisor and also data collection results, it was found that the production line produce 1040 mm paper box record the highest number of defects at glue workstation. It recorded 23 defects follows by slitter and partition workstations with 7 and 6 defects respectively as shown in Table 1. When Pareto Chart was plotted in Figure 1, it shows that the glue workstation has the highest frequency of defect with 23 defects or 46% of the defects that occurs on the whole process. Results from both Table 1 and Figure 1 shows that there is a problem occurs at the glue workstation.

Table 1. Number of defect occurs at each Workstation.

Defect	Pallet					Frequency
	1st	**2nd**	**3rd**	**4th**	**5th**	
Glue	////	///	~~////~~	~~////~~	~~////~~ /	23
Partition	//		/	/	//	6
Stitching		//		/		3
Slitter		//	/		////	7
Print	/				/	2
Slotter				//		2
Die Cut	/		/		//	4
Tie		/	/		/	3

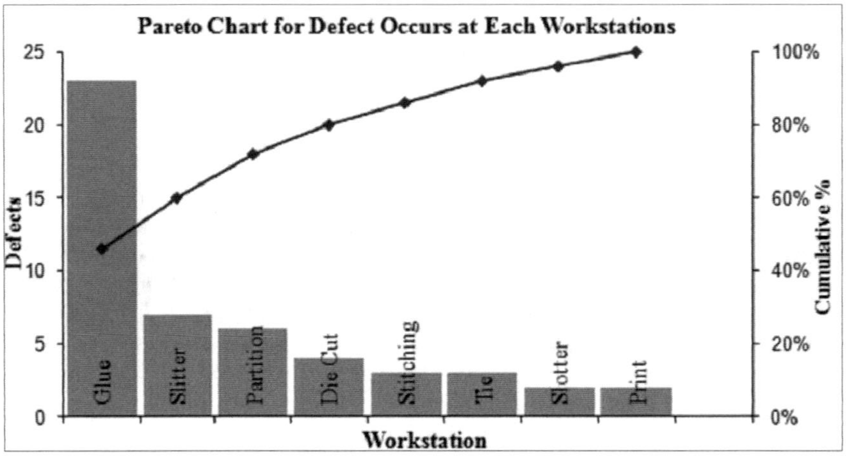

Figure 1. Pareto chart for defect occurs at each workstation.

From the Figure 1; we can see that glue workstation has the highest number of defects which can cause the company unable to meet the customer demand. At the same time, company daily output drop due to the error occurs in glue workstation. Hence improvement should be done at glue workstation to minimize the error. Next, root cause analysis is done by using cause and effect analysis (fish bone) diagram. It is divided into 4 main categories; Man, Machine, Method and Material. Since the workstation is manually operated, there is no analysis for machine. Feedback from operator during the interview shows that the box paper usually sticks one to another due to stacking the box paper immediately before the glue dried. Besides that, lack of rolling and pouring glue skills on the board is one of the root causes of this problem. Another problem is that, some of the operators were tired due to long working hour since they need to do overtime to fulfill the customer demand.

When looking at the method used in this workstation, the current stacking method that they used is not sufficient enough. After finish rolling the glue at the paper board, they stack it immediately before it dried. Besides that, the way they used to roll the roller in the glue tray is not sufficient because the operator arm wrist can injured easily caused by wrong position of body posture and at the same time doing repetitive work again and again. Some of the operator did not roll the roller first in the tray to minimize the glue quantity that will be rolled at the paper box. This leads to using an excessive of glue and it will take a longer time to dry.

When looking at the material side, the design of the box paper and material used to make the box can be improved to minimize the assembly time and manufacturing costs. Besides that, the type of glue used can be change to more environmental friendly and odourless glue which do not harm the operator. The summary of the root cause problem for glue workstation are shown in Figure 2.

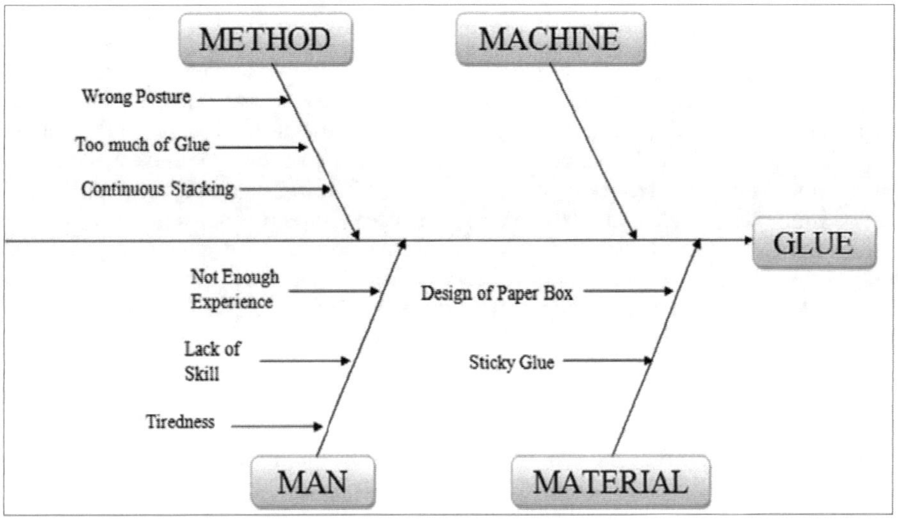

Figure 2. Summary of root cause analysis of glue workstation.

To solve this problem, there are three solutions have been proposed in this project. They are divided into three solutions. The first solution is divided the paper box into three groups before stacking up, followed by the usage of electric fan or airy place for glue workstation, and the third solution is to use table lamp or hair dryer to dried out the glue. The existing method shows that the defects occurs on glue workstation is due to the glue not dry in time and at the same time the operator straight away stacked the paper box on top of it and it caused the new paper box to stick together. Hence the alternative way to solve this problem is to propose not to stack the glued paper box together but divide it into three groups such in Figure 3. For an example, when the paper box is glued, the 1st paper box is place at the slot A while the 2nd paper box is place at slot B and the 3rd paper box placed at the C slot. Hence, the paper box got more time to expose to the air and the possibility to dry is much higher than just place the paper box at a place in a time.

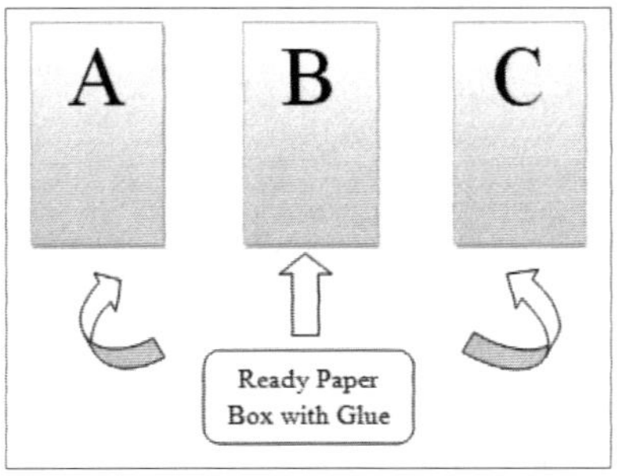

Figure 3. Grouping paper box into three groups.

Standing fan can be used by placing it in front of the paper box after it is glued. By switching on the fan, the wind that blows to the paper box can let the glue dry. Besides using the fan to dry out the glue at the paper box, the glue workstation can be relocate to an open area which has a good air circulation to make sure the paper box is dry. However, there is a concern where the wind that blows might be too strong and blow away the entire paper box. Hence, a weight need to place on top of the paper box to make sure it will not fly away.

Besides that, a small table lamp or hair dryer can be use to dry the glue. The heat transfer concept is applied where the heat of the lamp or hair dryer will dried out the paper box. Paper box can be place under the lamp or hair dryer and the air is blowing for a few seconds before stacked the paper box at semi-finish place. However, the table lamp that can be used is limited to the table lamp with long bulb. This is because the round bulb can only heat a small portion of the paper box while long bulb can heat longer portion of the paper box and reduce the production time.

A few trials were conducted to see the performance of the proposed solutions. After perform a trial for a few days, number of defect products was collected and the defect cost was calculated. The summary of the results are shown in Table 2. Here, we can see that the current glue workstation produce average of 123 defects, followed by table fan 100 defects and standing fan 91.2 defects, 3 table lamps 86 defects and lastly, by using three hair dryer 6.4 defects. By looking at the results, it shows a tremendous decreasing which is good for the company. This is due to the new arrangement of standard operation procedures to reduce the amount of glue used and give extra time to let the glue dry before stacking it.

Next, when we looked at the cost of defect per hour, the highest defects cost is the existing glue workstation which record RM 0.403932. Table fan is in the second place with RM 0.33283 and Standing fan in the third place with RM 0.3045533. Fourth place is 3 Table lamps with RM 0.2833786 and the last place which records the lowest defect costs is 3 hair dryers with RM 0.0224365. Based on the results, it is shown that by using 3 hair dryers, the number of defects can be reduce dramatically due to the usage of hair dryer to dry the glue. Besides that, the arrangement of the ready paper box with glue give an interval time to allow the glue to dry.

Table 2. Summary of experiment results between the existing workstation and proposed solutions.

Method	Trial					Average	Total Defect Costs (RM) / hour
	1	2	3	4	5		
Existing Glue Workstation	120	145	140	100	110	123	0.403932
3 Table Lamp	99	90	85	80	76	86	0.2833786
Table Fan	120	101	105	94	80	100	0.33283
Standing Fan	105	99	93	87	72	91.2	0.3045533
3 Hair Dryer	10	8	4	5	5	6.4	0.0224365

4. Conclusions

This paper describes about the using of industrial engineering tools to reduce number of defects and at the same time increase production line productivity. Form the experiment that have been done, the main objective to reduce number of defects product has been achieved. From all four options that have been proposed, it is found that the best solution is using three hair dryer which record 6.4 defects in one hour or total defects costs of RM 0.0224.

References

[1] Kapyla J, Jaaskelainen A and Lonnqvist A 2009 Identifying future challenges for productivity research: evidence from Finland *International Journal of Productivity and Performance Management.* . **59**(7): 607-623

[2] Heap J, 2007 Stormy productivity ahead? *International Journal of Productivity and Performance Management.* **56**(2): 170-177

[3] Lo V H Y and Sculli D 1995 Industrial Engineering and TQM *Training for Quality.* **3**(3): 4-7

[4] Jafri M R and Chan K T 2001 Improving Quality with Basic Statistical Process Control (SPC) Tools: A Case Study *Jurnal Teknologi.* **35**(A): 21-34

[5] Reed, D. 1992 *TQC Tools in Review*, New Jersey: Printed Circuit Fabrication

[6] Winco K C Y 1996 An Integrated Model for Manufacturing Process Improvement. *Journal of Materials Processing Technology* **61**: 39-43

[7] Paliska G, Pavletic D, Sokovic M 2007 Quality tools – Systematic use in Process Industry. *Journal of Achievements in Materials and Manufacturing Engineering.* **25**(1): 79-82

[8] Ishikawa K 1985 *What is Total Quality Control.* Prentice Hall. Eaglewood Cliff, N.J

[9] Pimblott J G 1990 Managing Improvement – Where to start, Quality Forum. *The TQM Magazine.* **16**: 165-173

[10] Jay H and Barry R 2008 *Principles of Operations Management.* 7th ed. New Jersey: Pearson Prentice Hall, Inc

[11] Pierchala C E, and Jyoti S B S 1999 Control Charts as a Tool in Data Quality Improvement. Technical Report Washington: NHTSA/ DOT HS 809 005

[12] Naser A A 2007 Application of quality tools by the Saudi food industry *The TQM Magazine.* **19**(2): 150-161

Design and development of an Integrated Slasher (Pulverizer) for Sweet Potato Harvester: A Review

Amer N N Kakahy[1], D Ahmad , M D Akhir[2] , A Ishak[3] and S.Sulaiman[4]
[1]Department of Biological & Agricultural Engineering, Faculty of Engineering, University Putra Malaysia , 43400 UPM, Serdang, Selangor, Malaysia.
[2]Mechanization & Automation Research Center, MARDI, Serdang, Selangor, Malaysia.
[3]Department of Electrical and Electronics Engineering,
[4] Department of Mechanical and Manufacturing Engineering, University Putra Malaysia

E-mail[1]:amer_kakahy@yahoo.com

Abstract. In this study information on the development of combined sweet potato harvester is presented. The study focuses on the design, manufacture and choice of the best slasher for cutting and fragmentation of stems and leaves (vegetative portion) for the crop to facilitate the process of harvesting and extraction of tubers. A survey on various types and forms of cutters (mowers) and knives normally used for cutting vines and leaves was conducted and concluded that the vertical type and flail type mowers are the most effective in removing of the vegetative growth. In addition there is a need to design an integrated harvester, which combines both the cutting and digging in a single pass in order to reduce the time taken for harvesting, the fuel consumption and the number of machines used since this will give some economical advantages. In addition to the design requirements, it is recommended that number of tests should be conducted to select the suitable speed for the tractor, the velocity for cutters and the length of knives for both sides of the ridges to avoid damage to the tubers.

1. Introduction

Sweet potato *Ipomoea batatas (L.) Lam.* belongs to *convoluvlaceae* family and is an important and valuable staple crop worldwide (FAO, 1987). It is a nutritious and generous food source for humans and animals as well as a raw material for starch, sugar, alcohol and so on (Kozai et al., 1997; Saiful Islam et al., 2002). The crop grows throughout the tropics and subtropics (Scott, 1992), especially in Asian and African countries where account for about 95% of the world's sweet potato production (Mok, Zhang and Carey, 1997). Sweet potato has high yield even though it was cultivated in adverse situations (Simonne et al., 1993).

Malaysian Ministry of Agriculture and Agro-based Industries has set a target to expand the production of sweet potato from 14,500 tons in 2000 to 300,000 tons in 2010. Sweet potato can be transformed from a mere cash or subsistence crop to a raw material for Agro-based industrial use such as flour. Mean yields targeted to increase from 7.9 to 30 t/ha to be achieved through varietal improvement, better crop management practices as well as reduced crop losses (Tan et al., 2007).
Harvesting is one of most important issues and therefore it will be necessary to have the equipment that facilitate the operations in short time and less quantity as production increases. The only

alternative is to replace labor with machines since agricultural sector is managed by old aged farmers and limited manpower. Farmers have to make the most of available resources. Bris soils are preferred for planting sweet potatoes because of their yield potential in addition to being an alternative crop to tobacco. Because of the increasing global demand for sweet potato and even from Malaysian, steps should be taken to reduce sweet potato production cost will need manpower increased from the estimated labor up to 750 man-hours needed per hectare of crops, although, current estimated labor cost production of sweet potato is RM 3700 to RM 5500 per hectare, depending on the type of soils. Increasing sweet potato production is associated with higher labor requirement, particularly for planting, crop maintenance and harvesting. In-field mechanization presents an attractive solution to this predicament. Machines can speed up operation times and also sustain and increase productivity. (Md. Akhir, 2010).

One of the main factors is the speed of the cutting blades that allow cutting without a counter edge and this depends on the diameter of the trajectory, the inertia, the stem's resistance to flexion and on the maximum cutting force of the apparatus. Also, the energy and the cutting power of mowers and shredders depend on the knives speed, on the feed speed, on the crop properties, on the knives' cutting width and on the number of blades cutting simultaneously. Flow of the material is the most important factor in the power requirements of the shredder which depends linearly on the shredder's advancement speed. The energy demand will be increased in the tractor Power-take-off (PTO) with increasing speed (and flow). The speed of the cutting blades has a considerable effect on the energy of the conveyance of the material, which depends mainly on the flow of feeding but also on the energy losses due to friction and to air resistance during circulation. As the speed of the blades increases, in flail shredders, the height and the distance to which the residues are thrown increases and so do the friction with the air and the demand for power. (Jorge et al., 2009)

Research shows a blade angle in the range 30–40° with a rake angle of 10–20°, are used for forage chopping having cutting speeds of up to 30 m/s and optimum radius of the cutting edge is approximately 0·05 mm. The review concludes that there is a need for more basic information and a better understanding of cutting mechanics. Several measurements are required of cutting forces, together with studies of the effects of cutting speed and providing mechanical support to stems during impact cutting. The physical properties of bulk forage materials are relevant to the cutting process. There is a need to establish a model of cutting mechanics and to relate knife parameters to forage material properties. Further information is required on the chop length distribution of different forms of cutting mechanism. (O'Dogherty, 1982)

Sweet potato Harvester requires suitable equipment for commercial production. Before harvest most of the top growth needs to be removed or it will become entangled in the digging machine. Vine removal is best done with a flail pulveriser where the flails are shaped to the contour of the bed. A standard slasher or pulveriser cannot remove material between the rows and causes chopping into the top of the hill and damaging the roots. As a result, the remaining vines can be cut on both sides of the hill with large sharp coulters mounted on a tool bar. This vine removal should be done a week before digging to toughen the skin of the roots. (Traynor, 2005).

The most common method to mechanically kill cover crops and leave their residues as surface mulches is by mowing. Because of affectivity making leaves cover crop residues uniformly distributed on the soil surface, flail mower is most desirable over a bush hog or rotary type mower. However, a mower may not be the best option to suppress weeds. Mowers generate small pieces of residue which decompose rapidly (Creamer and Dabney, 2002; Amanda, 2005).

Before harvesting the massive amount of intertwined vine material produced by sweet potatoes must be destroyed. Vine-potato detachment increases harvesting efficiency and decreases damage to the potato skin by reducing the amount of handling some mechanical harvesters require that the vines be removed prior to digging, (Smith and Wright, 1994). The rotary mower can be used, but it does not cut the vines between the rows of some varieties. Flails are contoured to fit the rows. Several types of mechanical diggers exist and still are being used to lift sweet potatoes out of the ground. The simplest machines are patented after a low flat-bed type potato digger, (Misener, et al., 1984).

There are three basic types of harvesting machines generally used for forage crops, namely, cutter-bar mowers, rotary mowers having knives on a vertical axis (disc and drum type) and rotary knives on a horizontal axis (flail type). Fine chopping of the mown crop requires a second cutting operation which is typically achieved with a precision cut cylinder. However, mowing and coarse chopping is obtained in a single operation with the flail harvester. Flail-type machines employ high speed rotating knives about a horizontal axis which cut the standing crop by impact. After cutting at a certain stubble height, the whole crop is chopped by repeated cutting, is lacerated against a shear bar and is then blown through the chute into an accompanying trailer or forage wagon. This type of harvester usually produces a comparatively long chop of non-uniform length.

The energy required to operate a flail-type forage harvester is consumed in (1) cutting the forage; (2) blowing the air; (3) moving the chopped particles through the cylinder housing and chute; and (4) deflecting the particles into the collecting forage wagon. In an upward cutting direct cut-and-throw flail forage harvester, the plant is cut and accelerated by the flail knives. The acceleration energy imparted to the chopped forage by the flail knives along with the air stream inside the housing supply the necessary energy to the chopped material for its conveyance. (Chattopadhyay and Pandey1999)

However there are many types of mower such as, rotary mowers, roller mowers, flail mowers, cylinder mowers, sickle bar mowers, hay/silage mowers, zero turn mowers and ride-on mowers. Standard flails are shaped like an extruded "T" or "Y" and the chain attaches to the bottom. There are also proprietary flails with different shapes for shredding larger brush and others that leave a smooth, finish cut.

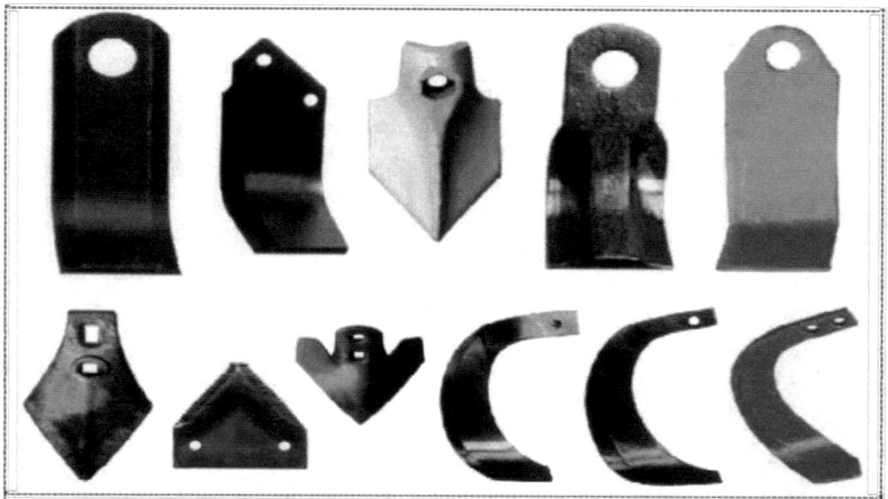

Figure 1. Some types of blades.

Series Flail Mower Specifications			
MODEL	FL-100	FL-125	FL-150
CUTTING WIDTH	39" (100cm)	49" (125cm)	60" (150cm)
CUTTING HEIGHT	1"-6"	1"-6"	1"-6"
ROTOR DIAMETER	15"	15"	15"
ROTOR SPEED	1780rpm	1780rpm	1780rpm
TRACTOR HP	18-30hp	20-35hp	25-40hp

1.1 Problem Statement

Until now, sweet potato is mostly harvested manually. The manual labor cost constitute about 30 - 40% of the total operational cost. Rotary mower cannot cut the vines between the rows of some varieties, another vine removal machine that works very well is a flail chopper they contoured to fit the rows (Md. Akhir, 2010).

Though often overlooked, so to benefit from the advantages such as, reduced labor requirements, time savings, reduced machinery wear, fuel savings, and decreased soil compaction. However, the possible impact of this research is that farmers can benefit from advantages of combining slasher that increase their income and decrease the outcome cost for sweet potato production.

The study focuses on the design, manufacture and choice of the best machine for cutting and fragmentation of stems and leaves (vegetative portion) for the crop to facilitate the process of harvesting and extraction of tubers. The speed of the cutting blades is one of the main factors that allow cutting without a counter edge and this depends on the diameter of the trajectory, the inertia, the stem's resistance to flexion and on the maximum cutting force of the apparatus.

A survey on various types and forms of cutters (mowers) and knives normally used for the cutting vine and leaves was conducted and concluded that the vertical type and flail type mowers (rotary knives on a horizontal axis) are the most effective in removing of the vegetative growth. In addition there is a need to design an integrated harvester, which combines both the cutting and digging in a single pass in order to reduce the time taken for harvesting, the fuel consumption and the number of machines used. This will give some economical advantages. In addition to the design requirements, it is recommended that number of tests should be conducted to select the suitable speed for the tractor and the velocities for cutters and the knives length for both sides of the ridges so that it does not damage the tubers.

1.2 Objectives

The General objectives in this study is to design and develop an integrated slasher (Pulverizer) for sweet potato harvester (two in one). The specific objectives in this study will be focused on:
1- Designing and choosing perfect blades for the slasher.
2- Finding out the best speed for cutting and effect of performance efficiency of cutter.
3- Designing and fabricating the slasher to be fixed to the harvester.
4-Measuring and evaluating the costs.
5- Measuring the power requirements and sweet potato yield by using different mower

1.3 Scope of Work

The research will focus on measuring the performance and costs of integrated slasher for sweet potato harvester used in the field.

2. Materials and Method

2.1 Parameters Determination for Machine Design

Parameters such as soil properties, crops characteristic, types of knives, will be determined by using the lab equipment and measuring tools. AutoCAD design software will be used for drawing the new designed knives .The completed blue print will be sent to the fabricator to fabricate the prototype combine harvester for tuber crops and the completed machine will be tested in the field for observing the performance.

2.2 The materials and methods that will be used in this research

1-Three speeds of tractors to pull the three different slashers (different knives) and three heights of cutting vine
2- Measuring equipment (Tachometer, momentum, time clock and weight meter.) and necessary instrumentation to measure the cutting torque and cutting force.
3- Preparation for pulverizing, designing the research and collecting the data.

This study will consist of three main factors: -

Three speeds of tractor to pull three different slashers (different knives) and three depths of cutting with six replicates and effects on power requirements, plant product and economics. The experiment will focus on the effect of Slasher types, forward speed and height of cutting on Sweet potato yield and their effects on the power requirements (Traction force, Fuel consumption and Power lost with slippage total traction efficiency.)

After gathering all required information from text books, magazines and internet, the test will be designed (RCBD design) based on the following steps:-
1-Testing the new Integrated Slasher
2-The first salsher (combined with harvester) will be pulled by Tractor at three different speeds .
3- Second and third cutting height will be used with the first slasher.
4- Second and third Slasher will be used to complete the treatments on the same above steps.
5- Collecting the data and analyzing the data.

In flail mowers the supplied power N is used to overcome the cutting resistance Nc; ventilation resistance – blast produced by the revolving flails Nw; resistance of the thrown-up cut material Ne; friction resistance in bearings Nt; and transmission gear Ni; or (Kanadyjski, and Karwoski .1974):

$$N = Ne + Nw + Ne + Nt + Ni \quad (1)$$

In the given power consumption balance the first two members are of decisive importance. The value of Nc at mowing vegetable material of specified of specified parameters increases, as is known, with increasing machine working speed, but the same time decreases with increasing cutting speed, or the peripheral speed of the flail blades. On the other hand, however, with an increase in the peripheral speed of the flails ventilation resistance increases

3. Experimental Design

ANOVA test at 95 – 99% confidence levels will be carried out to determine the interaction between the test variables. Data obtained will be subjected to appropriate statistical analysis and test, (t, f, chi-square, etc) to determine their mean differences and differences in their variances. Lastly regression models representing the outcomes of the test will be formed and the significance of each variable will be tested based on the set null and the alternative hypothesis.

4. Results and discussion

This research would contribute to the literature in similar work on design, development, performance and costs of the equipment and tractors used in the sweet potato field. It is expected that the perfect slasher would assist greatly in reducing labor requirement and expanding the new idea on using new technology and work successfully on the field.

A survey on various types and forms of cutters (mowers) and knives normally used for the cutting vine and leaves was conducted and concluded that the vertical type and flail type mowers (rotary knives on a horizontal axis) are the most effective in removing of the vegetative growth. In addition there is a need to design an integrated harvester, which combines both the cutting and digging in a single pass in order to reduce the time taken for harvesting, the fuel consumption and the number of machines used. This will give some economical advantages. In addition to the design requirements, it is recommended that number of tests should be conducted to select the suitable speed for the tractor and the velocities for cutters and the knives length for both sides of the ridges so that it does not damage the tubers.

A flail pulveriser is the best vine removal but the Slasher or pulveriser is not able to remove material between the rows and causes chopping into the top of the hill and damage the roots The new flails are shaped to the contour of the bed. As a result, the remaining vines can be cut on both sides of the hill with large sharp coulters mounted on a tool bar. The other option is to add longer knives on both sides of the flail mowers.

5. Conclusion

The review of research on design and development of an integrated slasher (Pulverizer) for sweet potato have been summarized, and required information to develop the sweet potato harvester like types of knives, slashers and harvester have also been discussed.

Acknowledgments

Thanks are due to the members of the Department of Biological and Agricultural Engineering, UPM and staff of Agricultural Mechanization & Automation Center,MARDI for their generous help and advice during the preparation of this paper.

References

[1] Amanda, L. S. (2005). Development of a more sustainable Sweet potato production system. A thesis submitted to the graduate faculty of Auburn university in partial fulfillment of the requirements for the degree of master of science Auburn, Alabama december 16, 2005

[2] Cz, Kanadyjski, and T. karwoski (1974). Agricultural machines, theory and construction, foreign scientific publications department of the national center for scientifical and economic information, warsaw, poland, vol. 2,pp 465-466

[3] Creamer, N. G. and S. M. Dabney(2002). Killing cover crops mechanically: review of recent literature and assessment of new research results. Amer. J. Of alternative agri. 17(1):32-40

[4] Food agriculture organization (1987). Roots and tubers: their role in food security courier, 10, 62

[5] Jorge, S. P. C. F; F. G. Pegna; C. I. Coronel; F. G. Reina and P. Spugnoli (2009). Power demand of a flail mower during pineapple field demolition. Cien. Inv. Agr. 36(1):59-68

[6] Kozai, T.; C. Kubota; J. Heo; C. Chun; K. Ohyama and G. Niu (1997). Towards efficient vegetative propagation and transplant production of sweet potato (Ipomoea batatas (L.) Lam.) under artificial light in closed ecosystems. In Proceeding of the international workshop on sweet potato production system toward the 21st century (pp. 201–214). Kyushu National Agricultural Experiment Station, Miyazaki 885-0091, Japan

[7] M. Traynor (2005). Sweet potato production guide for the top end. Information booklet 1.department of primary industry, fisheries and mines crops, forestry and horticulture division northern territory government, 2006

[8] Md. Akhir H. (2010). Design and development of a digging device for harvesting sweet potato, Unpublished PhD thesis Submitted to the School of Graduate Studies, Universiti Putra Malaysia

[9] Misener, G. C.; C. D. McLeod, and L. P. McMillan (1984). Evaluation of a prototype potato harvester. Transactions of the asae,: 24-28

[10] Mok, I G. ; Zhang, D. E., and Carey, E. E. (1997). Sweet potato breeding strategy of CIP. In L. LaBonte et al. (Ed.) Proceeding of the international workshop on sweet potato production system toward the 21st century (pp. 9–27). Kyushu National Agricultural Experiment Station, Miyazaki 885-0091, Japan.

[11] O'Dogherty, M. J. (1982). "A review of research on forage chopping." Journal of Agricultural Engineering Research 27(4): 267-289

[12] Chattopadhyay, P. S. and K. P. Pandey (1999). Effect of knife and operational parameters on energy requirement in flail forage harvesting. . Agric. Engng res. (1999) 73, 3}12 article no. Jaer.1998.0395

[13] Saiful Islam, A. F. M.; Kubota, C.; Takagaki, M., and Kozai, T. (2002). Sweet potato growth and yield from plug transplants of different volumes, planted intact or without roots. Crop Science, 42, 822–826

[14] Scott, G. J. (1992). Transforming traditional food crops: Product development for roots and tubers. In G. J. Scott, S. Wiersema, & P. I. Ferguson (Eds.). Product development for roots and tuber crops (Vol. 1, pp. 3). Lima, Peru: Asia. International Potato Center (CIP)

[15] Simonne, A. H.; Kays, S. J.; Koehler, P. E. and Eilenmiller, R. R. (1993). Assessment of bcarotene content in sweet potato breeding lines in relation to dietary requirements. Journal of Food Composition and Analysis, 6, 336–345

[16] Smith, S. L. and M. E. Wright (1994). A Preharvest vine disposal machine for sweet potatoes. Applied Engineering in Agriculture, American Society of Agricultural Engineers. VOL. 10(3): 321-326

[17] Tan, S. L. ; Abdul Aziz, A.M.; Zaharah, A.; Salma, O. and Khatijah, I. (2007). Selection of sweetpotato clones with high β-carotene for processing of nutritious food products. Journal Trop. Agric. and Food. Scince. Vol.35 (2): 213-220

Non-velocity Based Analysis of Passive Ultrasonic Signal for Source Location Detection in Composite Plates: A Pilot Study

Z M Hafizi[1,2], J Epaarachchi[1], C K E Nizwan[2] and K T Lau[1]
[1]Centre of Excellence in Engineered Fibre Composites University of Southern Queensland, Toowoomba QLD 4350, Australia
[2]Faculty of Mechanical Engineering, Universiti Malaysia Pahang, 26600 Pekan, Pahang, Malaysia

E-mail: MohdHafiziBin.Zohari@usq.edu.au, hafizi@ump.edu.my

Abstract. Acoustic Emission (AE) evaluation is one of the fast growing non-destructive techniques, owing to its ability to reveal in advance of any impending failure of a building structure. This capability makes the so called passive ultrasonic technique a very good tool in structural health monitoring; especially for composite structures. In metallic structures, AE technique is currently well established and able to provide accurate and consistent result. However, in composites, the challenge for a reliable AE results is huge due to the orthotropic behaviour of the materials. The present study investigates the energy attenuation of AE signals in thin composite plate and utilized the attenuation pattern into non-velocity based source location detection. Standard Hsu-Nielsen source location testing was applied on a glass fibre epoxy resin laminate and a single channel AE system was used to acquire AE signals with the support from *AEWin* software for signal analysis. A linear source location algorithm utilizing AE signals energy attenuation patterns was developed and tested for the composite specimen. The results revealed that the source location algorithm provides reasonably accurate results of source location for glass fibre epoxy resin laminate.

1. Introduction

Ultrasonic technique is one of the most popular non-destructive testing in current decades. The fast growing of the ultrasonic technology is owing to the recent achievement and development of computer technology which support the complex and tedious signal processing and analysis. The application of ultrasonic evaluation is numerous in Structural Health Monitoring (SHM). With more complicated design of current structures, the evaluation method for health monitoring also needs to be advance. Acoustic emission (AE); a technique that utilize the ultrasonic range of acoustic sound is also increasingly popular the last few decades, especially in the health monitoring of structures such as buildings, bridges, wind turbines and transport vehicles. It also can be called passive ultrasonic method, where the ultrasonic wave; can be continuous or burst signals, released from any source (e.g. flaws, cracks) travelled in medium of solid materials (e.g. concrete, metals) and can be easily captured by AE sensory systems.

AE technique differs from commercial ultrasonics method in one major aspect; it doesn't require the ultrasonic wave as input [1]. The energy released is totally from the test object, but this lead to the main drawback in AE method where the sources can contain a lot of information and a complex signal analysis need to be done to interpret the data. However, modern computer technology helps solving

Published under licence by IOP Publishing Ltd

this problem. The active sources of acoustic emission can be investigated in two main areas. Some researchers focused in statistical analysis of AE data [2-7] while the rest more interested in doing the wave analysis or modal analysis of AE signals [8-13].

One of the main advantages of AE technique is it able to reveal in advance of any impending failure of a building structure. In metallic structures, AE technique consistently performs well and able to give accurate result. But in composites, the challenge for a reliable AE results is huge due to the orthotropic behaviour of the materials. A lot of recent studies are done to make AE technique for health monitoring of composite structures become more quantitative, leading to more general result and not case specific [14, 15].

Location detection is very important especially when monitoring the progressive failure. Accurate location observing can give the precise information on the crack velocity [16], size of delamination and its orientation and many more. Location detection or damage localization in composite materials pulls a lot of study since commercials AE tools always give relatively bigger errors and inconsistent result. This is due to the anisotropic characteristic of composites.

There are a lot of source location methods for acoustic emission technique. One of the easiest ways of source location finding is by using zonal method where it utilizes the attenuation characteristic of acoustic waves. The theory is simple; AE sensor (in an AE sensor array) that nearest to the active sources will capture the highest AE signal amplitude. However, it needs to have a lot of sensor to get good accuracy [17, 19]. The other source location technique is the Time of Arrival (TOA) of AE wave at different sensors [18, 19]. Through this method, two AE sensors is required for linear location detection, while at least three AE sensors is needed for two dimensional detection or source point on a plate. AE sources must be inside or near the sensors array to get good result. Another source location method is by using modal analysis of AE waves [9-12] where it is one of the most recent developments. The technique applies the knowledge of two major wave modes in AE signals; asymmetric and symmetric wave mode. Both wave modes have different wave velocity and arrive at sensor at different arrival time. This makes single sensor linear source location detection possible [11, 12].

In both TOA and modal analysis methods, the information of wave velocity is compulsory for accurate result. However, AE waves didn't travel with a constant velocity in composite materials [20]. This present study describes an alternative concept for identification of damage sources in composite materials using non-velocity approach; signal energy attenuation. Also includes is the modal study of energy signal attenuation.

1.1. Signal Energy Attenuation
In ultrasonic wave study, it is necessary to include the effect of wave signal attenuation. Generally, sound wave attenuates with this relation [21],

$$A = A_0 e^{-\alpha z} \tag{1}$$

where A_0 is the amplitude of the propagating wave at some location or defect source. The amplitude A is the decreased amplitude as the wave travelled into distance Z from source or initial location. α value refers to attenuation coefficient of the wave travelling in z-direction and its dimension is *nepers/length*. The same model is applied to the signal energy attenuation in AE signals.

The main factors that weaken the AE signals energy is the combined effect of dispersion, diffraction and scattering. When AE wave travels through any medium, its signal energy will drop with the increase of distance and can be expressed with a universal model,

$$E_i = E_o e^{-\beta(|x_i - x_o|)} \tag{2}$$

In equation (2), x_o is referred to the location of source, x_i is the location of sensor i, E_o is the energy at source, E_i is energy at sensor i and β is the decay constant. By using equation (2), a set of two sensors can be used for linear source location detection. In the case of β value is unknown, three sensors are needed. Its relation can be expressed as,

$$\left(\ln\frac{E_3}{E_2} \Big/ \ln\frac{E_2}{E_1} \right) = \frac{|x_2 - x_0| - |x_3 - x_0|}{|x_1 - x_0| - |x_2 - x_0|} \qquad (3)$$

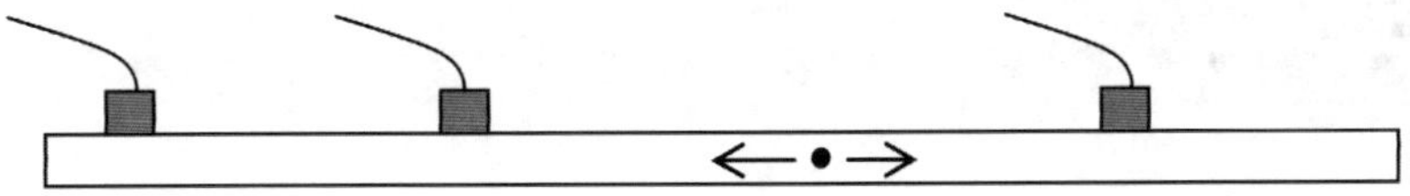

Figure 1. Source location using three sensors for pipe leak detection.

Energy attenuation source defect localization is mainly applied in linear structures such as pipe leak detection (figure 1) and tubes defect. However, for 2D planar source location, the concept of energy attenuation source location also can be applied with appropriate number of sensors.

One of the main benefits of using this method compared with conventional wave velocity dependent method is; no other input is required for source location calculation. Attenuation rate of AE wave will change with the age of the structure and also with different lay-up configuration. In the case of linear source location using three channels, the decay constant or attenuation constant doesn't need to be included into calculation because the number of sensors itself allow the in-situ attenuation rate determination; using equation (3). This is different with velocity dependent approach where AE wave velocity need to be calculated for each measurement.

2. Experimental details

An unidirectional laminate; 28.7 cm × 2.5 cm fibreglass epoxy resin plate and thickness 2.64 mm with stacking sequence $[0^o]_4$ was fabricated using the hand lay-up method. AE data acquisition used for experimentation was a single channel Physical Acoustic Corporation AE Node Systems, with AE sensor NANO-30 (working frequency 100 kHz to 700 kHz) and support by *AE Win* software. All the acquired signals were analogue filtered to the range of 20 kHz – 1000 kHz. The sampling rate for acquisition was set up to 5 Mega sample per second and threshold was set to 45 dB. All acquired waveforms were stored in mini notebook for further analysis.

2.1. Attenuation behaviour

The plate's signal energy attenuation rates were plotted for four different angles; [0°], [45°], [60°] and [90°]. It was done by breaking the pencil lead; a standard Hsu-Nielsen source testing at several points along the given angles in straight line, starting from one edge of the composite specimen towards other edge. Angle [0°] referred to a line along the glass fibre. An AE sensor was located at point C (figure 2) and captured all the AE signals. The signal's energy was calculated automatically using the AE Win software.

2.2. Source location

Only single channel of AE system were used for the linear source location detection testing, where the sensor was placed at three different positions; A, B and C as shown in figure 2, at different time, while breaking the pencil lead in fixed location. In other words, pencil break test at same location were done three times in order to get AE readings at position A, B and C. Source location computation for [0°], [45°], [60°] and [90°] were done by using an algorithm, developed base on equation (3).

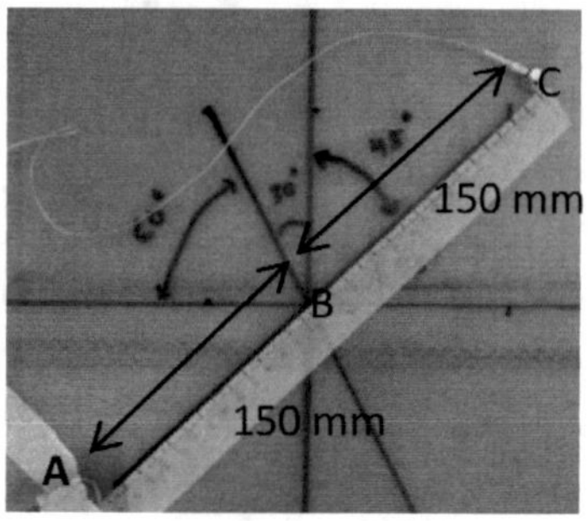

Figure 2. Unidirectional Glass Fibre Reinforced Polymer (GFRP) composite plate with attached AE sensor.

3. Results and discussions

Figure 3 shows the decay rate of AE signals energy in composite specimen for three different configurations. It can be seen that, the AE signals energy attenuation for the specimen, follow the general model of ultrasonic wave attenuation. The R-square show quite high values (near to 1.0) and the average decay constant for all configurations were 0.006.

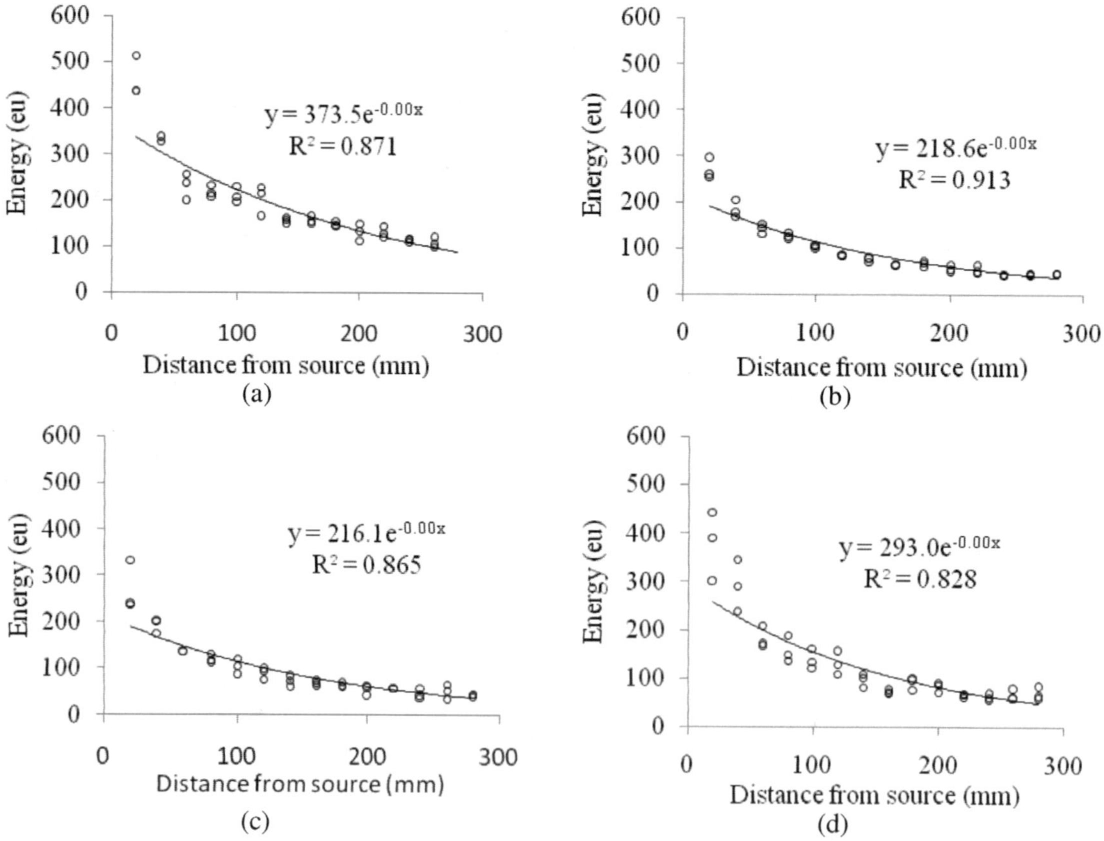

Figure 3. (a) Attenuation rate for [0°], (b) attenuation rate for [45°], (c) attenuation rate for [60°] and (c) attenuation rate for [90°].

Samples of typical AE signals and it frequency spectrums at 100 *mm*, 200 *mm* and 280 *mm* from AE sensor for [45°] are as shown in figure 4. Two major wave modes exist which are asymmetric (flexural) mode and symmetric (extensional) mode. Extensional wave mode usually has higher frequency and arrived earlier to sensor [13]. It's clear that energy from extensional mode dominates the total signal energy closer to the source but it attenuated very fast and shows a significant reduction with the distance. Meanwhile, the flexural wave energy was remained the same or reduced very little with the distance.

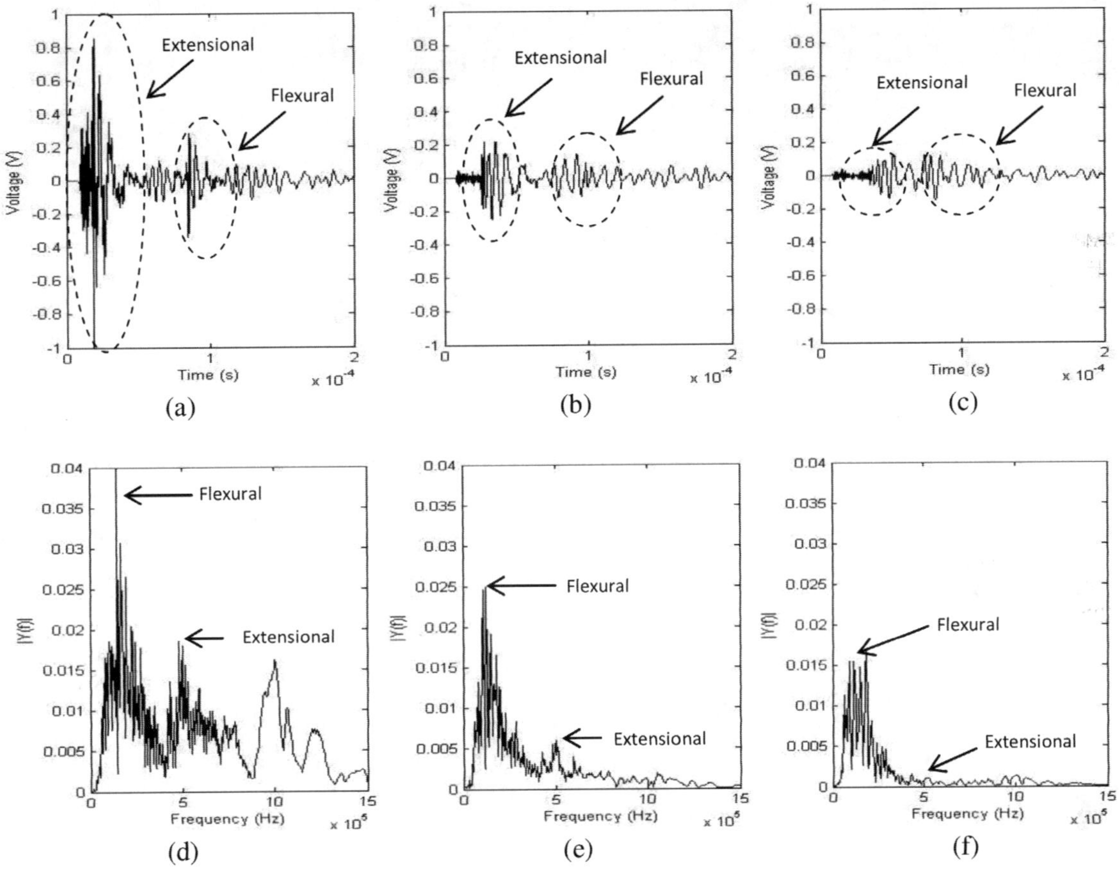

Figure 4. AE signals and its frequency spectrum at 100 *mm* (a), 200 *mm* (b) and 280 *mm* (c) from AE sensor for configuration [45°].

Test result for unidirectional composite plate with [45°] configuration source location determination using energy attenuation algorithm can be shown as in figure 5. Meanwhile, figure 6 shows test result for [0°] and [90°]. All plots randomly scattered very close to the actual source (triangle). The average errors can be concluded by table 1.

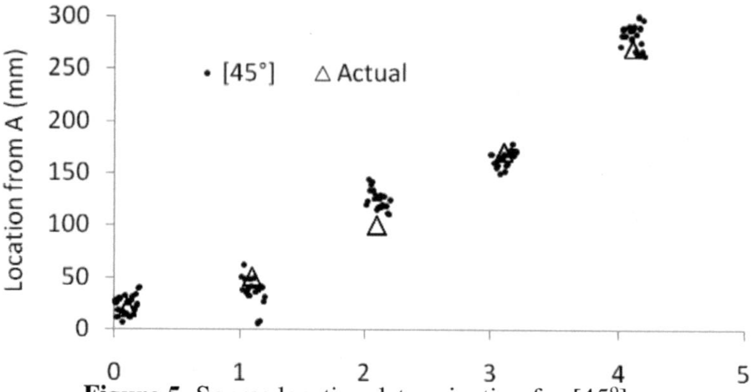

Figure 5. Source location determination for [45°].

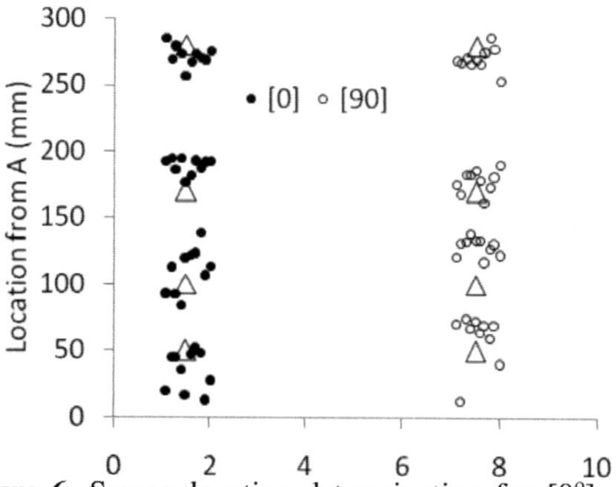

Figure 6. Source location determination for [0°] and [90°].

Table 1. Comparison between averages test results and it actual value.

	Calculated location (mm)	Actual location (mm)	Error (mm)
0°	34.97	50.00	15.03
	110.37	100.00	10.37
	188.91	170.00	18.91
	271.72	280.00	8.28
45°	36.96	50.00	13.04
	125.27	100.00	25.27
	163.78	170.00	6.22
	283.58	280.00	3.58
90°	59.98	50.00	9.98
	128.66	100.00	28.66
	178.28	170.00	8.28
	269.90	280.00	10.10

4. Conclusions

The experimental evaluation of linear source location detection using energy attenuation of acoustic emission signal has been presented. It can be concluded that the non-velocity dependent source location algorithm utilizing exponential model of AE signals energy attenuation shows a prospective approach for damage localization for composite materials.

Further investigation will be done using a more practical testing such as a standard tensile test for GFRP to further study the applicability and practicality the proposed algorithm. In addition, instead of only linear source location, 2D damage mapping of GFRP composite plate using the algorithm also will be investigated.

Acknowledgements

The author acknowledges and wishes to thank Faculty of Engineering & Surveying and Centre of Excellence for Engineered Fibre Composites, University of Southern Queensland for providing the financial support and excellent facilities for the research study. The writer also acknowledges the Faculty of Mechanical Engineering, Universiti Malaysia Pahang, for research collaborations through internal grant, *RDU100323*.

References

[1] Miller R K and McIntire P [Nondestructive Testing Handbook Second Edition, Volume Five, Acoustic Emission], American Society for Nondestructive Testing Inc., USA, 12 (1987).

[2] Godin N, Huguet S, Gaertner R and Salmon L. 2004. Clustering of acoustic emission signals collected during tensile tests on unidirectional glass/polyester composite using supervised and unsupervised classifiers. *NDT&E International* **37**, 253–64

[3] Johnson M. 2002. Waveform based clustering and classification of AE transients in composite laminates using principal component analysis. *NDT&E International* **35**, 367-76

[4] De groot P J, Wijnen P A M and Janssen R B F. 1995. Real-time frequency determination of acoustic emission for different fracture mechanism in carbon/epoxy composites. *Composites Science and Technology* **55**, 405-12

[5] Gutkin R, Green C J, Vangrattanachai S, Pinho S T, Robinson P and Curtis P T. 2011. On acoustic emission for failure investigation in CFRP: Pattern recognition and peak frequency analyses. *Mechanical Systems and Signal Processing*, **25** 1393–407

[6] Kostopoulos V, Loutas T and Dassios K. 2007. Fracture behavior and damage mechanisms identification of SiC/glass ceramic composites using AE monitoring. *Composites Science and Technology*, **67** 1740–46

[7] Oliveira R d and Marques A T. 2008. Health monitoring of FRP using acoustic emission and artificial neural networks. *Computers and Structures*, **86** 367–73

[8] Morscher G N. 1999. Modal acoustic emission of damage accumulation in a woven SiC/SiC composite. *Composites Science and Technology*, **59** 687-97

[9] Jiao Jing-Pin, He Cun-Fu, Wu Bin, Fei Ren-Yuan and Wang Xiu-Yan. 2006. A new acoustic emission source location technique based on wavelet transform and mode analysis. *Front. Mech. Eng. China* **3** 341−45

[10] Jiao Jingpin, Wu Bin and He Cunfu. 2008. Acoustic emission source location methods using mode and frequency analysis. *Struct. Control Health Monit.* **15** 642–51

[11] Jingpin Jiao, Cunfu He, Bin Wu, Renyuan Fei and Xiuyan Wang. 2004. Application of wavelet transform on modal acoustic emission source location in thin plates with one sensor. *International Journal of Pressure Vessels and Piping* **81** 427–31

[12] Oskouei A R and Ahmadi M. 2008. Using wavelet transform to locate acoustic emission source in composite plate with one sensor. *Acoustics 08, Paris, 29 June – 4 July 2008* p 2211 - 15

[13] Surgeon M and Wevers M. 1999. Modal analysis of acoustic emission signals from CFRP laminates. *NDT&E International*, **32** 311–22

[14] Scholey J J, Wilcox P D, Wisnom M R and Friswell M I. 2010. Quantitative experimental measurements of matrix cracking and delamination using acoustic emission. *Composites: Part A*, **41** 612–23

[15] Scholey J J, Wilcox P D, Wisnom M R and Friswell M I. 2009. A practical technique for quantifying the performance of acoustic emission systems on plate-like structures. *Ultrasonics* **49** 538–48

[16] Bohse J. 2000. Acoustic emission characteristics of micro-failure processes in polymer blends and composites. *Composites Science and Technology* **60**, 1213-26

[17] Miller R K. 1987. *Nondestructive Testing Handbook Second Edition, Volume Five, Acoustic Emission* ed McIntire P. (USA: ASNT Inc) p 137-8

[18] Miller R K. 1987. *Nondestructive Testing Handbook Second Edition, Volume Five, Acoustic Emission* ed McIntire P. (USA: ASNT Inc) p 145-53

[19] Prosser W H. 2002. *Nondestructive Evaluation: Theory, Techniques and Applications* ed Shull P J. (USA: Marcel Dekker Inc) p 410-8

[20] Prosser W H. 2002. *Nondestructive Evaluation: Theory, Techniques and Applications* ed Shull P J. (USA: Marcel Dekker Inc) p 388-90

[21] Prosser W H. 2002. *Nondestructive Evaluation: Theory, Techniques and Applications* ed Shull P J. (USA: Marcel Dekker Inc) p 107-14

Comparative Study of Biofuel and Biodiesel Blend with Mineral Diesel Using One-Dimensional Simulation

Rafidah Rahim, Rizalman Mamat, Mohd Yusof Taib
Faculty of Mechanical Engineering, Universiti Malaysia Pahang,
26600 UMP, Pekan, Pahang, Malaysia.
Phone: +6094242303 Fax: +6094242202

E-mail: rafidah.rahim@yahoo.com

Abstract - This study is intended to perform one-dimensional simulation for four cylinders diesel engine by using various type of fuels and blend. The testing of biofuels properties conducted according to ASTM standards. The physical properties of the fuel are investigated in chemical laboratory which comprises of flash point, kinematic viscosity, density, cloud & pour point, acid value and moisture content. There are three types of fuels used throughout the study, which are straight vegetable oil (SVO), biodiesel 20% blend (B20) and biodiesel 5% blend (B5). Then, the properties data from the experiment will be used in the simulation GT Power software. Simulation tests have been run with the aim of obtaining comparative measures of torque, power, specific fuel consumption and volumetric efficiency. The results is use to evaluate and analyze the performance of diesel engine running with the mentioned fuels above. The comparison performances for each fuel have been discussed. There is no significant difference in the engine performance when fueled with B5 and diesel. There is only about one percent lower of B5 and four percent higher of B20 and SVO compare to diesel fuel.

Keywords: *Biodiesel, diesel, simulation*

1. Introduction

Global warming and green house effects nowadays give a high impact to environmental problem. Due to the environmental policies to reduce carbon dioxide emission, taking of biodiesel as an alternative and renewable source to replace fossil diesel becomes increasingly important. Palm oil has been reported to be the most interesting option instead of consideration various oil sources to be the feedstock to biodiesel production plants. All fuels have properties that we can use to identify them. The more properties that we can identify, the better we know the fuel. To faithfully predict alternative fuel combustion, accurate prediction of the physical properties of alternatives fuels is critical in the representation of identification for each fuel.

Biodiesel is the general word for all types of fatty acid methyl esters (FAMEs) or ethyl ester made from different raw materials and used as fuels. It is produced from transesterification process of vegetables oils or animal fats with the addition of methanol [1]. The transesterification (alcoholysis) process is a probable method for biodiesel production. This process is a chemical reaction between triglycerides and alcohol in the presence of alkaline liquid catalyst, usually sodium or potassium methoxide. The alcohol reacts with the fatty acids to form the methyl ester (biodiesel) and glycerol [2].

Published under licence by IOP Publishing Ltd

Physically and chemically, all vegetable oils can be used to produce [3]. Commonly, the liquid has similar composition and characteristics such as cetane number, energy content, phase changes and viscosity compare to petroleum-derived diesel. So it can be used in any CI diesel engine without any modification when it blended together with petroleum-derived diesel. Biodiesel have enabled to become one of the most familiar biofuels in the world compare to petroleum-derived diesel because some of its distinct benefits such as lower greenhouse gases emissions, higher lubricity and cetane ignition rating [1].

Malaysia and Indonesia are respectively largest and second largest producers of palm oil in the world, jointly they produces 85% of world's palm oil [4]. Progressively growing domestic palm-oil production will provide the basis for a rapid growth of the biofuel industry during the coming decade. Production is expected to increase at a rate of about 10 percent annually, reaching 1.1 billion litres by 2017. In Southeast Asia (SE Asia) biodiesel production is drastically growing due to its high potentiality and yield factor of palm [4]. Tropical climate and cheap man power of this region is another beneficial point for growing of this plant [5]. The industry will be predominantly export oriented, with the EU as its target market [6].

Four methods to reduce the high viscosity of vegetable oils to enable their use in common diesel engine without operational problem such as engine deposits have been investigated: blending with petro diesel, pyrolisis, micro emulsification (co solvent blending), and transesterification [7]. In the early phases of starting biodiesel projects, it can be observed that simple process technologies and basic purification do not achieve the required high quality needed for the modern diesel engine [8]. This paper highlighted the new data of biodiesel in GT-Power that do not have in any other analysis of engine performance instead of commonly used the diesel data. The objective of this research is to study the effect of temperature of biodiesel fuel as an alternative fuel for the same diesel engine specification. The properties of diesel fuel are varied significantly with the properties of biodiesel fuel. The vapor of diesel and biodiesel fuel properties from GT-POWER is shown in Table 1. The performance of brake power, brake torque, brake specific fuel consumption (BSFC), brake mean effective pressure (BMEP), volumetric efficiency, and brake efficiency were being discussed in this research.

Table 1: Vapor fuel properties of diesel and biodiesel

Vapor Fuel Properties	Density (g/cm^3)	Energy content (MJ/kg)
D2	0.837	49.9618
B5	0.839	49.8964
B100	0.878	39.9161
SVO	0.917	39.8887

2. Model Setup

In general, a one dimensional (1D) simulation of an engine model consists of intake system, exhaust system, compressor and variable geometry turbocharger system (VGT), common rail fuel injection systems, exhaust gas recirculation systems, engine cylinders and valve train. The development of the four cylinder modeling in one-dimensional simulation for four-stroke direct-injection (DI) diesel engine was presented in this paper. Figure 1 below shows the diesel engine modeling using GT-POWER software.

Optional of the first step of building engine model in GT-POWER is to modeling the intake system. For the selected diesel engine, the intake system has a few components, size and different data. The system is started from environment till the intake valve. The intake system components in the GT-POWER model are environment, intrunner, inport, intvalve. Figure 1 shows the intake system components for one cylinder only. The other three cylinders have the same configurations of the intake system as for the one cylinder. The components in this system require a few data to complete the data form before running the model.

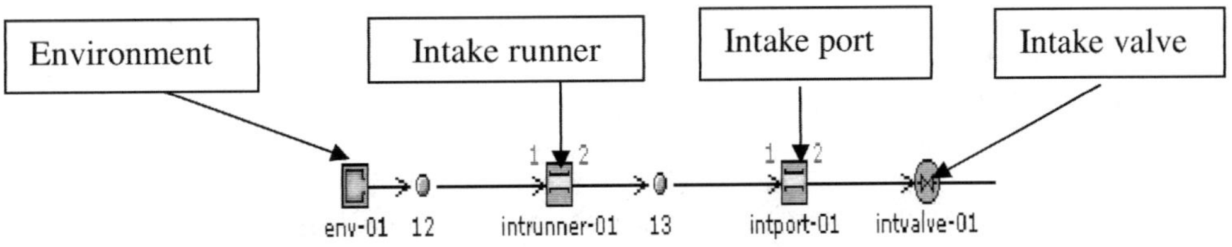

Figure 1. Intake system components

Data in environment panel are pressure, temperature, pressure flag and composition. For this study, the environment pressure is at standard atmospheric pressure which is 1 bar. The same goes for environment temperature, which is assumed to be 298 K. Initial fluid compositions is assumed to be fresh air and by neglecting the existing of NO, NO_2 and CO concentration.

The engine cylinder and fuel injection system are focused in the engine cylinder performance supports diesel fuel from fuel injection system, fresh air intake system and exhaust gas to exhaust system. There are many components in the engine cylinder and fuel injection system of the diesel engine. However the basic for all diesel engines are the same components. Then the components, size and data must be recorded and inserted into the GT-POWER form.

The engine cylinder and fuel injection system component are injector, cylinder and engine. Every component in this system needs any data to complete the data form and running the model. Figure 2 shows the engine cylinder and fuel injection system components for the selected diesel engine. It shows only for one cylinder and the rest three cylinders share the same configuration instead of engine. Engine is only one, which represents overall of the cylinders.

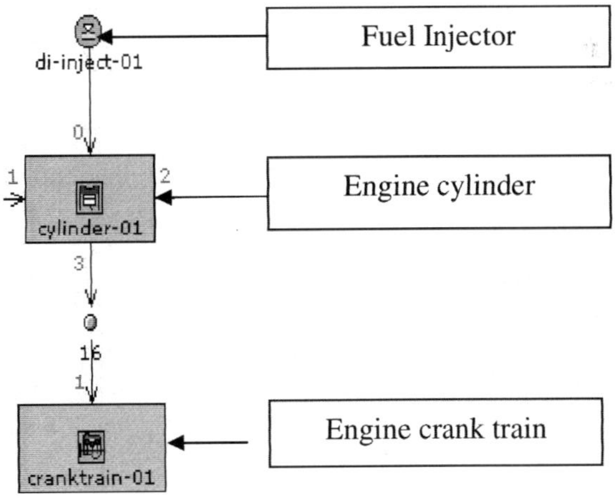

Figure 2. Engine cylinder and fuel injection system components

Engine cylinder input panel consists of various attributes such as the start of cycle, cylinder geometry object and initial state name. Figure 2 showed the input panel for it. Start of the cycle is also known as crank angle at intake valve close. Crank angle is at which each cylinder's cycle begins. This value does not affect the simulation predictions; it only specifies the starting and ending angle within a cycle over which integrated and averaged predictions are measured. This attribute should usually be set to "def" which sets the value to be equal to the value specified in 'Engine Crank Train' for Start of Cycle (CA at IVC). The dimension of bore, stroke and connecting rod which are measured corresponding to the real engine were inputted in this general engine panel. Other components, size and data must be recorded and inserted to the GT-POWER template library input panels. Data in the engine cylinder geometry are bore, stroke, and wrist pin to crank offset, compression ratio, TDC clearance height and connecting rod length.

The input panel for the template of engine crank train consists of number of cylinders, configuration of cylinder and engine type. For the selected diesel engine, the exhaust system has a few components, size and different data. The system is started from the intake valve till environment. The exhaust system components in the GT-POWER model are environment, exhrunner, exhport, and exhvalve. Figure 3 showed the exhaust system components configuration for the four cylinders. The components in this system require a few data to complete the data form before running the model.

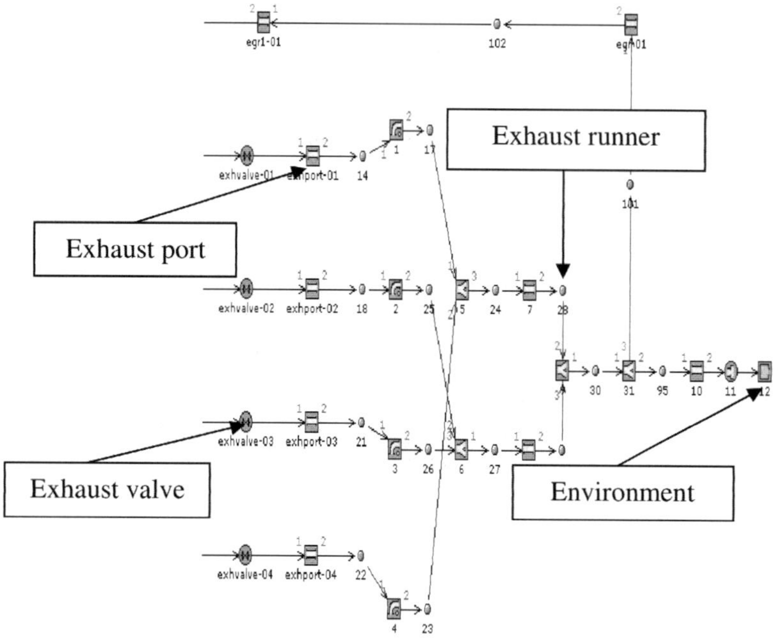

Figure 3. Exhaust system components

Figure 4 shows complete model of the engine. After completing the simulation model, Plot Options requesting were made. Plot Options are plots of results over the course of a single engine cycle (the last engine cycle in multi-cycle simulations) which indicated which interesting parameters of the study that are going to be observed. Plot Options may be requested by selecting the appropriate plot from the plot panel within each part. If the model is ready for simulation, GT-POWER simulation may be started and this will start the simulation running. By running the solver, a shell will open, noting the version of the solver being used. Windows will show the progress of simulation in the form of scrolling text. Once the input has been read successfully, it is all set to create the requested data for post-processing in GT-POST.

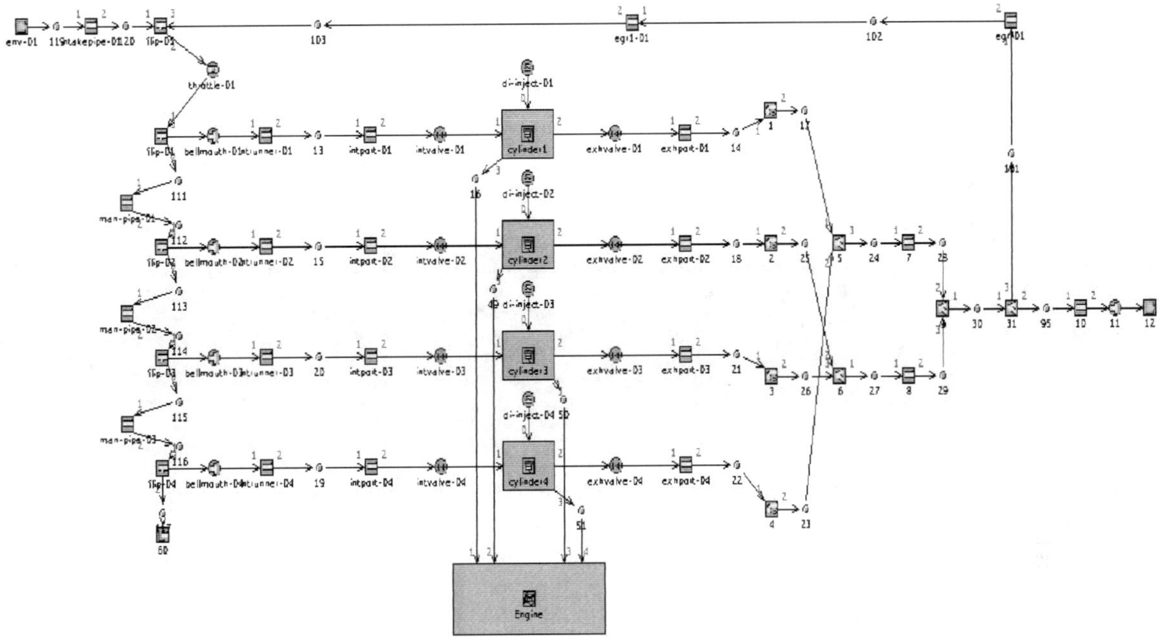

Figure 4. Four cylinders diesel engine modeling using GT-POWER

In GT-POWER, there is no biodiesel fuel in its Fuel Properties panel. In GT-POWER, a gas reference object is commonly described by its C: H: O: N compositions. For incompressible liquid properties, information about density, vapor fluid object and heat of vaporization are the necessary input to GT-POWER. Typically every liquid reference object must be associated with a gas reference object so that the properties of the liquid will be known if the fluid evaporates. Then, the gas/vapor properties of biodiesel need to be determined which it consists of molecular weight, lower heating value, and number of atoms per molecule which is carbon, hydrogen, nitrogen and oxygen.

In the selected diesel engine, the intake system has a few component, size and different data. The system is started from environment till the intake valve. All of the intake system components in the GT-POWER model are environment, intake runner air filter, air filter, intake runner, intake port, intake valve. The components in this system need a few data to complete the data form and running the model.
Engine cylinder and fuel injection system that focused in engine cylinder performance support diesel fuel from fuel injection system, fresh air intake system and exhaust gas to exhaust system. The components, size and data must be recorded and inserted to the GT-POWER form.

All of the engine cylinder and fuel injection system component are injector, cylinder and engine. Exhaust system is the last system in the diesel engine. The system starts from exhaust valve and finishes in the environment. The GT-POWER components in the exhaust system are exhaust valve, exhaust port, exhaust runner, muffler, exhaust runner exit, and environment. The details of the engine parameters used in this model are described in Table 2 below.

Table 2: Detail parameters of a four cylinder diesel engine

Parameter	Value
Bore (mm)	82.7
Stroke (mm)	93
Compression ratio	22.4
Displacement (cc)	500
Number of Cylinder	4
Connecting Rod Length (mm)	150
Piston Pin Offset (mm)	1
Intake Valve Open (°CA)	351
Intake Valve Close	-96
Exhaust Valve Open	125
Exhaust Valve Close	398

In this section, some basic parameters that are commonly used to characterize engine operation are developed. These include the mechanical output parameters of work, torque, and power; the input requirement of air, fuel and combustion; efficiencies; and emission measurement of engine exhaust [9-10].

4. Results and Discussion

Biodiesel operates in compression ignition engines with little or no modification just like petroleum diesel. Figure 5 below shows the variation fuels for brake mean effective pressure. The mean effective pressure is quite often used to calculate the performance of an internal combustion engine. The maximum brake mean effective pressure is essentially for well establish a good engine designs. The diesel fuel and B5 shows the highest BMEP compare to other two fuels, B100 and SVO. As shown, B100 and SVO significantly similar value, while trend for D2 quite similar to B5

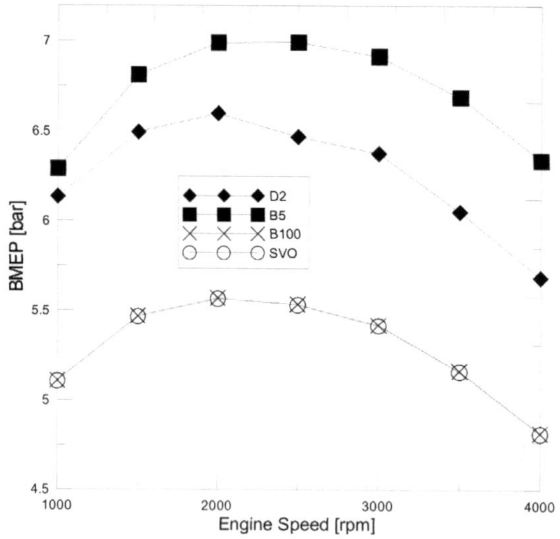

Figure 5. Effect of engine speed variation on BMEP

Brake specific fuel consumption is measured as a flow rate-mass flow per unit time. From figure 6 below, the analysis results for brake specific fuel consumption are well in good agreement. As shown in

IOP Conf. Series: Materials Science and Engineering **36** (2012) 012009 doi:10.1088/1757-899X/36/1/012009

figure, B5 and D2 shows the lowest BSFC while B100 and SVO shows the highest BSFC. For compression ignition engines, the best performance for BSFC is lowest value, it is obviously desirable [9]. The higher BSFC value in the case of SVO and B100 is due to the lower energy content as depicted in Table 1 above. This results cause the engine to inject more fuel to gain equal brake torque.

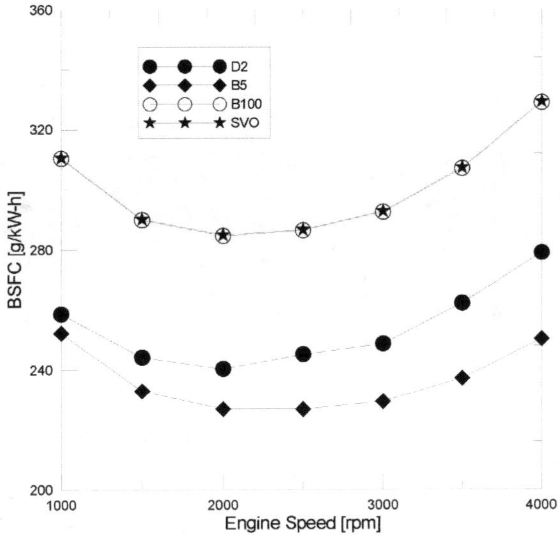

Figure 6. Variation of brake specific fuel consumption against engine speed

The effect of variation fuels on brake engine torque for various speeds is shown in figure 7 below. The torque is a function of engine speed [11]. At low speed, torque increases as the engine speed increase, reaches a maximum and then, as engine speed increase further, torque decreases as shown in figure below. The torque decreases because the engine is unable to ingest a full charge of air at the higher speed. Figure 8 below clearly shows the gap between D2 & B5 and B100 & SVO. The maximum reduction brake engine torque recorded between these fuels when engine speed achieved at 2000 rpm. The performance of the fuels same with other performance whereas B15 and B20 is the highest value of brake torque while the lowest value are SVO and B100 fuels.

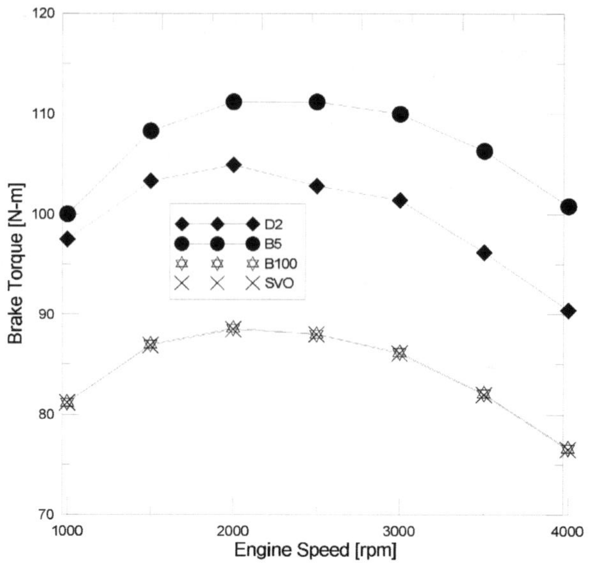

Figure 7. Variation of brake torque against engine speed

Brake power is defined as power that developed at the crankshaft or flywheel. Figure 8 below shows that SVO and B100 have a lower value of brake power compare to B5 and D2. It is resulting due to the lower energy level of fuel that causes some reduction in the engine power when it is used in diesel engine without any modification [12]. A close resemblance occurred at low speed representing small discrepancy in output between the fuels. The maximum brake power is 42.2 kW for B5 while the minimum value is 8.5 kW posses by B100 and SVO. It is well known that the heating value of the fuel affects the power of an engine. The lower energy level of the B100 and SVO fuel causes some reductions in the engine power when it is used in diesel engines without any modifications [12].

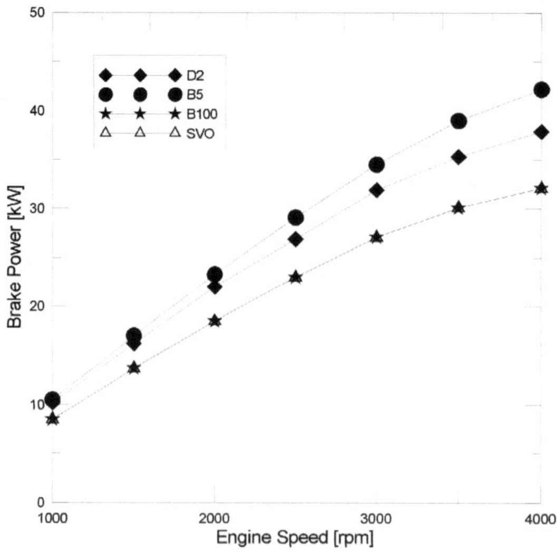

Figure 8. Variation of brake power against engine speed

Volumetric efficiency is one of the most important parameters which determine the performance level of a four-stroke engine. VE is the measure of success with which air supply is inducted into an engine. Figure 9 shows the effect of volumetric efficiency with respect to engine speed. It can be seen that the maximum efficiency achieved at 1000rpm of engine speed. Engine with higher volumetric efficiency will generally be able to run at higher speeds and produce more overall power due to less parasitic power loss moving air in and out of the engine. Volumetric efficiency was decreasing sharply as the engine speed increase further. The sharp decrease happens because of higher speed is accompanied by some phenomenon that have negative influence on η_v. These phenomenon's include the charge heating in the manifold and higher friction flow losses which increase as the square of engine speed [13].

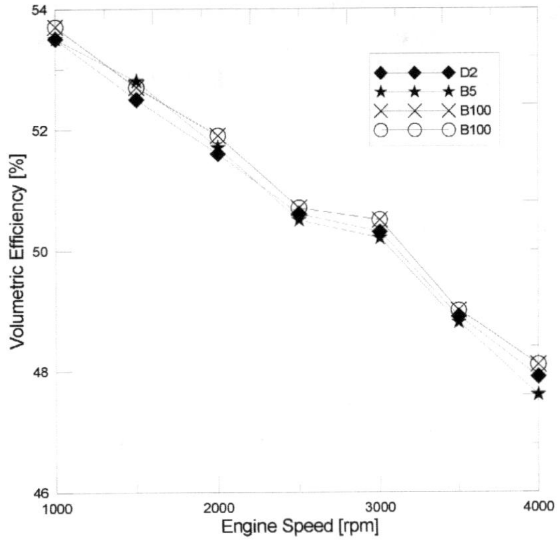

Figure 9. Effect of engine speed variation on volumetric efficiency

5.0 Conclusion

The performance of power, torque, specific fuel consumption, brake mean effective pressure and volumetric efficiency are compared. Analyzing results shows that the performance of diesel engine running with B5 and Diesel fuel has no significant difference in the engine performance. Inversely with the biodiesel fuel blended 100% (B100), it shows slightly difference performance for diesel engine. It highlights a lower performance and need some modification of engine to level up the performance as equal to B5 and diesel fuel.

References

1. Lim, S. and L.K. Teong, *Recent trends, opportunities and challenges of biodiesel in Malaysia: An overview.* Renewable and Sustainable Energy Reviews, 2010. **14**: p. 938-954.
2. Mamat, R., *Performance and Emission Characteristics of an Automotive Diesel Engine Using Biodiesel Fuel with The Influence of Air Intake Variables*, in *School of Mechanical Engineering*. 2009, The University of Birmingham Birmingham.
3. A.Z.Abdullah, et al., *Current status and policies on biodiesel industry in Malaysia as the world's leading producer of palmoil.* Energy Policy, 2009. **37**: p. 5440-5448.

4. Jayed, M.H., et al., *Environmental aspects and challenges of oilseed produced biodiesel in Southeast Asia.* Renewable and Sustainable Energy Reviews, 2009. **13**: p. 2452-2462.
5. Tan, K., et al., *Palm oil: addressing issues and towards sustainable development.* Renewable and Sustainable Energy Reviews, 2009. **13**(2): p. 420-427.
6. Wiebe, K.A., et al., *The State of Food and Agriculture.* Rome: FAO Publications, 2008.
7. Knothe, G., J.V. Gerpen, and J. Krahl, *The Biodiesel Handbook*, ed. L.o.C.C.-i.-P. Data. 2005.
8. Korbitza, W., et al., *Worldwide review on biodiesel production.* IEA Bioenergy Task 39, Subtask „Biodiesel", 2003.
9. Heywood, J.B., *Internal Combustion Engine Fundamentals* 1988: McGraw-Hill 930.
10. Pulkrabek, W.W., *Engineering Fundamentals of the Internal Combustion Engine*. Second ed. 2004: Pearson-Prentice Hall. 478.
11. Abu-Zaid, M., *Performance of single cylinder, direct injection Diesel engine using water fuel emulsions.* Energy Conversion & Management, 2004. **45**: p. 697-705.
12. Can, Ö., I. Çelikten, and N. Usta, *Effects of ethanol addition on performance and emissions of a turbocharged indirect injection Diesel engine running at different injection pressures.* Energy Conversion and Management, 2004. **45**(15-16): p. 2429-2440.
13. Rahman, M.M., M.K. Mohammed, and R.A. Bakar, *Effects of Air Fuel Ratio and Engine Speed on Engine Performance of Hydrogen Fueled Port Injection Engine.* American Journal of Scientific Research, 2009(1): p. 23-33.

Development of a Solar Assisted Drying System Using Double-Pass Solar Collector with Finned Absorber

M S M Azmi[1], M Y Othman[2], K Sopian[3], M H Ruslan[3], Z A A Majid[4], A Fudholi[3] and J M Yasin[5]

[1,2] School of Applied Physics, Faculty of Science and Technology, Universiti Kebangsaan Malaysia 43600 UKM Bangi, Selangor Darul Ehsan, MALAYSIA
[3] Solar Energy Research Institute (SERI), Universiti Kebangsaan Malaysia 43600 UKM Bangi, Selangor Darul Ehsan, MALAYSIA
[4] Kulliyyah of Allied Health Sciences, International Islamic University Malaysia 25200 Bandar Indera Mahkota, Kuantan, Pahang Darul Makmur, MALAYSIA
[5] OPF FELDA Factory, Felda Bukit Sagu 2, 25150 Kuantan, Pahang Darul Makmur, MALAYSIA

E-mail: syah_0602@yahoo.co.uk

Abstract. The Solar Energy Research Group, Universiti Kebangsaan Malaysia, International Islamic University Malaysia and Yayasan FELDA has designed and constructed a solar assisted drying system at OPF FELDA Factory, Felda Bukit Sagu 2, Kuantan, Pahang. The drying system has a total of six double-pass solar collectors. Each collector has a length of 480 cm and a width of 120 cm. The first channel depth is 3.5 cm and the second channel depth is 7 cm. Longitudinal fins made of angle aluminium, 0.8 mm thickness were attached to the bottom surface of the absorber plate. The solar collectors are arranged as two banks of three collectors each in series. Internal manifold are used to connect the collectors. Air enters through the first channel and then through the second channel of the collector. An auxiliary heater source is installed to supply heat under unfavourable solar radiation condition. An on/off controller is used to control the startup and shutdown of the auxiliary heater. An outlet temperature of 70–75 °C can be achieved at solar radiation range of 800–900 W/m^2 and flow rate of 0.12 kg/s. The average thermal efficiency of a solar collector is approximately 37%.

1. Introduction

Malaysia lies entirely in the equatorial region. On average, Malaysia receives of 4.21 to 5.56 kWh/m^2 of solar radiation a year. Solar energy has been identified as one of the most potential alternative energy resources as Malaysia receives abundant sunshine all year long [1]. This solar energy has great potential for low temperature applications, particularly for drying of agricultural products. Various types of solar systems for drying agricultural products have been reviewed and studied in the literature [2–6]. The solar

[1]Correspondence author: M. S. M Azmi
Email address: syah_0602@yahoo.co.uk

Published under licence by IOP Publishing Ltd

assisted drying systems mainly for agricultural produce are developed with compact collector design, high efficiency, integrated storage, and long-life drying system.

One of the most important components of a solar dryer system is the solar collector. Various designs of solar collectors have been proposed including multi-pass collectors. Satchunanathan and Deonarine [7], Satchunanathan and Persad [8] and Persad and Satchunanathan [9] performed the experiments and conducted analytical studies involving double-pass solar air collector. They concluded experimentally and analytically that the double-pass design with air flow between the two glass covers and under the absorber plate had a thermal efficiency about 10-15% higher than the single-pass design. They also concluded that reduced top loss improves the performance since the outer glass cover separating the collector from ambient temperature is cooled by the working fluid.

In addition, the use of a double-pass resulted in an increase in the pressure drop a cross the collector. Mohamad [10] conducted theoretical studies of a double-pass solar collector with porous media in the second channel. Sopian et al. [11] performed the experimental studies on the performance of a double-pass solar collector with porous media in the second or lower channel. The second or lower channel of the solar collector is filled up with porous media which acts as heat storage system. This will increase the outlet temperature and the performance of the system.

However, the low thermal conductivity of air results in poor heat transfer between the absorber plate and the flowing air; hence, the air collector efficiency is low. Several designs have been suggested to increase the heat transfer coefficient between the absorber and the flowing air. These include finned absorber [12,13], corrugated absorber [14], multiple passes [15] air collector, etc.

In this work, we have investigated the performance of a double-pass solar collector in which fins are introduced in the second channel flow passage parallel to the length of the collector. Air enters through the first channel formed by the glass cover and the photovoltaic panel and is heated directly by the sun. Next it enters the second channel formed by the back plate. The fins on the back plate will increase the heat transfer to the air and enhance the efficiency of the system.

A solar assisted drying system that incorporated the double-pass solar collector with fins has been designed, constructed and tested in the OPF Factory, Felda Bukit Sagu 2, Kuantan. The followings present the design and preliminary performance results of the solar assisted drying system.

2. The Solar Assisted Drying System

The main components of the solar assisted drying system are (a) the collector array, (b) blower, (c) auxiliary electric heater with on/off controller, (d) the drying chamber, and (e) data acquisition system. Figure 1 shows the schematic of the solar assisted drying system. Air flows into the collector by using blower into the drying chamber. An auxiliary heater is used to supply heat when the drying chamber temperature reached a lower setting temperature.

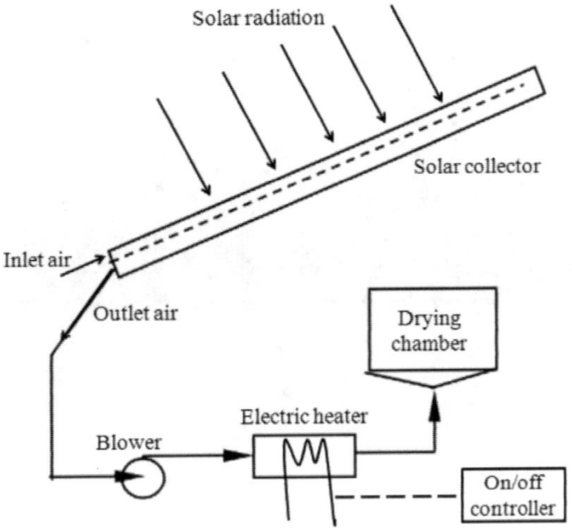

Figure 1. Solar assisted drying system.

This double-pass solar collector with fins consists of the glass cover, absorber plate, finned absorber and insulator. The upper part of the collector was covered by tempered glass, 5 mm thickness. The absorber plate was made of aluminuim sheet, 2 mm thickness and was painted black. Fins are used to increase heat removal from first channel to the second channel and improved the overall efficiency of collector. The side walls and the bottom floor of the collector were thermally insulated with 2.5 cm thick fiberglass to minimize heat losses. The schematic of the double-pass solar collector with fins is shown in figure 2.

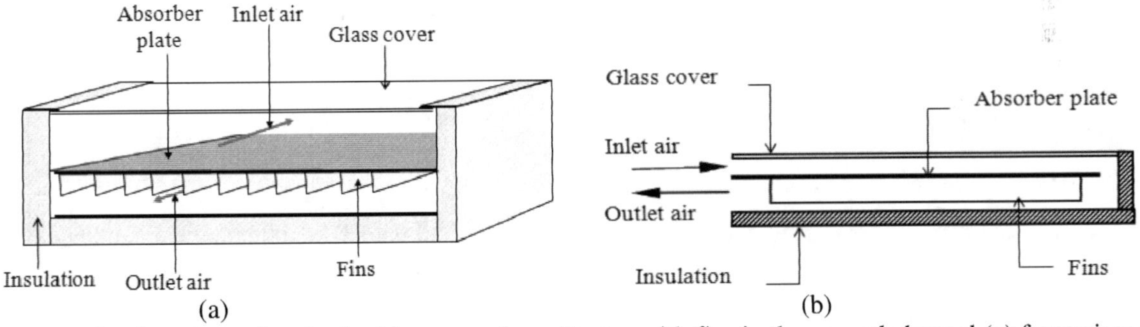

Figure 2. The schematic of a double-pass solar collector with fins in the second channel (a) front view (b) side view.

This solar collector was fabricated with a double channel configuration. The length of the collector is 480 cm and the width is 120 cm. The first and second channel depths are 3.5 cm and 7.0 cm respectively. These collectors have been arranged to form an array. The array has a total six collectors. Figure 3 shows the array for the solar drying system.

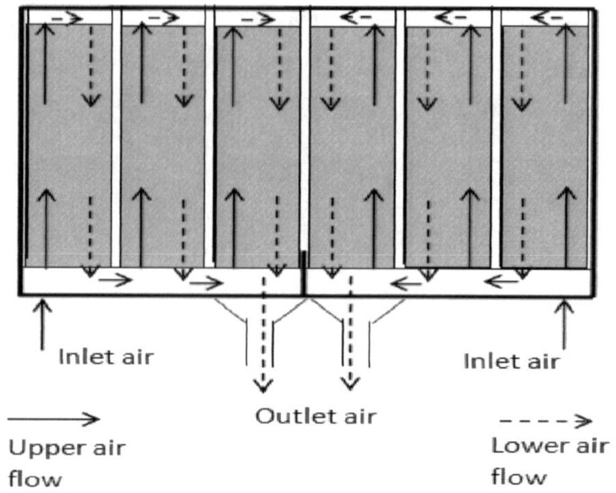

Figure 3. The array for the solar drying system.

The blower is used for sucks the hot air from the collector, because it is no natural convection flow in the double-pass solar collector. By using blower, the air flow may be varied. The electric heater is used for supply heat if the drying chamber temperature falls below a set temperature of 50 °C. Using the on/off controller, the electric heater is on if the temperature of drying chamber lower than 40 °C and off if higher than 50 °C. The electric heater is finned air heater with power supply is about of 5 kW. The drying chamber is batch type, it is used for dry oil palm frond, with capacity of about 400 kg wet dried. Data acquisition system, the data acquisition or computer interface system feed data from the thermocouple and pyranometer to the computer. The main components of the data acquisition system for the experimental setup consists of the following (a) the personal computer, (b) the data acquisition card, and (c) the personal computer application program.

3. Data Analysis

The thermal efficiency of a solar collector, η_c is the ratio of useful heat gain to the solar radiation incident on the plane of the solar collector. It is defined as

$$\eta_c = \frac{mC(T_o - T_i)}{A_c I} \times 100\%$$

where,
m = mass flow rate (kg/s)
C = specific heat of air (J kg^{-1} °C^{-1})
A_c = collector area (m^2)
T_i = inlet air temperature (°C)
T_o = outlet air temperature (°C)
I = solar radiation intensity (W/m^2)

4. Results and Observations

The solar drying is tested under varying environmental conditions for several days. The system is activated at about 9:00 in the morning and terminated at around 5:30 in the evening. Global solar radiation is measured by pyranometer. The inlet, outlet and ambient temperatures are measured by type-K thermocouple wire. These sensors are connected to the data acquisition system. The velocity in the outlet duct is recorded using a hand held air flow meter and the flow rate can be determined. The data are recorded and averaged for a period of one minutes and stored in the hard drive of the personal computer. The data can be retrieved using a spreadsheet program.

During five days of the experiment, the diurnal variation of solar radiation (I), ambient temperature ($T_{ambient}$), inlet air temperature (T_{inlet}), and outlet air temperature (T_{outlet}) for solar drying system were plotted and are shown in figure 4–8. An output temperature range of 40–60 °C can be obtained at solar radiation levels of 500–600 W/m^2. Figure 4 is a typical result for a cloudy day. It is also observed that the finned absorber in the second channel of the collector acts as the storage medium. This is also obvious from the fact that for the cases where the solar radiation falls to a lower level, the output temperature still be maintained. Figure 7 shows a typical result for a clear day in the morning and afternoon but rainy in the evening. The average solar radiation level from 9:00 am to 5:30 pm is about 540 W/m^2. The maximum outlet air temperature is about 76 °C for inlet temperature and air flow rate of 35 °C and 0.12 kg/s respectively.

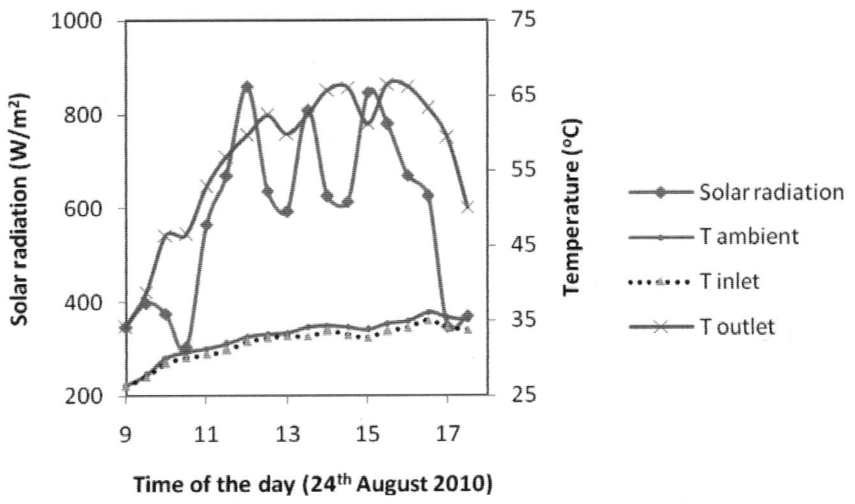

Figure 4. The solar radiation, ambient and inlet temperatures and the resulting outlet temperature for 24th August 2010.

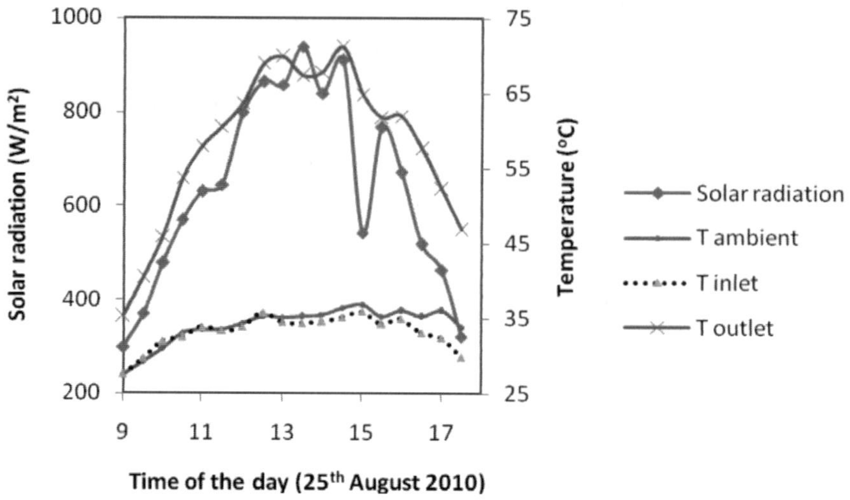

Figure 5. The solar radiation, ambient and inlet temperatures and the resulting outlet temperature for 25th August 2010.

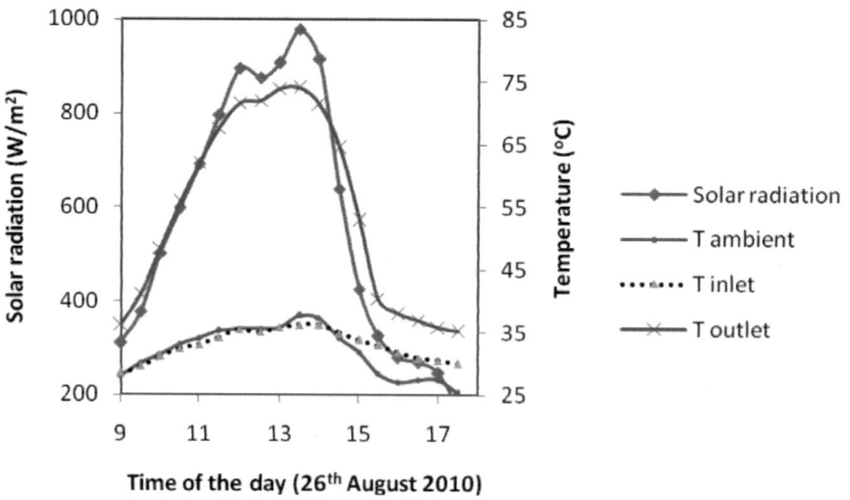

Figure 6. The solar radiation, ambient and inlet temperatures and the resulting outlet temperature for 26th August 2010.

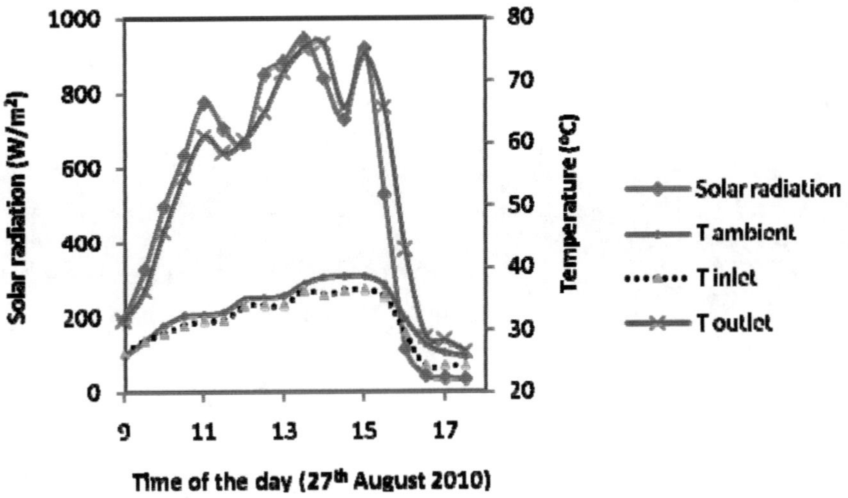

Figure 7. The solar radiation, ambient and inlet temperatures and the resulting outlet temperature for 27th August 2010.

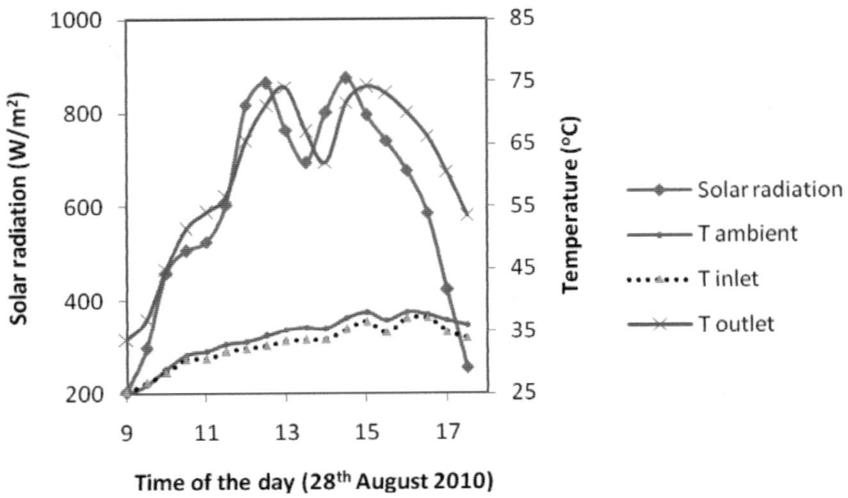

Figure 8. The solar radiation, ambient and inlet temperatures and the resulting outlet temperature for 28th August 2010.

Figure 9–13 shows the results of solar radiation and thermal efficiency of solar collector for the day. The average thermal efficiency of a solar collector varies from 25% to 37%. The higher thermal efficiency of a solar collector shows the result of the experiment for 27th August 2010. The fins acts as increase heat removal from first channel to the second channel and improved the overall efficiency of collector.

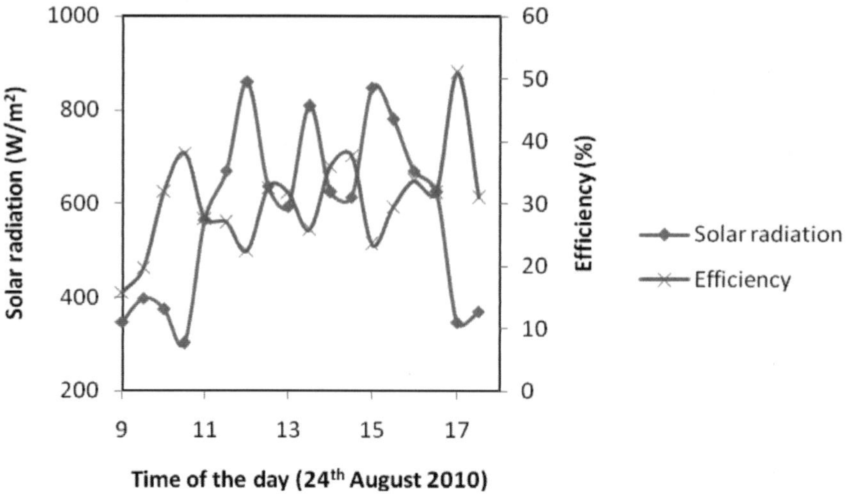

Figure 9. The solar radiation and thermal efficiency of solar collector for 24th August 2010.

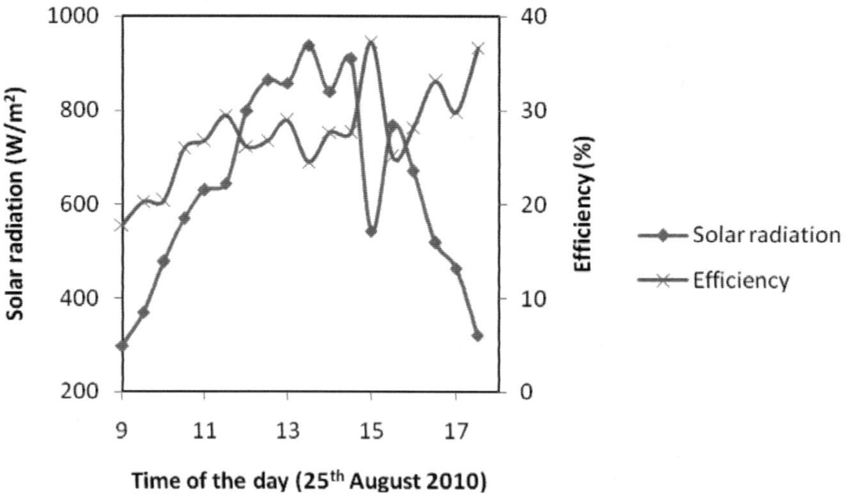

Figure 10. The solar radiation and thermal efficiency of solar collector for 25th August 2010.

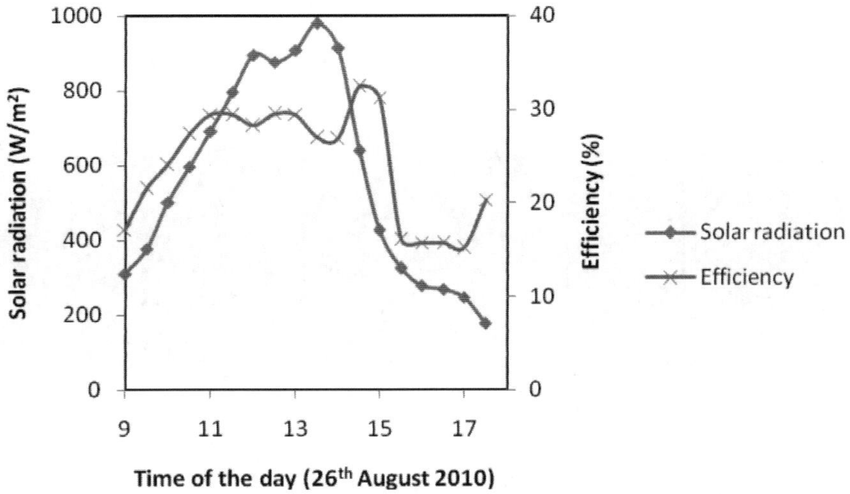

Figure 11. The solar radiation and thermal efficiency of solar collector for 26th August 2010.

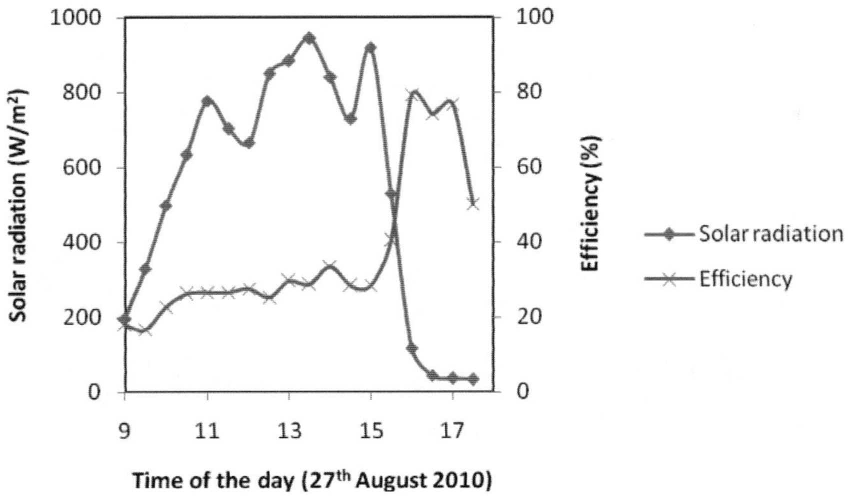

Figure 12. The solar radiation and thermal efficiency of solar collector for 27th August 2010.

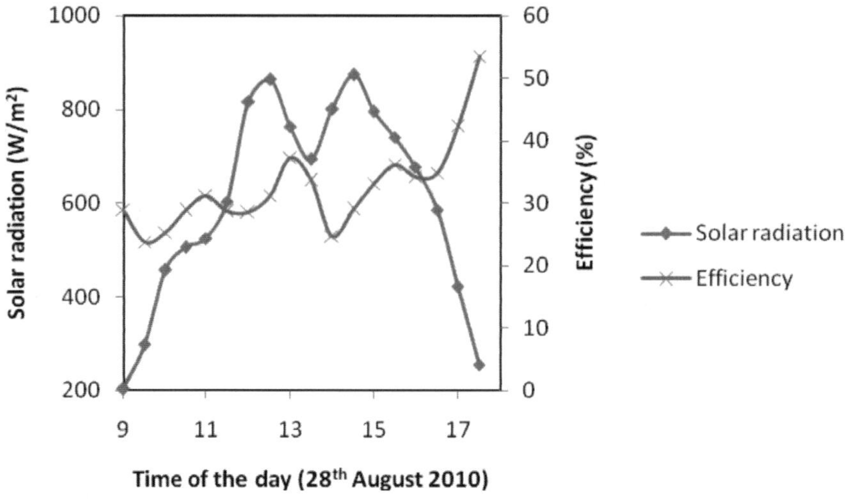

Figure 13. The solar radiation and thermal efficiency of solar collector for 28th August 2010.

5. Conclusions

The solar assisted drying system using the double-pass collector has been constructed and tested. The six collectors with a length of 4.80 m and width of 1.20 m is connected using internal manifold. Fins are added in the second channel of the collector. The fins acts as increase heat removal from first channel to the second channel and improved the overall efficiency of collector. Therefore, the higher and steady outlet temperature can be obtained even under intermitted solar radiation conditions. The outlet air temperature can reached 75 °C for average 800 W/m^2 and air flow rate of about 0.12 kg/s. The average thermal efficiency of a solar collector is 37%.

6. Acknowledgement

The authors would like to thank to the Yayasan FELDA for financial assistantship for this research project.

References

[1] Azhari A W, K Sopian, A Zaharim and M Al Ghoul 2008 A new approach for predicting solar radiation in tropical environment using satellite images-case study of Malaysia. *WSEAS. Trans. Environ. Dev.* **4** 373-377

[2] Garg H P and Prakash J 2000 *Solar energy fundamentals and applications* (New Delhi, India: Tata McGraw-Hill Publishing Company Limited)

[3] Ekechekwu O V and Norton B 1999 Review of solar-energy drying systems II: an overview of solar drying technology *Energy Conver Manage* **40** 615–55

[4] Koyuncu T 2006 Performance of various design of solar heaters for crop drying applications *Renewable Energy* **31** 1073–88

[5] Ekechekwu O V and Norton B 1999 Review of solar-energy drying systems III: low temperature air-heating solar collectors for crop drying applications *Energy Conver Manage* **40** 657–67

[6] A Fudholi, K Sopian, M H Ruslan, M A Alghoul and M Y Sulaiman 2010 Review of solar dryers for agricultural and marine products *Renewable and Sustainable Energy Reviews* **14** 1-30

[7] Satchunanathan S and Deonarine S 1975 A two pass solar air heater *Solar Energy* **15** 41-49

[8] Satchunanathan S and Persad P 1980 Experimental evaluation of the multipass solar air heaters *In: Proceedings of the Third Miami International Conference on Alternative Energy Sources, Miami Beach* 375-382

[9] Persad P and Satchunanathan S 1983 The thermal performance of the two pass cover solar air heaters *Journal of Solar Energy Engineering* **105** 254-258

[10] Mohamad A A 1997 High efficiency solar air heater *Renewable Energy* **60** 71–6

[11] K Sopian, Supranto, M Y Othman, W R W Daud and B Yatim 2007 Double-pass solar collectors with porous media suitable for higher-temperature solar-assisted drying systems *Journal of Energy Engineering* **133** 1–18

[12] H P Garg, G Datta and B Bandyopadhyay 1983 A study on the effect of enhanced heat transfer area in solar air heaters *Energy Convers. Manag.* **23** 43–49

[13]B E Cole Appel and R D Haberstroh 1976 Performance of air-cooled flat-plate collectors Proc. *Conf. On Sharing the Sun, Solar Technologies in the Seventies, Winnipeg* **2** 94–106

[14] C Choudhury, S L Andersen and J Rekstad 1988 *Sol. Energy* **40** 335

[15] A K Bhargava, H P Garg and V K Sharma 1985 Theory of multiple-pass solar air heaters *Energy* **10** 589–599

Effects of heat input on mechanical properties of metal inert gas welded 1.6 mm thick galvanized steel sheet

M.I. Rafiqul[1], M. Ishak[1, 2], M. M. Rahman[1, 2]
[1]Faculty of Mechanical Engineering, University Malaysia Pahang
[2]Automotive Excellent Centre, University Malaysia Pahang
Pekan, Kuantan, Pahang, Malaysia
Email: rafiqme99buet@gmail.com
mahadzir@ump.edu.my

Abstract. It is usually a lot easier and less expensive to galvanize steel before it is welded into useful products. Galvanizing afterwards is almost impossible. In this research work, Galvanized Steel was welded by using the ER 308L stainless steel filler material. This work was done to find out an alternative way of welding and investigate the effects of heat input on the mechanical properties of butt welded joints of Galvanized Steel. A 13.7 kW maximum capacity MIG welding machine was used to join 1.6 mm thick sheet of galvanized steel with V groove and no gap between mm. Heat inputs was gradually increased from 21.06 to 25.07 joules/mm in this study. The result shows almost macro defects free welding and with increasing heat input the ultimate tensile strength and welding efficiency decrease. The Vickers hardness also decreases at HAZ with increasing heat input and for each individual specimen; hardness was lowest in heat affected zone (HAZ), intermediate in base metal and maximum in welded zone. The fracture for all specimens was in the heat affected zone while testing in the universal testing machine.

1. Introduction

Galvanized steel is widely employed as structural material in various industries. It can be found almost everywhere. It is employed in different constructions such as bridges, pipelines, cooling towers, street light post, home appliances, railroads, electrification works, highway's divider etc [1]. For countless outdoor, steel frame house, steel parts in car that allow it to emerge from rainstorms unharmed, metal roofing, marine or industrial applications, galvanized steel is an essential fabrication component. Besides being inexpensive and effective, this metal is popular because it can be recycled and reused multiple times [2].

The principal method of making steel resist corrosion is by alloying it with zinc .This is why galvanized steel is gone through a chemical process to keep it from corroding. When steel is submerged in melted zinc, the chemical reaction permanently bonds the zinc to the steel through galvanizing. Therefore, the zinc isn't exactly a sealer, like paint, because it does not just coat the steel; it actually permanently becomes a part of it. The zinc goes through a reaction with the iron molecules within the steel to form galvanized steel. The most external layer is all zinc, but successive layers are a mixture of zinc and iron, with an interior of pure steel [1].

This steel with zinc at most outer layers will not be attacked by rust because of this protective metal. The multiple layers are responsible for the amazing property of the metal to withstand corrosion inducing circumstances, such as saltwater or moisture. In brief, whenever steel is exposed to most atmospheric corroding environments, a zinc coating deposited by the hot dip process is recommended as an effective and economic mean to protect the steel substrate. Zinc coatings prevent rusting of steel by acting as a barrier against the environment and by sacrificially corroding themselves as sacrificial layer to provide protection as cathode. If, for some reason, rust does take hold on the surface of

Published under licence by IOP Publishing Ltd

galvanized steel, the zinc will get corroded first. This allows the zinc that is spread over the breach or scratch to prevent rust from reaching the steel [1]. In addition, corrosion rate of zinc is lower than that of steel, thus improving the durability of the system. Zinc dissolution is also restrained by its own corrosion products which tend to block active sites hindering the electrochemical reactions [1]. Hence even if a galvanized steel surface is scratched down to the bare steel; the galvanizing coating will prevent the steel from rusting. Therefore, Galvanized steel is a superior product to steel with any other type of coating on it since it protects the steel even when the coating is damaged in handling or in service [3].

However it is not always possible or practical to fully assemble components before they are galvanized and in some cases the complete welded assembly may be too large to fit into the molten pool of zinc to galvanize or too difficult to handle and transport. In these cases too, it will be necessary to weld after galvanizing. This is why we have to apply welding in galvanized sheets, plates, pipes, bars and poles etc to make large structured complex shapes for outdoor application where corrosion is most [4].

But, while welding, the heat vaporizes the protective zinc coating near the weld as zinc has low melting point. Zinc melts at about 482°C and vaporizes at about 906°C. Since steel melts at approximately 1510°C [5] and the welding arc temperature is 8300 to 11090°C, the zinc that is near the weld does not stand a chance. It is vaporized. By the time the weld pool freezes, the zinc is gone [6].

Even though the remaining very little zinc continues to provide some protection to the zinc-free areas, the appearance is poor, and the zinc-free areas will rust when exposed to the environment. Paints which are high in elemental zinc will effectively restore full corrosion protection to the weld areas if properly applied. These paints are available in either spray cans or in containers suitable for brush or spray application [6, 7]. These paints can be applied to the weld after sand blasting or wire brushing to remove all welding slag, zinc oxide and zinc rich sublimate caused by burnt of zinc followed by wiping the weld clean with a rag. Thermal sprayed zinc is also effective in restoring corrosion resistance, but the surface has to be sufficiently roughened, usually by sand blasting or coarse abrasive conditioning to enable thermal sprayed zinc to stick properly [8,9].

Galvanized steel can be readily welded using stainless steel electrodes or filler. But stainless steel filler is electropositive (Cathodic) to zinc. This means that, in the presence of moisture, the zinc immediately next to the weld metal will corrode to protect the stainless steel. This is why steel products were always galvanized after fabrication because there used to be no practical way to restore the effectiveness of galvanizing after welding [1].

Nevertheless, in this research work it has been assumed that the properties degradation by loss of zinc and formation of zinc oxide or zinc rich sublimate in the welding area in the process can be restored by painting mentioned above in this literature and tried to find out an alternative and new approach of welding of galvanized steel with alternate filler material, stainless steel electrodes and MIG method. Mechanical properties like ultimate tensile load of the welded joint and hardness of the base metal, heat affected zone and welded zone were also studied.

2. Experimental

All the specimens were cut from the same sheet and tested in Chemical Analyzer to find out the chemical composition. The result found for the compositions is given below in the **table 1**. According to the compositions found, the steel is AISI 1012. The chemical composition for the filler, ER 308L is also given in **table 2**.

Table 1. Chemical Composition of the galvanized steel (Base Metal) (wt %)

Fe	C	Si	Mn	P	S	Cr	Mo	Ni	Al	Co	Cu	Nb	Ti	V	W	Pb
97.6	0.14	0.07	0.4	0.09	0.11	0.06	0.03	0.13	0.74	0.01	0.12	0.06	0.02	0.01	0.32	0.05

Then the specimens were hold tight enough by jigs in an automatic welding table and welded by using MIG method. Two specimens of same dimension were joined with no gap between. The shielding gas used was 98% CO_2 with 2% Argon at a pressure of 950 psi. The filler metal used was ER 308L; a stainless steel filler with 1 mm diameter. The specimen speed was kept 2 meter/min. The angle of the welding gun was 90°. The distance of the welding gun was kept 3 mm from the surface of the specimens. The parameters like voltage, current, feed rate of filler for each specimen were varied.

Table 2. Chemical Composition of ER 308L Filler [10]

C	Cr	Ni	Mn	Mo	Si	P	S	Cu	Co	V	N	Ti	Nb	Cr_{eq}[a]	Ni_{eq}[b]	Ferrite Content (FN)
0.015	20.0	9.8	1.4	<0.17	0.36	0.02	0.007	0.18	0.06	0.06	0.055	<0.08	<0.07	20.74	12.6	5.4

[a]Cr_{eq} =Cr +Mo + 1.5Si + 0.5Nb.
[b]Ni_{eq} =Ni + 0.5Mn + 30C+ 30N.

In **figure 1 and 2**, the automatic welding table with specimen gripper and welded specimen are shown respectively.

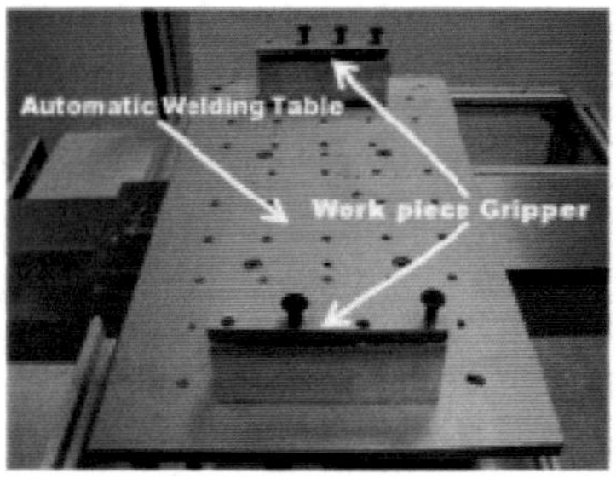

Figure 1. Automatic welding table with gripper where work piece can be gripped by tightening with screws.

Figure 2. Welded specimen

After welding, the specimens were cooled in normal room temperature at 20°C and cut into standard size using cutting machine to test for the tensile strength on universal testing machine. Again welding portion of each specimen were cut into dimension of (6×30×1.6) mm by using sectioning cut off machine for hot mounting to test in Vickers hardness tester and metallurgical microscope.

The net Heat input, H can be calculated from the below equation as the ratio of the total input power, P of the heat source to its travel speed or welding speed, V.

$$H= P/V \tag{1}$$

Where, H is in watt-seconds or joules per mm or inch, P in watts and V in mm per second or inches per minute. For an electric arc, P= EI where E = volts, I = amperes [11].

In **table 3** the variable parameters: voltage, current, feed rate of filler in the MIG welding machine and calculated heat input for each specimen are given.

Table 3. Voltage, Current, Heat Input and Feed Rate of filler and work piece for Different Specimens

Specimen No.	Voltage	Current	Heat input *(Joules/mm)*	Feed rate of filler *(m/m)*	Feed rate of specimen *(m/m)*
1	13.0	54	21.06	2.5	2
2	13.1	55	21.62	2.6	2
3	13.2	56	22.18	2.7	2
4	13.3	57	22.75	2.7	2
5	13.4	58	23.32	2.7	2
6	13.5	59	23.90	2.8	2
7	13.6	60	24.48	2.8	2
8	13.7	61	25.07	2.9	2

3. Results and Discussions

3.1.Strength of Joint

The results in this research work showed that with increasing heat input of the welding machine the ultimate tensile strength decreases gradually.

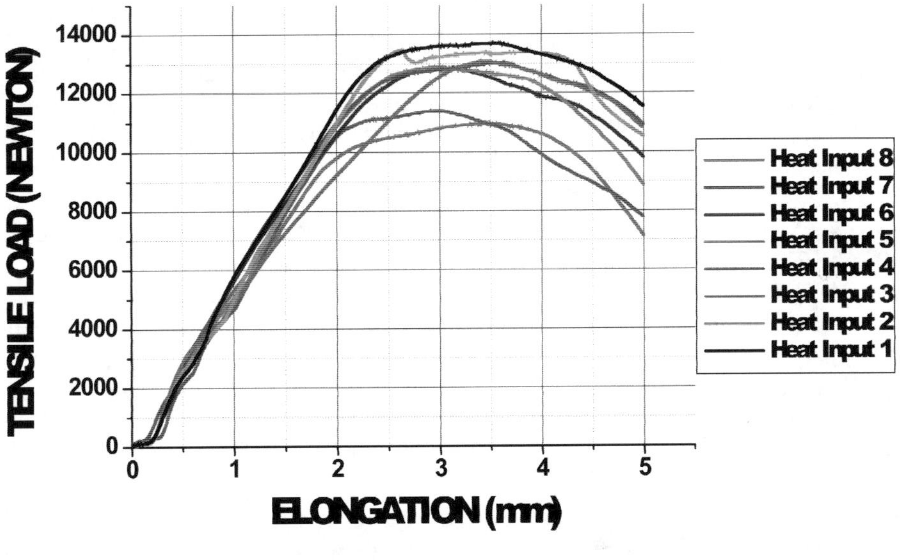

Figure 3. Comparison of ultimate tensile strength for all specimens showing the lowest heat input has highest strength and highest heat input has lowest strength because the higher the heat input, the higher the softening and degradation of the base metal quality at HAZ by burning and thinning the zinc coating and at the same time leaving higher thermal residual stress.

From **figure 3**, it is found that the ultimate tensile strength 13.69kN (342.41 MPa) for lowest heat input (Heat Input 1) was maximum and it continued to decrease. The ultimate tensile strength for highest heat input (Heat Input 8) was minimum and it was 10.95kN (273.81 MPa). This is because, the higher the heat input, the higher the softening and degradation of the base metal quality at HAZ by burning and thinning the zinc coating and at the same time leaving higher thermal residual stress. If the thermal residual stress is higher at HAZ, the joint strength is weaker. That is why the ultimate tensile strength of the joints decreases with increasing welding heat input. The ultimate tensile strength for the base metal was 14.77kN (369.27MPa) while tested in universal testing machine. In **table 4** below, it shows the ultimate tensile strength for base metal and welded joints, joint efficiency for various heat inputs (**Calculated**) and ultimate tensile strength of the base metal from reference [12].

Table 4. Comparison of Ultimate Tensile Strength of welded joints with base metal and Joint Efficiency [12]

Specimen No.	Size of Specimen *(mm × mm)*	Heat input *(Joules/mm)*	Maximum Force of Welding *(N)*	Ultimate Tensile Strength of Welding *(MPa)*	Ultimate Tensile Strength of Base Metal *(MPa)*	Joint Efficiency *(%)*
Base Metal	25×1.6	-	14770.95N	-	369.27(Calculated)	
1	25×1.6	21.06	13696.35N	342.41		92.54
2	25×1.6	21.62	13463.05N	336.57		90.96
3	25×1.6	22.18	13085.60N	327.14		88.41
4	25×1.6	22.75	13030.13N	325.75	370	88.04
5	25×1.6	23.32	12899.57N	322.50		87.20
6	25×1.6	23.90	12862.78N	321.57		86.91
7	25×1.6	24.48	11390.26N	284.76		76.96
8	25×1.6	25.07	10952.38N	273.81		74.00

From the **table 4**, we can see that the ultimate tensile strength for heat input 1 is 342.41Mpa with 92.5% joint efficiency. The next 5 heat inputs have also closer value of ultimate tensile strength to base metal except heat input 7 and 8 which have the value of 284.76 and 273.81MPa with joint efficiency 76.96% and 74.00 % respectively. So, it means that with low heat input higher strength and higher joint efficiency of the joint can be achieved.

The reason is, in Metal Inert Gas welding method the temperature rise for melting the ER 308L filler is 1510° which is very high and with increasing heat input the heat affected zone of galvanized steel is softened and affected extremely by burning and thinning the zinc coating and at the same time leaving higher thermal residual stress. Zinc has lower melting point than steel and higher thermal residual stress makes the material weaker at HAZ. This is why, increasing heat input has more influence in softening the base metal at HAZ gradually resulting weakening the strength of joint as well as joint efficiency and causing failure at heat affected zone.

3.2.Hardness Test

All the specimens were tested in Vickers Hardness testing machine. Before testing in the hardness tester all the specimens were mounted by hot mounting process and polished for smooth mirror surface. The hardness was tested across the welding seam for every 0.5 mm interval starting from the centre of the welded zone followed by heat affected zone and base metal. The load in the hardness tester was 100gf and dwelling time was 10s and the lens used to see that indentation was M50/0.68.

From **table 5** below, it can be seen that, the hardness was minimum at heat affected zone and maximum at welded zone and intermediate at base metal measured for each heat input and with increasing welding heat input from 1 to 8, hardness decreased for HAZ. The hardness for the welded zone and base metal is more or less unchanged.

Table 5. Vickers hardness number for various specimens in 0.5mm interval along the welded zone, heat affected zone and base metal.

	Vickers Hardness Number							
Distance from the centre of weld (mm)	Heat Input 1	Heat Input 2	Heat Input 3	Heat Input 4	Heat Input 5	Heat Input 6	Heat Input 7	Heat Input 8
0.5	493.61	489.56	483.64	480.71	478.48	479.07	487.55	488.65
1	489.28	490.46	487.55	485.83	487.87	484.02	480.23	479.34
1.5	487.35	487.02	493.03	483.36	481.54	473.86	479.08	485.85
2	350.19	347.55	344.67	337.34	328.02	326.29	324.55	321.66
2.5	347.92	343.89	340.71	334.93	332.49	330.26	325.59	320.86
3	330.03	317.32	309.07	298.07	289.08	285.55	279.93	271.24
3.5	400.77	396.09	393.98	390.76	388.06	379.07	375.36	370.13
4	403.23	400.54	400.98	393.29	395.04	393.49	386.45	390.66
4.5	404.79	399.29	389.92	387.59	385.02	394.32	388.33	392.06
5	399.28	397.58	387.85	387.09	394.42	393.53	387.89	379.91

This is because the deposition of ER 308L stainless steel filler which has tensile strength of 590 MPa [13, 14], higher than the ultimate tensile strength of the base metal and more over for very high temperature generation at the time of welding and for slow cooling at room temperature of 20°C, the hardness was highest in the welded zone i.e. fusion zone. There is absence of high strength filler material at HAZ and also it became soft by very high temperature. This is why the hardness number for heat affected zone was lower than welded zone and base metal. And with increasing heat the softening, degradation and formation of thermal residual stress at HAZ was more for consecutive specimens resulting the decreasing in hardness number. The hardness number of the unaffected base metal remained unchanged but was lower than the welded zone.

Both tensile strength and hardness are indicators of a metal's resistance to plastic deformation. Consequently, this is is roughly proportional for tensile strength as a function of the hardness for cast iron, steel and brass. The same proportionality does not hold for all metals [15]. This is why with increasing or decreasing Vickers hardness numbers the tensile strength of material increases or decreases. Hence, the fracture for the specimens was in lowest strength zone i.e. the heat affected zone while testing in the universal testing machine showed in **figure 4,** picture taken by optical microscope. The heat affected zone has the lowest strength as it has the lowest hardness number.

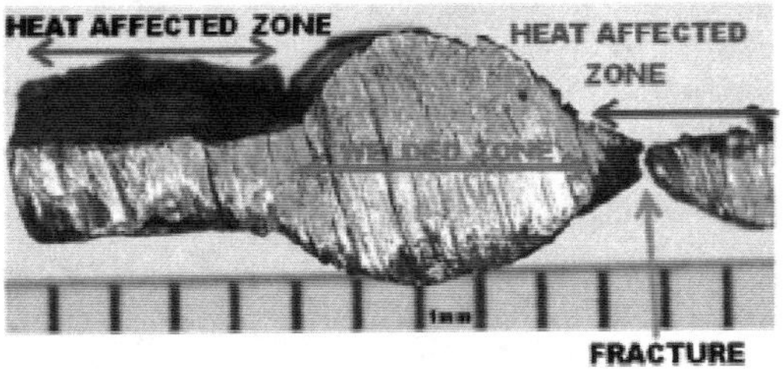

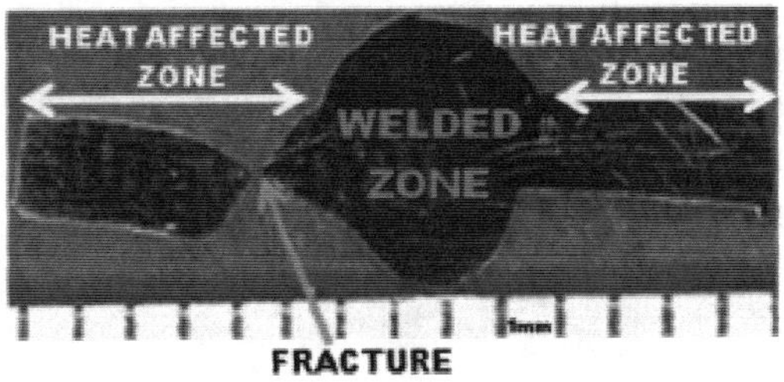

Figure 4. Picture taken from optical microscope shows the fracture is in heat affected zone, HAZ because of the absence of high strength filler in this zone and the high temperature rise at welding time has softened the HAZ and degraded its quality with leaving thermal residual stress.

IOP Conf. Series: Materials Science and Engineering **36** (2012) 012011 doi:10.1088/1757-899X/36/1/012011

From the graphs below in **figure 5** drawn using the data from **table 5** we can see how the hardness varies for each heat input along across the welded zone, heat affected zone and unaffected base metal.

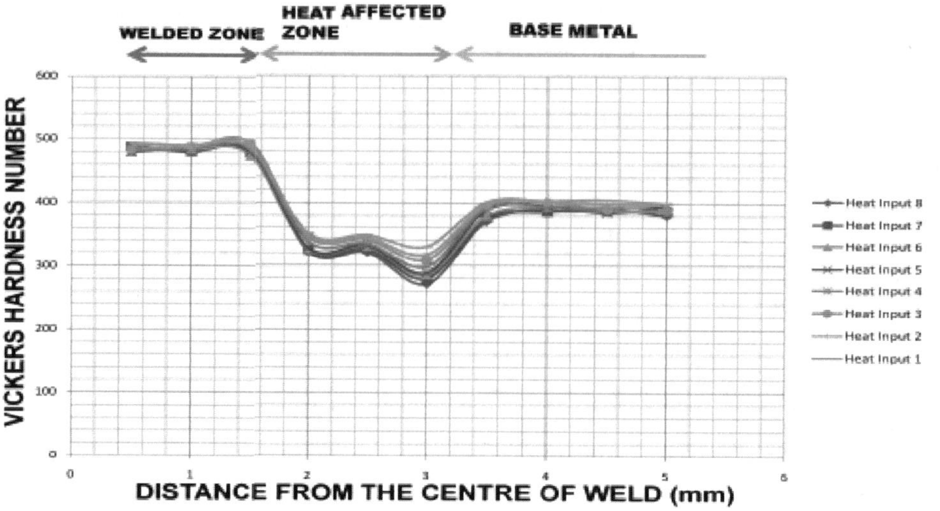

Figure 5. Comparison of Hardness number between all heat iputs measured starting from the centre of welded zone followed by heat affected zone and base metal showing highest hardness at WZ and lowest at HAZ and intermediate at base metal becauuse the zone nearest to the welding seam known as HAZ is mostly affected by heat softening, degrading the quality and leaving residual stress plus the absence of high strength filler material. The higher the heat input, the higher the softening, quality degradation and growth of residual stress and lower the hardness number and strength.

From **figure 4** and the graph in **figure 5** we can calculate the width of welded and heat affected zone and it can be seen that the width of the welded zone is about 3-4 mm and the width of the heat affected zone is about 1-2 mm. If we compare **figure 4 and 5**, it can be proved that the fracture was in the heat affected zone.

3.3 Macro Defects in the Weld

While testing for the hardness in the Vickers hardness machine with M50/0.68 lens, almost no macro defects were seen in the heat affected zone (HAZ) and welded zones for all the specimens. The surface of the specimens was also observed in the optical microscope, showed in **figure 6** and no macro defects were found for welded zone and HAZ. Further, the surface of the specimens at welded zone was observed in metallurgical microscope with M50/0.75 lens, showed in **figure 7** and almost no macro defects were found. It means that almost defects free welding was achieved.

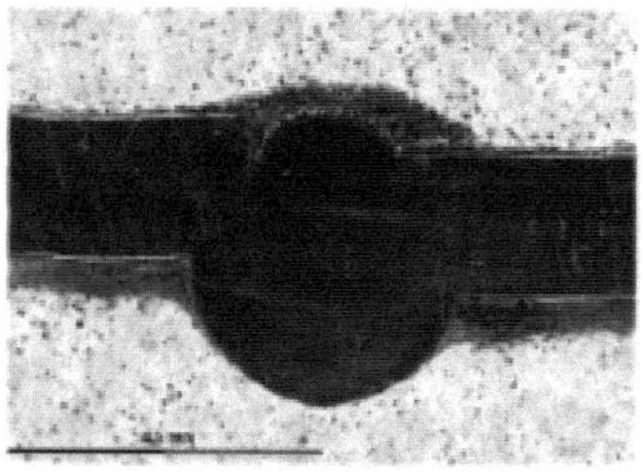

Figure 6.Welding pool and HAZ surface viewed on optical microscope **s**howing almost macro defects free area

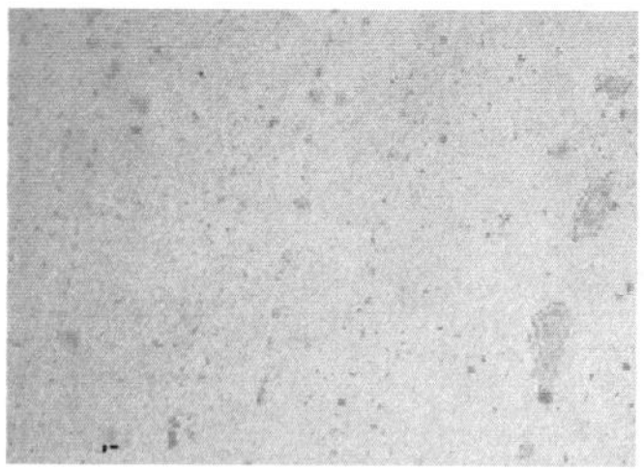

Figure 7.Surface at welded area viewed on metallurgical microscope shows almost no macro defects in the area

4. Conclusion

The following conclusion can be drawn from this research work:

- Very good quality and macro defects free welding was obtained in butt welding of galvanized steel sheet of 1.6 mm thickness using MIG method with ER 308L, a stainless steel filler material.

- The ultimate tensile strength for the joint with lowest heat input was highest but with increasing heat input the joint strength continued decreasing. For all heat inputs the fracture was in the heat affected zone.

- The filler ER 308L can be readily used as an alternative electrode for welding galvanized steel as it showed good mechanical properties in the welding joint.

- MIG method is suitable for welding galvanized steel as no macro defects were found in the welding joint and the method is mechanized, simple, easy and quick to operate.

Acknowledgement

The authors gratefully acknowledge the Research Management Centre for grant; CIM, Welding, Materials and Metrology Lab, Faculty of Mechanical Engineering, University Malaysia Pahang for helping with the experiments and completing this research work.

References

[1] Simoes A M P, Carbonari R O, Di Sarli A R, del Amo B, Romagnoli R 2011 An environmentally acceptable primer for galvanized steel: Formulation and evaluation by SVET *J. Corrosion Science* **53(1)** 464-472.
[2] http://www.wisegeek.com/what-is-galvanized-steel.htm.
[3] ASM International Committee 1992 *ASM Handbook, Metals Handbook* **13** 9th ed. (USA Corrosion) 432.
[4] Leidheiser Jr. H, Zuzuki I 1980 Towards more corrosion resistant galvanized steel *J. Corrosion (NACE)* **36 (12)** 701–702.
[5] Chen W, Ackerson P, Molian P 2009 CO_2 laser welding of galvanized steel sheets using vent holes *J. Materials and Design* **30** 245–251
[6] Welding Galvanized Steel Safely 1997 *The Fabricator*.
[7] Eijnshergen J F H V 1994 Duplex systems. Hot dip galvanizing plus painting: Pre-treatments for galvanized steel, Elsevier Amsterdam The Netherlands **57**
[8] Werner R 1989 Chemical surface treatment of steel coated with zinc and zinc alloys, Keynote Lecture: Proceedings of the international conference of zinc and zinc alloy coated steel sheet (GALVATECH) *The iron and steel institute of Japan* Tokyo 197–205.
[9] ASM International Handbook Committee 1994 Surface Engineering *ASM Handbook* 5, **3rd Printing** (USA).
[10] Das C R, Bhaduri A K, Srinivasan G, Shankar V, Mathew S 2009 Selection of filler wire for and effect of auto tempering on the mechanical properties of dissimilar metal joint between 403 and 304L(N) stainless steels *J. of materials processing technology* **209** 1428–1435.
[11] Energy for welding *Principles of Welding Processes, Physics, Chemistry and Metallurgy* **134**
[12] MatWeb: Material property data *2011http: www.matweb.com, October.*
[13] Singha P J, Guhab B, Achar D R G 2003 Fatigue tests and estimation of crack initiation and propagation lives in AISI 304L butt-welds with reinforcement intact *J. Engineering Failure Analysis* **10** 383–393
[14] Techalloy Welding Products, 2009 *www.techalloy.com*
[15] Mechanical Properties of Metals *Material Science and Engineering* **8th ed.** 179.

IOP Conf. Series: Materials Science and Engineering **36** (2012) 012012 doi:10.1088/1757-899X/36/1/012012

Fatigue behavior of Austenitic Type 316L Stainless Steel

K. A. Mohammad[1], Aidy Ali[1], B. B. Sahari[2] and S. Abdullah[3]
[1]Department of Mechanical and Manufacturing Engineering, Universiti Putra Malaysia, 43400 UPM, Serdang Selangor, Malaysia
[2]Institute of Advanced Technology (ITMA), Universiti Putra Malaysia, 43400 UPM, Serdang Selangor, Malaysia
[3]Department of Mechanical and Materials Engineering, Universiti Kebangsaan Malaysia, Bangi, 43600 UKM Bangi, Selangor, Malaysia

E-mail[1]: k.amohd@yahoo.com

Abstract. The aim of this work is to determine the fatigue life of 316L stainless steel. The mechanisms of fatigue of 316L stainless steels were investigated and discussed. The fatigue tests were carried out at constant-amplitude cyclic loading with load ratio R=0.1. Scanning Electron Microscope (SEM) is then used to examine the fracture surface. The results show that the fatigue limit of 316L stainless steel was 146.45 MPa.

1. Introduction

In leading-edge toward global market, many engineering components experienced fatigue mechanism in daily life and it is one of the oldest problems which concern scientists and engineers. Plenty of steel in the world has been investigated for a lot of application in the science and technology market. Fatigue caused the structure failed. Due to the complicated nature of fatigue mechanisms and the large number of factors that influence fatigue life, there is so far no unified approach that can treat all fatigue problems [1]. For instant, engineering components used in aviation, nuclear industry, transportation, including in oil and gas plant may be contain defect due to imperfection of manufacturing process. Examples of these applications include the utilizing of industrial piping in petrochemical refineries, aerospace, ship building industries, construction, repair nuclear power facilities, and ground vehicle industries. Safety and predicted life are the most significant in these industries. Therefore damage tolerant design can be used to predict residual life of components.

During the last decades, lots of designers, engineers and researchers investigated and explored to develop prediction models for high cycle fatigue (HCF) life, as since it take enormous time and efforts to construct a stress life (S-N) curve [2]. According to Landgraf [3], the HCF is recognized as highly influenced by microstructural variables such as grain size, the volume fraction of secondary phase and the amount of solute atoms or precipitates. In the other hand, many investigator have examined the fatigue crack initiation and propagation modes where manipulate the failure of component to breakdown [4-6]. From the authors Chen et al. [7], mentioned that the starting point of failure under low cycle fatigue is

Published under licence by IOP Publishing Ltd

mostly related to the geometrical discontinuities on the specimen surface and furthermore of creep-fatigue-environment may enhance the cracking problem.

Due to its application for daily use, life prediction of 316L steel is one element of the complex and crucial in designing a component into tubular or cylinder shape working in various of alternative loading. Up till now, a small number of researches work have been carried out in investigating and study the mechanism of tubular components cause by complexity of geometry to compare with simple plate geometry in order to predict the lifetime of steel before rupture.

In this paper, the proposed of this work is characterizing the fatigue life of austenitic stainless steel 316L at room temperatures which is subjected to variable loading.

2. Materials

In this study, austenitic type 316L stainless steel was used and studied and it chemical compositions are given in Table 1. Meanwhile Table 2 represents the mechanical properties for hourglass specimen of austenitic steel. Austenitic steel in form of tubular specimens were provided by local supplier, S.N Machinery Services Sdn. Bhd. that is designed and fabricated carefully in accordance with ASTM E606. In fatigue tests, the specimen image with its dimensions of austenitic Type 316L stainless steel specimen designs which was tested in transverse direction of axial fatigue testing as shown in Figure 1.

Table 1. Chemical compositions of Type 316L stainless steel [8]

Element (%)	C	Ni	Cr	Mn	P	S	Si	Mo	N
316L	0.020	11.21	17.38	1.86	0.027	0.0054	0.51	2.36	0.038

Table 2. Mechanical Properties of Type 316L Stainless Steel

Mechanical Properties	Type 316L stainless steel
Yield Point, MPa	332
Tensile strength, MPa	673
Modulus of Elasticity, GPa	165
Strength at break, MPa	586
Elongation at break, mm	35.5

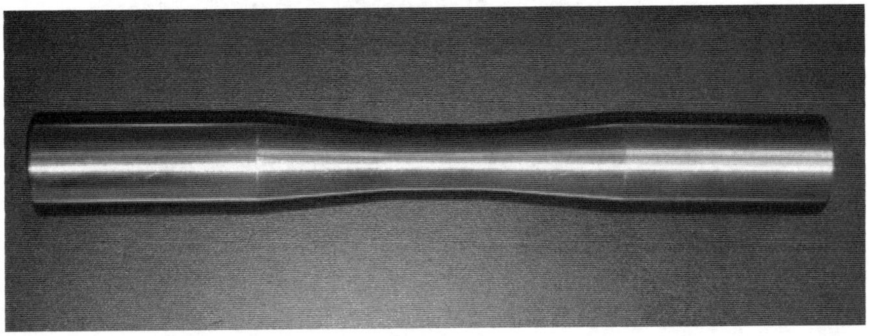

(a)

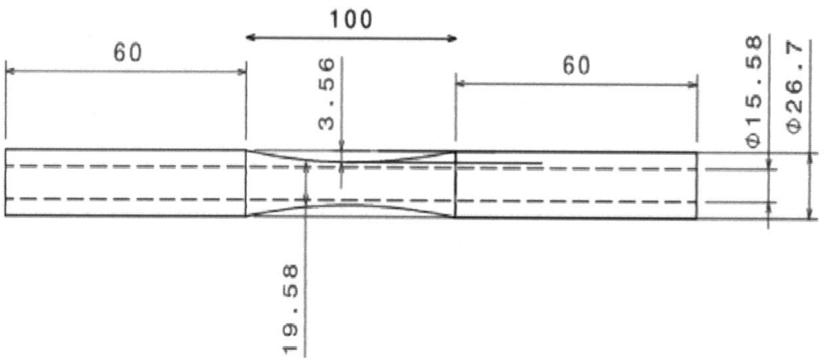

(b)

Figure 1. (a) Austenitic Type 316L Stainless Steel and (b) its dimensions

3. Jig Design

The jig is made out of carbon steel 45SC in this fatigue test as shown in Figure 2. The designed jig was accommodated to hold and grip specimens for fix it into the actuator of machine as additional supporter bar. In addition, use of jig also preventing specimen from slips and scratched.

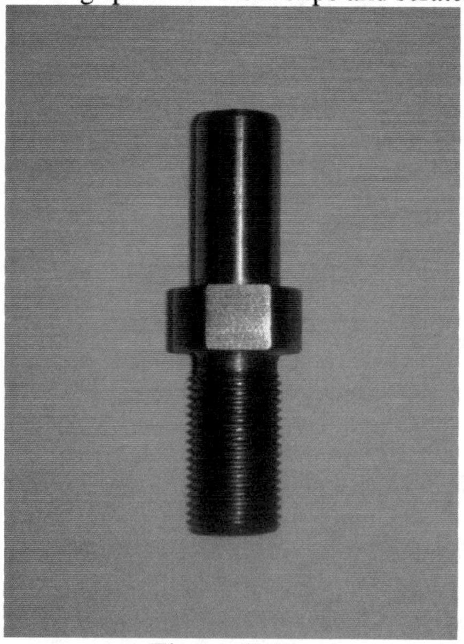

Figure 2. Jig

4. Experiment

The fatigue specimens were tested in push-pull fatigue test using Fast Track Hydraulic Universal Testing Machine Instron 8802 of 250 kN load capacity as shown in Figure 3. The machine is attached with computed Waveform Generator Software to record the fatigue data. The tensile test was carried out before commencing the fatigue test as guidance in order to determine the maximum load that the specimen could sustained. All fatigue tests for austenitic steel were performed in constant load amplitude, constant frequency of 5 Hz and a stress ratio, R equal to 0.1. Figure 4 shows the specimen undergoing fatigue test. Fatigue test results were plotted in graph in term of stress versus life data. Best fit curve were plotted for data analysis.

Figure 3. Fast Track Hydraulic Universal Testing Machine Instron 8802 of 250 kN load

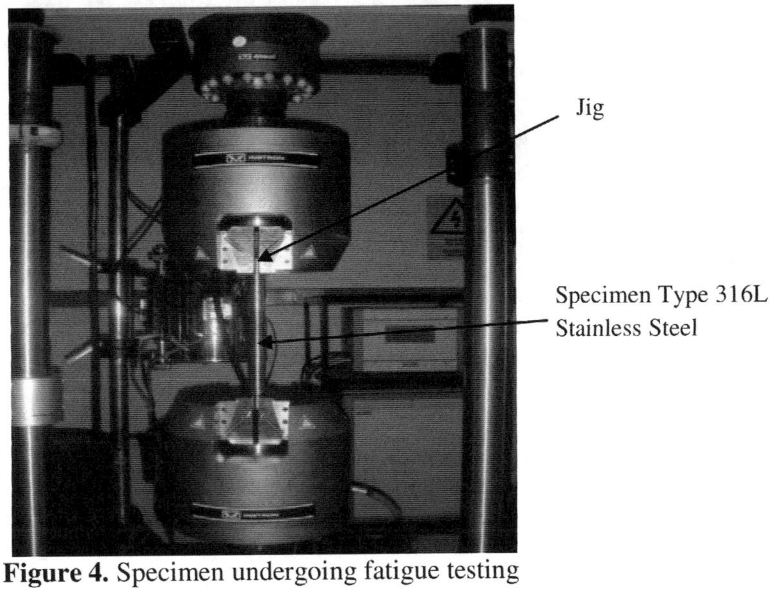

Figure 4. Specimen undergoing fatigue testing

5. Results and Discussion

In experimental, the fatigue tests were conducted using different loadings. The fatigue life of specimen obtains as shown in Table 3 in terms of number of cycles to failure. With considering eight's value of means and amplitude stresses used for carried out in tension-tension fatigue by applied loads on specimens in order to get life time at variable amplitude loading. From the results in Table 3, stress amplitude versus fatigue life is plotted to view the significant difference of fatigue life at high stress compared to lower stress.

Table 3. Stainless Steel Type 316L in term of Number of Cycles.

Load Ratio (R)	Stress Amplitude (MPa), σ_a	Life (Cycles) $2N_f$
0.1	334.00	4,628
	290.93	17,340
	275.20	55,478
	234.33	164,938
	220.15	450,447
	180.11	1,033,948
	160.69	4,832,284
	146.45	7,893,764

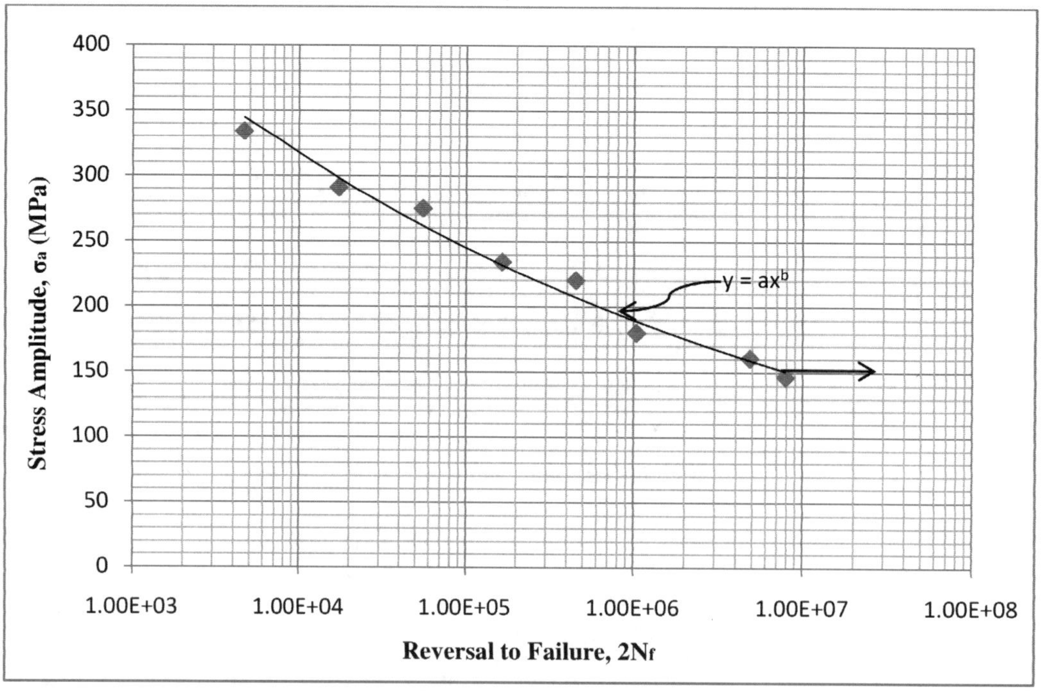

Figure 5. S-N curve

Fatigue data at different load were acquired with a data acquisition system. The results are plotted in S-N curve is shown in Figure 5. From the figure shows that variation of curve of fatigue test will decrease as different loading applied on specimen till fatigue failure. It is mentioned that type 316L stainless steel undergoes harden under cyclic loading [9]. The data collected from the test have been fitted by power law curves in order to describe the specimen fatigue behavior. The equations of power law curves are given by:

$$y = ax^b \tag{1}$$

This equation shown the relation between fatigue life and applied load is called Basquin relation. The Basquin relation is the general equation represents typical S-N curve and its expression developed from log–log S-N graphs. It is the most widely used equation. This equation is usable in the stress-based approach to fatigue analysis and design. The relation is given by:

$$S_a = S_f' \left(2N_f\right)^b \tag{2}$$

where S_a is stress amplitude, S_f' is the fatigue strength coefficient and b is the fatigue strength exponent. The fatigue experimental data was fitted with equation of Basquin Equation whereby the value of b exponent obtains from the slope of line in log-log scale. It is show that behavior of fatigue strength on type 316L stainless steel was hardening and followed by softening at surrounding condition [10].

From the experimental results, it can be observed that the data is decrease as applied load at different variable into fatigue endurance zone. At range of 10^3-10^4, it is categorized as low cycle fatigue where stress amplitude's value between 334 MPa – 290.93 MPa yields cycle ranging 4628 – 17340 measured for hazardous area where the fatigue life is very low. Meanwhile range of 10^5 and above could be categorized as high cycle fatigue has 55478 – 7893764 cycles and it is considered it will have a much longer fatigue life. Each life's component would be replaced by another one since it reaches at this stage more than a 1000000 cycles. The arrow represents the specimen beyond 1 millions cycles did not break even after applied many specimen throughout the tests. From observation the graph, it can be that fatigue limit is successfully characterized for type of 316L stainless steel in this research and found to be 146.45 kN at 7,893,764 cycles.

6. Fractographic Analysis of Fracture Surface

In order to investigate the morphological-microstructure relationship of failed specimen, fracture surface were investigated using SEM in terms of the pattern crack would be irregularities can degrade fatigue strength. The fracture initiate at surface irregularities such as voids and inclusions in according to Lah et al. [11]. This experiment was carried out by Scanning Electron Microscope (SEM - Hitachi S-3400N) as shown in Figure 6.

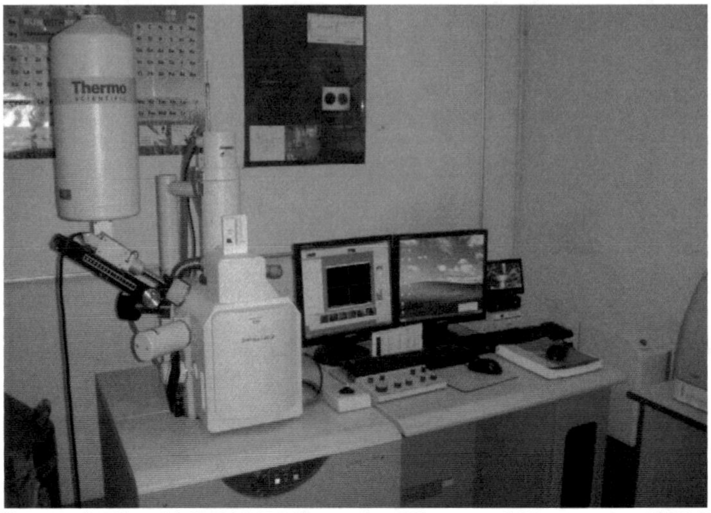

Figure 6. Hitachi S-3400N Scanning Electron Microscope (SEM)

In the Figures 7, 8 and 9 show SEM fractograph for Low Cycle Fatigue (LCF) regime of specimen tested at 334 MPa. The specimen shows evidence of beachmark due to fatigue failure. Crack initiation and early propagation exhibits the transgranular fracture mode in fracture surface specimens for first stage of fracture surface also shown in Figure 7. It shows the fracture strip and which reveals the shear surface morphology. Otherwise, from the figure revealed also the fracture surface has more than one porosity initiation site.

Figure 7. Nucleation Crack

At second stage, fracture surface of the specimen shows a propagation zone where mixed fracture path completely intergranular mode in fracture behavior of specimen as shown in Figure 8.

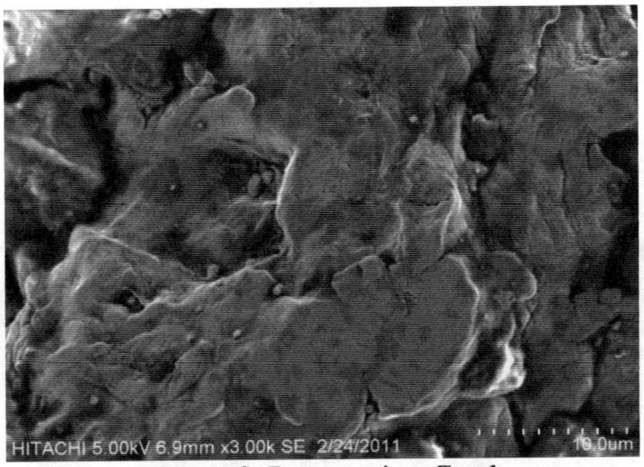

Figure 8. Propagation Crack

Figure 9 represents the captured picture on fracture surface at third stage before failed. Unstable of mode between intergranular fracture and transgranular fracture in this specimen tests was found in final stages (rupture zone). In this stage, fracture surface is look like changed from intergranular to transgranular before reach unstable stage to rupture.

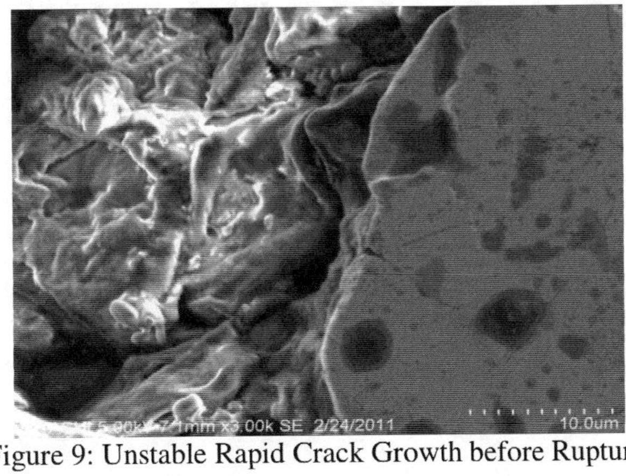

Figure 9: Unstable Rapid Crack Growth before Rupture

7. Conclusions

As conclusions, the fatigue life of austenitic stainless steel 316L is successfully characterized from the experiment conducted at ambient temperature. Since experimental results have higher loading causes the specimen to fail tremendously fast instead of lower load. Fatigue test imposed a limit of fatigue life of 10^7 cycles due to cost and time limitation. The fatigue limit from the result is 146.45 kN at 7,893,764 cycles for this specimen. Meanwhile with observation using SEM show that the nucleation crack was transgranular mode, intergranular in crack propagation, and last stage exhibits mixed of transgranular and transgranular mode in continuous cyclic loading in this research.

Acknowledgement

The author would like to thank to University Putra Malaysia for the scholarship under Special Graduate Research Assistant (SGRA) under grant University (Project No.: 91726) which made to pursuit of this research achievable.

References

[1] Weixing Y, *Stress Field Intensity Approach for Predicting Fatigue Life.* International Journal of Fatigue, 1993. 15(3): p. 243-246.

[2] Beden S M, Abdullah S, Ariffin A K, Al-Asady N A, Rahman M M, *Fatigue Life assesement of different steel based Shell Materils Under Variable Amplitude Loading.* European Journal of Scientific Research. ISSN 1450-216X, 2009. 29 No.1: p.157-169.

[3] Landgraf, R.W. Handbook, Fundamentals of Metal Fatigue Analysis Prentice Hall, 1979. *Fatigue and Microstructure.* In: Meshii M, edit ASM; p. 439-440.

[4] Marchionni, M., Osinkolu, G. A. and Maldini, M. *High Temperature Cyclic Deformation of A Directionally Solidified Ni-Base Superalloy.* Fatigue Fracture Engineering Structure, 1995. 19(8): p. 955-962

[5] Gayda, J. and Miner, R. V. *Fatigue Crack Initiation and Propagation In Several Nickel-Base Superalloys At 650°C.* International Journal Fatigue. 1983(5): p.135-143.

[6] Mcmahon, C. J. and Coffin L. F. *Mechanisms of Damage and Fracture In High Temperature, Low Cycle Fatigue Of A Cast Nickel- Based Superalloy,* Metallurgical and Materials Transactions B. 1970. 1, No 12: p.3443-3450

[7] Chen, L. J., Yao, G., Tian, J. F., Wang, Z.G., and Zhao, H.Y. *Fatigue and Creep Fatigue Behavior of A Nickel-Base Superalloy At 850° C,* International Journal Fatigue, 1998. 20 No.7: p. 543-548.

[8] Maeng W.Y., Kang, Y.H. *Creep-Fatigue and Fatigue Crack Growth Properties of 316LN Stainless Steel At High Temperature,* Transactions of the 15th International Conference on Structural Mechanics in Reactor Technology (SMiRT-15) Seoul, Korea, August 15-20, 1999.

[9] Kim, D.W., Chang, J.H., Ryu, W.S. *Evaluation of the Creep-Fatigue Damage Mechanism of Type 316L and Type 316LN Stainless Steel.* International Journal of Pressure Vessels and Piping, 2008. 85(6):p. 378-384

[10] Vogt, J.-B., Foct, J., Regnard, C., Robert, G., Dhers, J. *Low temperature fatigue of 316L and 316LN Austenitic Satinless Steel.* Metallurgical Transaction A, 1991. 22A :p. 2385-2392

[11] Lah, N.A.C, Aidy Ali, Ismail, N., Chai, L. P., Mohamed, A. A. *The Effect of Controlled Shot Peening on Fusion Welded Joints,* Materials and Design. 2010, 31: p. 312-324

IOP Conf. Series: Materials Science and Engineering **36** (2012) 012013 doi:10.1088/1757-899X/36/1/012013

Detection of defects of Kenaf/Epoxy by Thermography Analyses

M J Suriani[4], Aidi Ali[1], A Khalina[2], S M Sapuan[1], S Abdullah[3]
[1]Department of Mechanical and Manufacturing Engineering, Faculty of Engineering, Universiti Putra Malaysia, 43400 UPM, Serdang, Selangor, Malaysia.
[2]Department of Biological and Agricultural Engineering, Faculty of Engineering, Universiti Putra Malaysia, 43400 UPM, Serdang, Selangor, Malaysia.
[3]Department of Mechanical and Materials Engineering Faculty of Engineering and Built Environment, Universiti Kebangsaan Malaysia, 43600, Bangi, Selangor, Malaysia.
[4]Institut of Teacher Education, Darulaman Campus, 06000 Jitra, Kedah, Malaysia

E-mail[4]: surianimatjusoh @gmail.com

Abstract. There are quite There are quite a few defects can occur due to manufacturing of the composites such as voids, resin-rich zones, pockets of undispersed cross-linker, misaligned fibres and regions where resin has poorly wetted the fibres. Such defect can reduce the mechanical properties as well mechanical performance of the structure and thus must be determine. In this study, the defect of Kenaf/epoxy reinforced composite materials has been determined by thermography analyses and mechanical properties testing of the composites have been done by tensile test. 95% of the thermography analyses have proved that the defects occur in the composite has reduced the mechanical properties of the specimens.

1. Introduction

Kumar et al. reported that fibre reinforced polymer composite finds application from head gear to aircraft due to lightweight, higher strength to weight ratio and adaptability to customize the composite according to the required strength and functionality for which it is employed [1]. Consequently, it is necessary to for which develop a sophisticated and advanced tool or method of analyses such as non-destructive technique (NDT), i.e. Infrared thermal imaging to detect the defect of such lignocelluloses fibre reinforced polymer composites before its will be used in a variety industries. Infrared and thermal measurement and investigation techniques have been used in a variety of applications, i.e. inspection of subsurface defects and features, identification of thermo-physical properties, detecting of coating thickness and hidden structures [2].

Thermal imaging technique or thermography is a surface thermal radiation measurement technique that is used to detect spatial variations in the measured surface temperature pattern. Thermography reveals flaws by searching anomalous hot-spots after thermal excitation [3]. This robust methodology that applies will be used to monitor the integrity of composite structures before further applications and during their operational life-cycle and to anticipate failure before occurrence to ensure reliable and

Published under licence by IOP Publishing Ltd

safe operation. Knowledge of defects, their determination is also important in cutting manufacturing cost and maintenance operation.

2. Experimental procedures

2.1. Materials

The fibre used is treated kenaf bast fibre provided by Institute of Tropical and Forest Product (INTROP), UPM, Serdang, Selangor, Malaysia. Epoxy with density of 1.15 g/cm³ used as binding material. The raw kenaf bast fibres were combed in order to clean and untangle the strong bonding of individual fibres. Bernard et al. [4] reported that combed fibre has strong mechanical properties in comparison to uncombed fibre. Kenaf bast fibres were cut to 23 cm fibre length using scissors. Figure 1(a) and Figure 1(b) shows raw materials of treated kenaf bast fibres and unidirectional kenaf bast after have been combed and cut to 23cm.

Figure 1(a). Bulky of treated kenaf bast fibre.

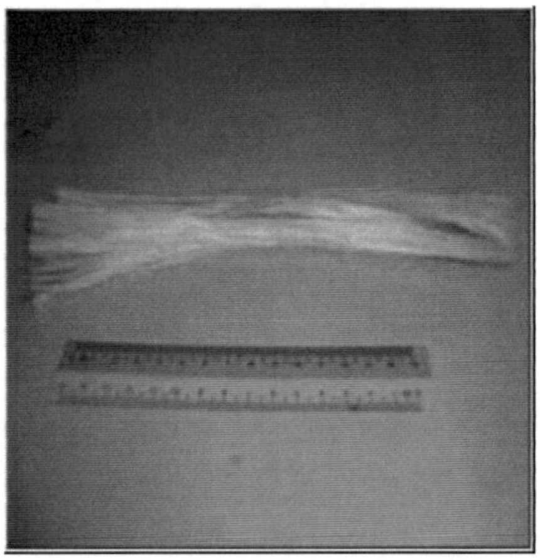

Figure 1(b). Combed Kenaf bast fibre.

2.2. Preparation of Epoxies Samples

Epoxies samples were produced in order to control and determine the optimum temperature and curing time for processing the kenaf/epoxy composite. The volume of epoxies used for the thickness of 0.3cm is 112.5cm^3 equal to 129.4g based on the volume of the mould, in accordance with ASTM D3039 [5].

2.3. Fabrication of Kenaf /Epoxy Composites

The composite specimens were prepared by combining the kenaf bast fibre as reinforcement and epoxy resin as a binder or matrix through hand lay-up and compress using hot-press compression moulding, using a mould similar to that used to fabricate epoxy samples. One of the main obstacles that need to be addressed in the function of kenaf fibre reinforced composites is uneven fibre distribution [6]. Kenaf fibres are difficult to manually separate and visually disperse evenly during manufacturing [7]. At 60:40 percent total volume with 0.1cm thickness, kenaf bast fibre to epoxy, 25.2g of kenaf bast fibre and 17.25g of kenaf bast fibre were prepared. At 60:40 percent total volume with 0.3cm thickness, kenaf bast fibre to epoxy, 75.6g of kenaf bast fibre and 51.8g of kenaf bast fibre were prepared. At 60:40 percent total volume with 0.5cm thickness, kenaf bast fibre to epoxy, 129.4g of kenaf bast fibre and 84.0g of kenaf bast fibre were prepared.

The composites were compressed using hot-press compressing machine at temperature of 70°C, pressure of 50 Bars. Under these circumstances, the mould was pre-heated for 3 minutes followed by pressing for 20 minutes and venting for 7 times to maintain uniform thickness and avoid air entering the mould. Figure 2(a) and Figure 2(b) shows the mould used to fabricate the kenaf/epoxy specimens.

Figure 2(a). Steel spacer.

Figure 2(b). Steel plate.

2.4. Testing

2.4.1. Thermal Imaging Technique

Infrared and thermal measurement and investigation techniques have been used to detect the defects on the surface of kenaf/epoxy composites. Figure 3 shows the IR Fluke Flexcam camera used in this study.

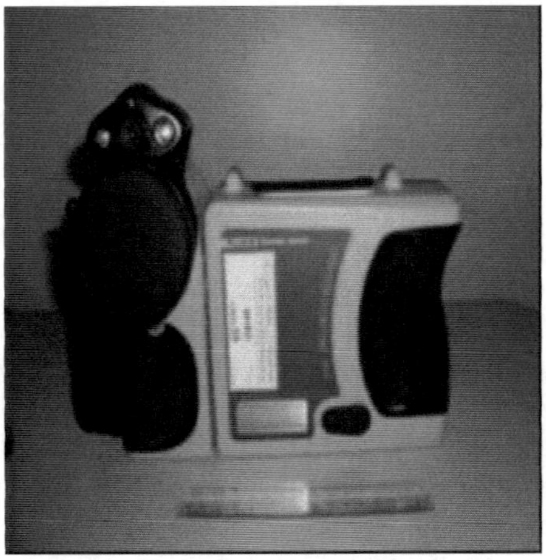

Figure 3. IR Fluke Flexcam camera

This investigation and measurements were using active thermography whereby kenaf/epoxy specimens were heated in the oven at temperature $100^{\circ}C$, for 60 minutes before thermography were recorded. In active thermography the surface of a sample is heated using external heat source and the surface is monitored [8]. The kenaf/epoxy specimens were place on the non-metal stand to avoid heat dissipate from the stand through conductivity. These experiments were carried on the emissivity of 0.92 and the room temperature is $29^{\circ}C$ with the relative humidity of 70%. Figure 4 shows the schematic diagram of the method used to detect the defect in this study.

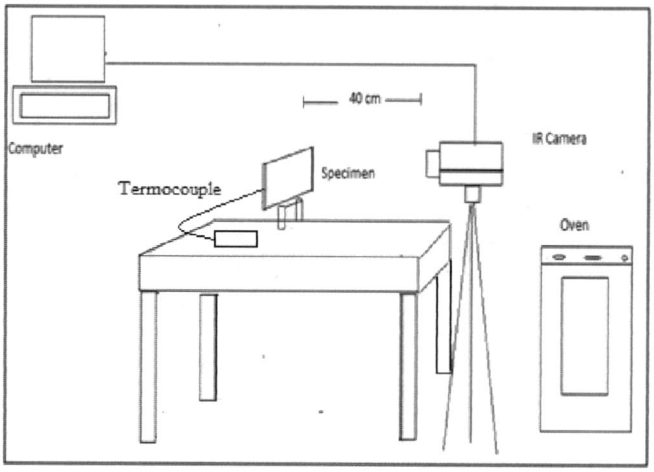

Figure 4. Schematic diagram of the method used to detect defects.

2.4.2. Tensile Test
Tensile test were done using an Instron machine at Universiti Putra Materials. Tensile test procedures are accordance with ASTM standard D 3039/ D 3039M [5].

3. Results and Discussion

3.1 Determination of healthy and faulty parts
An infrared camera scans surface of components and records any changes in temperature. In this study the healthy parts is declare as a part which have a good properties of the composite materials surface and was presented mostly as blue colour area. While the faulty part is declare as an area which contains most defects and mostly present in red colour.

3.2 Thermography analyses
Thermography results are presented in a contour of contrast colours. Thermography analyses of the kenaf/epoxy specimen reveal that the faulty part is showing higher temperature than the healthy part. Defects or inhomogeneities which lie below the surface affect the heat transfer rate when thermal energy is propagating into or out the structures [9]. These defects generally occur during the commercial production of composites, and can be due to batch-to-batch variation of the prepreg, as well as the sometimes manual construction known as 'lay-up'. Variability can arise from differences in the prepreg tack level during lay-up, variable resin content. Composites can contain a number of defects introduced during manufacturing, which can considerably increase their likelihood of the composite failure. Such defects include voids, resin-rich zones, and pockets of undispersed cross-linker., misaligned fibres and regions where resin has poorly wetted the fibres [10].

Table 1 shows the comparison between of the defects in the three specimens by thickness of 0.1cm, 0.3cm and 0.5cm at 45% kenaf to 55% epoxy, 60% kenaf to 40% epoxy and 75% kenaf to 40% epoxy composites respectively.

IOP Conf. Series: Materials Science and Engineering **36** (2012) 012013 doi:10.1088/1757-899X/36/1/012013

Table 1 Comparison between of the defects in the three specimens by thickness and fibre loading.

Thickness/ Fibre Loading	**0.1cm**	**0.3cm**	**0.5cm**
45% Kenaf to 55% Epoxy	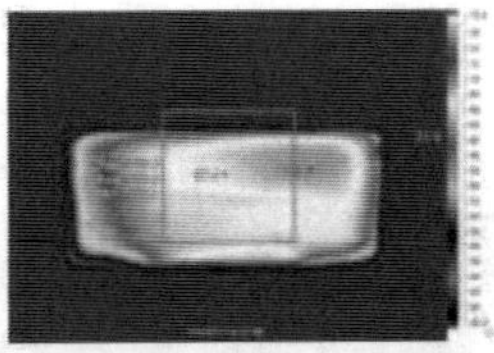		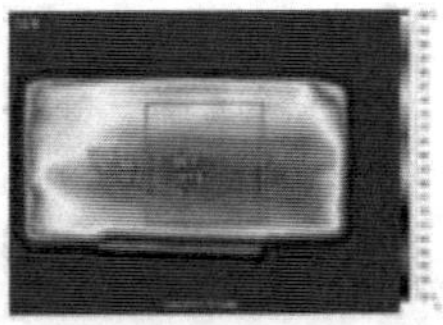
60% Kenaf to 40% Epoxy	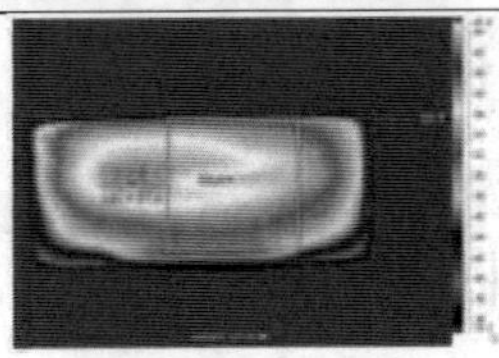	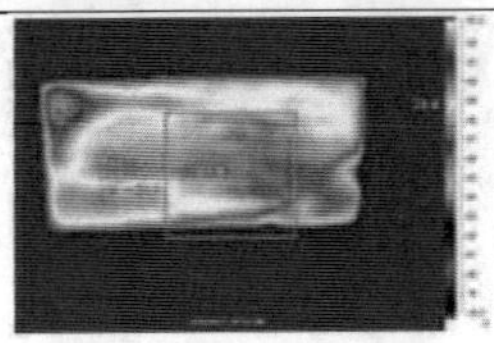	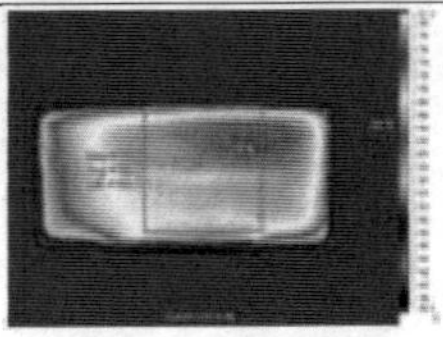
75% Kenaf to 15% Epoxy		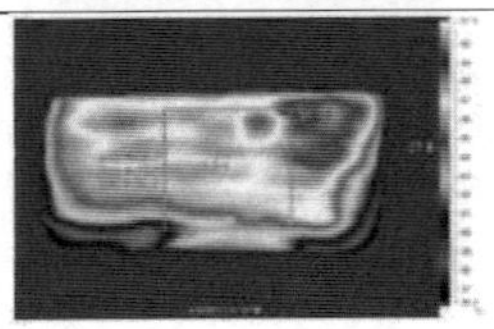	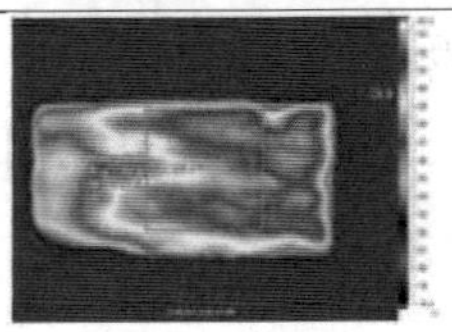

3.3 Results of tensile test

Specimen of 0.1cm thickness at 60% kenaf to 40% epoxy shows the highest ultimate tensile strength (UTS) of 186.55 MPa. Specimen of 0.5cm thickness at 75% kenaf to 40% epoxy shows the lowest UTS of 57.77 MPa. These results have proved that the specimens which contain most defects have reduced the mechanical properties of the composite materials. These results also claimed by Avdelidis et al., (2005) in their study pulsed thermography on aircraft materials [11] It also concluded that the mechanical properties of the specimens are decrease with the increase of specimen's thickness and percentage of fibre loading. Figure 5 shows the results of tensile test for specimen of 0.1cm, 0.3cm, and 0.5cm thicknesses of kenaf/epoxy specimens.

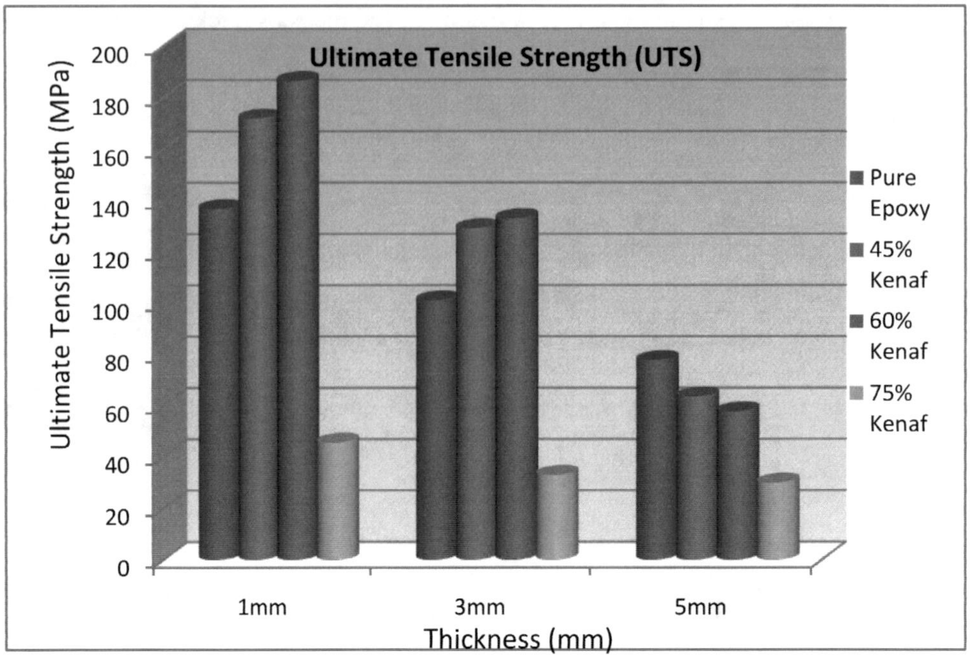

Figure 5. Result of Tensile test

4. Conclusions

It can be concluded that Infrared Thermal Imaging technique have been proved as a NDT to detect defect in Kenaf/epoxy composite materials. Thermography analyses of kenaf/epoxy have been matched with the result from tensile test. The most defect or faulty parts specimens shows the lowest value of ultimate tensile strength. It has been proved that the defects in composites will reduce the mechanical properties of the kenaf/epoxy specimens. Defects due to manufacturing were studies and found that kenaf/epoxy shows the faulty parts are mostly caused by contain highest quantity of void and pocket of undispersed cross-linker.

In spite of the indicated benefits, it should be reminded that this suffers from some disadvantages, which can be solved in the next research. The most important point is to ensure measurement of constant temperature during recorded thermography of the kenaf/epoxy specimen which is suggested here is by using thermocouple. The next part of this research also included the determinations of defects by thermography analyses and validations the results by using scanning electron microscopy (SEM) and optical miscroscope.

Acknowledgement

The authors would like to thank Department of Mechanical and Manufacturing Engineering, Universiti Putra Malaysia (UPM), Serdang, Malaysia and Institute of Tropical and Forest Product (INTROP), UPM, Serdang, Selangor, Malaysia for the assistance on this project.

References

[1] Kiran P K, Raghavendra N V, Srihara B K 2011 Optimazation of infrared radiation cure process parameters for glass fiber reinforced polymer composites. *Material and Design* **32** 1129-1137

[2] Maldague X 2000 *Applications of infrared thermography in nondestructive evaluation.* In: Rastogi P K, Inaudi D editors. Trends in optical nondestructive testing, Amsterdam: Elsevier Science 281-285

[3] Choi M, Kang K, Park J, Kim W, Kim K 2008 Quantitative determination of a subsurface defect of reference specimen by lock-in infrared thermography. *NDT & E International* **41** 19-124

[4] Bernard M, Khalina A, Janius R, Faizal M, Kamarudin KS 2009 *Effect of Processing Parameters on Mechanical Properties of Kenaf Fiber Plastic Composite.* Paper presented at the meeting of the International Advanced of Technology Congress (ATCi), Malaysia; November 3-5

[5] ASTM D-3039 2008 *Standard Test Method for Tensile Properties of Polymer Matrix Composite Materials.* American Society of Testing Materials

[6] Zampoli M., Pourboghrat F, Yankovich S A, Rodgers B N, Moore J, Drzal L T, Mohanty A K, Misra M 2007 Kenaf natural fibre reinforced polypropylene composites: A discussion on manufacturing problems and solutions. *Composites Part A* **38** 1569-80

[7] Joseph K, Thomas S, Pavithran C 1996 Effect of chemical treatment on the tensile properties of short sisal fibre-reinforced polyethylene composites. *Polymer* **37** 5139-49

[8] Maldague X P 2001 *Theory and Practice of Infrared Technology for Nondestructive Testing* New York John Wiley & Sons Inc. 356-361

[9] Aggelis D G, Kordatos E Z, Strantza M, Soulioti D V, Matikas T E 2011 NDT approach for characterization of subsurface cracks in concrete. *Construction and Building Materials* **25** 3089-3097

[10] Scheirs J. 2000 *Compositional and Failure Analysis. A Practical Approach.* Chichester John Wiley 458-476

[11] Avdelidis N P, Almond DP , Marioli-Riga Z P, Dobbinson A, Hawtin B C 2005 *Pulsed thermography: philosophy, qualitative & quantitative analysis on aircraft materials and applications.* Paper presented at the Proceeding Vth workshop, Advances in Signal Processing for Non Destructive Evaluation of Materials Quebe City Canada 2-4 August

Experimental Study on Revetec Engine Cam Performance

Maisara Mohyeldin Gasim[1], Lee Giok Chui, Khirul Azhar bin Anwar
Faculty of Mechanical Engineering, Universiti Malaysia Pahang, 26600, Pekan, Pahang, Malaysia

E-mail[1]: maisara@ump.edu.my

Abstract. In Revetec engine (three-lobed) cam replaces the crankshaft to convert the reciprocating motion of the engine piston, to a rotating motion in the drive line. Since the cam controls the piston movement, the cam profile has a great effect on engine performance. In this paper an experimental study was done to a (three- lobed) cam with Cycloidal motion profile but with different ratios between the base circle radius of the cam and the radius of the roller follower. DEWESoft was used to find the displacement and the vibration of the piston, and compare the actual results from the test with the theoretical results from the cam profile equation. The results showed that there is a periods of miss contact between the follower and the cam when the ratio between the base circle radius of the cam and the radius of the roller follower is less than a certain value, and also increasing of vibration . The suggested ratio between the cam and follower radius is to be more than 2:1.

1. Introduction

Efficiency of conventional IC engines is less than 50% percent due to heat and friction losses. About two- third of the input energy is lost in exhaust gases and water cooling [1], and the mechanical losses due to friction are about 8% [1].

Due to economic crisis and shortness of oil resources, efforts are done to increase IC engines efficiency by reducing the heat and friction losses.

All conventional IC engines used crankshafts to convert the piston reciprocating motion to a rotating motion in the drive line, but crankshafts are responsible of side thrust force, vibrations and also they are not efficient in transferring the power to the drive line.

Crank-less and free piston engines were a good example to overcome the crank shaft problems, but these new kinds of engines are facing a control and starting problems [2].

In 1996, an Australian engineer called Bradley Howell Smith, managed to produce a new mechanism to convert the reciprocating motion of the internal combustion engines pistons, to rotating motion in the drive line, by using a three-lobed counter rotating cams and called the new engine arrangement as Revetec engine [3].

In Revetec engine, shown in Figure 1, there are two opposite pistons controlled by two (three-lobed) cams. On both sides of each piston, there are two bearing which act as the follower for the cams. As each cam has three arms (three-lobed), meaning, in every two cycles of engine there are three power stokes compared to one power stroke in the conventional engines, and every stroke in Revetec engine completed in just 60 degrees instead of 180 degrees in conventional engines.

The cam profile contributes great effect in the engine performance, because the cam profile controls the piston motion, meaning that by changing the cam profile the piston motion will change too.

There is a wide variety of cam motion schemes, but the main motions for cams followers, are: simple harmonic, constant acceleration and Cycloidal motion [4]. Cycloidal motion is the best profile for high speed engines because it maintain the maximum cylinder pressure when compared with the other motion schemes [5].

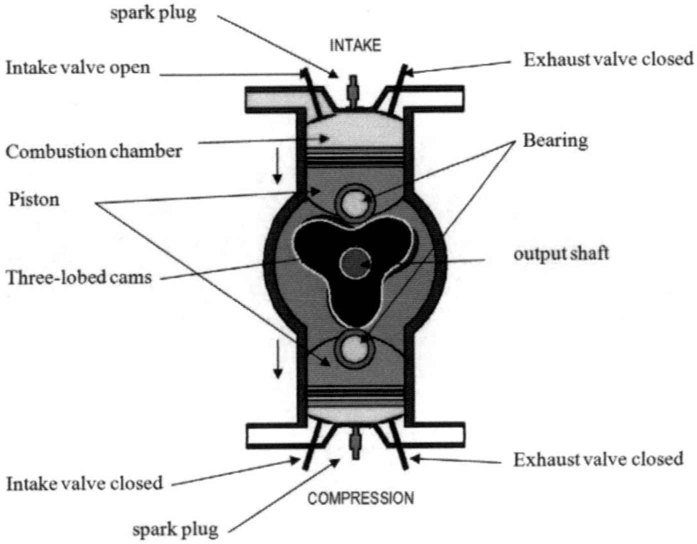

Figure 1: Revetec engine with two cylinders during intake-compression stroke

In this paper an experimental study done to calculate the displacement, speed and acceleration of the piston when connected to three lobed cam having a Cycloidal motion profile.

2. Cam design and manufacturing

By using CNC milling machine cams with different radii of the cam base circle were manufactured. Figure 2 shows one of these cams. The x and y coordinates of the milling cutter are given as:

$$C_x = -\left[R_f + R_b + \Delta R\right]\sin\phi + \left[R_c - R_F\right]\sin(\phi - \alpha) \tag{1}$$

$$C_y = \left[R_f + R_b + \Delta R\right]\cos\phi - \left[R_c - R_F\right]\cos(\phi - \alpha) \tag{2}$$

Where:
R_f = Radius of the roller follower
R_b = Radius of the base circle
ϕ = angle into rise or fall interval that defines the instantaneous follower properties.
R_c = Mill cutter radius
ΔR = Instantaneous follower displacement, which can be calculated from the equation:

$$\Delta R = H\left[\frac{\phi}{\beta} - \frac{1}{2\pi}\sin\left(\frac{2\pi\phi}{\beta}\right)\right] \tag{3}$$

H = Maximum follower displacement
β = Rotation angle of cam during the rise or fall interval.
α = the angle between the follower centerline and the cam contact point, which calculated from:

$$\alpha = \tan^{-1}\left[\frac{v\left(R_f+R_b+\Delta R\right)}{\omega\left(R_f+R_b+\Delta R\right)^2}\right] \tag{4}$$

ω = Rotational speed of the cam
v = instantaneous velocity of the cam follower at the cam angle ϕ, which obtained from the equation:

$$v = \frac{H\omega}{\beta}\left[1 - \cos\left(\frac{2\pi\phi}{\beta}\right)\right] \quad (5)$$

one of the cams designed depending on these equations and by using AutoCAD is shown in Figure 2. Then using CNC machine one of the manufactured cams is shown in Figure 3.

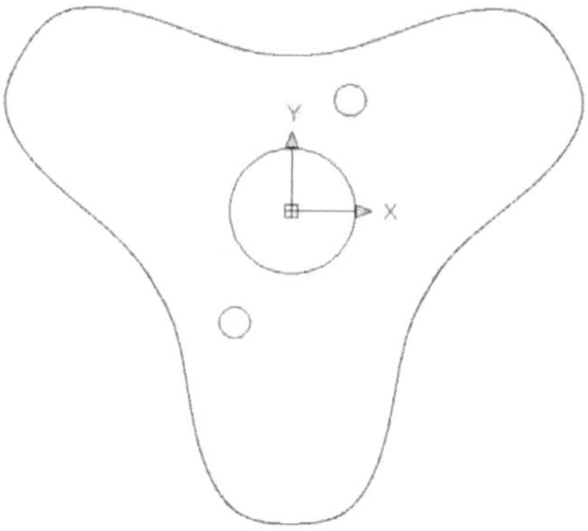

Figure 2: Three-lobed cam drawing

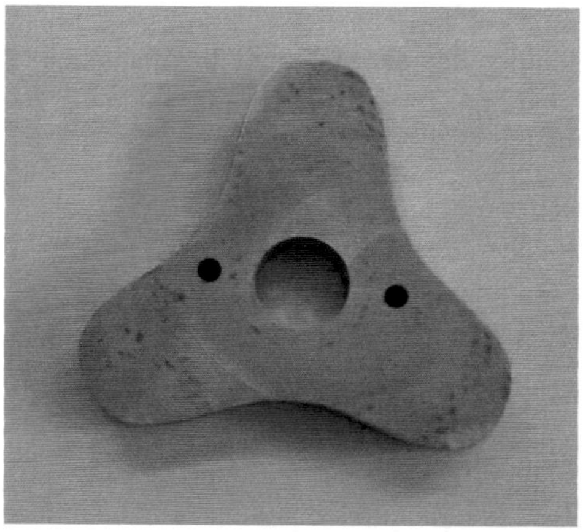

Figure 3: Three-lobed cam

2. Test rig layout

Figure 4 shows the test rig layout which consist of the hardware components. In the hardware there are: shaft (1) which drives the three lobed cam (2) using the electric motor (3). Three sensors exerted in different positions, sensor (4) measures the follower (5) displacement. The spring (6) maintain the contact between the follower and the cam.

The data from the sensor is transferred to DEWESoft which is data acquisition software. By using the software the data is recorded and illustrated in Figures 5 and 6.

Figure 4: Test rig layout

3. Results

Figure 5 illustrates the displacement and vibration of cam follower against the time. The graph shows unobtainable displacement at the beginning of fall intervals. This unobtainable displacement is due to large diameter of the roller follower or the- respectively- small base circle of the cam.

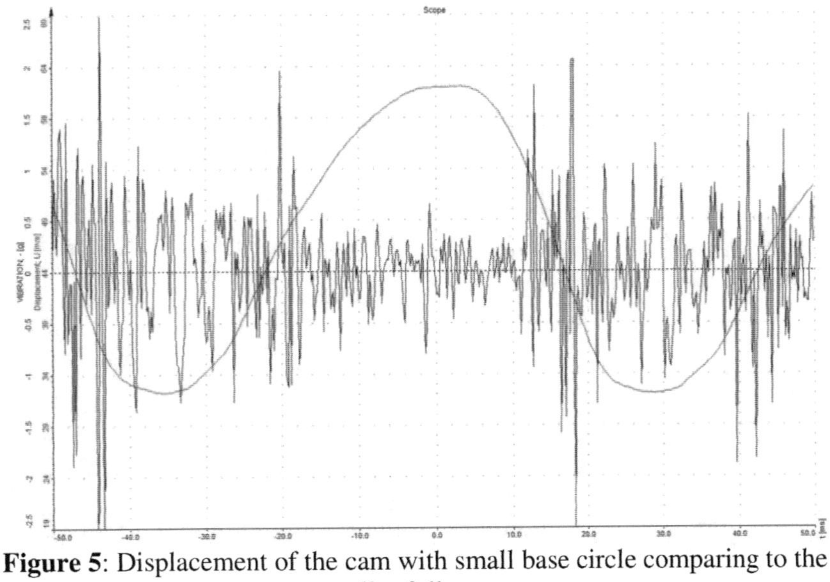

Figure 5: Displacement of the cam with small base circle comparing to the roller follower.

After increasing the diameter of the base circle the miss contact between the cam and the follower is disappeared, and also the vibration decreased, as shown in figure 6.

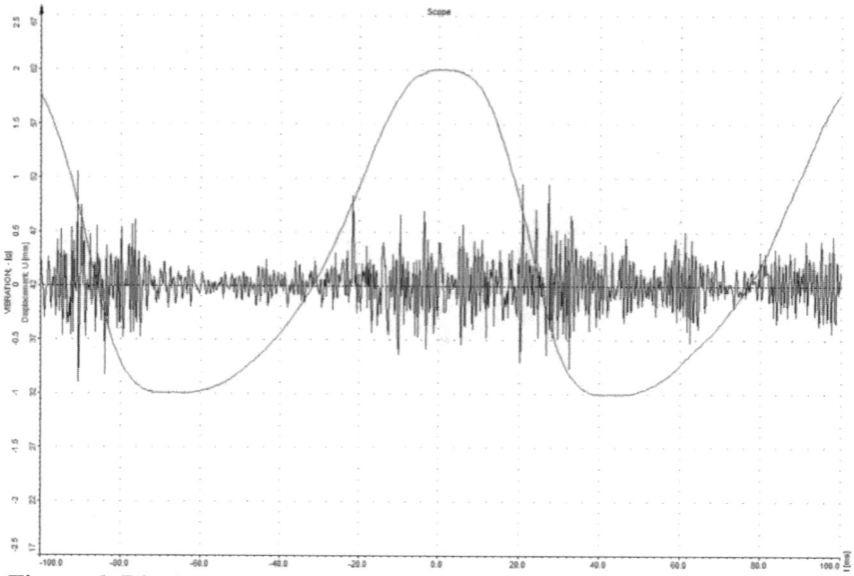

Figure 6: Displacement of the cam with large base circle comparing to the roller follower.

4. Conclusion

Three-lobed cam used in Revetec engine to convert the piston reciprocating motion to a rotating motion in the drive line. The cam profile used here was designed to give the piston a Cycloidal motion, which is proven to be the best motion to maintain the maximum cylinder pressure during combustion.

Different cam base circle radius used and tests were carried out measure the displacement using DEWESoft. The results from the test shows that the best ratio-to avoid miss contact between the cam and follower and to decrease the vibration- is more than 2:1.

References

[1] V. Pandiyarajan , M. Chinna Pandian , E. Malan , R. Velraj, R.V. Seeniraj, Experimental investigation on heat recovery from diesel engine exhaust using finned shell and tube heat exchanger and thermal storage system, 2010, Applied Energy 88 (2011)

[2] Mikalsen R., Roskilly A.P., Performance simulation of a spark ignited free-piston engine generator, Applied Thermal Engineering 28 (2008) 1726–1733

[3] Bradley Howell-Smith, Opposed Piston Combustion Engine, W.O., international Patent classification F02B 75/24 (2006.01)

[4] Maisara Mohyeldin Gasim, A. K. Amirruddin, A. Shahrani, The Effect of Cam Profile on Cylinder Pressure in Revetec Engine , 2nd National conference in mechanical Engineering for Research and Postgraduate studies, 3-4 Dec. 2010, University Malaysia Pahang.

[5] Myszka David H., Machines and Mechanism, Pearson Education International, 2005.

Computational Analysis on Performance of Thermal Energy Storage (TES) Diffuser

M.A.H.M. Adib[1], F. Adnan[2], A.R. Ismail[1], K. Kardigama[1], H.A. Salaam[1], Z. Ahmad[1], N.H. Johari[1], Z.Anuar[1] and N.S.N.Azmi[1]
[1]Sport and Human Engineering Group (SHEG), Faculty of Mechanical Engineering, Universiti Malaysia Pahang, 26600 UMP, Pekan, Pahang, Malaysia.
[2]Department of Applied Science and Technology, Universiti Kuala Lumpur, 32200 Lumut, Perak, Malaysia.

E-mail[1]: azrul@ump.edu.my

Abstract. Application of thermal energy storage (TES) system reduces cost and energy consumption. The performance of the overall operation is affected by diffuser design. In this study, computational analysis is used to determine the thermocline thickness. Three dimensional simulations with different tank height-to-diameter ratio (HD), diffuser opening and the effect of difference number of diffuser holes are investigated. Medium HD tanks simulations with double ring octagonal diffuser show good thermocline behavior and clear distinction between warm and cold water. The result show, the best performance of thermocline thickness during 50% time charging occur in medium tank with height-to-diameter ratio of 4.0 and double ring octagonal diffuser with 48 holes (9mm opening ~ 60%) acceptable compared to diffuser with 6mm ~ 40% and 12mm ~ 80% opening. The conclusion is computational analysis method are very useful in the study on performance of thermal energy storage (TES).

1. Introduction

Air conditioning system is essential for residential, commercial and also the industrial sector especially for countries with high ambient temperature such as Malaysia. The usage of conventional air conditioning system as what most household and building have, are generally being used more often during the day time especially during the afternoon when it is at the "peak-time". Peak time is when demand for electricity energy is high and to cater to this demand, utility company will have to generate more energy and this comes with a price hence contributes a big portion to the customers' monthly utility bill. The solution for this is to shift the energy or load needs during peak time to off peak. Off peak is when the electric energy demand is low and the tariff per kilowatt hour is less than that of the peak time. Thermal energy storage or TES for short is a storage system that can be use to shift the peak load demand to off peak load demand. TES is a concept where the ice or chilled water is produced and stored in storage during at night. The stored cooling medium is later used to meet a cooling load of the total integrated complexes during the day when the electricity rate is higher. The concept behind TES is simple. Storage medium is cooled by chillers during off-peak hours and stored in an insulation tank. Electrical costs peak during the day is very high, and are less during evening hours, because demand decreases. TES in naturally stratified tanks is used to reduce the operating costs and refrigeration plant capacity of many large cooling systems. TES takes advantage of natural

Published under licence by IOP Publishing Ltd

properties of water which cause it to naturally stratify at different temperatures. For this TES system was controlled with an inlet and outlet diffuser. By properly designing an inlet and outlet diffuser system, a tank can be used to store cold and warm water separated by a boundary layer knows as a thermo cline. In this study, the diffuser must be designed to minimize the effect of mixing (optimum diffuser design), so that the density differences of the cold and warm water will result in stratification.

2. Method

Selection design of diffuser will have to consider many criteria such as performance, efficiency and the design itself since different diffuser have different system setup for example radial and octagonal diffuser is more suitable for cylindrical tank and H-diffuser is suitable for rectangular tank. However, this study objective is to obtain the optimum diffuser design. With using CFD-Fluent will be carried out to fulfill the objective. The ratio of inertia force to kinematics force calls Reynolds number. This dimensionless number is use to contribute the inlet flow through the cylindrical storage tank. (Dorgan *et al.*, 1993) suggested using 200 Reynolds number. For the 5 meter length and above, 400 to 850 Reynolds number is suggested to use. Even though, the Reynolds number must less than 2000.

$$\mathrm{Re} = \frac{\rho v d}{\mu} \quad (1)$$

3. Parameters and Equation

From this project, all the parameters are considered from two sources of investigator, (J.E.B. Nelson *et al.*,1999; Mohd Shamsurie Anuar,1996; B.J. Sliwinski *et al.*,1978; C.R. Truman *et al.*,1989; Wildin, M. *et al.*,1990) experiment that have been done from their research of parametric studies on thermally stratified chilled water storage system.

Table 1. Parameter setting

Characteristic	Value
Warm water temperature	290 K
Chilled water temperature	277 K
Aspect Ratio (Ar)	3.0, 4.0, 5.0
Number of Diffuser	24 holes, 48 holes
Types of Diffuser (Octagonal)	Single Ring, Double Ring
Diffuser Size (Diameter)	6 mm, 9 mm, 12 mm
Flow rate	123 Liters / Hours
Reynolds Numbers	Re 200
Initial temperature different	277 K to 290 K

4. Result and Discussion

Comparison was made for thermocline thickness produced and good agreement was observed. Figure 1 and 2 below shows comparison made for charging process with the same inlet *Re* of 200. Study on shallow tank with HD 3.0, 4.0 and 5.0.

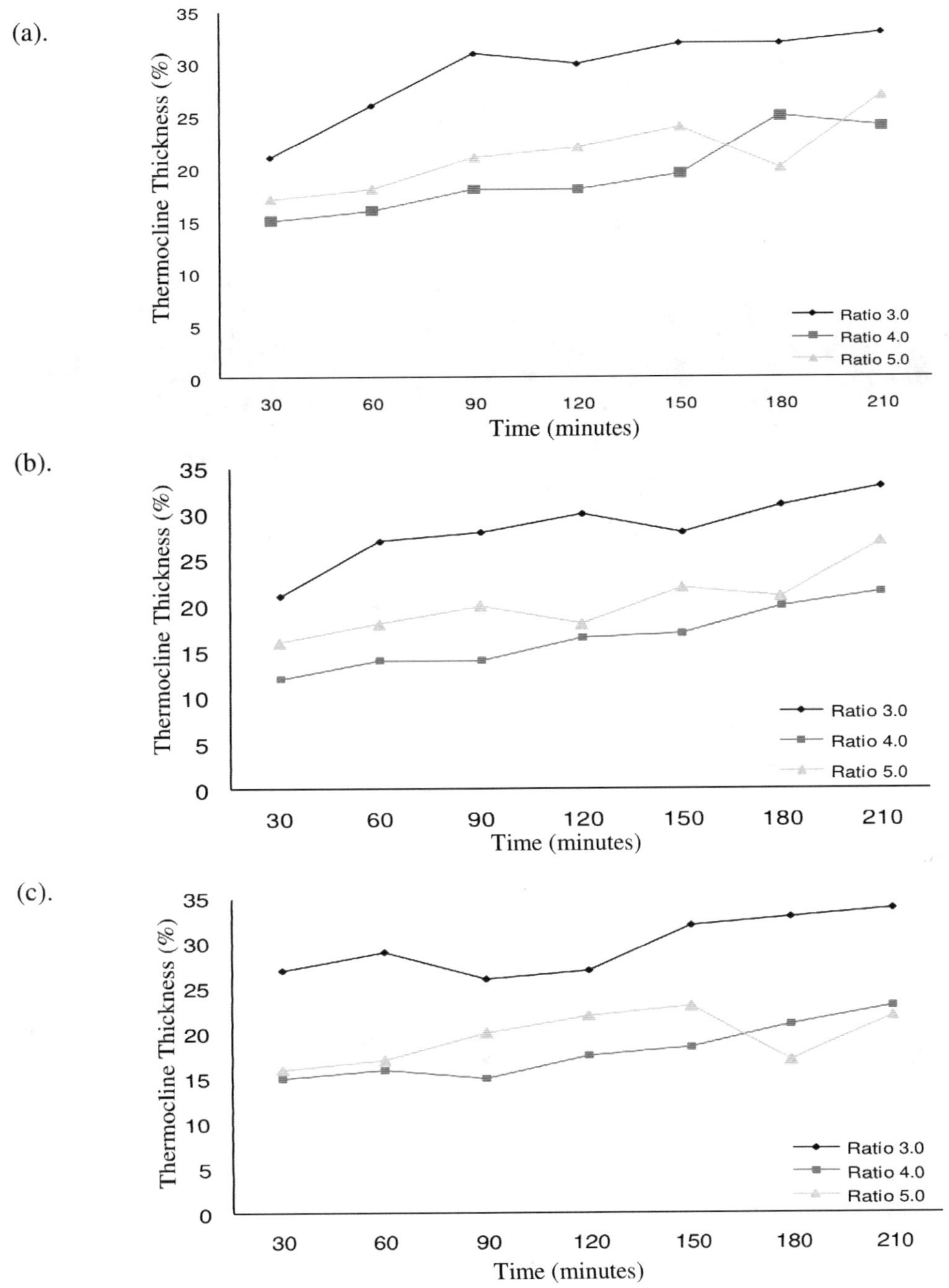

Figure 1. Thermocline thickness with Double Ring type for different diffuser size; a) 40% (6mm) ; b). 60% (9mm) ; c). 80% (12mm)

Chilled water at various Reynolds (*Re*) was charged into the tank and observations on the development of the thermocline were made and all the result with compare between diffuser size of Double Ring and Single Ring are shown in Figure 2 below.

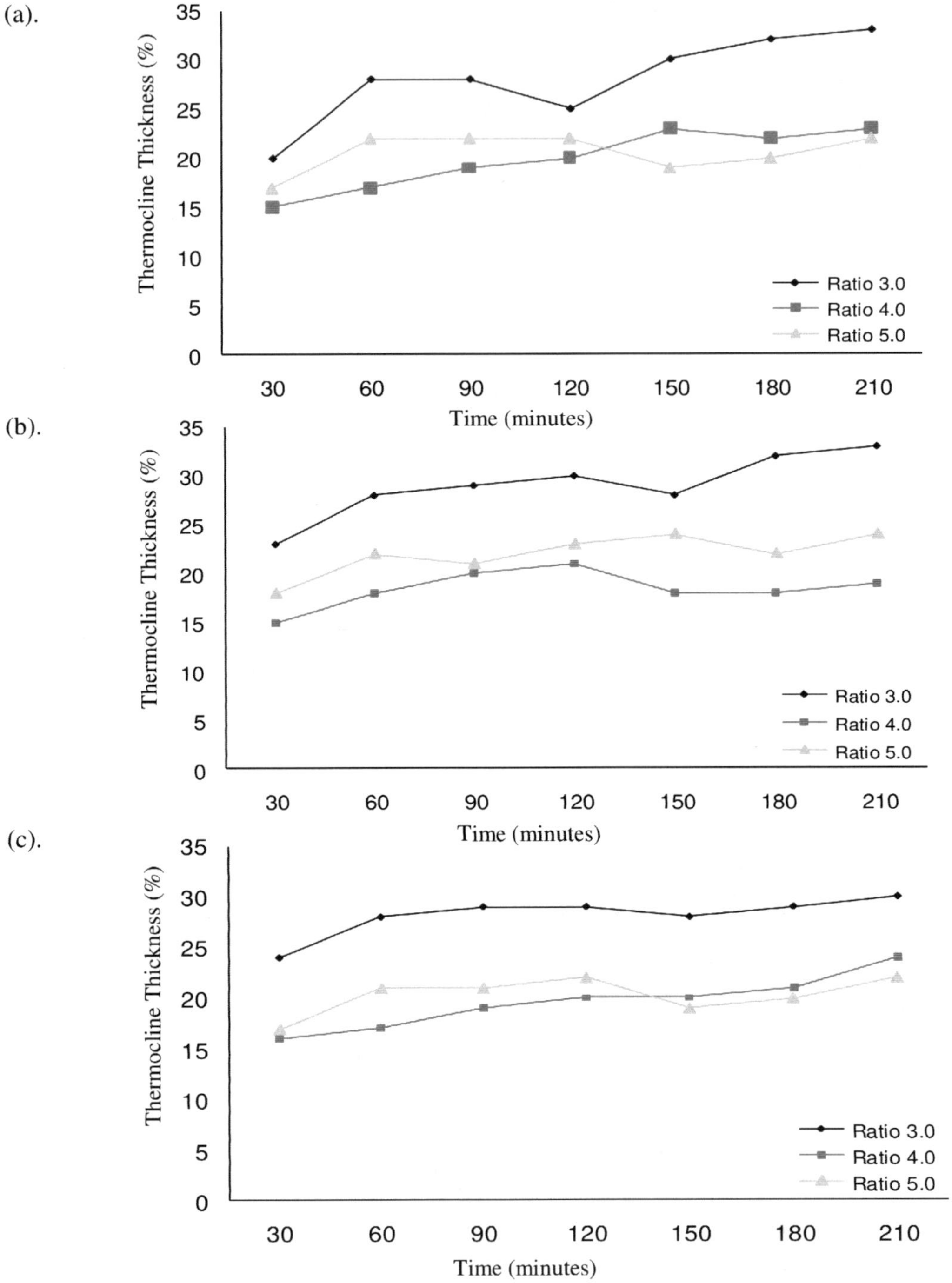

Figure 2. Thermocline thickness with Single Ring type for different diffuser size; a). 40% (6mm) ; b). 60% (9mm) ; c). 80% (12mm)

Results reveal that the thermocline degradation is more pronounced at low flow rates. Thermal diffusion, axial wall conduction and heat exchange with the ambient are rate process. The energy degradation due to these is directly proportional to the charging time. Hence the degradation of energy due to heat conduction across the thermocline and long the storage tank (axial wall conduction) increase with the increase in charging time. Stratification improves with increasing flow rate up to a certain value and there after remains constant. The rates of warming up due to energy transfer from the ambient decreases with decrease in the aspect ratio and also with the diffuser size (opening holes). The thickness of the thermocline is increased by heat conduction through the thermocline during the charging and discharging portions of cycle and this may obscure the initial thickness produced by mixing and heat transfer during thermocline formation. It has been found that mixing near the inlet diffuser may be minimized by introducing fluid through that diffuser so that it initially forms gravity current. Gravity current is a thin layer, composed almost entirely of incoming fluid that travels across the tank due to a density difference rather than inertia of the incoming flow. Such a current gently pushes the fluid previously in the tank out of the way so the mixing occurs only at the front of the current during the first pass of the current across the tank.

(a).

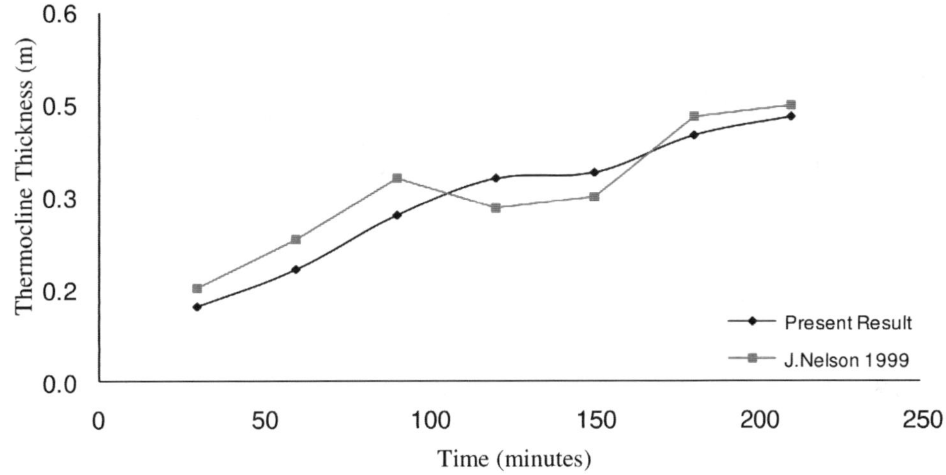

(b).

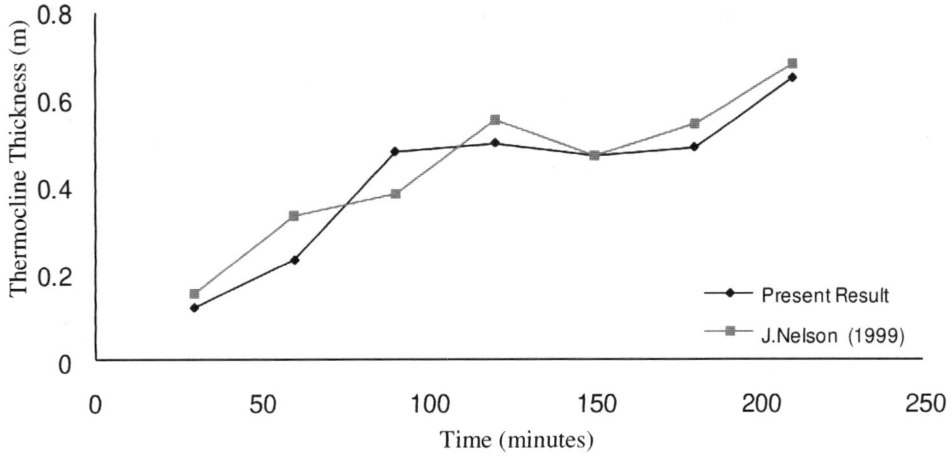

Figure 3. Verification Result for 3D model for thermocline thickness between; a). Double Ring with 48 holes ; b). Single Ring with 24 holes

IOP Conf. Series: Materials Science and Engineering **36** (2012) 012015 doi:10.1088/1757-899X/36/1/012015

Figure 3 shown verification result between double ring and single ring. For double ring result with 48 holes, the accuracy of the thermoclines thickness compare between present result and J.Nelson (1999) result achieved 86.3% and for single ring result with 24 holes, the accuracy of thermoclines thickness achieved 73.4%. Based on both 3D comparison result shown, it seen that the thermocline thickness can be predictable with a different diffuser size (hole size).

5. Conclusion

The need for ease and comfort when indoor have lead to building nowadays either it residential, commercial or industrial equipped with air conditioning system. Air conditioning systems have considerably contributed a big portion to monthly utilities bill hence, the need for a more efficient and reliable system is essential. Thermal energy storage, chilled water system to be more precise is undeniable a great solution to be used as air conditioning system considering it benefits and the finding on this study had shown that the best performance of thermocline thickness during 50% time charging occur at double ring type diffuser with holes diameter of 9 mm (60% opening) in range 16% and the good initial state of thermocline thickness range is between 15%~21%. Medium height-diameter (HD) ratio 4.0 simulation also shows the good thermocline behavior and clear distinction between warm and cold water regions.

Acknowledgement

The support of the Universiti Malaysia Pahang, under SHEG group is gratefully acknowledged. The authors of this paper also would like to express their gratitude to Universiti Kuala Lumpur (UniKL) for supporting these research activities.

References

[1] B.J. Sliwinski, A.R.Mech and T.S.Shih. 1978. Stratification in thermal storage during charging, Proc.Sixth Int. Cof. Heat Tranfer, American Society of Mechanical Engineers, Vol. 4, pp. 149-154.

[2] C.R. Truman, M.W. Wildin. 1989. Finite Difference Model for Heat Transfer in a Stratified Thermal Storage Tank with Through Flow, Numerical Heat Transfer with Personal Cumputers and Super Computer, ASME/Ache National Heat Transfer Conference.

[3] Dorgan and J.S. Elleson. 1993. Design Guide for Cool Thermal Storage, Atlanta, Atlanta Ga, America Society of Heating, Refrigerating and Air-Conditioning Engineers, Inc. ASHRAE.

[4] J.E.B. Nelson A.R. Balakrisnan, S. Srinivasa Murthy. 1999. Experiments on Stratified Chilled Water Tanks, Int. Journal of Refrigeration.

[5] Mohd. Shamsurie Anuar. 1996. Penyimpanan Tenaga Terma Air Dingin: Proses Simulasi Tangki Penyimpanan Silinder Menegak, Universiti Teknologi Malaysia.

[6] Wildin, M. 1990. Diffuser Design for Naturally Stratified Thermal Storage. ASHRAE Transactions 96(1)1094-1102.

[7] Wildin, Yoo, and C.R.Truman.1986. Initial formation of a thermocline in stratified thermal storage tanks. ASHRAE Transactions 92(2): 280-92.

RSM model to evaluate material removal rate in EDM of Ti-5Al-2.5Sn using graphite electrode

Md. Ashikur Rahman Khan[1], M.M. Rahman[1,2,3], K. Kadirgama[1] and A.R. Ismail[1]
[1]Faculty of Mechanical Engineering
Universiti Malaysia Pahang, 26600 Pekan, Pahang, Malaysia
[2]Automotive Engineering Centre
Universiti Malaysia Pahang, 26600 Pekan, Pahang, Malaysia

E-mail[3]: mustafizur@ump.edu.my

Abstract. The usage of electrical discharge machining (EDM) is increasing gradually owing to its capability to cut precisely, geometrically complex material regardless hardness. Many process parameters greatly affect the EDM performance and complicated mechanism of the process result the lag of established theory. Hence, it becomes important to select the proper parameter set for different machining stages in order to promote efficiency. In view of these barriers, it is attempted to establish a model which can accurately predict the material removal rate (MRR) of titanium alloy by correlating the process parameter. Effect of the parameters on MRR is investigated as well. Experiment is conducted utilizing the graphite electrode maintaining negative polarity. Analysis and modelling is carried out based on design of experiment as well as response surface methodology. The agreeable accuracy is obtained and thus the model can become a precise tool setting the EDM process cost effective and efficient. Moreover, high ampere, short pulse-off time and low servo-voltage combined with about 250 μs pulse-on time generate the highest MRR.

1. Introduction

In EDM, the tool (anode) and the work piece (cathode) are immersed in a dielectric medium separated from each other by a small gap of the order of about 5-100 μm [1]. A controlled spark is generated between the two electrodes by applying a voltage (~200V) which breaks down the dielectric medium causing the voltage falls to about 25-30 V (discharge voltage) and the current to rise to a constant value set by the operator [2]. During machining electrons start flowing from cathode to anode which ionizes the dielectric medium and form a plasma channel between the cathode and anode. Some of the electrodes materials are melted and vaporized resulting material removal owing to intense heat generated in the plasma. The growth of the plasma channel, energy sharing between electrodes, process of vaporization, formation of recast layer, plasma-flushing efficiency, and temperature sensitivity of thermal properties of the work material are a few physical phenomena that render the machining process highly complex and stochastic [3]. The mathematical consideration of all these complex phenomena is very difficult. Therefore, mathematical prediction of material removal rate when compared with the experimental results shows wide variation.

---

* To whom any correspondence should be addressed.

Published under licence by IOP Publishing Ltd

Many process parameters greatly affect the electrical discharge machining performances. Consequently, it becomes important to select properly the process parameter set for different machining stages in order to promote efficiency [4]. Although, the desired process parameters can be determined based on experience or handbook values, it is a great challenge to ensure that the selected parameters are perfect for a particular material and environment.

In EDM process, parameter selection are often far from the optimum and at the same time selecting optimization parameters is costly and time consuming [5]. To improve the production rate and to decrease the dependence on experience, it is necessary to establish an optimization model in EDM process. A number of works has been carried out in modelling division using different types of material such as copper–steel (EN-8), die Steel, CK45 steel, C40 steel, Tungsten Carbide, tool steel 1.2714, AISI D2 tool steel, Ti-6Al-4V, Ti-15-3, etc. [5-16]. However, until now no study is noticed in electrical discharge machining considering titanium alloy Ti-5Al-2.5Sn workpiece and graphite tool.

In these circumstances, this paper aims to develop a mathematical model for material removal rate of titanium alloy Ti-5Al-2.5Sn based on design of experiment (DOE) and response surface methodology. The selection of this material was made taking into account its wide range of applications in airframes, jet engines, steam turbine blades, aircraft engine, compressor blades, missile fuel tanks and structural parts, etc. [17]. Four process parameters namely, peak current, pulse-on time, pulse-off time and servo-voltage are taken into account as input parameters of the model. The experimental work is performed as DOE utilizing graphite electrode as tool with negative polarity. Besides, the influence of the process parameters on material removal rate is investigated as well as attempt is made to estimate the maximum MRR.

2. Experimental procedure

The experiments were carried out utilizing a numerical control programming electrical discharge machine known as ''LN power supply AQ55L''. The EDM has the provisions of movement in three axes such as longitudinal (X-axis), lateral (Y-axis) and vertical direction of electrode (Z-axis) and has also a rotary U-axis with maximum rpm ±40. In this research, titanium alloy Ti-5Al-2.5Sn was taken into consideration as workpiece material and cylindrical graphite tool was applied in order to machine the workpiece. Process parameters namely, peak current, pulse-on time, pulse-off time and servo-voltage were selected in accordance with the literature review as well as preliminary experiment to connect these parameters with output criteria MRR. Pulse-on time refers the duration of time in which the current is allowed to flow whilst pulse-off time is the duration of time between the sparks per cycle [18] and [19]. Each cycle has an on-time and off-time that is expressed in units of microseconds. Since all the work is done during on-time, the duration of these pulses and the number of cycles per second (frequency) are important.

Peak current is the amount of power used in discharge machining, measured in units of amperage, and is the most important machining parameter in EDM [20]. Peak current (I_p) is the maximum current during spark. Servo-voltage is a reference voltage that can be set by operator. If the average gap voltage is higher than the servo reference voltage the feed speed increases. In contrast, the feed speed decreases or the electrode is retracted when the average gap voltage is lower than the servo reference voltage. MRR is an important indicator of the efficiency and cost effectiveness of the electrical discharge machining technique [9]. Material removal rate (mm^3/min) is calculated by measuring the average amount of material removed and the machining time as follows [6]:

$$\text{MRR } (\text{mm}^3/\text{ min}) = \frac{\text{reduction in weight of workpiece(g)}}{\text{density of workpiece } (\text{g/mm}^3) \times \text{machining time(min)}} \tag{1}$$

It can also be expressed as

$$MRR = \frac{1000 \times W_w}{\rho_w \times t} \tag{2}$$

$$W_w = W_1 - W_2 \tag{3}$$

where W_w is the weight loss of the workpiece in gm
W_1 is initial weight of work piece
W_2 is final weight of work piece
ρ_w is the density of the workpiece material (density of Ti-5Al-2.5Sn is 4.41 g/cm^3)
t is the machining time in minutes.

A well-designed experimental plan can substantially reduce the number of experiments [7]. This is the why; experimental design via central composite design of response surface methodology (RSM) was developed with an attempt to formulate the mathematical relations using smallest number of experiments possible. The coded levels for all process parameters used are displayed in table 1. The listing of experimental parameters is scheduled in table 2. Each experiment was conducted for fixed period, 40 minutes. A new set of the workpiece and graphite tool were applied for each run and total 62 experimental run, including one replication were carried out according to the design of experiment. The other process parameters were kept constant during experiments. In order to evaluate the amount of material removal the workpiece was weighed before and after machining using a digital single pan balance (maximum capacity=210 gm, precision=0.1 mg).

Table 1. Machining parameters and their levels.

Process parameters	Designation	Level 1	Level 2	Level 3	Level 4	Level 5
		-2	-1	0	1	2
Peak Current, I_p (A)	X_1	1	8	15	22	29
Pulse-on time, T_{on} (μs)	X_2	10	95	180	265	350
Pulse-of time, T_{off} (μs)	X_3	60	120	180	240	300
Servo-voltage, S_v (V)	X_4	75	85	95	105	115

Table 2. Experimental settings.

Working parameters	Description
Work piece material	Ti-5Al-2.5Sn
Size of work piece	22 mm × 22 mm × 20 mm
Electrode material	Graphite
Size of electrode	ϕ 20 mm × 50 mm (length)
Electrode polarity	Negative
Dielectric fluid	Commercial Kerosene
Applied voltage	120 V
Flushing pressure	0.15 MPa
Machining time	40 Minute

3. Modelling using RSM

The design of experiments (DOE) technique is a powerful work tool which allows us to model and analyse the influence of determined process variables over other specified variables, which are usually known as response variables. These response variables are unknown functions of the former design variables, which are also known as design factors. Response surface methodology is an interaction of mathematical and statistical techniques for modelling and optimizing the response variable models involving quantitative independent variables [8]. Through the use of the response surface methodology of DOE and applying regression analysis, the modelling of the desired response to several independent input variables can be gained.

In RSM, the independent process parameters can be represented in quantitative form as:

$$Y = f(X_1, X_2, X_3, \dots X_n) \pm \varepsilon \tag{4}$$

where Y is the response MRR, f is the response function, ε is the experimental error, and X_1, X_2, X_3, . . ., X_n are independent variables.

The form of f is unknown and may be very complex. Therefore, RSM aims at approximating f by a suitable lower ordered polynomial in some region of the independent process variables. If the response can be well modelled by a linear function of the independent variables, the equation (4) can be written as:

$$Y = \beta_0 + \beta_1 X_1 + \beta_2 X_2 + \dots\dots\dots + \beta_n X_n + \varepsilon \quad (5)$$

In the case of present study whilst the number of the independent variables is four, the linear equation is formed as follows:

$$Y = \beta_0 + \beta_1 X_1 + \beta_2 X_2 + \beta_3 X_3 + \beta_4 X_4 \quad (6)$$

On the other hand, the second-order model is normally used when the response function is nonlinear. However, if the model is not well fitted by the linear function then a higher order polynomial such as the quadratic model can be used. In the present study, both the linear model and second-order (quadratic model) model have been studied. The experimental values are analysed and the mathematical model is then developed that illustrate the relationship between the process variable and the output.

The mathematical model based on a second-order polynomial is given as follows:

$$Y = \beta_0 + \sum_{i=1}^{n} \beta_i X_i + \sum_{i=1}^{n} \beta_{ii} X_i^2 + \sum_{i,j=1, i \neq j}^{n} \beta_{ij} X_i X_j + \varepsilon \quad (7)$$

where Y is the corresponding response, X_i is the input variables, X_i^2 and X_iX_j are the squares and interaction terms, respectively, of these input variables. β_o, β_i, β_{ij} and β_{ii} are the unknown regression coefficients. In this work, equation (7) can be rewritten according to the four variables used as:

$$\begin{aligned} Y = {} & \beta_0 + \beta_1 X_1 + \beta_2 X_2 + \beta_3 X_3 + \beta_4 X_4 + \beta_{11} X_1^2 + \beta_{22} X_2^2 + \beta_{33} X_3^2 + \beta_{44} X_4^2 \\ & + \beta_{12} X_1 X_2 + \beta_{13} X_1 X_3 + \beta_{14} X_1 X_4 + \beta_{23} X_2 X_3 + \beta_{24} X_2 X_4 + \beta_{34} X_3 X_4 \end{aligned} \quad (8)$$

where X_1, X_2, X_3, and X_4 are four input variables as peak current (I_p), pulse-on time (T_{on}), pulse-off time (T_{off}) and servo-voltage (S_v), respectively.

4. Results and discussion

4.1 Mathematical model

In this work, RSM is utilized for determining the relations between the various EDM process parameters with the various machining criteria and exploring their effects on the material removal rate. After knowing the values of the measured response, the values of the different regression coefficients of mathematical equation are estimated. The mathematical model based on the response surface methodology is developed by utilizing test results obtained through the entire set of experiments.

Both linear and non-linear regression models were examined and the acceptance was based on high to very high coefficients of correlation (R) calculated as well as model adequacy. The first-order and second-order mathematical equation are executed using the same experimental data. It is essential to check the adequacy of the fitted models, because an incorrect or under-specified model can lead to confusing evaluation of the response. It can be verified whether the model is under specified by checking the fit of the model. The model adequacy checking includes the test for significance of the regression model, model coefficients, and lack of fit. Analysis of variance (ANOVA) as shown in table 3 and table 4 was performed in support of both linear and non-linear model owing to the aforesaid motives.

It is observed that the standard error, S (0.206950) for quadratic equation is smaller than the value of S (2.04202) for linear equation. Then again, the value of R^2, R^2-adjusted and R^2-predicted in the table 4 are greater than that of table 3. The lower value of S and higher value of R^2 shows the adequacy of the model. Consequently, the regression model with quadratic terms is more adequate and significant over the model with linear terms.

Furthermore, if the p-value of residual error is less than the α-level, it means that the model does not accurately fit the data. The p-value for the lack-of-fit in the case nonlinear model is 0.095, which is larger than 0.05 (i.e. α=0.05, or 95% confidence). Hence, the lack-of-fit term is insignificant as it is desired. The fit summary recommended that the second order model is statistically significant for analysis of MRR.

Table 3. Analysis of variance for single order model of MRR.

Source	Degree of freedom	Sum of squares	Mean squares	F-ratio	P
Regression	14	1025.74	256.434	61.50	0.000
Linear	4	1025.74	256.434	61.50	0.000
Residual error	16	108.42	4.170		
Lack-of-Fit	10	108.30	5.415	285.59	0.000
Pure Error	6	0.11	0.019		
Total	30	1134.15			
Standard deviation (S) = 2.04202					
R^2= 90.44%					
R^2-adjusted = 88.97%					
R^2-predicted = 85.41%					

Table 4. Analysis of variance for second order model of MRR.

Source	Degree of freedom	Sum of squares	Mean squares	F-ratio	P
Regression	14	1133.47	80.962	1890.38	0.000
Linear	4	1025.74	256.434	5987.49	0.000
Square	4	89.36	22.339	521.59	0.000
Interaction	6	18.38	3.063	71.51	0.000
Residual error	16	0.69	0.043		
Lack-of-Fit	10	0.57	0.057	3.01	0.095
Pure Error	6	0.11	0.019		
Total	30	1134.15			
Standard deviation (S) = 0.206950					
R^2= 99.94%					
R^2-adjusted = 99.89%					
R^2-predicted = 99.70%					

Finally, the second order model of MRR can be obtained as follows putting the values of the coefficients that obtained through response surface methodology.

$$\begin{aligned} MRR = {} & 4.93853 + 1.28459 I_p + 0.0480985 T_{on} - 0.0591216 T_{off} - 0.127086 S_v \\ & + 0.0321069 I_p^2 - 8.58835\times10^{-5} T_{on}^2 + 6.58753\times10^{-5} T_{off}^2 + 8.94884\times10^{-4} S_v^2 \\ & + 3.80572\times10^{-4} I_p T_{on} - 8.90482\times10^{-4} I_p T_{off} - 0.0131805 I_p S_v + \\ & 3.31059\times10^{-5} T_{on} T_{off} - 2.31268\times10^{-4} T_{on} S_v + 3.28692\times10^{-4} T_{off} S_v \end{aligned} \quad (9)$$

4.2 Confirmation test

The developed model is also validated through confirmation test and the results are presented in table 5. It is apparent that the error between the observed value and predicted value of MRR is in the range of 1.82-9.19% having a mean error as 4.29%. Accordingly, the developed regression model is demonstrated to be a practical and effective way for the evaluation of tool wear rate in EDM process.

Table 5. EDM conditions in verification test.

Peak current (A)	Pulse-on time (μs)	Pulse-off time (μs)	Servo-voltage (V)	MRR (mm^3/min) Experimental	Predicted	Error (%)
15	180	120	85	9.7939	9.9726	-1.82
29	320	120	85	31.0053	31.5841	-1.87
5	95	120	85	1.5230	1.3830	9.19
					Average	4.29

4.3 Effect of process parameters on MRR

An attempt is also made to investigate the effect of the input parameters as peak current, pulse-on time, pulse-off time and servo-voltage on material removal rate. The figure 1–3 shows the influences of the process parameters. It is obvious in the figure 1 that the increase of peak current increases the material removal rate. This is due to the fact that increasing peak current increases the discharge energy which results more material removal. Thus, material removal more increases with peak current and the similar observation was found in the research of Lee and Li [21] and Singh et al. [22]. The material removal rate increases as the pulse-on time increase up to certain value of on time (250 μs) hereafter the MRR decreases with on time. At too long pulse duration, the spark intensity is decreased in the discharge spots because of the expansion of the plasma channel [23]. Accordingly, the long pulse-on time lowers the material removal rate.

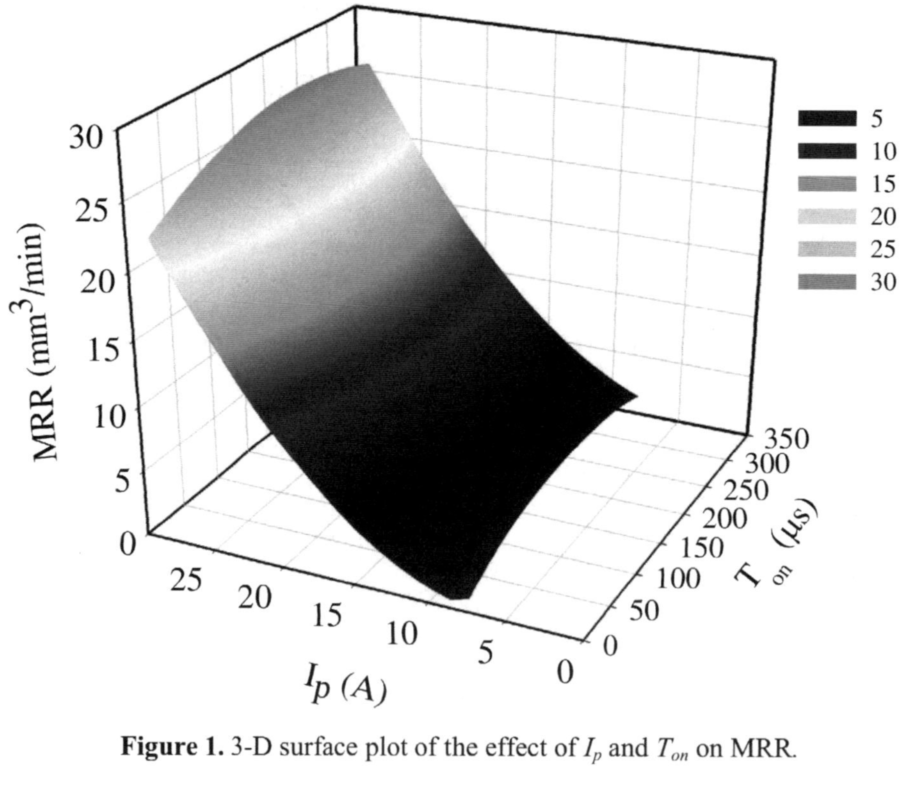

Figure 1. 3-D surface plot of the effect of I_p and T_{on} on MRR.

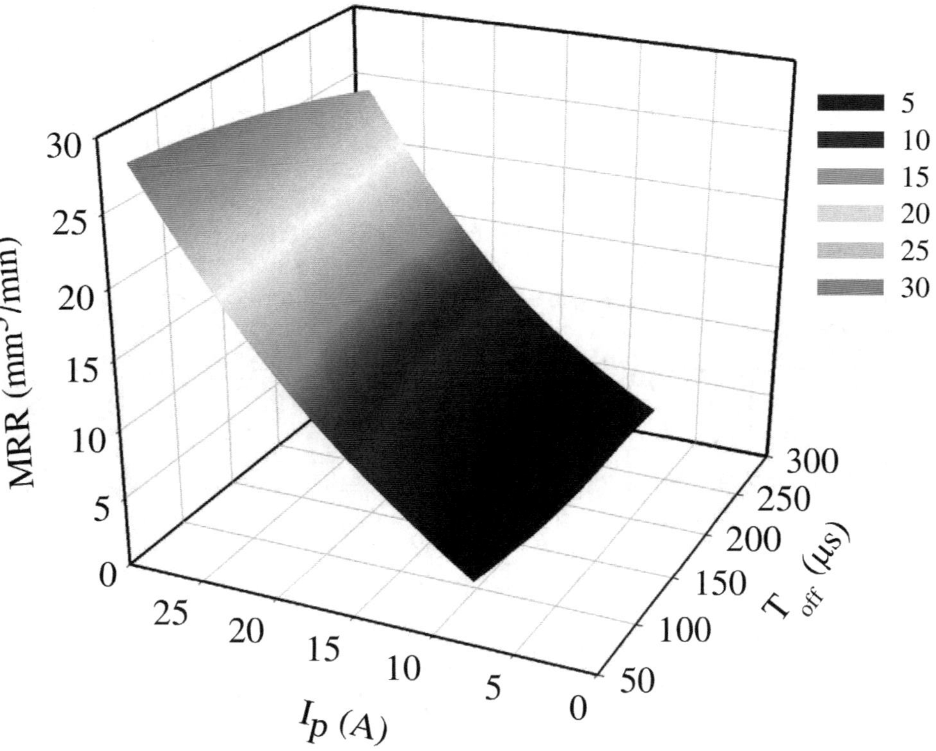

Figure 2. 3-D surface plot of the effect of I_p and T_{off} on MRR.

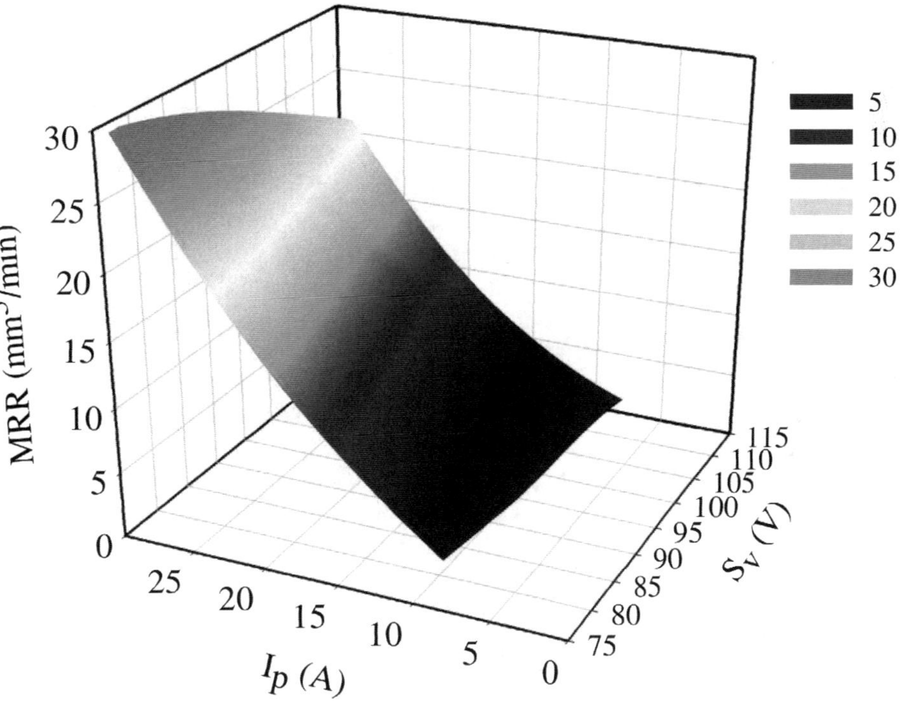

Figure 3. 3-D surface plot of the effect of I_p and S_v on MRR.

It is perceived from figure 2–3 that the increase of pulse-off time as well as servo-voltage reduces the material removal rate. The pulse off time is the time during which no energy is applied to the workpiece surface. Therefore, the short pulse off time create accessibility the application of heat energy on the workpiece surface for long duration and consequently the material is eroded at faster rate [24]. The short pulse-off time combined with lower servo-voltage generates the maximum material removal rate. Electrode's speed towards the workpiece together with gap width depends on servo-voltage. Longer ignition delay is occurred when the average gap voltage is higher than the servo reference voltage. On the contrary, when the average gap voltage is lower than the servo reference voltage results a smaller ignition delay. In another word, the high servo-voltage causes longer ignition delay and lower servo-voltage reduces the ignition delay. Therefore, as servo-voltage increase the ignition delay increases reducing the material removal rate and the controversy is also true. Han and Kunieda [25] and Kunieda et al. [20] also found the similar effect of servo-voltage on material removal.

5. Conclusions

This study is accomplished for the development of MRR model aiming to make the EDM cost effectiveness. The developed model is checked through the ANOVA henceforward confirmation test is performed to test the validity of the model. The result from analysis of variance exhibits that the model having satisfactory fitness. The confirmation test reveals the error from 1.82% to 9.19 with a mean error as 4.29% which is acceptable as less than 10%. Thus, the RSM model can be used tool to predict the MRR successfully making the EDM technique as cost effective and efficient. Moreover, the influences of the parameter is investigated and observed that peak current and pulse-on time have almost the similar effect and the converse response is apparent in the case of pulse-off time and servo-voltage. High ampere combined with short off-time and low servo-voltage yield maximum MRR while the pulse-on time is about 250 μs.

Acknowledgements

The authors would like to acknowledge the support of Universiti Malaysia Pahang for funding under University grant number RDU 100108 and GRS 090335.

References

[1] Lin Y C, Chen Y F, Wang D A and Lee H S 2009 *J. Mater. Process. Technol.* **209** 3374–83
[2] Joshi S N and Pande S S 2010 *J. Manuf. Process.* **12** 45–56
[3] Panda D K and Bhoi R K 2005 *Mater. Manuf. Processes.* **20**(4) 645–72
[4] Su J C, Kao J Y and Tarng Y S 2004 *Int J Adv Manuf Technol.* **24** 81–90
[5] Gao Q H, Zhang Q, Su S P and Zhang J H 2008 *J Zhejiang Univ-Sci A.* **9**(1) 104–08
[6] Chattopadhyay K D, Verma S, Satsangi P S and Sharma P C 2009 *J. Mater. Process. Technol.* **209** 1454–65
[7] Bhattacharyya B, Gangopadhyay S and Sarkar B R 2007 *J. Mater. Process. Technol.* **189** 169–77
[8] El-Taweel T A 2009 *Int J Adv Manuf Technol.* **44** 100–13
[9] Tomadi S H, Hassan M A, Hamedon Z, Daud R and Khalid A G 2009 *Proceedings of the International MultiConference of Engineers and Computer Scientists* Hong Kong **2** pp 1803–08
[10] Zarepour H, Tehrani A F, Karimi A D and Amini S 2007 *J. Mater. Process. Technol.* **187–188** 711–14
[11] Pradhan M K and Biswas C K 2009 *Int. J. Eng. Appl. Sci.* **5**(5) 346–351
[12] Rahman M M, Khan M A R, Kadirgama K, Noor M M and Bakar R A 2010 *Int. J. Mech. Mater. Eng.* **1**(3) 135–40
[13] Rahman M M, Khan M A R, Kadirgama K, Maleque M A and Bakar R A 2011 *Com. Sim. Modern Sci.* ed N Mastorakis, M Demiralp, V M Mladenov and A Zaharim USA WSEAS Publishers 5 pp107–13

[14] Rahman M M, Khan M A R, Kadirgama K, Noor M M and Bakar R A 2011 *Adv. Mater. Res.* **213** 402–08
[15] Rahman M M, Khan M A R, Kadirgama K, Noor M M and Bakar R A 2011 *Adv. Mater. Res.* **152–153** 1595–1602
[16] Khan M A R, Rahman M M, Kadirgama K, Maleque, M A and Bakar R A 2011 *WASET* **74** 121–25
[17] Rahman M M, Khan M A R, Kadirgama K, Noor M M and Bakar R A 2010 in: *Advances in Control, Chemical Engineering, Civil Engineering and Mechanical Engineering*, ed V Mladenov USA WSEAS Publishers pp 34-37
[18] Kumar S, Singh R, Singh T P and Sethi B L 2009 *J. Mater. Process. Technol.* **209** 3675–87.
[19] Rahman M M, Khan M A R, Kadirgama K, Noor M M and Bakar R A 2011 *J. Appl. Sci.* **11**(3) 549–54
[20] Kunieda M, Lauwers B, Rajurkar K P and Schumacher B M 2005 *CIRP Annals-Manuf. Technol.* **54**(2) 64–87
[21] Lee S H and Li X P 2001 *J. Mater. Process. Technol.* **115**(3) 344–58
[22] Singh S, Maheshwari S and Pandey P C 2004 *J. Mater. Process. Technol.* **149**(1-3) 272–77
[23] Hascalik A and Caydas Ulas 2007 *Appl. Surface Sci.* **253** 9007–16
[24] Kansal H K, Singh S and Kumar P 2008 *Math. Comp. Model.* **47** 1217–37
[25] Han F and Kunieda M 2004 *Precision Eng.* **28** 65–72

Performance of Process Damping in Machining Titanium Alloys at Low Cutting Speed with Different Helix Tools

M.A.Shaharun[1], A.R.Yusuff , M.S.Reza, K.A.Jalal
Faculty of Mechanical Engineering, Universiti Malaysia Pahang, 26600 Pekan, Pahang, Malaysia

E-mail: [1]adib@ump.edu.my

Abstract. Titanium is a strong, lustrous, corrosion-resistant and transition metal with a silver color to produce strong lightweight alloys for industrial process, automotive, medical instruments and other applications. However, it is very difficult to machine the titanium due to its poor machinability. When machining titanium alloys with the conventional tools, the wear rate of the tool is rapidly accelerate and it is generally difficult to achieve at high cutting speed. In order to get better understanding of machining titanium alloy, the interaction between machining structural system and the cutting process which result in machining instability will be studied. Process damping is a useful phenomenon that can be exploited to improve the limited productivity of low speed machining. In this study, experiments are performed to evaluate the performance of process damping of milling under different tool helix geometries. The results showed that the helix of 42° angle is significantly increase process damping performance in machining titanium alloy.

Key Words: Titanium, Process Damping, Helix Angle, Cutting Force, Frequency Response Function

1. Introduction

Titanium's physical qualities of high strength, toughness, durability, low density, corrosion resistance and biological compatibility have made it useful in a variety of applications. Titanium is currently used in aerospace applications, automobiles, prosthetics, buildings, and sporting equipment [1]. Titanium, which weighs of forty percent less than carbon steels, can be strengthened by alloying it with elements such as aluminum and vanadium. Titanium is nonmagnetic and possesses good heat transfer properties. It has a corrosion resistance to acids, nontoxic and biocompatible [2, 3]. These properties make titanium and its alloys useful in a wide range of structural, chemical, petrochemical, marine and biomaterial applications. The most widely used titanium alloy, Ti-6Al-4V, is present in forty-five percent of industrial applications. The unique combination of this alloy's physical and mechanical properties such as workability, fabricability, along its good production experience and as well as its commercial availability allows it to be economically useful. Therefore the efficiency of titanium machining is one of the major challenges in production engineering.

The common cause of instability in metal cutting is the regenerative chatter, and this often limits the productivity of machining operations. The regenerative chatter stability boundary is known as the stability lobe diagram and is a function of depth of cut and spindle speed, as shown schematically in figure 1. It has been reported that regenerative chatter is caused by interaction between the structural dynamics of the machine tool and the dynamics of the cutting process [4]. The requirement for high productivity of machining leads to a desire to suppress chatter and various methods have been proposed. This includes adding passive or active damping [5, 6], spindle speed manipulation or variation [7,8] and alternative methods [9,10,11]. Besides the damping produced from the structure of machine tools, the machining process itself can create damping to the system through a phenomenon known as process damping. The term of process damping or also known as resistance force was introduced by Tobias and Fishwick [12] and happens when the tool's flank face or rake angle rubs against the wavy workpiece surface at low spindle speeds. At high spindle speeds, the lobbing effect can be observed, and this allows high productivity cutting to be performed on easy-to-machine materials such as aluminium alloys.

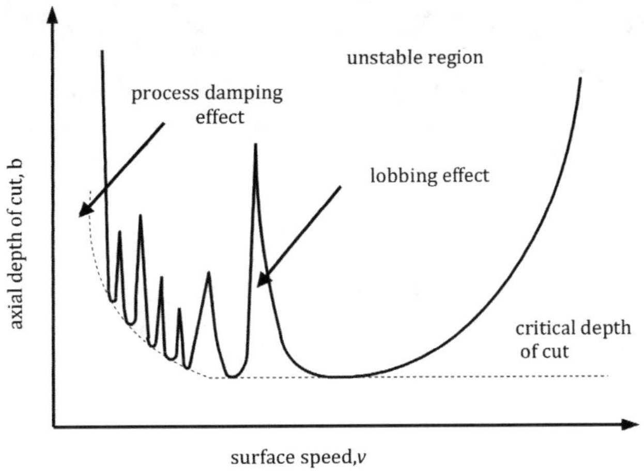

Figure 1. Process damping in stability lobe.

Unfortunately the resulting high surface speeds are incompatible with more difficult to- machine materials such as titanium and nickel based alloys. Previously, the practitioner is limited to low spindle speeds, where the chatter stability is strongly influenced by the process damping phenomenon. However, understanding of process damping remains as an unsolved problem in chatter research [13]. Most of the researcher agreed with the effect of low speed spindle to damping process however none has been reported the best tool application such as Montgomery and Altintas [14] used a model-based approach to investigate ploughing forces. Delio et al. [15] considered the wavelength of chatter vibration and the loss of process damping behavior at higher spindle speeds. Elbestawi et al. [16] modeled process damping effects when cutting aluminium and showed that the model could produce additional damping forces due to the tool flank / workpiece interference. Ranganath et al. [17] also developed a time-domain model of process-damped milling and compared the results with experimental data from an aluminium alloy workpiece. Huang and Wang [18] proposed a model that considered the consequences of chatter vibration on the effective rake and relief angles. They included additional empirical parameters in their model so that the cutting stiffness became a function of these effective angles, and thereby produced a process damping effect.

Until recently, Yusoff et al. [19] identified the significant contribution of variable helix/pitch compare to rake, relief angles and edge radius to increase process damping performance. From a practical and industrial point of view, understanding the effect of these variable helix and pitch on process damping is clearly of great importance, because it allows the practitioner to choose better tooling for specific machining problems. Therefore, the aim of the present study is to perform experimental milling experiments so that different tool helix geometries can be ranked in terms of their positive influence on process damping in milling.

2. Theory

Chatter is produced from self-excited vibration during cutting resulting in a high amplitude unstable motion. The amplitude of this motion is limited by nonlinearities such as tool loss of contact, nonlinear cutting force coefficients and nonlinear stiffness of the machine tool structure. The chatter frequency is close to a natural frequency of the system. At low speeds, the wavelength λ of these surface waves is much smaller since the wavelength is proportional to surface velocity v and inversely proportional to regenerative vibration frequency f_c , as shown by equation (1):

$$\lambda = \frac{V}{f_c} \tag{1}$$

As the spindle speed (and hence surface velocity) is reduced, the process damping phenomenon becomes sufficient for the regenerative chatter to be stabilized or suppressed as shown in figure 1. The corresponding surface vibration wavelength is given by equation (1) and is referred to as the process damping wavelength λ. The commonly proposed mechanism of process damping is shown schematically in figure 2. As each tooth removes the chip from the wavy surface, process damping forces are generated that act on the structure. The damping force corresponds to inference between tool flank face and wavy surface, where more damping force occurs at point 'B'. A ploughing force can be produced from the workpiece being deformed by the tool, whilst the surface angle changes the effective shear angle of the tool. Interference is minimized when the tool travels upwards on the wave (position 'D' in Figure 2) due to the positive slope of the machined surface. According to this concept, low relief angles should produce high ploughing forces from the engagement between tool and workpiece. Consequently, different helix angles should be considered in the evaluation of process damping.

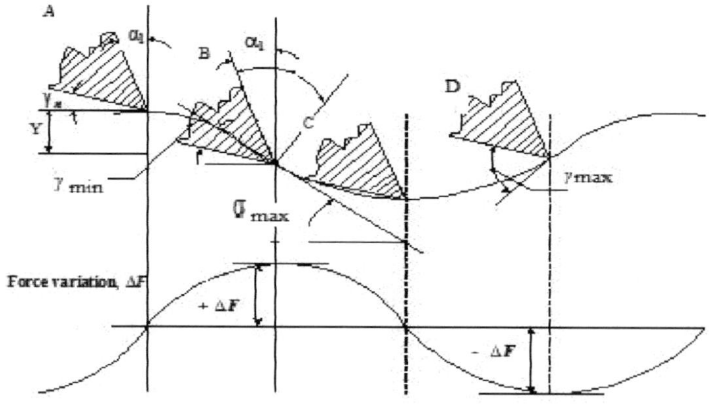

Figure 2. Process damping mechanism γ.

In the stability of high speed milling, axial depth of cut b is the most influential factor since the cutting forces are often considered in cutting force model, as shown in figure 3. This is to be given by the relationship

$$F = K_s\, b\, (Yo - Y) \tag{2}$$

where F is the cutting force, b is the axial depth of cut, K_s is the empirical cutting stiffness, and $(Yo - Y)$

is the change in surface position between current and previous cuts. In theory, the stability boundary is then independent of the feed rate despite the influence of the feed rate on the mean chip thickness. In practice, the empirical cutting stiffness K_s changes with the feed rate so that the feed rate does have some influence on the overall stability. K_s depends on K_t and K_n based on below formula

$$K_s = \sqrt{K_t^{\,2} + K_n^{\,2}} \tag{3}$$

$$\theta = tan^{-1}\left(\frac{K_t}{K_n}\right) \tag{4}$$

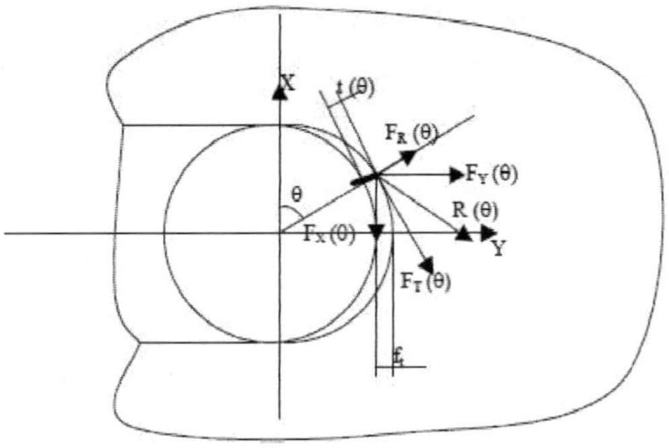

Figure 3. Model of cutting forces in milling operation [20].

This tangential cutting forces F_t, represent a resistance to workpiece rotation and constitute approximately more than twice normal cutting force. This trend will be occur in every experimental and different tools have different cutting force stiffness. Cutting force is very important thing that we need to be considered in machining process. The cutting forces that are developed during the milling process can directly or indirectly estimate process parameters such as tool wear, tool life, surface finish and also production time. Therefore it is necessary to identify the relationship between the cutting forces and uncut chip area that can be expressed as a product of the axial depth of cut and feed per tooth in order to predict cutting behaviour. During one tooth period, the average milling force in x and y direction is given by equation (5) [20].

$$\begin{aligned}\bar{F}_x &= \left\{\frac{Nbc}{8\pi}\left[K_t\cos(2\theta) - K_n[2\theta - \sin 2\theta]\right] + \frac{Nb}{2\pi}\left[-K_{te}\sin\theta - K_{ne}\cos\theta\right]\right\}_{\theta st}^{\theta ex}\\ \bar{F}_y &= \left\{\frac{Nbc}{8\pi}\left[K_n\left[2\theta - \sin(2\theta)\right]\right] + K_t\,cos(2\theta) - \frac{Nb}{2\pi}\left[K_{te}\cos\theta - K_{ne}\sin\theta\right]\right\}_{\theta st}^{\theta ex}\end{aligned} \tag{5}$$

Where K_{te} and K_{ne} represent the tangential and normal edge cutting force coefficient, while K_t and K_n represent the tangential and normal cutting force coefficient, respectively. The entry and exit angles of the cutter are $\theta_{st} = 0$ and $\theta_{ex} = \pi$, respectively. These cutting coefficients can be determined through cutting measurement with a force dynamometer and cutting certain conditions. Therefore, the average force per tooth period to be found

$$\begin{aligned}\overline{F_x} &= -\frac{mb}{4}K_n c - \frac{mb}{\pi}K_{ne}\\ \overline{F_y} &= \frac{mb}{4}K_t c - \frac{mb}{\pi}K_{te}\end{aligned} \tag{6}$$

In function of chip load *c*, equation (6) can be written as

$$\bar{F}_q = \bar{F}_{qc}C + \bar{F}_{qe}(q = x, y, z) \tag{7}$$

The experiment completes for a multiple cutting test with a range of chips load to recorded the cutting force values in *x* and *y* direction. A linear regression was then performed on the mean force values to determine cutting coefficient in equation (3) and (4) as given:

$$K_t = \frac{4F_{yc}}{Nb}, \quad K_n = \frac{4F_{xc}}{Nb}, \quad K_{te} = \frac{\pi F_{ye}}{Nb}, \quad K_{ne} = \frac{-\pi F_{xe}}{Nb} \tag{8}$$

F_{ye} and F_{xe} are determined from intersection at force axis for *x* and *y*- force data direction, while F_{yc} and F_{xc} are determined from the gradient at force axis for *x* and *y* – force data directions. However, under process-damped cutting conditions, the effect of feed rate has been observed to be significant [17]. It is useful to express the feed rate in terms of the maximum chip thickness h_{max} [18]:

$$h_{max} = fpt\sqrt{\frac{4r}{D} - \left(\frac{2r}{D}\right)^2} \tag{9}$$

Here, *r* is the radial immersion of the tool, and *D* is the tool diameter. The feed per tooth *fpt* is related to the machining feed rate *f*, number of teeth *m* and spindle speed *n* as shown by equation (10).

$$f = m \times fpt \times n \tag{10}$$

Using a high depth of cut at low cutting speed results in the chatter stability being dominated by process damping effects as shown on the left side of the stability diagram in figure 1. Meanwhile, using a low radial immersion helps to reduce the total machining forces and improves tool life. This approach will be employed in the present study in order to determine the process damping wavelength λ_c under different tool geometry and feed rate conditions.

4. Experimental Procedure

4.1. Frequency Response Function (FRF)

The experimental approach involved of the impact hammer model in which the testing applied was to measure flexure frequency response function and cutting tool frequency response function. The data acquisition system apparatus used was a Bruel & Kjaer model type 7539A 5/1 channel. Meanwhile, for an impact testing, a normal force created by using a 2302-10 Meggit hammer with vinyl tip was applied at the tool tip. The acceleration response was then captured by a 4507B Bruel & Kjaer accelerometer which was located opposite to the hammer impact point. From the frequency response function analysis, the wavelength of the vibration has been obtained.

4.2. Cutting force determination

The cutting force coefficient tests were conducted by using Haas 3 Axis CNC milling machine TM2. The average helix angles of 4 tools were measured by using an optical comparator (Mahr-MM320). Meanwhile, the force response of the cut titanium workpiece was obtained by using a Kistler Multi-Compenent Dynamometer type (9257B). The cutting condition was varied between 95mm/min and 222mm/min. The parameters used in experiment are listed in table 1. The feed rate values are dependent to the changing values of feed.

IOP Conf. Series: Materials Science and Engineering **36** (2012) 012017 doi:10.1088/1757-899X/36/1/012017

Table 1. Parameter set up in cutting force experiment.

Condition	Feed	Spindle speed (rpm)	Axial depth of cut(mm)	Radial Emmersion(%)	Feed rate (mm/min)
1	0.03	796	2	100	95
2	0.04	796	2	100	127
3	0.05	796	2	100	189
4	0.06	796	2	100	191
5	0.07	796	2	100	222

4.3. Process Damping

Process damping experiment was performed by using Haas 3 Axis CNC milling Machine TM2, as shown in figure 4. From the frequency response function , the resonant frequency obtained was used to choose a starting value for spindle speed, so that the expected wavelength of vibrations was $\lambda = 0.1$ mm. This was achieved by using both equation (1) and the relationship between tool diameter, spindle speed and surface speed. Based upon previous study [19], this initial wavelength was reported to be below the process damping wavelength λ_c. Meanwhile, the desire maximum chip thickness, the feed per tooth and hence the initial feed rate were then determined by using equations (9) and (10). A low radial width of cut ($r = 1$ mm) and large axial depth of cut ($b = 7$ mm) were used to minimize forced vibration as well as to reduce tool wear and also to prevent further damage to the tool if severe chatter occurred. The spindle speed, n and feed rate, f were incremented simultaneously by 10 percent in order to maintain constant fpt and h_{max}, until chatter was detected. Process damping performance was then evaluated in terms of λ_c from equation (1). Here, the chatter frequency was obtained from Fourier analysis of the vibration signal and the surface speed, v was determined based upon the spindle speed at which chatter occurred. The procedure was repeated for five h_{max} values in the range of 0.04 and 0.12 mm for each tool. Here, a block of titanium was mounted on a flexible structure (as shown in figure 4) and it was adapted from Yusoff et al [19]. During cutting, the vibration signal was recorded by using an accelerometer type (352C33).

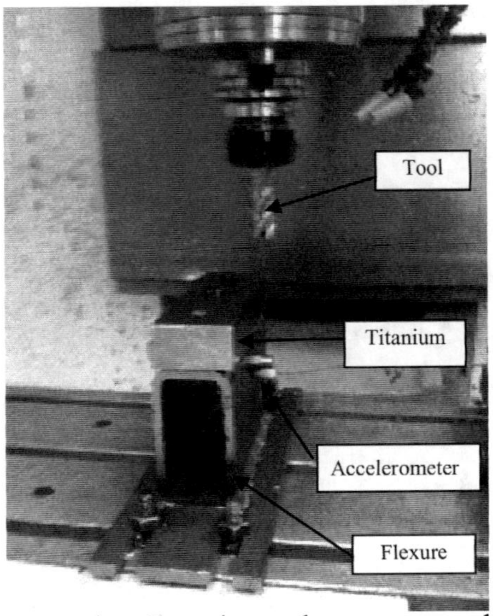

Figure 4. Experimental process damping arrangement

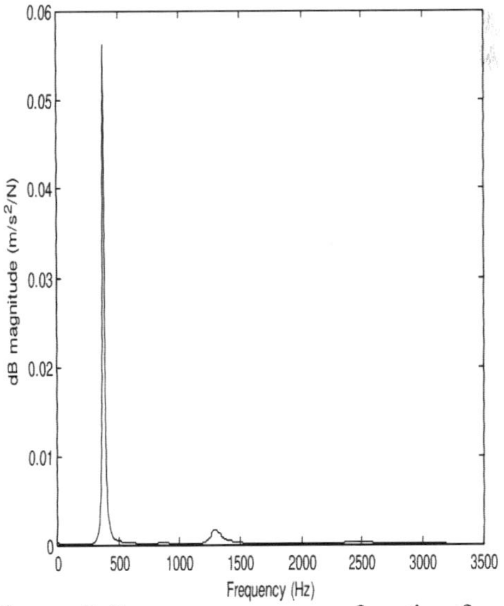

Figure 5. Frequency response function for Flexure.

5. Result and Discussion

5.1. Frequency Response Function (FRF)

Figure 5 shows the frequency response function flexure obtained from the conducted experiment. It shows that the dominant frequency of flexure was 367 Hz. A huge distinction of frequency between tools and flexure in the frequency response function experiments is significant in order to get a dominant damping frequency from workpiece.

5.2. Cutting force experiment

In this experiment, a cutting force coefficient for each tool was measured in order to provide a clear understanding of the influence of helix geometry would have on cutting force process. The cutting force coefficient investigation was also intended to explore whether the cutting force coefficient were linear during low feed rates and also to understand whether an increase in cutting force coefficient is due to damping wavelength. From the result obtained, the average force corresponded to a particular feed rate for each point (figure 6). By using relation in equation (8) and linear regression, the cutting force coefficient values can be calculated and were summarized in table 2.

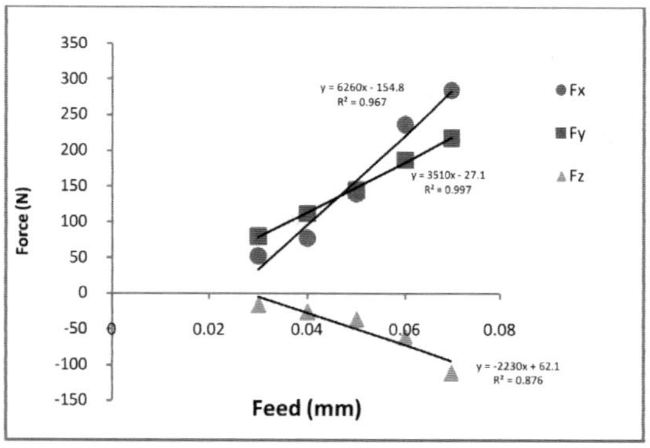

a) T1

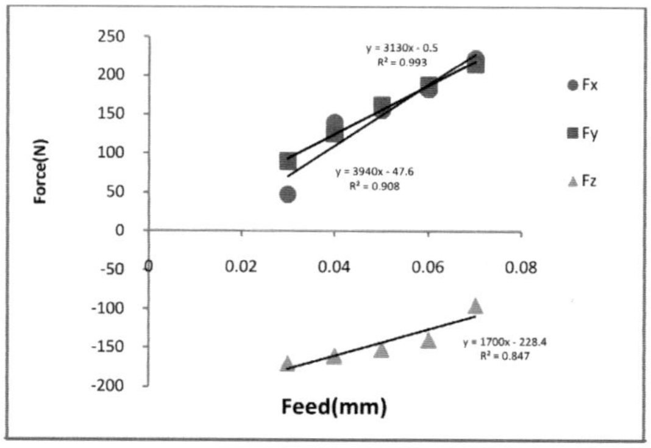

b) T2

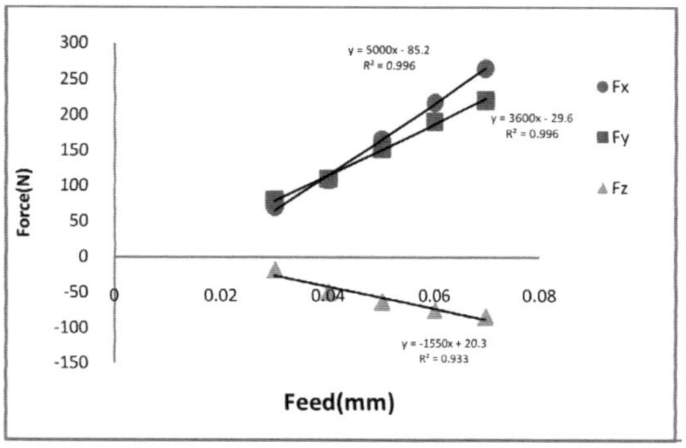

c) T3

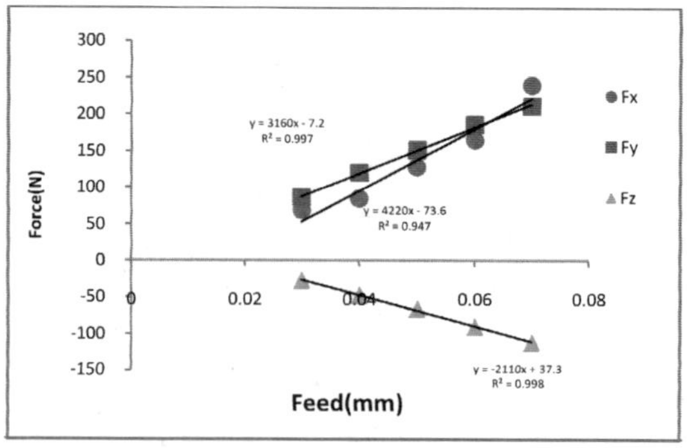

d) T4

Figure 6. Cutting force coefficient linearity for each tool.

Table 2. Cutting Force Coefficient for Each Tools

Tool	Helix Angle (°)	K_t (MN/m^2)	K_n (MN/m^2)	K_s (MN/m^2)	θ (°)
T 1	37°	3510	6260	7177	29
T 2	45°	3130	3940	5032	39
T 3	42°	3600	5000	6161	36
T 4	52°	3160	2110	3799	56

5.3. Analysis of Process Damping Wavelength

Figure 7 presents the repeatability test of T1 on process damping wavelength. It can be seen that the tools indicate a repeatability error is less than 10 percent between the initial test and repeat test. By performing the repeated test after the other experiment means that, there would be a slight amount of tool wear, so the influence of this wear could be compared to the influence of the other process parameters. It is clear that even considering this repeatability error, the maximum chip thickness has very significant effect on process damping wavelength.

Figure 8 shows the comparison of the damping wavelength obtained in this study for each tool. It can be seen that tools with different helix angle have different process damping performance. Generally, the damping wavelengths are gradually increase for all tool from h_{max} 0.04 to 0.12. The lower and higher helix angle causes low damping performance at high h_{max}.

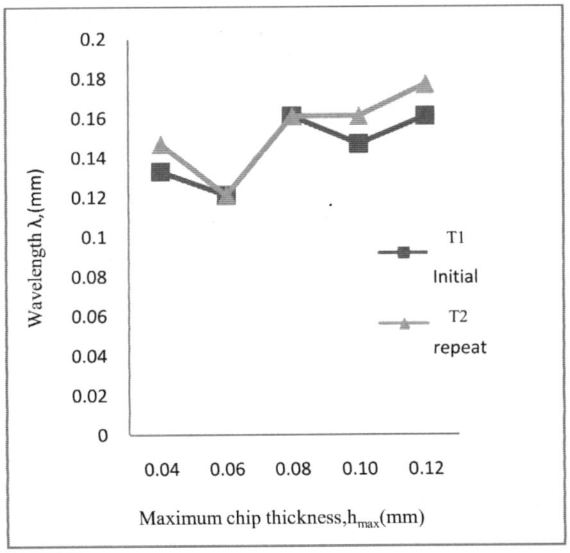

Figure 7. Repeatability of T1 on process damping wavelength.

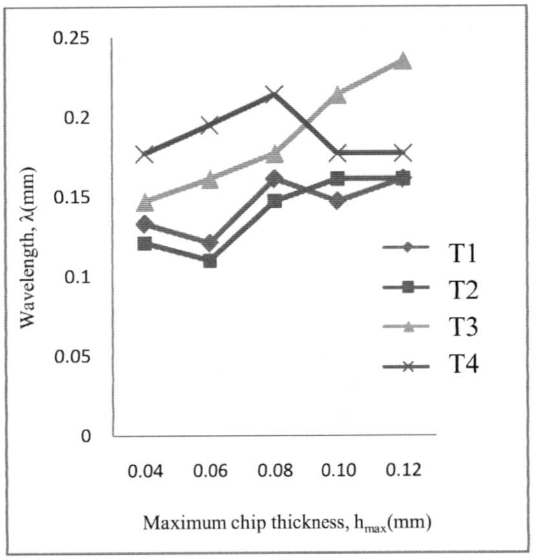

Figure 8. Comparison damping wavelength for each tool.

In Figure 8, the best damping performance was T3. T3 was able to achieve the highest damping wavelength at h_{max} 0.10 and h_{max} 0.12 compared to T1, T2 and T4. This tool also has a steady process damping performance from h_{max} 0.04 to 0.12. It is useful to relate this result to cutting force coefficient. According to the damping model as shown in figure 2, the force coefficient on the tangent component is important to increase process damping performance. In this cases where T3 have the highest K_t compare to T1, T2 and T4, and second high K_s. This tool geometry needs high K_s and K_t to increase process damping performance. In conclusion, high process damping performance depends on cutting force coefficient and helix angles.

6. Conclusion

This paper presents a relationship between the cutting force coefficient and damping wavelength. From the data gathering in all experiments, the results have showed that this is useful relationships between damping wavelength and cutting force coefficient. The experimental data showed that T3 have the highest value of K_t, according to that T3 have the highest damping wavelength compare to the others. The results showed that the helix of 42° angle most significantly increase process damping performance in machining titanium alloy. Further work will be conducted to study the effect process damping performance due on variable helix, variable pitch and variable helix and pitch tools.

Acknowledgement

The author gratefully acknowledges the financial support from University Malaysia Pahang (UMP) under UMP RDU Short Grant (RDU110305). Furthermore, the equipment are the courtesy Faculty of Mechanical Engineering,

References

[1] D.C Kirk, *Tools and Dies for Industry*, Proc. Conf, 7677. Met. Soc. London, 1976-7, 77-98.

[2] E.O.Ezugwu and Z.M.Wang (1977). *Titanium alloys and their machinability*, Journal of Material Processing Technology.Vol.68, 262-274.

[3] E.O.Ezugwu and J.Bonney and Y.Yamane(12002). *An overview of the machinability of aeroengine alloysTitanium alloys and their machinability*, Journal of Material Processing Technology.Vol134, 233-253.

[4] Merritt, H.E. *Theory of Self-excited machine tool chatter, contribution to machine-tool chatter research-I.* Transaction ASME Journal of Engineering Industry, 1965, 87(4), 447-454

[5] Zhang, Y. and Sims, N.D. *Milling workpiece chatter avoidance using piezoelectric active damping: a feasibility study.* Smart materials and structures, 2005, 14, 65-70.

[6] Ganguli, A., Deraemaeker, A. and Preumont, A. *Regenerative chatter reduction by active damping control.* Journal of Sound and Vibration, 2007, 300(3-5), 847.

[7] Soliman, E. and Ismail, F. Chatter suppression by adaptive speed modulation. International Journal of Machine Tools and Manufacture, 1997 37(3), 255

[8] Liao, Y.S. and Young, Y.C. *A new on-line spindle speed regulation strategy for chatter control.* International Journal of Machine Tools and Manufacture, 1996, 36(5), 651.

[9] Olgac, N. and Sipahi, R. *Dynamic and stability of variable pitch milling. Journal of Vibration and Control*, 2007, 13(7), 1031-1043.

[10] Sims, N.D., Mann, B. and Huyanan, S. *Analytical prediction of chatter stability for variable pitch and variable helix milling tools.* Journal of Sound and Vibration, 2008, 317(3-5), 664-686.

[11] Yusoff,A.R., Turner, S., *Optimisation of Variable Helix Tool Geometry for Regenerative Chatter Mitigation.* International Journal of Machine Tools and Manufacture, 2011, 51(2) 133-141

[12] Tobias, S.A. and Fishwick, W. *Theory of regenerative machine tool chatter.* The Engineer, 1958, 205, 199-203.

[13] Altintas, Y. and Weck, M. *Chatter Stability of Metal Cutting and Grinding.* CIRP Annals - Manufacturing Technology, 2004, 53(2), 619-642.

[14] Montgomery, D. and Altintas, Y. *Mechanism of cutting force and surface generation in dynamic milling.* Transaction ASME Journal of Engineering Industry, 1991, 113, 160-168.

[15] Delio, T., Smith, S., Tlusty, J. and Zamudio, C. *Stiffness, Stability, and Loss of Process Damping in High Speed Machining* : , Fundamental Issues in Machining, Vol PED 43, Dallas, TX, 1990. pp. 171-191. ASME, New York, NY. Precision Engineering, 1991, 13(3), 236.

[16] Elbestawi, M.A., Ismail, F., Du, R. and Ullagaddi, B.C. *Modelling machining dynamics including damping in the tool-workpiece interface.* Transaction ASME Journal of Engineering Industry, 1994, 116(4), 435-439.

[17] Ranganath, S., Narayanan, K. and Sutherland, J.W. *The role of flank face interference in improving the accuracy of dynamic force predictions in peripheral milling.* Journal of Manufacturing Science and Engineering, 1999, 121, 593-599.

[18] Huang, C.Y. and Wang, J.J.J. *Mechanistic Modeling of Process Damping in Peripheral Milling.* Journal of Manufacturing Science and Engineering, 2007, 129(1), 12-20.

[19] Yusoff, A.R., Turner, S., Taylor, C.M. and Sims, N.D. *The role of tool geometry in process damped milling.* International Journal of Advanced Manufacturing Technology, 2010(85), 883-895.

[20] Altintas, Y, *Manufacturing Automation; Metal Cutting Mechanics, Machine Tool Vibrations and CNC Design,* 2004, Press Syndicate of University of Cambridge.

Open loop control by equalization for an analog Duffing system

J A Vazquez Feijoo, PT Matadamas Ortiz, V J Rios Olivera
Instututo Politecnico Nacional, CIIDIR-Oaxaca, academia de ingeniería,Hornos #1003, Col. Noche Buena, Sta. Cruz Xoxocotlan, Oacaca, México

E-mail: javazquezfeijoo@yahoo.com.mx

Abstract. The inverse Volterra operators have been proposed as an open look control for sensors and actuators systems. Large mathematical development has been done in the area, however the implementation of a continuous time model of the inverse Volterra may be difficult. That is why in this work a discrete version of the Associated Linear Equations for the inverse operator is proposed. The control strategy is tested on an analog electronic Duffing oscillator system.

1. Introduction

Associate linear equations [1] are explicit models of the Volterra operators [3]. They have been used for analysis, identification [2] and control [4] for continues and time invariant nonlinear systems. The control is proposed in open look by connecting in tandem the nonlinear system with the inverse Volterra operator (expressed by ALEs). The present work presents a linear Autoregressive with eXogeneus inputs (ARX) discrete model for the Volterra operators. They may work better for digital treatment that the analytical explicit model. The development of the direct and inverse discrete ALEs for an analog Duffing Oscillator of third power is done here. After that an equalization strategy is carried out with the inverse Volterra to retrieve the input signal. A measurement system may be beneficiated with this equalization as the reading is not affected by the dynamics of the measurement system. The inverse Volterra produce a theoretical infinite band width, also compensate losses and eliminates the system inertia. The work is organized as follows, next two sections treat the general theory of the ALEs and the Inverse Volterra, The fourth section the discrete version of the ALEs for the inverse Volterra are developed. The fifth section presents the characteristics of the analog Duffing oscillator system. The sixth section deals with the identification by a NARMAX model of the Duffing oscillator. The last section shows the results of applying the equalization strategy to the Duffing oscillator.

2. The Associated Linear Equations (ALEs)

In [1], a general second order equation is given as,

$$\ddot{y}(t)+\dot{y}(t)+y(t)=\sum_{i=1}^{N}k_j y^j(t)+\sum_{j=1}^{P}c_j x^j(t)+\sum_{s=1}^{Q}g_s(t) \tag{1}$$

[1]Correspondence author: J.A Vazquez Feijoo
Email address: javazquezfeijoo@yahoo.com.mx

Published under licence by IOP Publishing Ltd

which includes most of the Volterra systems. In this equation $g_n(t)$ is a not necessarily a continuous function. The Volterra series [Schetzen] break down the signal into an infinite number of terms called Volterra operators,

$$y(t)=\sum_{i=1}^{\infty} y_n(t) \tag{2}$$

Each operator is obtained by the Associated Linear Equations (ALEs). From equation (1) a Duffing oscillator of third order can be obtained as,

$$\ddot{y}+2\zeta\omega_n\dot{y}+\omega_n^2 y+k_2 y^3(t)=x(t) \tag{3}$$

From which according to [2], even operators are zero, and the first four odd operators ALEs are,

$$\begin{aligned}
&\ddot{y}_1+2\zeta\omega_n\dot{y}_1+\omega_n^2 y_1=x(t)\\
&\ddot{y}_3+2\zeta\omega_n\dot{y}_3+\omega_n^2 y_3=k_3 y_1^3(t)\\
&\ddot{y}_5+2\zeta\omega_n\dot{y}_5+\omega_n^2 y_5=-3k_3 y_3(t)y_1^2(t)\\
&\ddot{y}_7+2\zeta\omega_n\dot{y}_{77}+\omega_n^2 y_7=-3k_3\left(y_3^2(t)y_1(t)+y_5(t)y_1^2(t)\right)
\end{aligned} \tag{4}$$

Each equation represents a linear system that depends on the *n*th- harmonic order of its input that is conformed by products of lower order operators. The analogic electronic system used in this work is a third degree Duffing oscillator as represented in equation (3).

3. Postinverse Volterra operators by ALEs

The Volterra inverse can be considered a system **K** that is connected in tandem with the system **H** to obtain a system **W** that is equivalent to an element of unitary gain [5] as shown in figure 1, without a DC (case treated in [4]). For the specific case of the Duffing oscillator, it can be written,

$$w(t)=x(t)+\sum_{i=1}^{N} w_j(t)=x(t) \tag{5}$$

where N is the maximum power of the output signal $y(t)$ in equation (1). A Duffing oscillator of only a third degree term needs only the first and the third inverse operators ($\mathbf{Z}_1$ and $\mathbf{Z}_3$), where,

$$Z_1=H_1^{-1} \tag{6}$$

and

$$Z_3(t)=-Z_1(Y_3[Z_1(t)]) \tag{7}$$

for the post inverse case one has,

$$w(t)=z_1(y(t))-z_1(y_3[z_1(y(t))])=x(t) \tag{8}$$

as can be seen in figure (2).

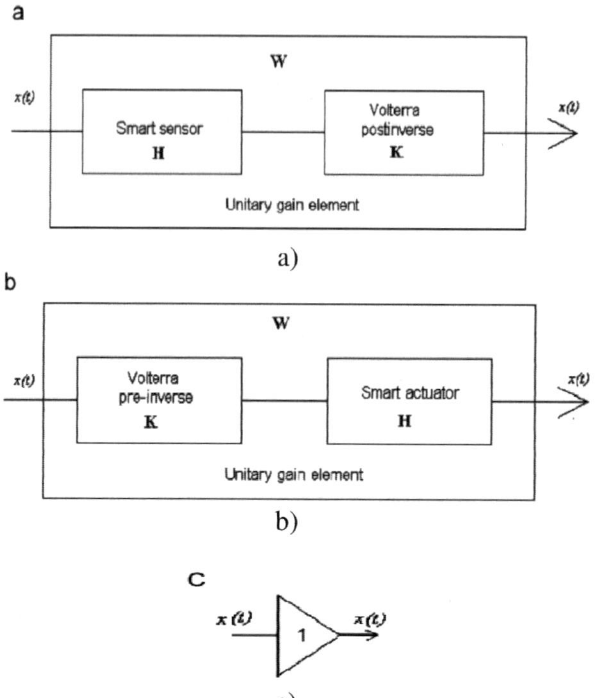

Figure 1. The inverse Volterra operator transform a system in an element of unitary gain. a) Equalization for sensors, b)Predistortion for actuators. c) The overall system

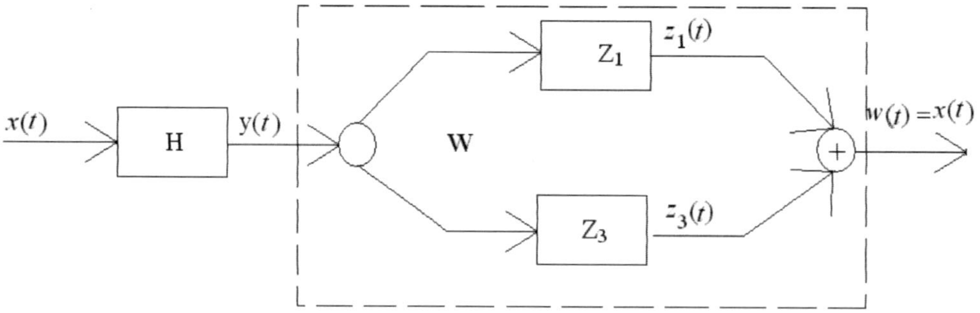

Figure 2. The full postinverse of a Duffing oscillator with only a third order term in the output.

4. Autoregressive with eXogenous inputs (ARXs) inverse models

Continuous nonlinear systems, may be properly modelled by discrete models called Nonlinear Autoregressive Moving average with eXogenous inputs [6], that in its most general case one has,

$$y(i)=\sum_{j=1}^{K}k_j y(i-j)+\sum_{m=1}^{P}p_j x(i-m)+\sum_{r=1}^{Q}q_r e(i-s)+\sum_{s1=1}\sum_{s2=1}\sum_{s3=1}u_{s1,s2,s3}y(i-s1)y(i-s2)y(i-s3) \quad (9)$$

where the argument is the instant i and the lags j, m, s. $e(i)$ are the model of additive noise. In [7] a cubic Duffing oscillator is modelled as,

$$y(i)=k_1y(i-1)+k_2y(i-2)+p_1x(i-1)+p_2x(i-2)+q_1e(i-1)+q_2e(i-2)+u_{1,1,1}y^3(i-1)+u_{2,2,2}y^3(i-2)$$
$$+u_{1,1,2}y^2(i-1)y(i-2)+u_{1,2,2}y^2(i-2)y(i-1) \quad (10)$$

As it was done before for continuous systems by the Associated Linear Equations ALEs, the Duffing oscillator discrete model can be decomposed also in ARX models. A discrete Volterra series can be described as,

$$y(i-j)=\sum_{i=1}^{\infty} y_i(i-j) \quad (11)$$

On substitution on (8) and separating the response in harmonic orders one has,

$$y_1(i)=k_1y_1(i-1)+k_2y_1(i-2)+p_1x(i-1)+p_2x(i-2) \quad (12a)$$

$$y_3(i)=k_1y_3(i-1)+k_2y_3(i-2)+y_1^3(i-1)+y_1^3(i-2)$$
$$+y_1^2(i-1)y_1(i-2)+y_1(i-1)y_1^2(i-2) \quad (12b)$$

$$y_5(i)=k_1y_5(i-1)+k_2y_5(i-2)+3u_{1,1,1}y_1^2(i-1)y_3(i-1)+3u_{2,2,2}y_1^2(i-2)y_3(i-2)$$
$$+3u_{1,1,2}y_1^2(i-1)y_3(i-2)+3u_{1,2,2}y_1^2(i-2)y_3(i-1) \quad (12c)$$

They become then the first, third and fifth harmonic order discrete (ARX)ALEs.

5. Discrette inverse Volterra

As discussed in section two, the first inverse operator is the inverse function of the original first order operator of the system, as in equation (6). The z-transform of equation (12a) is,

$$H(j)=\frac{p_1z^{-1}+p_2z^{-1}}{1-k_1z^{-1}-k_2z^{-2}} \quad (13)$$

The inverse function of an operator in the time domain is in the z-transform the inverse multiplicative, therefore,

$$H(j)=\frac{1}{W(j)} \quad (14)$$

So, the first inverse operator has as a z-transform,

$$W(j)=\frac{1-k_1z^{-1}-k_2z^{-2}}{p_1z^{-1}+p_2z^{-1}} \quad (15)$$

Back to the discrete time domain it is obtained,

$$w_1(j-1)=\frac{1}{p_1}\left(y_1(j)-k_1y_1(j)(j-1)-k_2y_1(j)(j-2)-p_2w_1(j-2)\right) \quad (16)$$

For the case of post inverse Volterra, equation (16) may be arranged as,

$$w_1(j-1)+\frac{p_2}{p_1}w_1(j-2)=\frac{1}{p_1}\left(y_1(j)(j)-k_1y_1(j)(j-1)-k_2y_1(j)(j-2)\right)$$

The left hand side is proportional to the first derivative of the input, one may define,

$$wd_1(j-1)=\frac{1}{p_1}(y_1(j)-k_1y_1(j-1)-k_2y_1(j-2)) \tag{17}$$

Though the response possesses a delay of one time interval, the value at the present time may be predicted by the following equation,

$$wd_1(j)=3wd_1(j-1)-3wd_1(j-2)+wd_1(j-3) \tag{18}$$

$z_1(t)$ in Figure (2) belongs to the post inverse Volterra, then the real input into $\mathbf{Z}_1$ is $y(t)$. The delayed output of this operator is obtained from equation (16) with the proper input, i.e.,

$$wd_1(j-1)=\frac{1}{p_1}(y(j)-k_1y(j-1)-k_2y(j-2)) \tag{19}$$

then, from equation (10) one obtains,

$$wd(j-1)=\frac{1}{p_1}\begin{pmatrix}k_1y(i-1)+k_2y(i-2)+p_1x(i-1)+p_2x(i-2)+q_1e(i-1)+q_2e(i-2)+u_{1,1,1}y^3(i-1)+u_{2,2,2}y^3(i-2)\\+u_{1,1,2}y^2(i-1)y(i-2)+u_{1,2,2}y^2(i-2)y(i-1)\\-k_1y(j-1)-k_2y(j-2)\end{pmatrix}$$

which reduces to,

$$wd_1(j-1)=x(i-1)+\frac{p_2}{p_1}x(i-2)+\frac{1}{p_1}\begin{pmatrix}q_1e(i-1)+q_2e(i-2)+u_{1,1,1}y^3(i-1)+u_{2,2,2}y^3(i-2)\\+u_{1,1,2}y^2(i-1)y(i-2)+u_{1,2,2}y^2(i-2)y(i-1)\end{pmatrix} \tag{20}$$

This is the first inverse Volterra operator. To calculate the second operator, equation (7) is used. The output of the first inverse operator $\mathbf{Z}_1$ is the input into $\mathbf{Y}_1$, which is

$y_1(i)=k_1y_1(i-1)+k_2y_1(i-2)+p_1wd_1(j-1)$

together with equation (20) yields,

$$\begin{aligned}y_1(i)&=k_1y_1(i-1)+k_2y_1(i-2)+p_1x(i-1)+p_2x(i-2)+q_1e(i-1)+q_2e(i-2)\\&+u_{1,1,1}y^3(i-1)+u_{2,2,2}y^3(i-2)+u_{1,1,2}y^2(i-1)y(i-2)\end{aligned} \tag{21}$$

which trivially is,

$$y_1(i)=y(i)$$

This is the input into $y_3(t)$, then,

$$\begin{aligned}y_3(i)&=k_1y_3(i-1)+k_2y_3(i-2)+u_{1,1,1}y^3(i-1)\\&+u_{1,1,2}y^2(i-1)y\ (i-2)+u_{1,2,2}y\ (i-1)y^2(i-2)+u_{2,2,2}y^3(i-2)\end{aligned} \tag{22}$$

This signal according with equation (7) is the input into wd_1(t) from equation (17),

$$wd_1(j-1)=\frac{1}{p_1}(y_3(j)-k_1y_3(j-1)-k_2y_3(j-2))$$

From the same equation (7), multiplying by -1 the third inverse operator $wd_3(j-1)$ is,

$$wd_3(j-1)=\frac{1}{p_1}(-y_3(j)+k_1y_3(j-1)+k_2y_3(j-2))$$

On substitution of equation (22) it is obtained,

$$wd_3(j-1)=\frac{1}{p_1}\begin{pmatrix}-k_1y_3(i-1)-k_2y_3(i-2)-u_{1,1,1}y^3(i-1)\\ -u_{1,1,2}y^2(i-1)y\ (i-2)-u_{1,2,2}y\ (i-1)y^2(i-2)-u_{2,2,2}y^3(i-2)+k_1y_3(j-1)+k_2y_3(j-2)\end{pmatrix}$$

$$wd_3(j-1)=\frac{1}{p_1}\begin{pmatrix}-k_1y_3(i-1)-k_2y_3(i-2)-y^3(i-1)-y^3(i-2)\\ -y^2(i-1)y\ (i-2)-y\ (i-1)y^2(i-2)+k_1y_3(j-1)+k_2y_3(j-2)\end{pmatrix}$$

reducing,

$$wd_3(j-1)=\frac{1}{p_1}\left(-y^3(i-1)-y^3(i-2)-y^2(i-1)y\ (i-2)-y\ (i-1)y^2(i-2)\right) \tag{23}$$

Adding together first an third order inverse Volterra operators (equations (20) and (23)), the output is,

$$wd_1(j-1)=x(i-1)+\frac{p_2}{p_1}x(i-2)+\frac{1}{p_1}(q_1e(i-1)+q_2e(i-2)) \tag{24}$$

As we know, the noise model from equation (10) can be additivately subtracted from the inverse output which is equation (24) and then if the noise model is exact,

$$wd_1(j-1)=x(i-1)+\frac{p_2}{p_1}x(i-2)$$

The Inverse Volterra operator w_1(j-1) is then obtained as,

$$w_1(j-1)=wd(j-1)-\frac{p_2}{p_1}w_1(j-2)$$

then as it should,

$$w(j-1)=x(i-1)$$

6. The analog Duffing oscillator system

The analog Duffing system considered in here (Single Degree Of Freedom SDOF), is constructed according with the following equation,

$$\ddot{y}(t)+2\zeta\omega_n\dot{y}(t)+\omega_n^2y(t)+\omega_o^2y(t)^3=Ax(t) \tag{25}$$

where ζ,ω and A are the linear parameters of the system. This equation can be also rewritten as in [8],

$$u(t)+2\zeta\omega_n\int u(t)dt+\omega_n^2\int\int u(t)dt+\omega_o^2\left(\int\int u(t)dt\right)^3=Ax(t) \tag{26}$$

To build a system from a model in terms of integrals instead of derivatives is reported to have more stability. Figure 3 shows the schematic of the system. That is obtained from the Fourier transformation,

$$U(\omega)=AX(\omega)-2\zeta\omega_n\frac{U(\omega)}{i\omega}-\omega_n^2\frac{U(\omega)}{(i\omega)^2}-\omega_o^2\left(\frac{U(\omega)}{(i\omega)^2}\right)^3$$

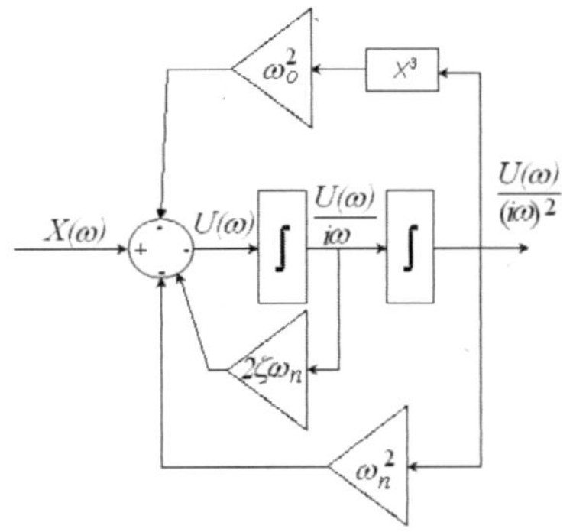

Figure 3. Block Diagram of the analog Duffing oscillatorsytem

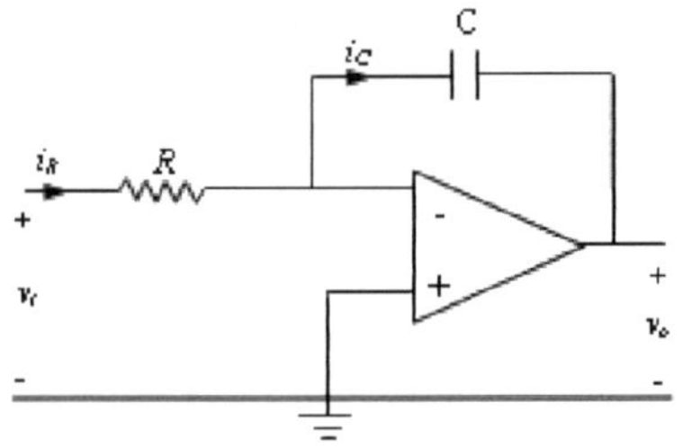

Figure 4. Integrator circuit with an Op amp 720

The integral blocks are constructed as traditionally (figure 4). The cubic operator is done by connect in cascade three multiplicative circuits AD633JN/AN. The electronic design can be seen in figure 5.

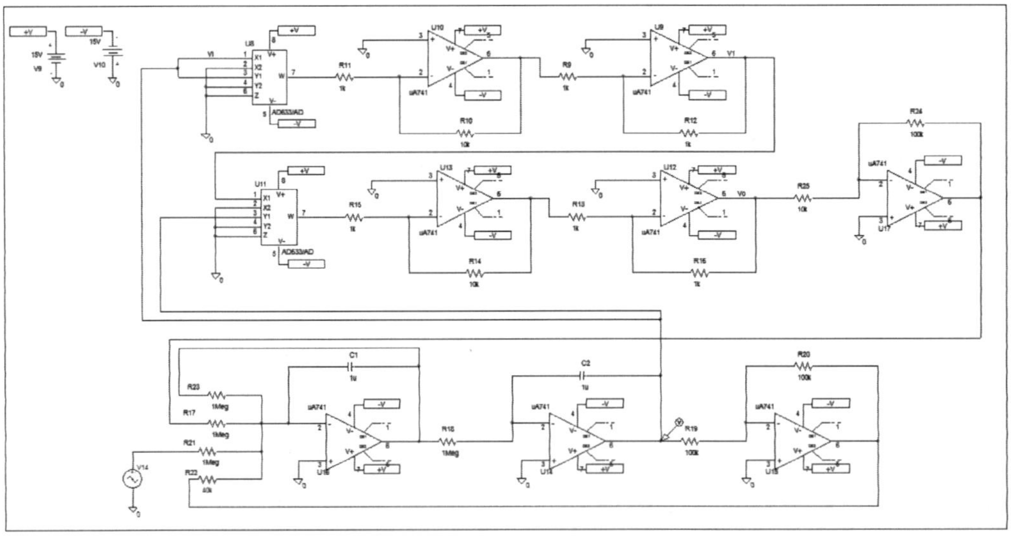

Figure 5. Analogic Duffing oscillator of therd negree

7. Modelling

The input into the analog system is a three harmonic signal, as shown in the figure 6. The output is obtained is presented in figure 7. The input frequencies are 6, 12, and 15 rd/sec. The output is filtered to eliminate signals with frequencies larger than 50 rd/sec, to prevent noise contamination. The data was taken at an interval of 0.02 sec.

The NARMAX model proposed for this system is,

$$y(i) = k_1 y(i-1) + k_2 y(i-2) + p_1 x(i) + u_{1,1} y^2(i-1) + u_{1,1,1} y^3(i-1)$$

This is because a second harmonic order was found. After filtering the ouput signal, the coefficients are obtained, giving the following model,

$$y(i)=0.9804y(i-1)+7.321\times10^{-6}y(i-2)+1.9\times10^{-3}x(i)+6.269\times10^{-4}y^2(i-1)+2.4\times10^{-3}y^3(i-1) \quad (27)$$

Figure 8 compares the simulation output with the original one. The analog system presents certain instabilities both in frequency and in amplitude. That implies that the model in equation (26) is valid only up to input amplitude of 13 V and a frequency of 60 rs/sec.

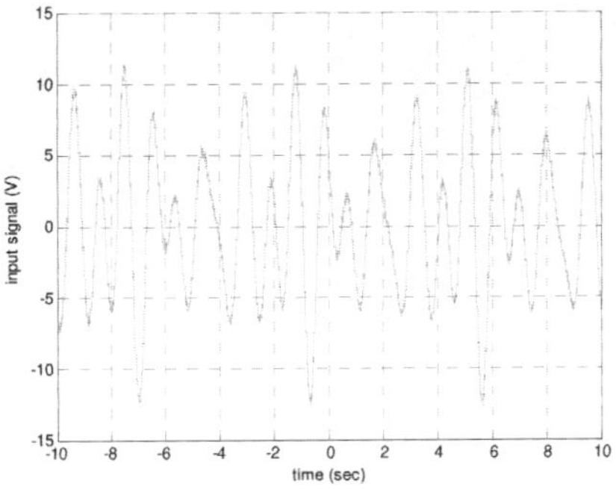

Figure 7. Input signal into the analogic Duffing oscillator

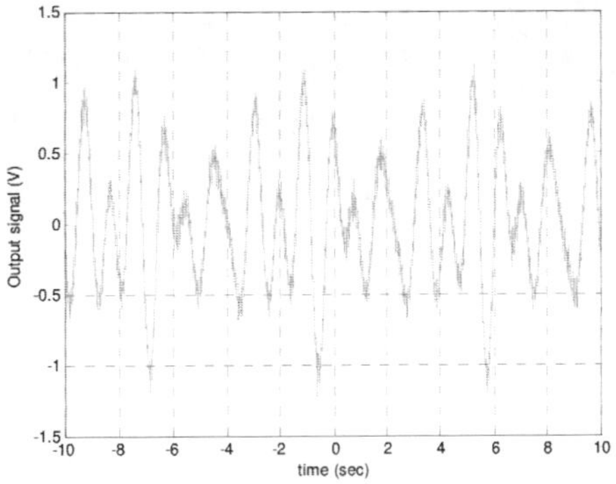

Figure 8. Output signal from the analogic Duffing *oscillator*

From equations (12), the first three non-zero ALE's operators are,

$$y_2(i)=0.9804y_2(i-1)+7.321\times10^{-6}y_2(i-2)+6.269\times10^{-4}y_1^2(i-1)$$
$$y_3(i)=0.9804y_3(i-1)+7.321\times10^{-6}y_3(i-2)+2.4\times10^{-3}y_1^{\,3}(i-1)+1.254\times10^{-3}y_1(i-1)y_2(i-1)$$

Only the first three inverse ALEs are obtained as the inverse Volterra just need to be built up the third order.

Figure 9 compares these ALEs with the filtered signal. The second order signal is obtains by subtracting the first order ALE from the filtered original signal. After this one may subtract the second order ALE and obtain the third order from the original signal. It is expected that higher orders from the original signal are insignificant.

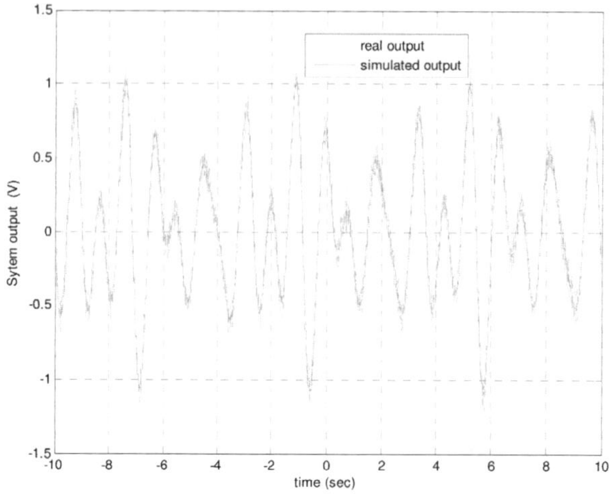

Figure 9. Comparing the output signal against the NARMAX model obtained for the analogic Duffing oscillator.

8. Inverse Volterra for the analog Duffing oscillator system

From equation (13) and the system model (27), the first inverse Volterra operator is obtained by the z-transform as,

$$W(j)=\frac{1-0.9804z^{-1}-7.321\times10^{-6}z^{-2}}{1.9\times10^{-3}} \tag{28}$$

$$w_1(j)=\frac{1}{1.9\times10^{-3}}y_1(j)-\frac{0.9804}{1.9\times10^{-3}}y_1(j-1)-\frac{7.321\times10^{-6}}{1.9\times10^{-3}}y_1(j-2) \tag{29}$$

Equation (7) is very close for the second order operator, i.e,

$$Z_2(t)=-Z_1(Y_2[Z_1(t)]) \tag{30}$$

the third order inverse Volterr operator is now,

$$Z_3(t)=-Z_1(Y_3[Z_1(t)])-2Z_1(Y_2(Z_1(t),Z_2(t))) \tag{31}$$

As the second order signal results not zero.

Figure 10 shows the plot of the second and third order Volterra inverse operators. In figure 11, the difference between the input signal and the inverse volterra as more inverse operators are added. Figure 12 contain the input signal and the full inverse Volterra operator.

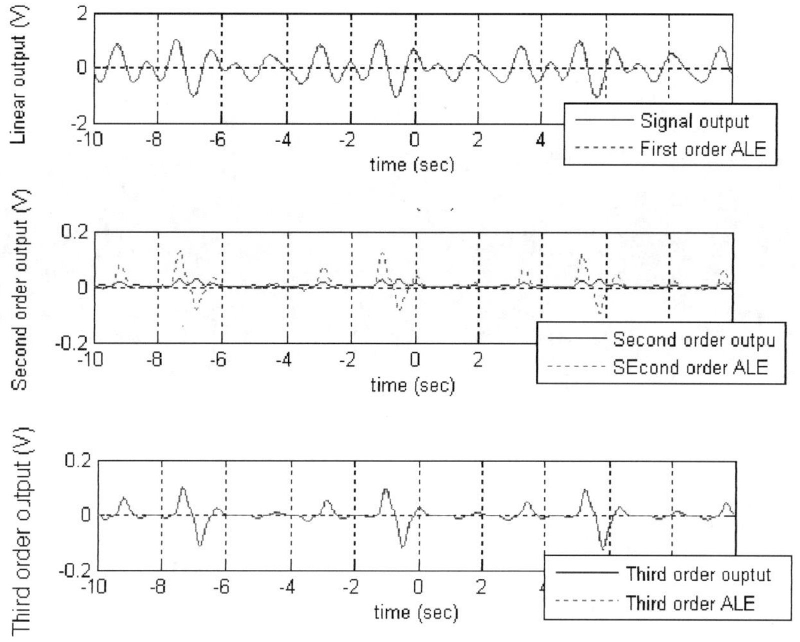

Figure 10. Plot of the ALEs against the corresponding order original system.

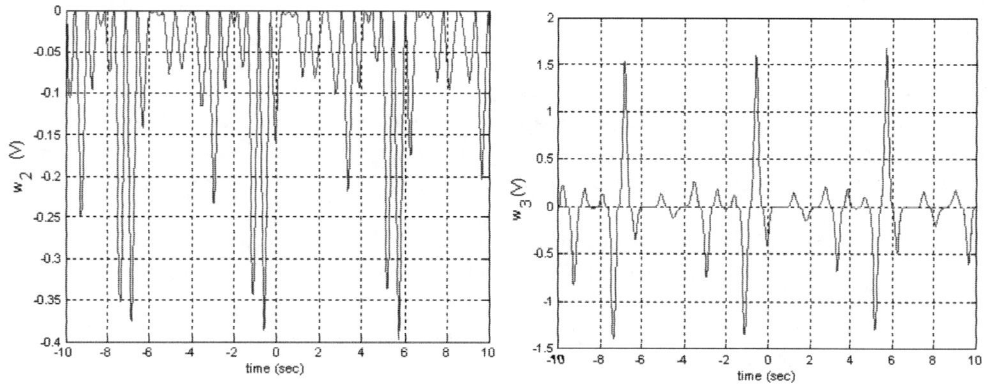

Figure 11. The inverse Volterra operator of second and third order.

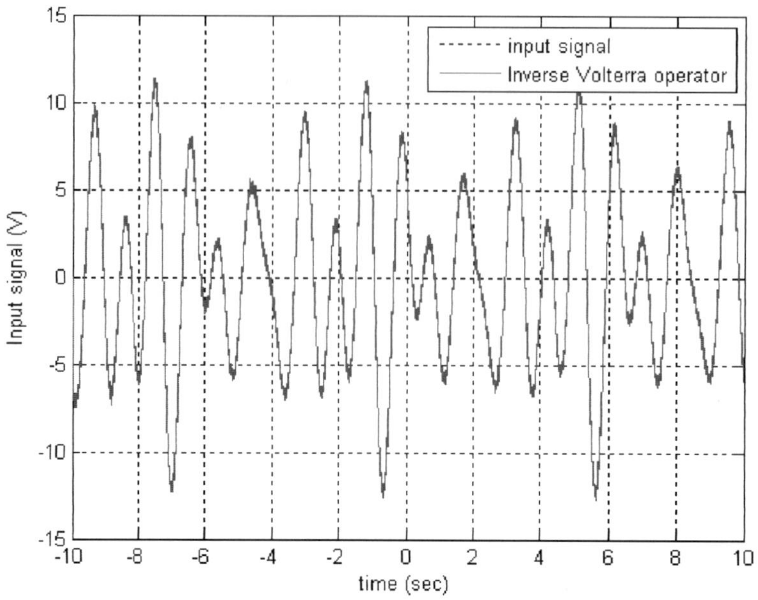

Figure 12. The system input against the inverse Volterra input.

9. Conclusions

The objective of the present work is to develop a discreet version of the Associated Linear Equations (ALEs).Then by obtaining the post inverse inverse discrete Votlerra system, it is intend to implement a open loop control of a Duffing oscillator analog system. Though the analog system exhibit certain instabilities, there where found a region in which the system is invariant regarding frequency and amplitude. The system is modelled by a NARX model and from there the inverse discrete models of the ALEs are found.

From the discrete Volterra operators, the inverse Volterra is obtained. It is found that the noise may be amplified by the inverse system and the original output may not be obtained. However, a proper filtering or noise model can help to make quite accurate the inverse prediction of the input.

It is very promising the inverse Volterra as a control tool after been applied to a system so noisy and instable as the present. The system inertia and system losses are compensated and harmonics are eliminated from the output. This implies the possibility of obtain more accurate measuring systems.

Though the inverse Volterra operator is said to have a infinity band width, because of the nature of the analog system it is not possible to demonstrate this.

References

[1] Vazquez Feijoo, J.A, Worden, K., Stanway, R. Analysis of time-invariant systems in the time and frequency domain by associated linear equations (ALEs). Mechanical Systems and Signal Processing. 896-919. 20 (2006)

[2] Vazquez Feijoo, J.A., Worden, K., Stanway, R. System identificationusign Associalted Linear Equations (ALEs), Mechanical Systems and Signal Processing. 431-455. 18 (2004)

[3] Schetzen, M, 1980, The Volterra and Wiener Theories of nonlinear systems, Wiley-Interscience Publications, NY, ISBN 0471044555.

[4] Vazquez Feijoo, J.A., Worden, K., Stanway, R. Associated Linear Equations for Volterra operators. Mechanical Systems and Signal Processing. 57-69. 19 (2005)

[5] Vazquez Feijoo, J.A., Worden, K., Stanway, R, Juarez Rodriguez, N, Transformation of a sensor or actuator system into a unitary gain element. Mechanical Systems and Signal Processing. 3088–3107. 21 (2007)

[6] S. Chan, S. A. Billings, C. F. N. Cowan, P. M. Grant, Practical identification of NARMAX models using radial basis functions, International Journal of control, 1327-1350, 52 (1990)

[7] H. K. Jang, K. J. Kim, identification of Cubic Stiffness Nonlinearity by Linearity-Conserved NARMAX Modeling, KSME Journal, Vol. 8, No., pp. 332-342,3-8 (1994)

[8] Bush, V., Hazen, H., (1931). "The differential analyzer: A new machine for solving differential equations," J. Franklin Inst., vol. 212, no. 4, pp. 447–488.

IOP Conf. Series: Materials Science and Engineering **36** (2012) 012019 doi:10.1088/1757-899X/36/1/012019

Study on Dynamic Behaviour of Wishbone Suspension System

M Kamal[1] and M M Rahman[1,2]
[1]Faculty of Mechanical Engineering
Universiti Malaysia Pahang
26600 Pekan, Pahang, Malaysia
Phone: +6094242246; Fax: +6094242202
[2]Automotive Engineering Centre
Universiti Malaysia Pahang
26600 Pekan, Pahang, Malaysia

E-mail: mustafizur@ump.edu.my

Abstract. This paper presents the characteristic model of the wishbone suspension system using the quarter car model approach. Suspension system in an automobile provides vehicle control and passenger comfort by providing isolation from road disturbances. This makes it essential that the detailed behavior of suspension should be known to optimize the performance. A kinetic study is performed using multi body system (MBS) analysis. The dirt road profile is considered as an applied loading. The spring constant, damping coefficient and sprung mass are studied on the performance of the suspension system. It can be observed that the spring constant is inversely related with time required to return to initial position and the amount of deformations. The damping ratio affects the suppression of spring oscillations, beyond a certain limit damping ration has the negligible effect. Sprung mass effected the equilibrium position of the suspension system with a small effect on its oscillation behavior. It is shown that the spring constant, damping ratio and sprung mass are significant parameters to design the suspension system. This study is essential for complete understanding of working of the suspension system and a future study with real geometries.

1. Introduction

Suspension system in its simplest form may be thought of as a linkage to allow the wheel to move relative to the body and some elastic element to support loads while allowing that motion. It is an assembly of springs, shock absorbers (dampers) and suspension arms that connects a vehicle to its wheels. In a running vehicle, the suspension system keeps the occupants comfortable and isolated from road noise, bumps, and vibrations. It also provides the vehicle, handling capabilities, allowing the driver to maintain control of the vehicle over rough terrain or in case of sudden stops. Additionally, the suspension system prevents the vehicle from damage and wearing [1-2].

Basically suspension system consists of spring, damper and structural components carrying the sprung mass (car chassis). The springs absorb impacts and provide cushioning when a wheel hits a bump in the road. The springs also resist the wheel's movement and rebounds, pushing the wheel back down, so to keep the control of vehicle by keeping the wheels in contact with the road. Shock absorbers (dampers) perform two functions. They absorb any larger than average bumps in the road so that the upward velocity of the wheel over the bump is not transmitted to the car chassis. Secondly,

[1,2]Correspondence author: M. Mustafizur Rahman
Email address: mustafizur@ump.edu.my

Published under licence by IOP Publishing Ltd

they keep the suspension at full as much as possible during the travelling for the given road conditions, in brief, they keep the wheels planted on the road [3-4]. Due to the criticality of suspension system performance related to ride comfort and vehicle control and passenger safety, the understanding of design variables and behavior of suspension system in severe or harsh loading cases should be well known and optimized. This paper analyze a characteristic model of wishbone type suspension system, with study the effects of spring constant, damping coefficient and supported sprung mass, which is not part of suspension system but can affect the response of suspension. As passenger comfort is one of the prime criteria for performance evaluation of suspension, the same is used to monitor the system under study by plotting the deflection of sprung mass when a dirt road profile is applied as loading.

2. Description of Basic Suspension System

Different types of suspension systems are being used in modern vehicles. Generally, the suspension systems can be divided into two groups: dependent suspension system and independent suspension system. The dependent suspension system has one or more solid axles that hold opposite wheels together, therefore the wheels cannot move independently. The independent suspension system does not have such a wheel axle so that the wheels are allowed to rise and fall without affecting the opposite wheel [5-6]. A suspension hierarchy is defined in figure 1 and schematic diagram of wishbone system is shown in figure 2.

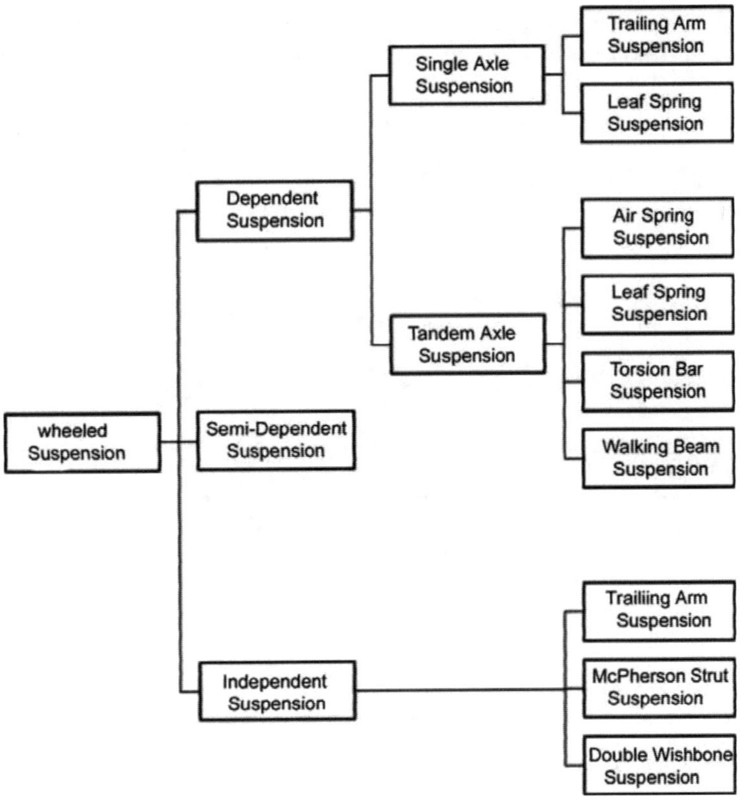

Figure 1. Hierarchy of popular suspension types.

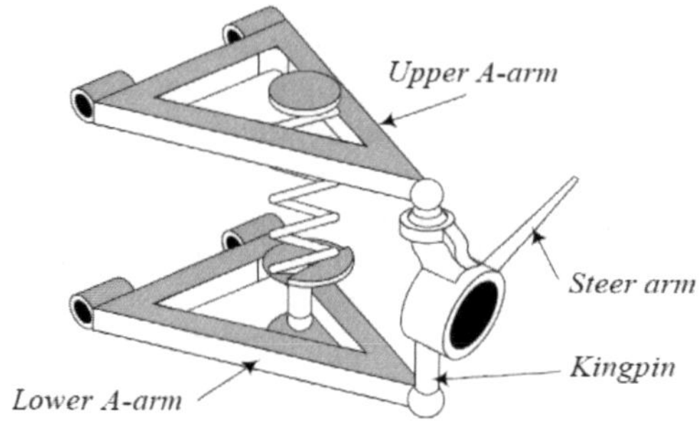

Figure 2. Wishbone suspension system.

3. Computational Model

A characteristic model of wishbone suspension system is developed in MSC ADAMS for the current study as shown in figure 3. The dimensions of components are taken from literature [7]. Material of connecting arms, and knuckle is aluminum (Density: 2700 Kg/m^3) (ASM Data sheet). Applied loading is in the form of dirt road profile, figure 4, induced in model through excitation platform [8]. The components are considered as rigid bodies. The model is constrained using revolute (one rotation D.O.F. allowed) and ball joints (three rotation D.O.F. allowed) (shown in figure 3), to properly simulate the suspension components motion. Due to simplified geometry the center of gravity of suspension arms and other components are not as in the real wishbone suspension. Tire is considered to be rigid, with weight assumed to be of a standard car tire (rubber + rim) taken as 11kg. Dirt road profile is converted to time history graph with respect to assumed speed of vehicle (30 km/hr). The values of spring constant and damping coefficient and sprung mass are assumed, table. 1. Spring is assumed to be pre-loaded, such that it is in equilibrium at initial condition.

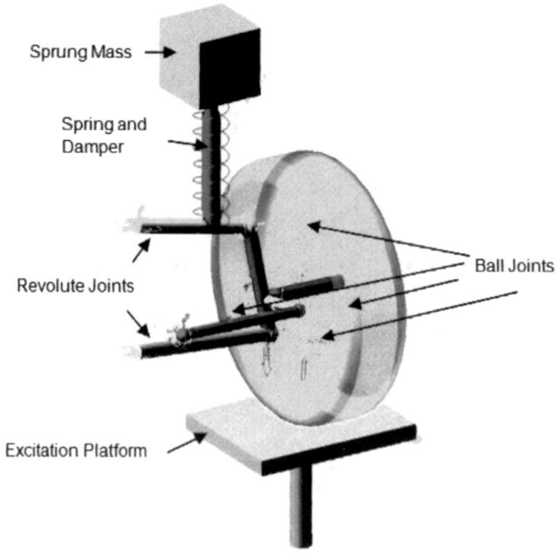

Figure 3. Characteristic model of wishbone suspension system.

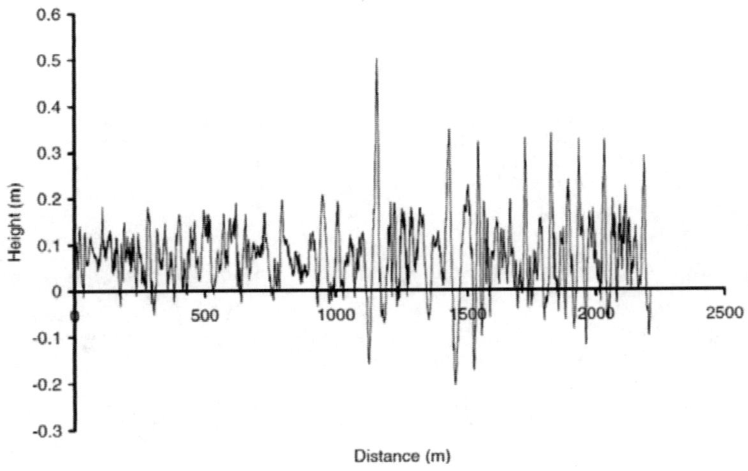

Figure 4. Dirt Road Profile [6].

Table 1. Assumed values of stiffness, damping and sprung mass.

Spring stiffness, *k* (N/mm)	Damping coefficient, *c* (N.s/mm)	Sprung mass (Kg)
0.1	0	600
1	2	1200
10	6	1800
100	10	2400
150	15	3500

4. Results and Discussion

The effect of spring stiffness, damping coefficient and sprung mass is studied separately with values mentioned in table 1. Following is the time history response of sprung mass to the dirt road profile input, for each value of variables under study.

4.1. Spring Stiffness Effects

Effects of stiffness change on the time history of motion of sprung mass is shown in figure 5. Figure 6 shows the spring deformation time history with changed stiffness. From the sprung mass response time history it can be concluded that, with respect to the input profile the motion induced in the sprung mass has no noticeable change with changing stiffness of spring. This means that for every stiffness value the sprung mass response is similar without any noticeable difference. But the spring deformation in figure 6 with two different stiffness show that the difference is in the force present in spring. That is what makes the stiffness calibration important for the optimum behavior of suspension system. As this is the force responsible to keep the tires to remain in contact with the road and minimize the return time of suspension system to its normal position.

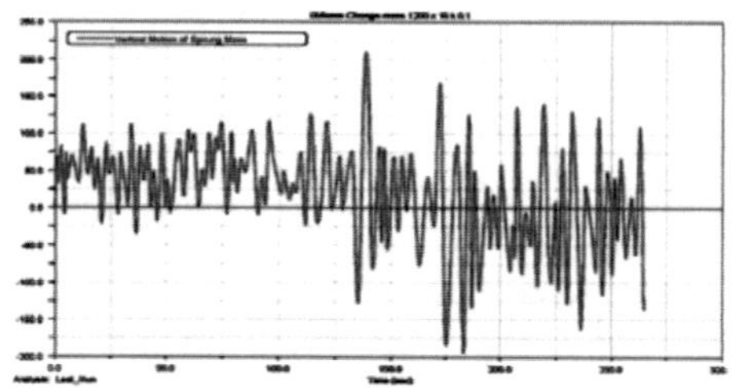

(a) k = 0.1 N/mm

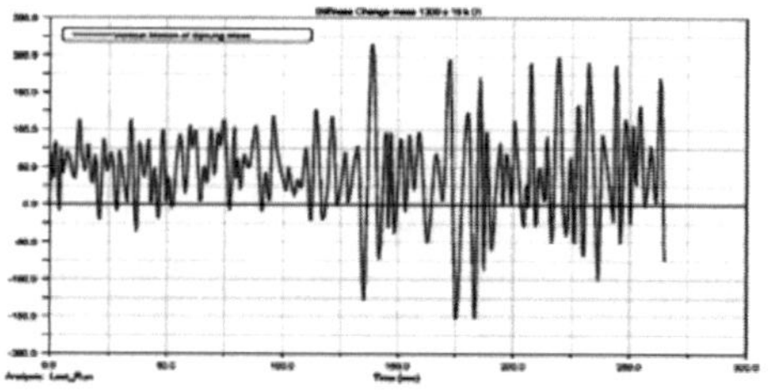

(b) k = 1.0 N/mm

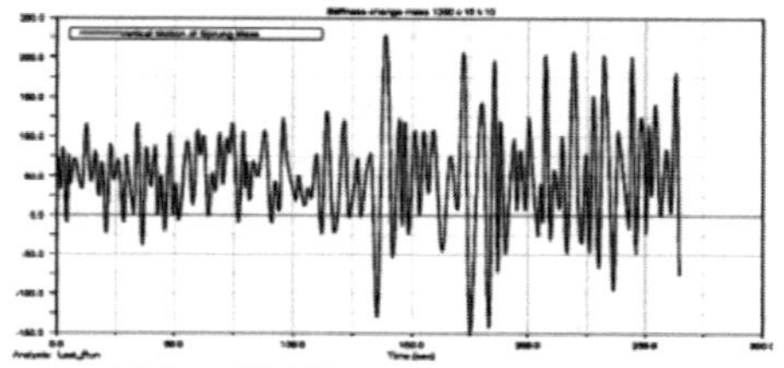

(c) k = 10 N/mm

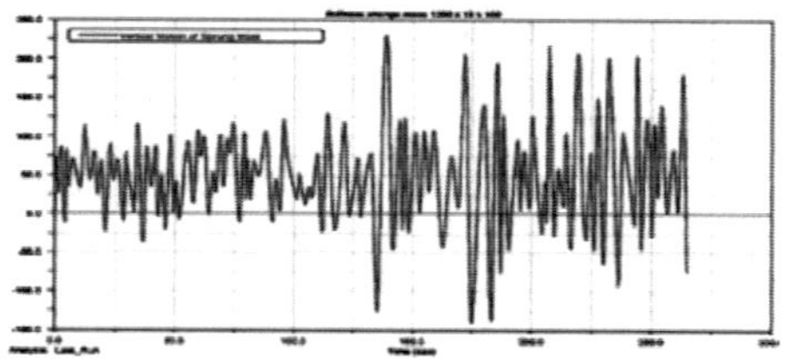

(d) k = 100 N/mm

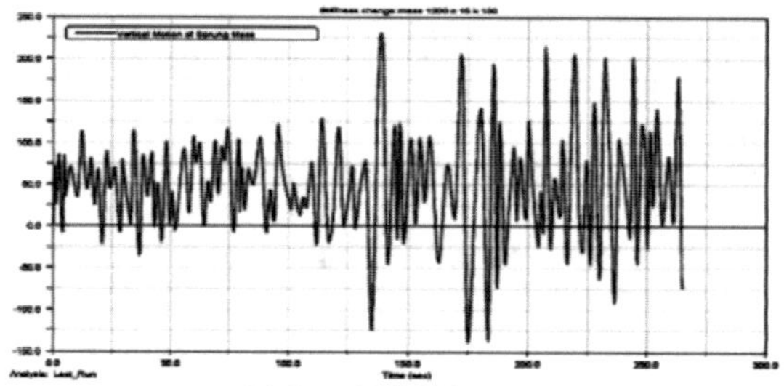

(e) k = 150 N/mm

Figure 5. Time history response of sprung mass with various spring constants.

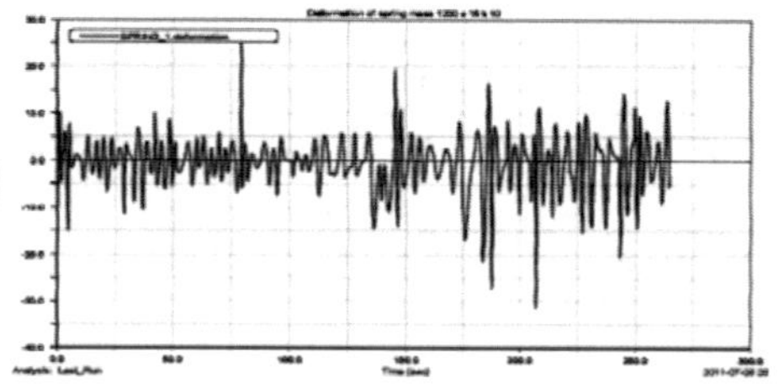

(a) k = 10 N/mm

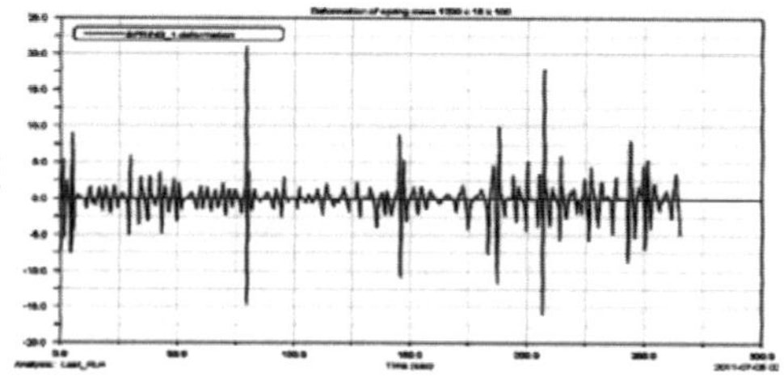

(b) k = 100 N/mm

Figure 6. Deformation change with change in spring stiffness.

4.2. Damping Coefficient Effects

The effect of damping coefficient is illustrated in figure 7. After comparing these sprung mass response time histories, it can be concluded that damping control the oscillations in spring. The response figure 7(a) is without damping and spring oscillations are clearly visible. For vehicle control these oscillations should be damped as soon as possible so suspension is ready to counter next road bump in minimum possible time and improve the handling of vehicle. Figure 7(b-e) shows the effect of damping as it is introduced in the system, the spring oscillations are damped and the resulting response shape is very close to the induced profile.

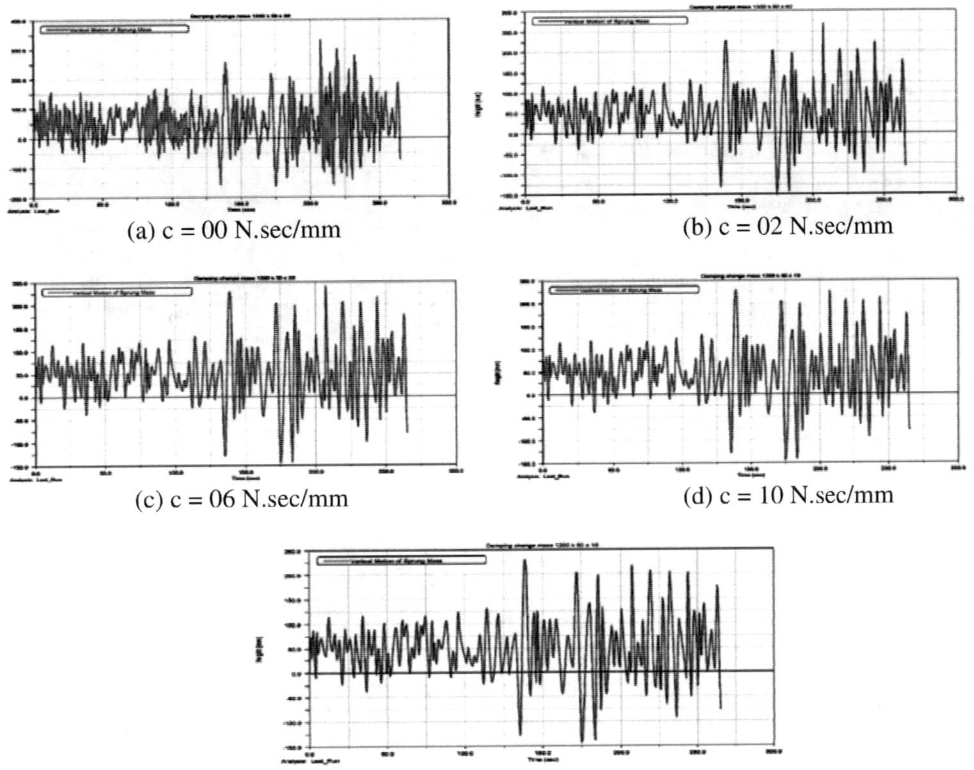

(a) c = 00 N.sec/mm
(b) c = 02 N.sec/mm
(c) c = 06 N.sec/mm
(d) c = 10 N.sec/mm
(e) c = 15 N.sec/mm

Figure 7. Time history response of sprung mass with various damping ratios.

4.3. Sprung Mass Effects

The effect of sprung mass change is reflected in figure 8. It can be clearly deduced that increase in mass only change the equilibrium position of the sprung mass. This makes the quantity of sprung mass important, such that when it is at equilibrium position it should leave enough deformation space for spring to counter the road disturbances and not reach its solid length.

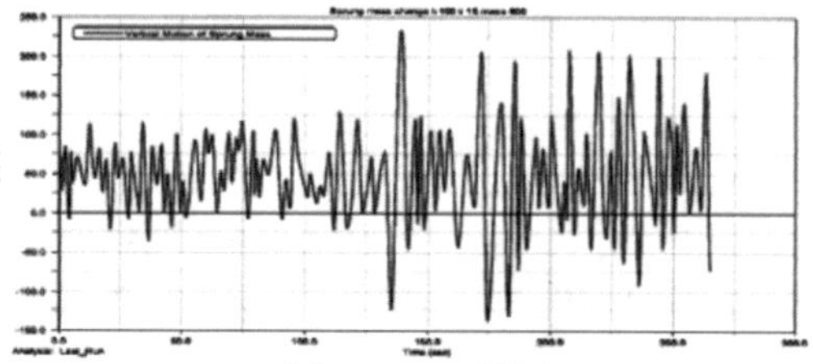

(a) mass = 600 kg

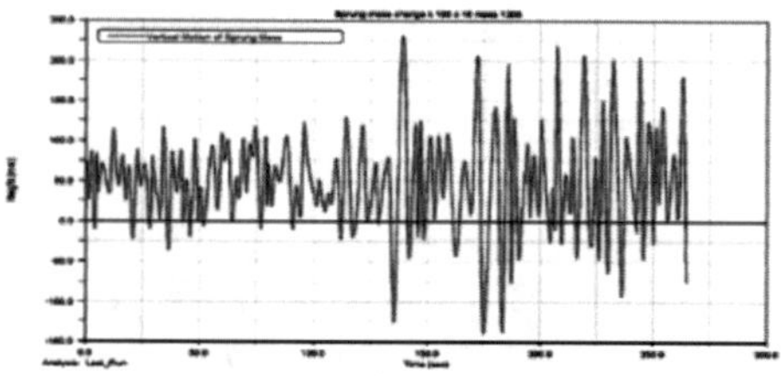

(b) mass = 1200 kg

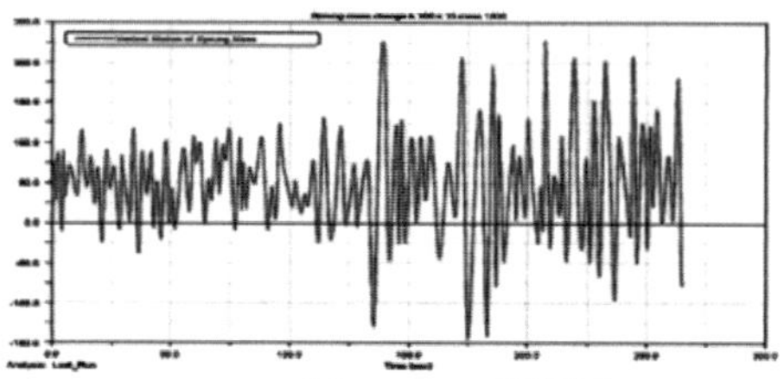

(c) mass = 1800 kg

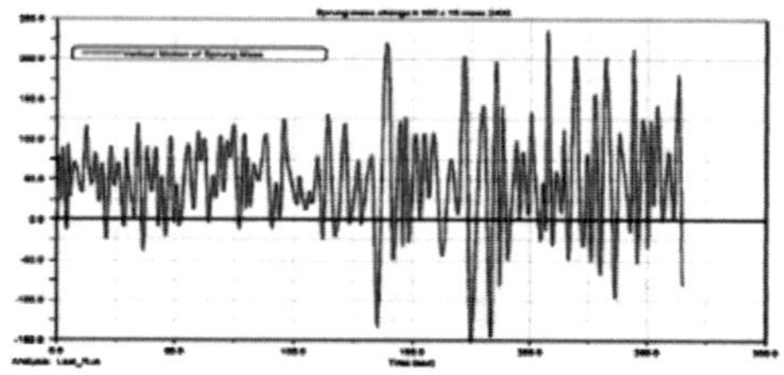

(d) mass = 2400 kg

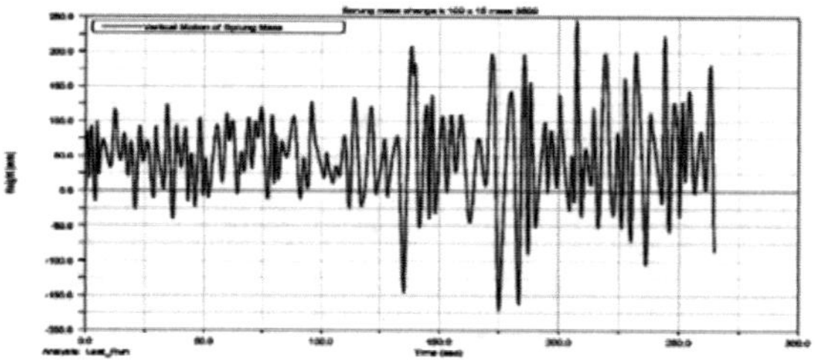

(e) mass = 3500 kg

Figure 8. Time history response of sprung mass with various mass values.

5. Conclusions

The present study provides the initial understanding of the effects of some important parameters of suspension system performance. The spring and damper characteristics are the most vital for the optimum behavior of suspension and should be thoroughly studied. The sprung mass effects must be known so the correct combination of spring damper combination can be selected and calibrated according to the design requirements. A study with real geometry of suspension system components is suggested to fully capture the response behavior. Tire should be modeled as flexible to capture the real forces transmitted to the suspension system.

Acknowledgments

The authors would like to thank Universiti Malaysia Pahang and financial support.

References

[1] Reimpell J, Stoll H and Betzler J W 2002 The Automotive Chassis: Engineering Principles, Second ed. (Warrendale: SAE).

[2] Blundell M and Harty D 2004 Multibody Systems Approach to Vehicle Dynamics (Burlington: Elsevier).

[3] Gillespie T D 1992 Fundamentals of Vehicle Dynamics (Warrendale: Society of Automotive Engineers, Inc).

[4] Nasir M Z M 2006 Analysis of Double Wishbone Suspension System Components. MSC Dissertation in Automotive Engineering, Coventry University, School of Engineering.

[5] Jazar RN 2008 Vehicle Dynamics: Theory and Applications (New York: Springer).
[6] Dixon J C 2009 Suspension Geometry and Computation (Chichester, New York: John Wiley & Sons Ltd).
[7] Jihui L and Ke C 2007 MSC.ADAMS Basic Training and Examples (Shenyang Ligong University).
[8] Snyman J A and Naude A F 2003 Appl. Mathematical Model. 27 263.

Preliminary results of impeller design modifications

Y C Yan, C H Sam, S E Soo, X S Khoo, J K Lau, C H Khek and O Younis[1]
School of Engineering, Taylor's University,
No.1, Jalan Taylor's, 47500, Subang Jaya,
Selangor Darul Ehsan, Malaysia.

E-mail: obai.younis@taylors.edu.my

Abstract. Power consumption in the oil mixing process is very much affected by the impeller design. Minor modification on the impeller design might result in better mixing and less energy consumption. This paper reports an early stage of applying the observation technique to modify three standard impellers, namely the Pitched blade turbine impeller, the CHEMSHEAR impeller and the Vertical Flat-Blade Turbine Impeller. The advantages and disadvantages of these three impellers were analyzed; accordingly minor modification on the impeller design is adopted. The CFD tool is used to evaluate one of the modified impellers; results indicated that the modified impeller achieves better fluid dispersion.

1. Introduction

Liquid substances have always been used as lubricants to minimise the friction, head, and wear between mechanical parts that come in contact with each other. Today, the most commonly used lubricant is lubricant oil, because of its wide range of applications. Some of the main functions of lubricating oils and lubricants to:

1. Reduce friction between moving metal surfaces resulting in significantly less wear.
2. Cool mechanical parts by removing heat and also decreasing the amount of heat produced [1].
3. Ensure leak tightness between the mechanical components such as cylinders and pistons [1].
4. Keep engine components clean by removing impurities [1].

Lubricant oil always undergoes a mixing process. Mixing is essentially the reduction of the inhomogeneity of a system in order to achieve a desired process result. The inhomogeneity is usually in terms of concentration, phase or temperature. Secondary effects such as mass transfer, reaction, and product properties are usually the critical objectives [1]. Mixing has always been an energy consuming activity of which impeller geometry can play a very important role. In view of the recent trends towards greenhouse gases, the successful design of a stirrer has become extremely important due to the insight of its energy -saving potential [2]. Numerous researchers have investigated the impeller shape effect on power consumption [3-5].

In this paper, the authors utilized the observation technique to propose minor modification on three different standard impellers namely the Pitched blade turbine impeller, the CHEMSHEAR impeller, and the Vertical Flat-Blade Turbine impeller, to achieve better mixing with less energy consumption. The paper only reports the on early stage of the work, which includes the proposed modification and the preliminary CFD results for one of the modified impellers.

[1]Correspondence author: Obai Younis
Email address: obai.younis@taylors.edu.my

Published under licence by IOP Publishing Ltd

2. Methodology

In the current early stage study, mainly two basic methods were applied, observation techniques and CFD. The observation technique was used for the design stage; the standard impellers used in the industry were studied and classified. The chosen impellers for this study were analyzed and the special attention was paid to the strengths and weakness of the design. The numerical simulation CFD was used for performance testing. Numerical simulations of the standard and modified impellers were carried out to compare the performance before and after the suggested modifications.

3. Discussion

3.1. Pitched Blade Impeller (Standard)

This impeller has substantial axial flow with a significant radial component which is designed for wide changes in the viscosity process [1]. The Pitched blade impeller is a type of traditional impeller for larger volume applications. It has a specific role in applications where a degree of fluid shear is beneficial to the overall process result [7]. This impeller has a generally good shearing action at high speed, but normally it does not operate under such conditions. It is power economical, but not efficient when it encounters viscous fluids.

3.2. Pitched Blade Impeller Modification

Mainly, the standard pitched blade impeller is modified by adding attaching baffles to lower surface of the blades. Baffles typically consist of solid surfaces positioned in the path of tangential flows generated by the rotating motion. It has a significant influence on the flow behaviour and resulting mixing quality. Baffles also increase the drag and power draw of the impeller [1]. As inertial forces are developed when the velocity of a fluid changes direction or magnitude due to the baffles, the turbulence of the flow increases. Fluid in motion tends to continue in motion until it meets a solid surface of another fluid moving in a different direction. Forces are developed during the momentum transfer that takes place [7]. Utilising this theory, baffles are attached to the lower surface of the impeller blades to disrupt the flow and cause the forces acting on the impeller blades to fluctuate in a random manner and increase the intensity of the turbulence to achieve better mixing.

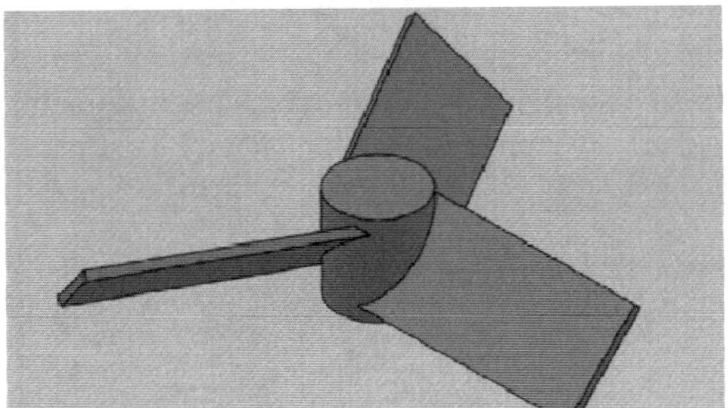

Figure 1. The Pitched blade impeller (standard).

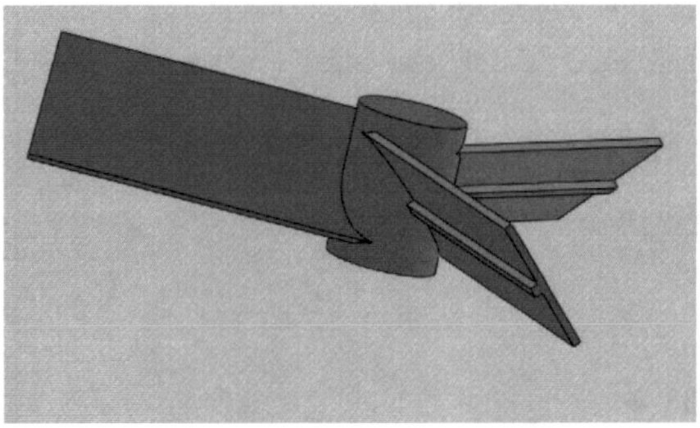

Figure 2. The modified Pitched blade impeller (modified).

3.3. CHEMSHEAR Impeller (Standard)

The ChemShear Impellers are made in 4 different styles. Style 1 has the widest blades and style 4 the narrowest blades. The one chosen for study is style 4 (Figure 3). Various processes require different ratios of shear and pumping [8]. The wider blades provide high pumping capabilities, while the narrower blades have the higher level of shear. This kind of impeller is suitable for applications which require moderately high shear but also some level of pumping. Most of the high shear devices are very low in pumping capability and require an axial impeller to provide turnover. The ChemShear impellers often overcome the need for an axial pumping impeller.

3.4. CHEMSHEAR Impeller Modification

As for this impeller, the focus was on further improving the pumping capabilities without sacrificing the high shear stress to achieve better circulation in the mixing process. The impeller's blades are twisted by a specified angle. Introducing such an angle will result in a changed blade pitched which directly results in changes in the pumping rate and the thrust [7]. However this design needs further investigation to study the affect of the angle on the pumping rate and to find the optimum angle.

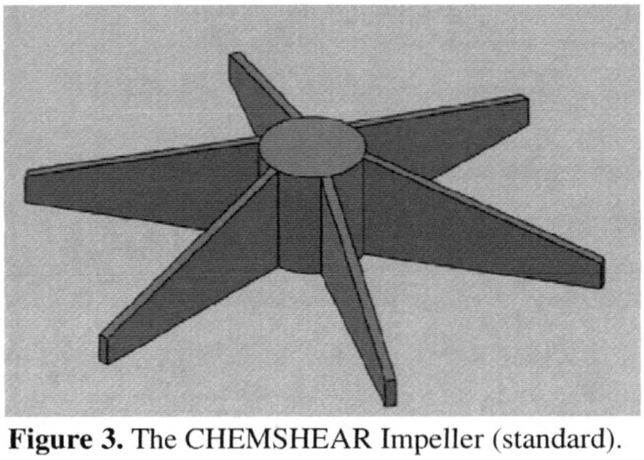

Figure 3. The CHEMSHEAR Impeller (standard).

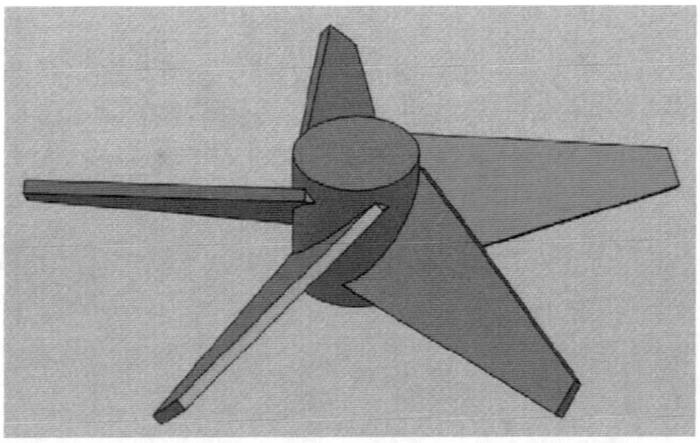

Figure 4. The CHEMSHEAR Impeller (modified).

3.5. Vertical Flat-Blade Turbine Impeller (Standard)

The Vertical flat-blade turbine impeller is a versatile impeller, used in a wide application range. This general paddle acts as horizontal circular flow at low speed and as radial flow at high speed. If there are baffles on the stagnant walls, it results in axial circular flow. It is suitable for mixing, homogenizing, dissolving, heat transfer and crystallization in a low viscosity medium. It also operates in laminar region in high viscosity medium. This design is used widely in low liquid levels or high shear applications with relatively high power consumption [6].

3.6. Vertical Flat-Blade Turbine Impeller Modification

Nozzle is a device designed to control the velocity of a fluid as it passes through the nozzle. The nozzle is often a tube of a varying cross sectional area and can be used to direct or modify the flow of a fluid. Nozzles are frequently used to control the rate of flow, speed, direction, mass, shape, and/or the pressure of the stream that emerges from them [7]. With the increased velocity, the turbulence intensity of the mixing increases. The particles of the fluid are also being dispersed into smaller particles to enhance particle-to-particle collision in mixing. With this ability, this impeller is modified by having the nozzle installed onto the blade to increase the fluid velocity as it emerges from the nozzle and to achieve better dispersion.

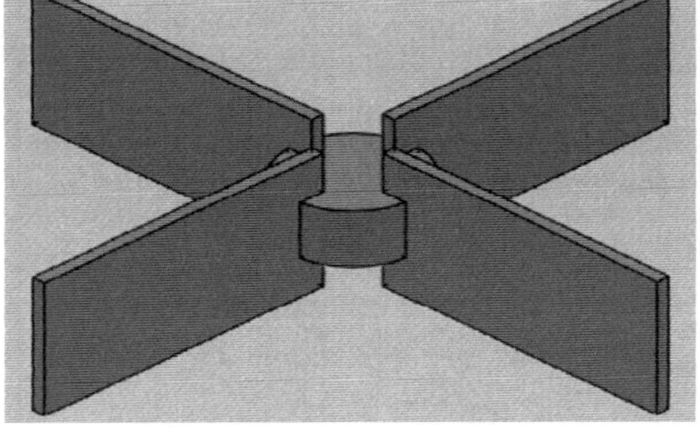

Figure 5. The Vertical Flat-Blade Turbine Impeller (standard).

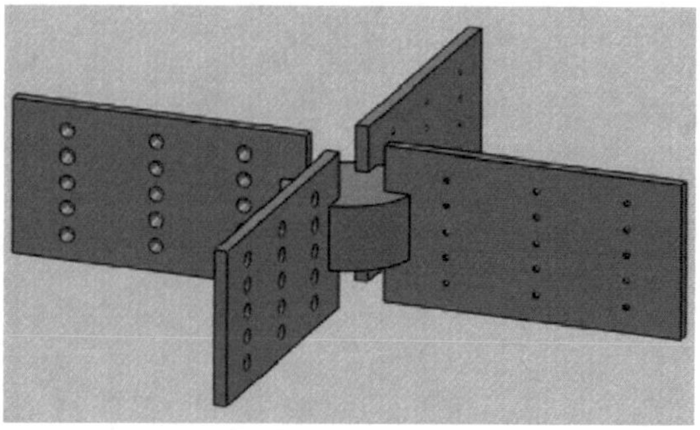

Figure 6. The Vertical Flat-Blade Turbine Impeller (modified).

3.7. Preliminary CFD results

Over the last two decades CFD has became an essential tool for design and testing. Advances in computer technology have led to considerable improvement in the modelling of the mixing process. However, as this paper reports the early stage of the work, only one impeller was simulated using the COMSOL software, namely the pitched blade turbine impeller. The authors mainly focused on the shear stress field as an indicator of fluid.

The shear stress distribution is used to study the radial powers of radial impellers, which is the ability to disperse fluid particles in a mixing process. In Figures 8 and 9, the intensity of shear stress is indicated by colours ranging from red to blue, with red being the highest intensity and blue being the lowest intensity.

It is clear that the intensity of the shear stress of the fluid produced by the modified pitched blade turbine impeller (figure 8) is higher than that of the standard pitched blade impeller (figure 7). It can then be concluded that the modified pitched blade impeller is able to disperse the fluid better than the standard impeller.

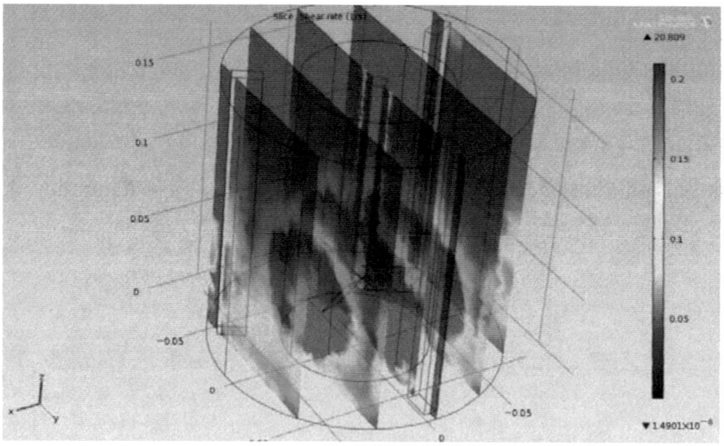

Figure 7. Shear stress distribution of the Standard Pitched Blade impeller.

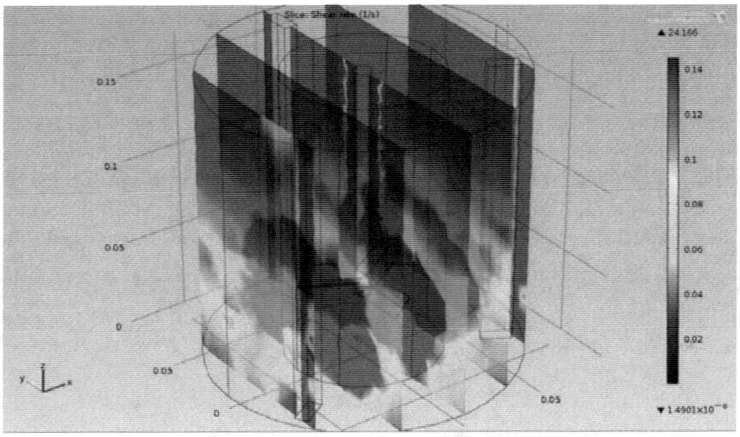

Figure 8. Shear stress distribution of the modified Pitched Blade impeller.

4. Conclusion

The advantages and disadvantages of three different impellers, namely the Pitched blade turbine impeller, The CHEMSHEAR impeller, and the Vertical Flat-Blade Turbine Impeller are analyzed. Accordingly minor modification on the impeller design are suggested and adopted to improve the mixing process as well as to reduce the power consumption. The CFD results of the modified Pitched Blade impeller indicated that the modification results in a better fluid dispersion compared to the standard impeller. However, as this paper only reports on the early stage of the work, more testing and deep analysis is under consideration.

References

[1] Edward L.Paul, Victor A. Atiemo-Obeng, Suzanne M.Kresta 2004 Handbook of Industrial Mixing*: Science and Practical* John Wiley 30.

[2] D.Georgiev, S. D. Vlaev 2006 Flow flied variation by minor modification of impeller blades, *12th Eurpoian on mixing, Bologna* 27-30.

[3] Orvalho S. C. P., Vasconcelos J. M. T., Alves S. S. 2000 Hold-up and gas-liquid mass transfer performance of modified Rushton turbine impellers. In: *Proceedings of 10th European conference on mixing* Delft 461-468

[4] Vlaeve S. D., Mavros P., Seichter P., Mann R. 2002 Operational characteristics of a new energy-saving impeller for gas liquid mixing *Journal of Chemical Engineering* **80** 653-659.

[5] Szalai E. S., Arratia P., Johnson K., Muzzio F. J. 2004 Mixing analysis in tank stirred with Ekato intermig impellers *Chemical Engineering science* **115** 3793-3805.

[6] Simanzhenkov V., Idem R. 2003 Crude Oil Chemistry *Marcel Dekker* New York 59-63.

[7] Coker. A. K 2007 Ludwig's Applied Process Design for Chemical and Petrochemical Plants 4th Edition, 445-447.

[8] Perry, R.H.; Green, D.W. Perry 1997 Chemical Engineer Handbook 7th edition, *McGraw-Hill.*

Four Decades of Utilizing Shadowgraph Techniques to study Natural Convection in Cavities: Literature Review

Syed Mohammed Aminuddin Aftab[1,2], Obai Younis[1] & Musthak Al-Atabi[1]

School of Engineering, Taylor's University, No 1 Jalan Taylor's 47500, Subang Jaya, Selangor, Darul Ehsan, Malaysia.

E-mail[2]: syedaftab2020@gmail.com

Abstract. Natural convection in cavities has been a field of interest to researchers over the past 50 years. One of the basic techniques used to investigate the natural convection in cavities is shadowgraph flow visualisation, used to experimentally observe the boundary layer growth, formation of double layer structures, intrusions and plume generation. This paper accounts for Various other observations also made when fins are placed along the side wall of the cavity, the effect due to change in shape and orientation of the cavity, how these changes effect the natural convection have also been discussed. The applications of various types of shadowgraph methods to understand the flow variation with density have also been included. The paper summarizes the literature of shadowgraph techniques in natural convection.

1. Introduction

Shadows are dark patterns which are cast on a screen when light falls on an object between the source of light and the screen. Shadows have fascinated mankind for ages. In the 17th century, Robert Hooke demonstrated shadowgraphy by showing the shadow of the convective plume of a lit candle on a screen using sunlight as the source [1] using two candles, a lens and an object. In the 20th century, researchers started using this technique to study mixing of water, the atmosphere and rising convection plume from a hand [1]. Schardin used this flow visualisation technique and called it as direct shadow method and later coined the term shadowgraph and the recorded image was called shadowgram [1]. Since then, shadowgraph techniques have been used by many researchers in wide range of applications such as aerospace, ballistics, crystals, explosions, blast, shockwaves, turbo machinery, supersonic jets, Heat ventilation and air conditioning, combustion, geophysics, biomedical, material processing, optical processing and many other applications [1] [2].

It has been observed that edge-tones produced by wood wind instruments can be observed using shadowgraph method [1]. This paper gives a brief insight into the use of shadowgraph in natural convection.

2. Natural Convection

Natural convection is one of the basic heat transfer mechanisms which can be observed in nature and in daily life. It depends on the various properties of fluids such as kinematic viscosity, density, thermal conductivity and specific heat as well as the velocity of the fluid as a result of which the convection heat transfer is more complex [3]. In natural convection the motion of the fluid is induced by buoyancy force, the variation in the fluid density due to the temperature variation makes the fluid lighter or heavier (depending on whether the fluid is being heated or cooled), as a result the lighter fluid moves up and the heavier (denser) fluid moves downward which results in circulation. The heat transfer coefficient which is velocity dependent is as low as the fluid velocities in natural convection which is normally around 1*m/s* [3]. As the velocity of the fluid increases the heat transfer rate increases.

Natural convection is more desired than forced convection in industrial applications as it minimises the operating cost. The main advantage of natural convection over forced convection is that it does not require a fan or pump to move the fluid. Natural convection plays an important role in the manufacture, design and use of electronic devices, heat ventilating and air-conditioning (HVAC) systems within buildings and power plants. In nature, natural convection helps in the heat transfer such mantel convection, oceanic and atmospheric motions are also due to natural convection.

3. Shadowgraph

Shadowgraph is mainly used to visualise the flow which is not possible to realize through the naked eyes, it is useful in visualising turbulence, shock wave formation in transonic flows and supersonic flows. In this method a couple of mirrors (Parabolic, Spherical, Plane First-Surface Folding Mirrors or Cleaning First-Surface Mirrors) or lenses (collimating or condenser) are used, focal length of a mirror or lens determines how strongly the light converges or diverges the beam consequently light source is placed at the focal length of the mirror to obtain parallel beams of light from a point light source. The parallel beam of light then passes through the transparent walls of the test section and is reflected again using a second mirror, the shadow is obtained on the screen or is captured using a CCD camera which is also located at the focal length. The images captured through the CCD camera are directly recorded on a pc and later processed using image processing software. The shadowgraph reveals the second order of the laplacian $\partial^2 y/\partial x^2$ of the refractive field and this shows the inhomogeneities in the best possible way. The shadow which is cast on the screen or CCD camera or photographic plate is known as shadowgram and it should not be confused with an image it is just mere shadow, and there is no 1:1 relation between the object and the shadow, as a result shadows are not true scale to the object.

3.1. Direct Shadowgraphy

Direct Shadowgraphy involves a tiny bright light source, the object and a screen for capturing the shadow. Light source can be a small candle or the one who lights the whole planet (sun), as stated earlier Hooke [1] used this method to see the rising plume from the candle. Direct shadowgraph is further divided into two types:

- Direct Shadowgraphy in Diverging light
- Direct Shadowgraphy in Parallel light

3.1.1. Direct Shadowgraphy in Diverging light

In this method light rays from a point source are used for casting the image of the object on the screen as shown in figure 1. The object is placed at a distance from the screen. The diverging light from the point source which is at a distance from the screen produces the shadow of the object on the screen.

3.1.2. Direct shadowgraph in Parallel light

This method is most widely used as the quality of beam is more superior and focused on the object. To obtain parallel beams of light a collimating lens or parabolic mirrors can be used. The light source and the camera are placed at the focal length of each of the spherical mirrors; the setup is in z-shape as shown in figure 2.

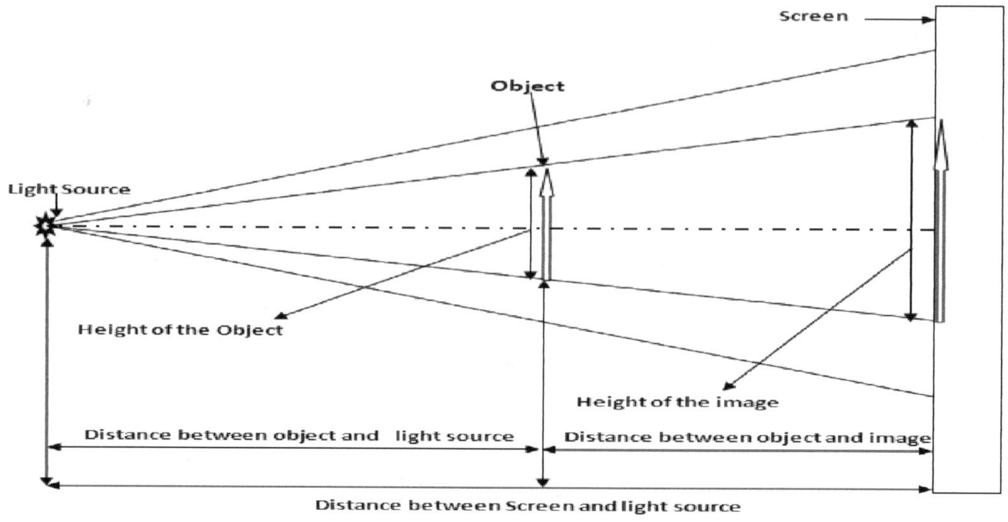

Figure 1. Direct Shadowgraph in diverging light

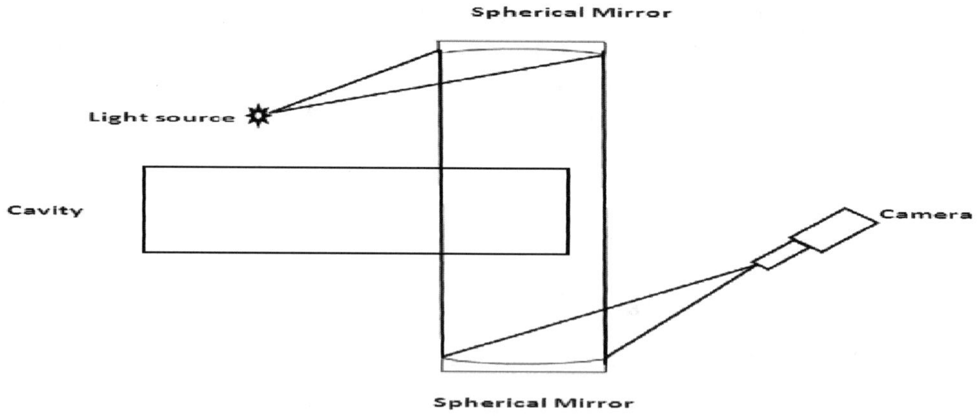

Figure 2. Z-type shadowgraph system

3.2. Focused Shadowgraphy

In this method one or more lenses are used to obtain a focused image, here the lenses are placed in between the object and light source as well as the object and the screen as shown in figure 3. The main advantage of focused Shadowgraphy is that it allows variable magnification of the shadowgram there are limitations of focused Shadowgraphy as well, such as it is limited by the field lens or mirror diameter.

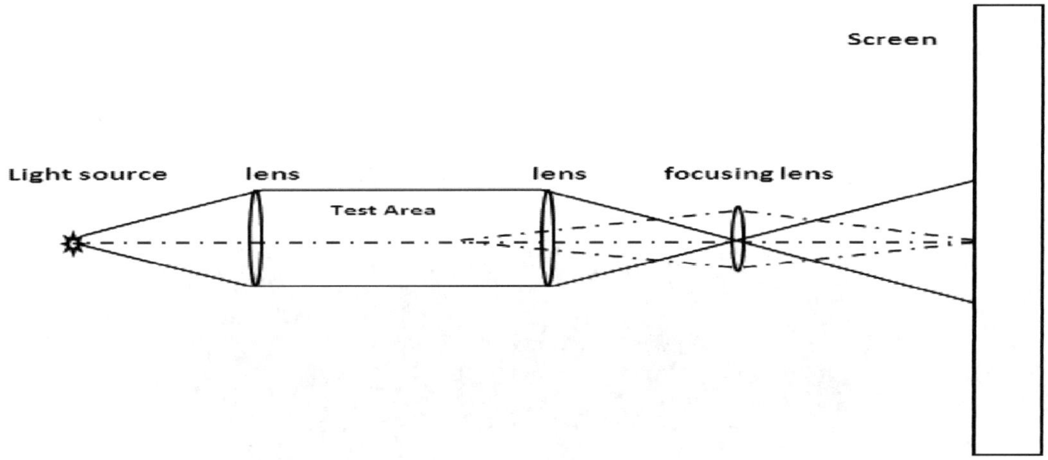

Figure 3. Focused Shadowgraphy

3.3. Distinctive Shadowgraph Techniques

Large scale Shadowgraphy is used for obtaining larger than life shadowgrams. Microscopic Shadowgraphy is used in the study of crystals. Stereoscopic Shadowgraphy is used for the study of the wake turbulence effect and also to visualise shockwaves, holographic Shadowgraphy in the study of aerodynamic testing and computed Shadowgraphy is used to generate images from CFD, they are known as synthetic shadowgrams and conical Shadowgraphy is used in supersonic flows about sharp streamlined bodies, i.e. it's application is mainly in conical flows [1].

4. Applications of shadowgraph in natural convection

This section deals with the shadowgraph in natural convection which includes various applications such as natural convection in differentially heated/cooled cavities and Rayleigh-Benard convection. Shadowgraph technique is used in visualising a water filled square cavity when it is suddenly heated and cooled on the opposite side walls, the top and bottom walls are thermally insulated [5]. In this experiment it was clearly observed for the first time the formation of travelling waves in the hot boundary layer region immediately after the experiment started, the horizontal intrusion and the interaction of these intrusions with the boundary layer was also visualised, the work describes the early stages of flow development [5]. Later further investigations showed the evolution of two travelling wave groups the interaction between group of travelling waves and how steady state is reached. Furthermore, the triggering of thermal stratification in cavity core and vertical boundary layer adjustment is also observed [6].

Work has been carried out on generating artificial shadowgraph images by numerically integrating the path of light through the flow field for comparison of numerical and experimental results, this method can be very helpful in qualitative interpretation of the experimental images, it was observed that the shadowgraph images which were artificially constructed were in good agreement with the experimental images for the experiment which was carried [7].

Comparative study of Schlieren, Shadowgraph and Interferometer in a differentially heated rectangular cavity where the top wall is cooled and bottom wall is heated and the side walls are adiabatic using air and water as the working fluids revealed that, at low temperature gradients the Schlieren results were much better than shadowgraph images and the interferometer. At high temperature gradients both Schlieren and shadowgraph images had good clarity than interferometer due to refraction error and at high Rayleigh number experiments the shadowgraph images were more clear than Schlieren and interferometer [8].

The transition of the boundary layer into double layer structure along the sidewall of a differentially heated cavity is studied. Flow behavior and heat transfer properties in double layer structure are investigated and clearly visualised using shadowgraph technique, the shadowgraph images as shown in figure 4. clearly show the strip formation at top and bottom portion of the wall the meeting of these layers at the mid height and increasing the thickness of the vertical boundary layer, both numerical and experimental results were found to be in agreement and all the three stages of boundary layer growth are studied [9].

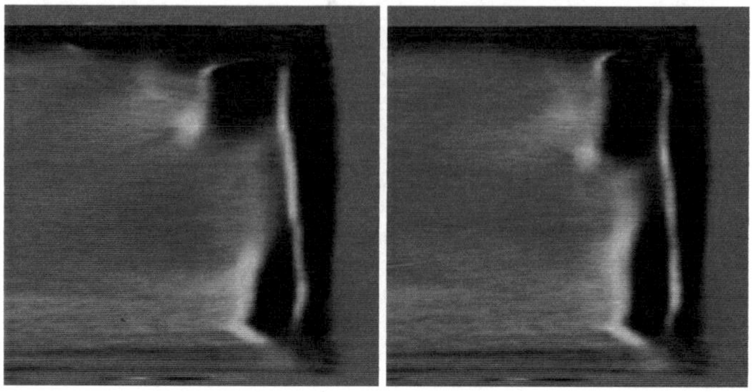

Figure 4. Development of the double layer structure [9]

A square fin is placed on the hot wall as shown in figure 5. inorder to study the effect of square obstruction on the boundary layer flow in a differentially heated cavity. The presence of intrusion (lower and upper) around the head of fin is observed. The boundary layer destabilisation and re-stabilisation due to the intrusion, the comparison of the thermal boundary layer properties with and without the obstruction on the side wall have also been reported, the important finding was the generation of plume head in presence of fin and its collapses and merger with secondary boundary layer [10].

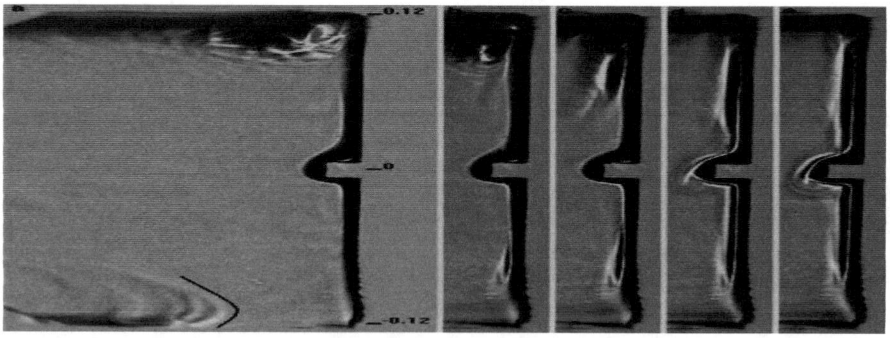

Figure 5. Square fin placed on the side wall [10]

A differentially side heated cavity is studied using shadowgraph, all the three transition stages (initial growth stage, entrainment development stage and quasi-steady stage) have been observed and the reverse flow in the double layer structure due to unstable travelling waves in the quasi-steady state is studied, the mechanism for the travelling waves in quasi steady thermal boundary layer [6] has been answered and demonstrated in this work [11].

The unsteady thermal flow around a fin on a differentially heated cavity is studied, the formation of gravity currents under the fin and plume formation on the fin are observed, the unsteady velocity scales of the lower intrusion, the oscillations of the thermal flow due to the fin has been measured using fast response thermistors and the sensitivity due to geometry of the fin, the enhancement in convection due to the fin which trigger instabilities downstream of thermal boundary layer have been reported [12].

The transient natural convection of variable and constant viscosity fluids in side cooled cavity is studied, sudden cooling of the walls of cavity shows new fluid behaviour in shadowgraph images which were not seen in differential heating as shown in figure 6. and new scaling analysis of the transient behavior is introduced to include the effect of variable viscosity and there was a good agreement between the scaling and the shadowgraph images [13].

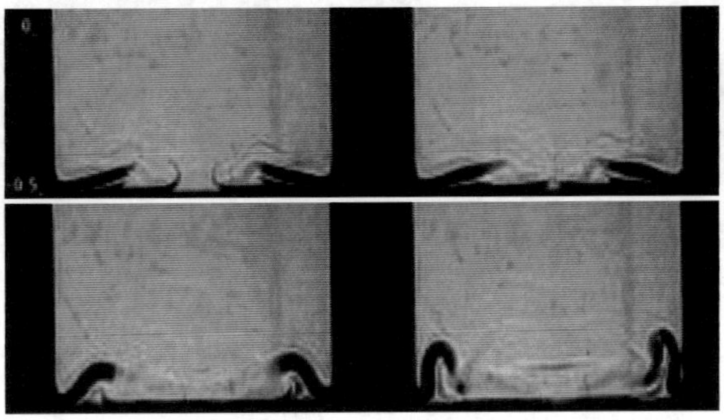

Figure 6. Showing development of horizontal intrusions at different time intervals [13]

Natural convection flow and transport process due to mass transfer in inclined enclosures at different angles is studied, the flow is visualised using particle tracer and shadowgraph, the inclination angle is from 30 to 90 degrees, as the inclination is varied and the aspect ratio of the enclosures is changed the results indicate the change in the flow structure and boundary layer [14].

The heat transfer through triangular cavity is examined in this study. Transient natural convection in water-filled isosceles triangular enclosure is considered with cooling on the inclined walls and heating at the bottom wall. The initial stage showed the development of thermal boundary layer adjacent to the surface, transitional stage showed the convective instabilities in the form of rising and sinking temp variation, the steady state at low Grashof number is characterized by symmetry flows and at higher Grashof number it results in pitchfork bifurcation [15].

5. Conclusions

The above review gives an insight into the area of shadowgraphy and its application in the field of natural convection. The various observations such as the formation of the travelling waves, transition of the boundary layer into double layer structure, transition of thermal boundary layer into three stages, the flow behaviour and heat transfer properties in the boundary layer, and the formation of the intrusions. Shadowgraphy has great contribution in the field of natural convection in cavities leading to new insights and better understanding of various phenomenas.

Shadowgraph is clearly one of the least expensive and easy to setup experimental methods for flow visualisation having wide range of applications.

Acknowledgement

The authors would like to thank the Malaysian government for providing financial support through FRGS (Fundamental Research Grant Scheme) and Taylor's University for sponsoring the PhD candidate.

References

[1] Settles, G. S., 2001, *Schlieren and shadowgraph techniques: visualising phenomena in transparent media*, Berlin: Springer-Verlag.

[2] Yahya, S.M., 2003, *Fundamentals of Compressible Fluid Flow - SI Units*. New age international publishers.

[3] Yunus A. Cengel., 2006, *Heat and Mass Transfer-A Practical Approach*-McGraw Hill.

[4] W.Schopf and J.C.Patterson, 1995, "Natural convection in a side-heated cavity visualisation of the initial flow features", *J. Fluid Mech*, **295**: 357-379.

[5] W.Schopf and J.C.Patterson, 1996, "Visualisation of natural convection in a side-heated cavity: transition to the final steady state", *Int. J. Heat Mass Transfer*, **39**: 3497-3509.

[6] W.Schopf, J.C. Patterson, A.M.H. Brooker, 1996, "Evaluation of the shadowgraph method for the convective flow in a side-heated cavity", *Experiments in Fluids*, **21**, 331-340.

[7] Atual Srivastava, Atanu Phukan, P.k. Panigrahi and K. Muralidhar, 2004 "Imaging of a convective field in a rectangular cavity using interferometry, Schilerene and shadowgraph", *Optics and Lasers in Engineering*, **42**: 469-485.

[8] Feng Xu , J.C. Patterson and Chengwang Lei, 2008, "On the double layer structure of the boundary layer adjacent to a sidewall of a differentially heated cavity", *Int. J. Heat and Mass transfer* **51**, 3803-3815.

[9] Feng Xu , J.C. Patterson and Chengwang Lei, 2006, "Experimental observations of the thermal flow around a square obstruction on a vertical wall in a differentially heated cavity", *Experiments in* Fluids, **40**, 364-371.

[10] Feng Xu, J.C. Patterson and Chengwang Lei, 2005, "Shadowgraph observations of the transition of the thermal boundary layer in a side-heated cavity", *Exeriments in Fluids*, **38**, 770-779.

[11] Feng Xu , J.C. Patterson and Chengwang Lei, 2008, "An experimental study of the unsteady thermal flow around a thin fin on a sidewall of a differentially heated cavity", *Int. J. Heat and Fluid Flow*, **29**, 1139-1153.

[12] Younis, F. Xu, J. C. Patterson, C. Lei, J. Pallares, F. X. Grau, "scaling of transient natural convection cooling in side-cooled cavity- the effect of variable viscosity". Proceedings of the ASME/JSME 2011 8th Thermal Engineering joint conference, Honolulu, Hawaii, USA.

[13] D.Z. Jeng, C.S.Yang and C.Gau, 2009, "Experimental and numerical study of transient natural convection due to mass transfer in inclined enclosures", *Int. J. Heat and Mass transfer*, **52**, 181-192.

[14] Chengwang Lei, Steven W. Armfield, J.C. Patterson, "Unsteady natural convection in a water-filled isosceles triangular enclosure heated from below", *Int. J. Heat and Mass transfer*, **51**, 2637-2650.

Response Surface Design Model to Predict Surface Roughness when Machining Hastelloy C-2000 using Uncoated Carbide Insert

N.H. Razak[1], M M Rahman[1,2] and K. Kadirgama[1]
[1]Faculty of Mechanical Engineering
Universiti Malaysia Pahang
26600 Pekan, Pahang, Malaysia
Phone: +6094242246; Fax: +6094242202
[2]Automotive Engineering Centre
Universiti Malaysia Pahang
26600 Pekan, Pahang, Malaysia

E-mail: mustafizur@ump.edu.my

Abstract. This paper presents to develop of the response surface design model to predict the surface roughness for end-milling operation of Hastelloy C-2000 using uncoated carbide insert. Mathematical model is developed to study the effect of three input cutting parameters includes the feed rate, axial depth of cut and cutting speed. Design of experiments (DOE) was implemented with the aid of the statistical software package. Analysis of variance (ANOVA) has been performed to verify the fit and adequacy of the developed mathematical model. The result shows that the feed rate gave the more effect on the both prediction values of Ra compared to the cutting speed and axial depth of cut. SEM and EDX analyses were performed in different cutting conditions. It can be concluded that the feed rate and cutting force give the higher impact to influence the machining characteristics of surface roughness. Thus, the optimizing the cutting conditions are essential in order to improve the surface roughness in machining of Hastlelloy C-2000.

1. Introduction

Surface roughness is a measure of the technological quality of a product and a factor that greatly influences manufacturing cost. It describes the geometry of the machined surface and combined with the surface texture, which is process dependent, can play an important role on the operational characteristics of the part (e.g. appearance of excessive friction and/or wear). Surface roughness is a commonly encountered problem in machined surfaces. Furthermore a good-quality machined surface significantly improves fatigue strength, corrosion resistance, and creep life [1]. Surface roughness is consisting of a multitude of apparently random peaks and valleys. When two rough surfaces are brought to be in contact, it is occurred in smaller area, which is called the real area of contact. This area is not only a function of the surface topography but also on the study of interfacial phenomena, such as friction and wears [2]. Lee and Ren [3] were explained that surface roughness plays an important role in affecting friction, wear, and lubrication of contacting bodies. Regardless of the method of production, all surfaces have their own characteristics, which are referred to as surface

[2]Correspondence author: M. Mustafizur Rahman
Email address: mustafizur@ump.edu.my

Published under licence by IOP Publishing Ltd

texture [4]. Surface texture is the pattern of the surface which deviates from a nominal surface. The deviations may be repetitive or random and may result from roughness, flaws, and waviness [5]. Therefore, the actual surface profile is the superposition of error of form, waviness, and roughness. Surface roughness is defined as closely spaced, irregular deviations on a scale smaller than that of waviness. Figure 1 shows standard terminology and symbols to describe surface roughness. The profile *p* is the contour of any specified section through a machined surface on a plane that is perpendicular to the surface. Roughness width cutoff *l* (i.e., sampling length) is included in the measurement of average roughness height. The mean line *m* of the profile *p* is located so that the sum of the areas above the line (within the sampling length *l*) is equal to the sum of the areas below the line.

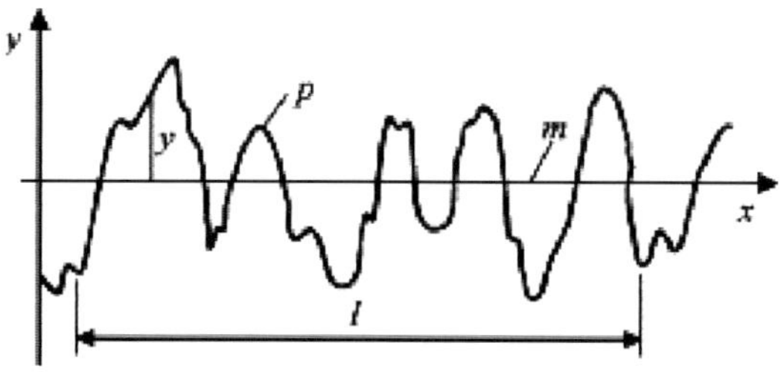

Figure 1. Surface roughness definition.

Despite the different surface finish parameters, the roughness average R_a is the most used international parameter of surface roughness. It is defined as:

$$R_a = \frac{1}{l}\int_0^l |y(x)| dx \qquad (1)$$

where *l* is the sampling length and *y* is the ordinate of the profile curve.For inspecting a surface, several commercially available instruments, called surface profilometers, are used. Lundberg [6] has investigated the effect of surface roughness on the lubricant film characteristics under conditions of combined normal and sliding motion. Besides that, Choudhury and El-Baradie [7] were mentioned that the effect of cutting variables such as speed, feed rate and depth of cut on surface roughness by considering one variable at a time .Nickel-based alloys play an extremely important role in gas turbine engines. In addition to their use in aircraft, marine, industrial and vehicular gas turbines, nickel-based alloys are now also used in space vehicles, rocket engines, experimental aircraft, nuclear reactors, submarines, steam power plants, petrochemical equipment and other high-temperature applications. Kwon et al. [8] proposed a model providing a relationship between surface roughness and tool wear. They concluded that this model can serve for a better utilization of tool in a way that tools can be employed to the fullest extent until they do not achieve the required surface quality. Sahin and Motorcu [9] established first-order and second-order equations (in which the independent variables are logarithmic transformations of speed, feed rate and depth of cut) using response surface methodology in order to predict surface roughness in machining mild steel and reached a conclusion that the main influencing factor on surface roughness is the feed rate. Nickel-based alloys are known as some of the most difficult-to-machinesuperalloys in order to satisfy production and quality requirement [10]. It has been reported that nickel-based alloys strengthened by heat treatment are very sensitive to microstructure change due to their high strength at high temperature, high ductility, high tendency to work hardening, etc [11]. Major changes in the machined surface layer may include: (i) residual stresses (tensile or compressive) induced in machining; (ii) changes in hardness of the surface layer due to work hardening; (iii) microcracking or macrocracking, particularly in grinding (iv) tears, laps and crevice like defects associated with the 'built-up edge' produced in machiningand severe flank wear; (v) plastic deformation as a result of hot or cold work; (vi) metallurgical transformations attributed to high temperature; and (vii) chemical change including high temperature oxygenation and diffusion action between the workpiece and tool materials. Response Surface Method (RSM) is a collection of statistical and mathematical methods that are useful for the modelling and optimization of the engineering problems. In this technique, the main objective is to optimize the responses that are influencing by various parameters [12]. RSM also quantifies the relationship between the controllable parameters and the obtained response. In modelling of the manufacturing processes using RSM, the sufficient data is collected through designed experimentation. In general, a second order regression model is developed because of first order models often give lack off fit. The study uses the Box-Behnken design in the optimization of experiments using RSM to understand the effect of important parameters. Box-Behnken design is normally used when performing non-sequential experiments. That is, performing the experiment only once. These designs allow efficient estimation of the first and second-order coefficient. Because Box-Behnken design has fewer design points, they are less expensive to run than central composite designs with the same number of factors. The RSM is practical, economical and relatively easy for use and it was used by lot of researchers for modeling machining processes [13-14]. Mead and Pike [15] and Hill and Hunter [16] reviewed the earliest study on response surface methodology.The aim of the present study is to develop a surface roughness prediction model for machining Hastelloy C-2000 using uncoated carbide based on response surface method with three cutting paramaters, which are feed rate, axial depth and cutting speed.

2. Experimental Setup

2.1. Design of Experiments

Design of experiment (DOE) is used to reduce the number of experiments and time. The study uses the Box-Behnken design because it has fewer design points and less expensive to run than central composite designs with the same number of factors. Three levels of cutting parameters were selected to investigate the machinability of this alloy which is covering the feed rate range for Hastelloy C-2000 workpiece, 0.1 mm/tooth, 0.15 mm/tooth, 0.2 mm/tooth, different values of depth of cut, such as 0.4 mm, 0.7 mm and 1.0 mm and various cutting speed, 15 m/min, 23 m/min and 31 m/min were selected. Various input parameters in the conduct of the experiments are listed in table 1 and design of experiment is also shown in table 2.

Table 1. Machining parameters and their levels

Destination	Process parameters	level		
		-1	0	1
X1	Feed rate (mm/tooth)	0.1	0.15	0.2
X2	Axial Depth (mm)	0.4	0.7	1
X3	Cutting speed (mm/min)	15	23	31

2.2. Workpiece and Cutting Tool Material

The chemical and physical properties of the workpiece material of Hastelloy C-2000 are given in table 3 and table 4 respectively. The constituents of the workpiece chromium (23%) and molybdenum (16%) being high, the material is hard to machine. Nickel consists of approximately 50%, which makes the alloy suitable for high temperature applications. The dimension of test specimen used in the conduct of the experiments was 46 mm × 120 mm × 20 mm. The test block was annealed and has Rockwell B90 hardness.

Table 2. Design values

Experiment No.	Feed rate (mm/tooth)	Axial Depth (mm)	Cutting speed (m/min)
1	0.15	0.4	31
2	0.15	1	15
3	0.1	0.7	15
4	0.2	1	23
5	0.2	0.7	31
6	0.15	0.7	23
7	0.15	0.7	23
8	0.2	0.7	15
9	0.1	0.4	23
10	0.15	1	31
11	0.15	0.4	15
12	0.1	0.7	31
13	0.1	1	23
14	0.15	0.7	23
15	0.2	0.4	23

Table 3. Chemical composition of workpiece material (Hastelloy C-2000)

Ni	Cr	Mo	Fe	Cu	Al	Mn	Si	C
BAL	23%	16%	3%	1.60%	0.50%	0.50%	0.08%	0.01%

Table 4. Physical properties of workpiece material (Hastelloy C-2000) at room temperature

Parameters and unit	Value
Density (g/cm^3)	8.5
Thermal conductivity (W/m°C)	9.1
Mean coefficient of thermal expansion (μm/m°C)	12.4
Thermal Diffusivity (cm^2/s)	0.025
Specific heat (J/kg°C)	428
Modulus of elasticity (GPa)	223

The experimental study was carried out in wet cutting conditions on a CNC milling machine by slotting machining equipped with a maximum spindle speed of 4000 rpm, feed rate of 5.1 m/min and a 5.6-kW drive motor. The cutting tool insert used to cut the material was uncoated carbide, grade designation K15, with 6 % composition of Co and the rest was tungsten (WC) and the grain size was 1 μm. The following are the details of the tool geometry of inserts when mounted on the tool holder: (a) special shape ; (b) axial rake angle: 19.5°; (c) radial angle: 5°; and (d) sharp cutting edge. The material was machined with 1 pass (120mm) in direction of *Y*and stop. Then the surface roughness of the material was assessed by perthometer manufactured by Mahr (Surf PS1). Six observations were taken for each sample and were averaged in order to get the value of roughness (Ra).CNC milling machine, workpiece set up in the machine and uncoated carbide insert are shown in figure 2.

(a) workpiece at CNC milling machine

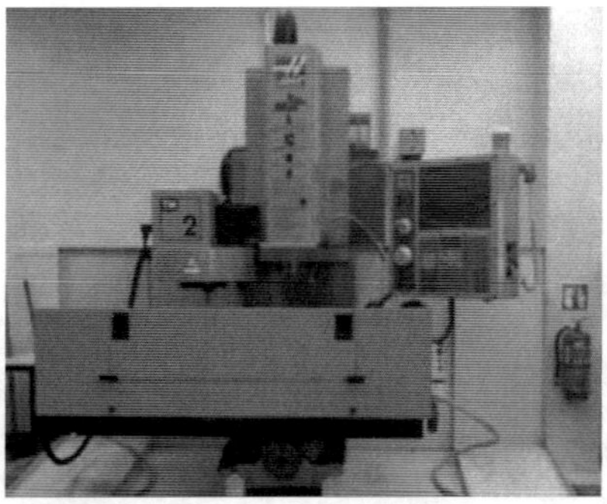

(b) CNC milling machine

(c) Shape of uncoated carbide

Figure 2. Experimental set up and shape of uncoated carbide insert.

2.3. Response Surface Method

The main objective is to optimize the response surface that is influenced by various process parameters. RSM is quantified the relationship between the input parameters and the obtained response surfaces [12,17]. The second-order polynomial mathematical model for surface roughness is developed as Eq. (2):

$$Y = C_o + \sum_{j=1}^{k} C_j x_j + \sum_{j=1}^{k} C_{jj} X_j^{\,2} + \sum_{i<} \sum_{j=2}^{k} C_{ij} X_i X_j \tag{2}$$

where Y is the corresponding response (surface roughness,SR) yield by the various variables and X_1(1,2,3…n) are coded levels of n quantitative process variables,the term C_0, C_j, C_{jj} and C_{ij} are the second order regression coefficients. Equation (2) can be written as Eq. (3) :

$$\begin{aligned} Y &= C_0 + C_1X_1 + C_2X_2 + C_3X_3 + C_{11}X_1^{\,2} + C_{22}X_2^{\,2} + C_{33}X_3^{\,2} + C_{12}X_1X_2 \\ &+ C_{13}X_1X_3 + C_{23}X_2X_3 \end{aligned} \tag{3}$$

where $X_{1,}X_2, X_3$ are feed rate (mm/tooth), axial depth (mm) and cutting speed (m/min) respectively. The equations of the fitted model for SR are represented in Eq. (4):

$$\begin{aligned} Y &= 1.42033 + 0.28350X_1 + 0.38988X_2 - 0.34488X_3 + 0.00583X_1^{\,2} - 0.63692X_2^{\,2} \\ &+ 0.20708X_3^{\,2} + 0.02875X_1X_2 - 0.28175X_1X_3 - 0.32300X_2X_3 \end{aligned} \tag{4}$$

The analysis of variance is presented in table 5. The adequacy of the second-order model is verified using ANOVA. At a level of confidence of 95%, the model is checked for its adequacy. Based on table 5, the model is adequate due to the fact that the P values of lack-of-fit are not significant. This implies that the model could fit, and it is adequate. Therefore, the model is acceptable and there is some indicator to measure the effectiveness of the model that built in the value of surface roughness prediction data.

Table 5. Analysis of variance for second order surface roughness (R_a)

Source of Variation	Degree of freedom	Sum of squares	Mean of squares	F ratio	P value
Regression	9	5.2950	0.58834	7.89	0.017
Linear	3	2.8105	0.93684	12.56	0.009
Square	3	1.7464	0.58213	7.81	0.025
Interaction	3	0.7382	0.24605	3.30	0.116
Residual Error	5	0.3729	0.07458		
Lack of fit	3	0.1546	0.05153	0.42	0.733
Pure error	2	0.2183	0.10914		
Total	14	5.6679			

4. Results and Discussion

Figure 3 shows the effect of surface roughness against the feed rate and cutting speed. It can be seen that the feed rate has the most dominant effect on the surface roughness, followed by the axial depth and the cutting speed. This situation can be explained by the decrease in cutting forces resulting from the decrease in feed rate. Smaller cutting forces cause less vibration and provide better surface finish (Figure 3a). It is clear from figure 3(b) that surface at roughness increases with the decrease in feed rate. A low feed rate uniform the outer surface thus increasing the surface finish. Another factor to consider is cutting speed. It is understood that an increase in cutting speed improves surface quality. This result supports the argument that high enough cutting speeds reduce cutting forces together with the effect of natural frequency and vibration, giving better surface finish [18]. Hence, a better surface roughness can be obtained by employing high cutting speed, low axial depth and low feed rate [19]. Figure 4 shows the predicted results are closely agree with the experimental values. Therefore, the model of the response surface method can be accepted as well.

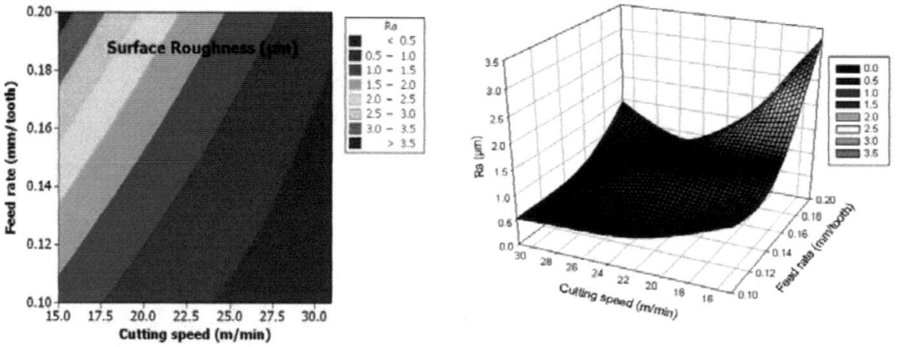

(a) Contour plot 2D (b) Surface plot 3D

Figure 3. Variation of surface roughness against cutting speed and feed rate (a) 2D contour ; (b) 3D surface plot

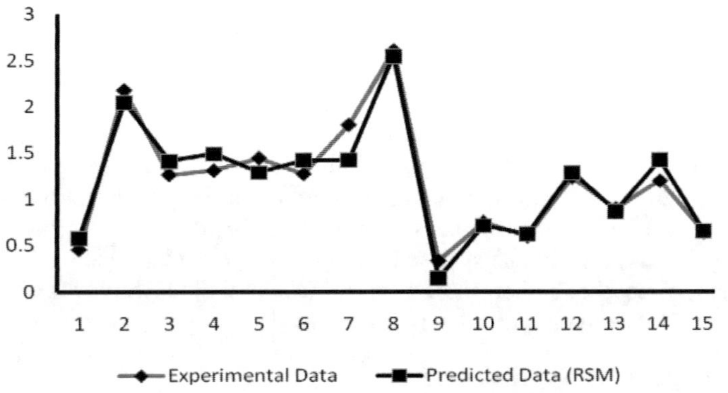

Figure 4. Comparison between the experimental and predicted results of surface roughness.

Figure 5 shows the SEM viewing and EDX with different magnification level at low cutting speed (15m/min) and medium feed rate (0.15 mm/tooth) and maximum axial depth (1.0mm). It can be seen for an experimental 10 where the low cutting speed (15m/min) and medium feed rate (0.15 mm/tooth) and maximum axial depth (1.0mm) produces inferior surface roughness (0.745 μm).

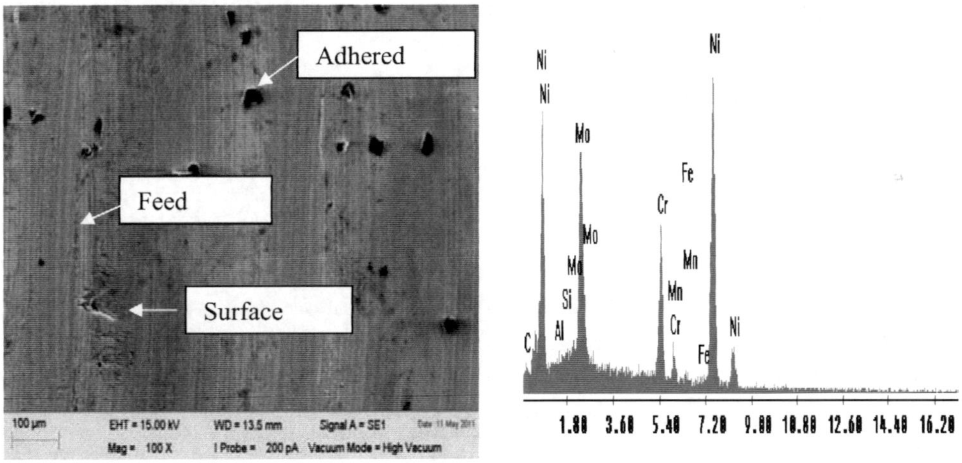

(a) SEM Viewing (b) EDX picture

Figure 5. SEM viewing for magnification level 100x and EDX result (Experiment no. 10) at low cutting speed (15m/min) and medium feed rate (0.15 mm/tooth) and maximum axial depth (1.0mm)

From figure 5, there are many surface defects on the surface texture such as feed marks, chip redeposition to the surface, adhered material particles and surface cavities. It has been reported that when the thermal softening of the material is increased, compressive stresses also increase and such surface flaws clear out of the machined surface, as well as enabling the workpiece near- surface to reconstruct itself easily [20]. These types of defects were observed by different researchers in many different nickel- and titanium-alloys such as Ni Cr20 TiAl [21], IN-718 [22-23]. Besides that, figure 6 (a) SEM viewing for 100x and (b) EDX result shows the surface texture of experiment 4 where the maximum cutting speed is 23 m/min, maximum feed rate is 0.2 mm/tooth and maximum axial depth is 1.0 mm produces 1.308 μm that is inferior than experimental 10.

IOP Conf. Series: Materials Science and Engineering **36** (2012) 012022 doi:10.1088/1757-899X/36/1/012022

Figure 6(a) shows the formation of carbide cracking, smearing, and distributing marks appear on the machined surface. This phenomenon is called carbide cracking, and it causes a sudden increase in the shear stress during cutting that leads to surface cavities due to plucking. This process causes residual cavities and cracks to be formed inside the machined surface, causing even further problems. As a result, carbide cracking can be a serious problem in terms of micro-scale surface integrity. Especially when the depth of cut and feed values are very small, the carbide particle sizes become too close to a concerning level that carbide cracking might gain significant importance in the surface of the end product. From the observations based on table 6, it shows that the the experimental 10 gives the superior value of surface roughness based on the feed rate value which is lower than experimental 4. The feed rate gives the higher impact to influence the machining characteristics of surface roughness, followed by axial depth and cutting forces.

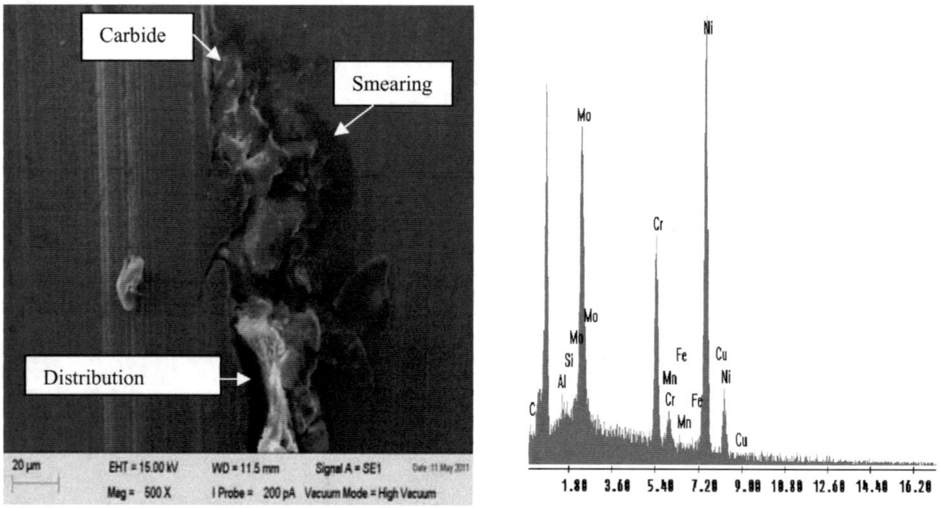

(a) SEM Viewing (b) EDX result

Figure 6. SEM viewing for magnification level 100x and EDX result (Experiment no. 4) at low cutting speed (23 m/min) and medium feed rate (0.2 mm/tooth) and maximum axial depth (1.0 mm).

Table 6. Result of surface roughness during the machining process.

Article I.No.of experiment	Feed rate (mm/tooth)	Axial Depth (mm)	Cutting Speed (m/min)	Surface Roughness (µm)
4	0.2	1.0	23	1.308
10	0.15	1.0	15	0.745

5. Conclusions

In this paper, RSM has been used to determine the prediction of surface roughness by machining Hastelloy C-2000 with uncoated carbide for various input parameters namely the feed rate, axial depth and cutting speed. The feed rate has the most dominant effect on the surface roughness, followed by the axial depth and the cutting speed. Higher value of feed rate decreases the surface quality of and may contribute to surface defect such as surface flaw and cavities. The RSM model can successfully relate the above process parameters with the response surface roughness. Thus, the optimizing the cutting conditions are essential in order to improve the surface roughness in machining.

Acknowledgments
The authors are grateful to the Faculty of Mechanical Engineering, Universiti Malaysia Pahang for financial support under Graduate Research Scheme (Project No. RDU090398).

References

[1] Stark G A and Moon K S 1999 *Trans. of the ASME, J. Manuf. Sci. Eng.* 121, 251.
[2] Bhushan B 1999 *Handbook of Micro-Nano Tribology*. CRC Press.
[3] Lee S C and Ren N 1996 *Tribol. Trans.* 39(1), 67.
[4] Kalpakjian S and Schmid S R 2001 *Manufacturing Engineering and Technology* (New Jersey: Prentice-Hall).
[5] Lou S 1997 *Development of Four in-Process Surface Recognition Systems to Predict Surface Roughness in End Milling,* Ph.D. Thesis, Iowa State University, Iowa.
[6] Lundberg J 1995 *Tribol. Int.* 28(5), 317.
[7] Choudhury I A and El-Baradie M A 1997 *J. Mater. Process. Technol.* 67(1-3), 55.
[8] Kwon Y, Ertekin Y and Tseng T 2004 *Mach. Sci. Technol.* 8(1), 39.
[9] Şahin Y and Motorcu A R 2005 *Mater. Des.* 26(4), 321.
[10] Field M 1968 *Machining Aerospace Alloys, Iron and Steel Institute,* Special Report 94, 151.
[11] Niemi R M 1971 *Integrity Prediction*, SME Technical Paper, Dearbon, Michigan.
[12] Montgomery C 1997 *Design and Analysis of experiments* (4th ed.) (New York: Wiley).
[13] El Baradie MA 2007 *J. Eng. Manufacture* 43.
[14] Hasegawa M A Seireg and Lindbergm R A 1976 *Trib. Int.* 285.
[15] Mead R and Pike D J 1975 *Biometrics* 31, 803.
[16] Hill W J and Hunter W G 1966 *Technometrics* 8, 571.
[17] Kwak J S 2005 *Int. J. Mach. Tools Manuf.* 45, 327.
[18] Sturesson P O, Hakansson H, Claesson L I 1997 *Mech. Syst. Signal Process.*11, 459.
[19] Ezugwu E O and Wang Z M 1996 *Progress of Cutting and Grinding* 111, 102.
[20] Pawade R S, Joshi S S, Brahmankar P K and Rahman M 2007 *J. Mater. Process. Technol.*159,192.
[21] Zou B, Chen M, Huang C and An Q 2009 *J. Mater. Process. Technol.* 209, 5802.
[22] Sharman A R C, Hughes J J and Ridgway K 2004 *Machining Sci. Technol.* 8(3), 399.
[23] Dudzinski D, Devillez A, Moufki A, Larrouquerre D, Zerrouki V and Vigneau J 2004 *Int. J. Machine Tools Manufac.* 44, 439.
[24] Che-Haron C H and Jawaid A 2005 *J. Mater. Process. Technol.* 166, 188.

Effect of Propellant Composition to the Temperature Sensitivity of Composite Propellant

Amir Aziz [1], Rizalman Mamat and Makeen Amin
Faculty of Mechanical Engineering, Universiti Malaysia Pahang, 26600 Pekan, Pahang.

Wan Khairuddin Wan Ali
Department of Aeronautical Engineering, Universiti Teknologi Malaysia 81310 UTM Skudai, Johor.Malaysia.

E-mail: amiraziz@ump.edu.my[1]

Abstract. The propellant composition is one of several parameter that influencing the temperature sensitivity of composite propellant. In this paper, experimental investigation of temperature sensitivity in burning rate of composite propellant was conducted. Four sets of different propellant compositions had been prepared with the combination of ammonium perchlorate (AP) as an oxidizer, aluminum (Al) as fuel and hydroxy-terminated polybutadiene (HTPB) as fuel and binder. For each mixture, HTPB binder was fixed at 15% and cured with isophorone diisocyanate (IPDI). By varying AP and Al, the effect of oxidizer- fuel mixture ratio (O/F) on the whole propellant can be determined. The propellant strands were manufactured using compression molded method and burnt in a strand burner using wire technique over a range of pressure from 1 atm to 31 atm. The results obtained shows that the temperature sensitivity, a, increases with increasing O/F. Propellant p80 which has O/F ratio of 80/20 gives the highest value of temperature sensitivity which is 1.687. The results shows that the propellant composition has significant effect on the temperature sensitivity of composite propellant

1. Introduction

The effect of propellant composition on burning characteristic of composite propellant has been long recognized. Bozic et al [1] reported that, the propellant composition is one of the several parameter that influence the temperature sensitivity. An understanding of this effect is important for improving the capability to the selection of the proper formulations to satisfy specific needs. The previous studies proved that temperature sensitivity can be significantly lowered by utilized larger percentage of aluminum [2]. Then, several researches [1, 3-4] mention that varying oxidizer-fuel mixture ratio will give a different value of temperature sensitivity. The objectives of this study are to determine the temperature sensitivity of four different compositions of aluminized AP/HTPB composite propellants. The scope included measuring burning rate over a range of pressure from 1atm to 31atm, while varying the percent of aluminum and ammonium perchlorate.

Published under licence by IOP Publishing Ltd

2. Methodology

2.1. Propellant selection composition

The specific impulse (Isp) and the burning rate are among the most important performance parameters of propellants. High Isp is required to obtain optimum performance of a propellant.The theoretical Isp was estimated using NASA CEC71 program [5], a computer program for rocket performance. The same program was also applied by several researchers [6-7] to estimates the performance characteristics of AP/HTPB based propellant. The propellant used in this experiment was composite propellant consisted of ammonium perchlorate (oxidizer), aluminium (fuel) and hydroxy terminated polybutadiene, (HTPB) binder (fuel). Due to the poor processibility and mechanical properties of the propellant formulation with higher solid loading fraction [3, 8] and reduced Isp of the propellant with low oxidizer fraction, the range of composition tested was limited. Only four composition were selected and finalized which have O/F; 60/40, 66/34, 74/36 and 80/20 as shown in Table I. For each of the compositions, the percentage of HTPB binder was set constant at 15%. Initially, the percentage of Al used was 25% in the mixture and subsequently decrease until it reached 5% with proportional increase in AP. The formulation codes were named according to the percentage of AP applied.

Table 1. Formulation of the propellant.

Formulation code	AP (%)	AL (%)	O/F
p80	80	5	80/20
p74	74	11	74/26
p66	66	19	66/34
p60	60	25	60/30

2.2. Propellant preparation

The first step in mixing process is to produce the binder by mixing HTPB with IPDI accordingly. The ingredients were mixed together using a glass stirring rod in an agitating and swirling motion, similar to the method reported by Matthew Stephens et al. [9]. Al was then added to the binder and blended together until all the Al powder was coated by the binder. Next, the AP was added according to the desired formulation and the mixture was again stirred until a uniform consistency was achieved. Without degassing, the propellant was compressed into a straw mould which has 70mm length and 5mm diameter. There were 2 main reasons for choosing these dimensions. Firstly, to minimize the pressure and temperature increase in the strand burner combustion chamber during testing. Depending on the composition of mixture, combustion of 2.54 cm (1 inch) strand could increases as much as 10-20% pressure inside the burner [10]. Secondly, this small size reduces material cost and improves safety. Each of 100 grams of mixtures can produce approximately 27 strands of propellants. The length of the strand burnt is the importance parameter to be measured for burning rate calculation, while the size and shape of the strand is less significant [10-11]. The strands were then transferred to the oven and cured at 64°C for 5 days. The strands were visually inspected and were rejected if cracks, pores, or shape irregularities were seen. A cross-sectional view of the propellant was observed under SEM photographs as shown in Fig. 1 in order to investigate the structure of propellant matrix.

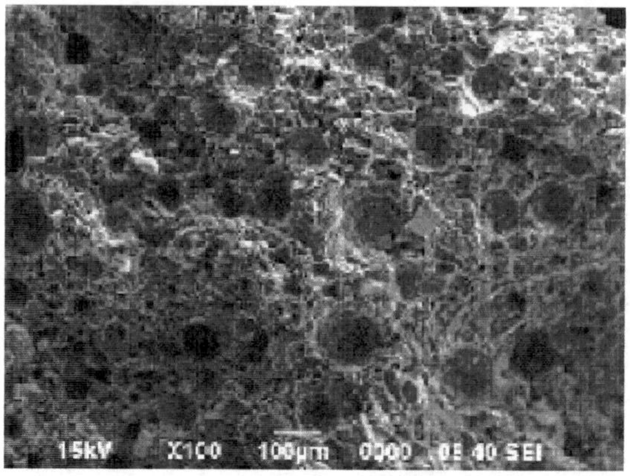

a) Magnified 100X.

b) Magnified 1000X

Figure 1. Cross sectional view of propellant p80

3. Burning rate measurement

The burning rate measurement was carried out with a strand burner which was pressurized by nitrogen gas. The strand burner was designed to handle test pressures up to 38atm (550 psi). The body, flange and both end cap are made of low carbon steel. The 23 cm long cylinder has an inner diameter of 10 cm and an outer diameter of 13 cm, offering a wall thickness of 1.5 cm thickness. Each end cap is 1.5 cm thick, making the overall length of the burner 26 cm as shown in Fig. 2(a). Both end caps are square with side length of 21 cm. The propellant strand was fixed to the end cap using strand holder made of 5mm low carbon steel nut. Nitrogen and combustion products escape to the atmosphere via a stainless steel proportional relief valve as shown in Fig. 2(b).

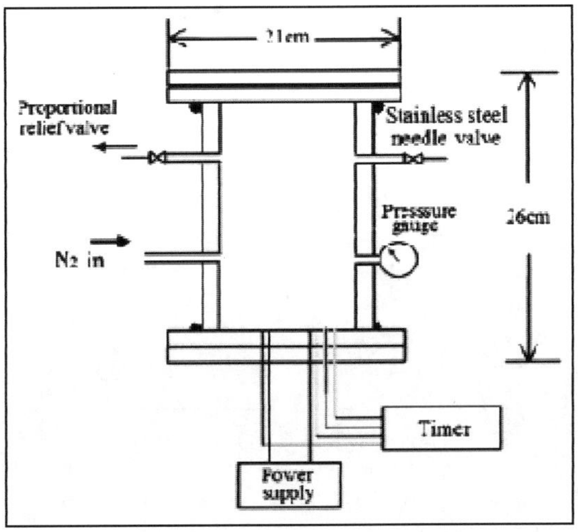

a) Assembly view.

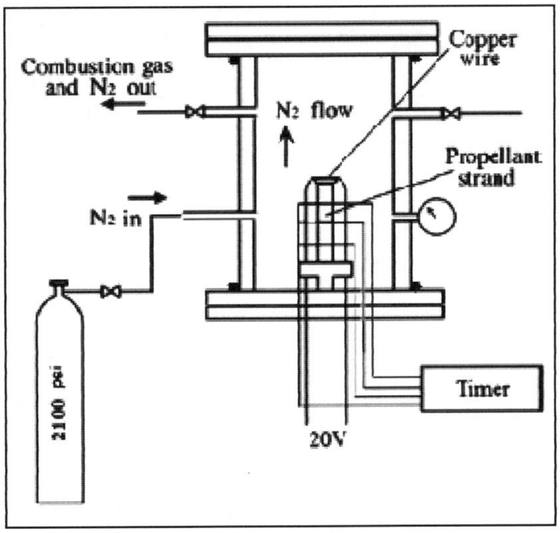

b) Schematic diagram.

Figure 2. The strand burner facilitie

Wire technique was used to measure the burning rate. Each strand was 5 mm diameter and 70 mm in length. Three small holes were accurately placed along the strand length using a needle. An igniter and two fuses wires were passed through these holes and connected to a power supply and electronic timer respectively. All wires were of the same type of 38 S.W.G. tinned copper wire with 0.152 mm thickness, the same type of igniter wire used by Rodolphe et al.[10].The strand is mounted vertically and is ignited at the top end using electrical current. The burning rate was measured for several combustion pressure ranging from 1 atm to 31 atm. It was determined from the period it took for both fuses separated at a distance of 50 mm apart to cut-off. Both ends were left with 10mm distance to avoid extinction transient [12]. Generally, three strands established the burning rate at a single pressure level and repeatability of the burning rates was observed within 5% and is acceptable according to K Jayaraman et al. [13]

4. Result and discussion

The data of temperature sensitivity for each of formulations were collected as shown in Table 2 and were used to plot the graph in Figure 3 and Figure 4.

Table 2. Temperature sensitivity of the propellant.

Formulation code	Temperature sensitivity, *a*
p80	1.687
p74	1.578
p66	1.466
p60	1.455

Figure 3 shows that with small quantity of Al, i.e. less than 19%, will has little effect on the combustion mechanism of propellant, further increasing the amount of Al has resulted in the decreasing in temperature sensitivity. These observation also confirm the observation made by T Godai and M Shimizu[8]. The discussion from previous study [14-15] reported that, the presence of aluminum particles in propellant act as heat sinks by absorbing energy released by gas phase reactions and decrease the burning rate. As a result, reduces the burn rate temperature sensitivity. It means that, the propellant becomes less sensitive to the temperature.

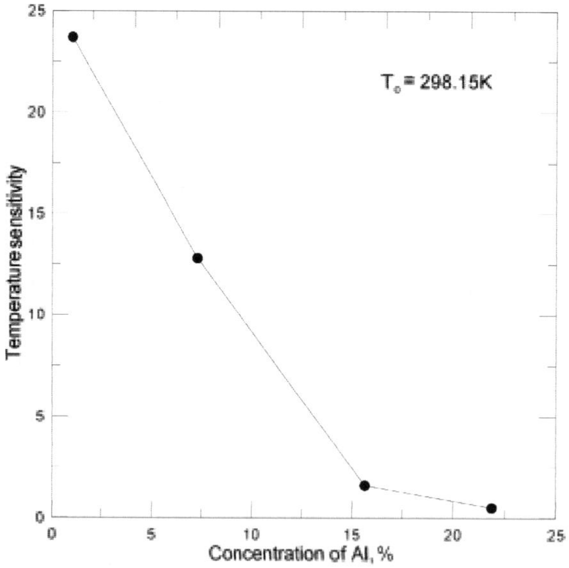

Figure 3. Effect of Al concentration on temperature sensitivity

Figure 4 shows that the temperature sensitivity will increase by increasing the percentage of ammonium perchlorate because of the fast reaction rate due to the high flame temperature of propellants containing higher loading of oxidizer [8]. A experimental result in Figure 4 shows that by increasing 20% of the oxidizer from p60 to p80, increase the pressure sensitivity by 12% which is similar with the study made by Makoto and Hirotatsu [16].

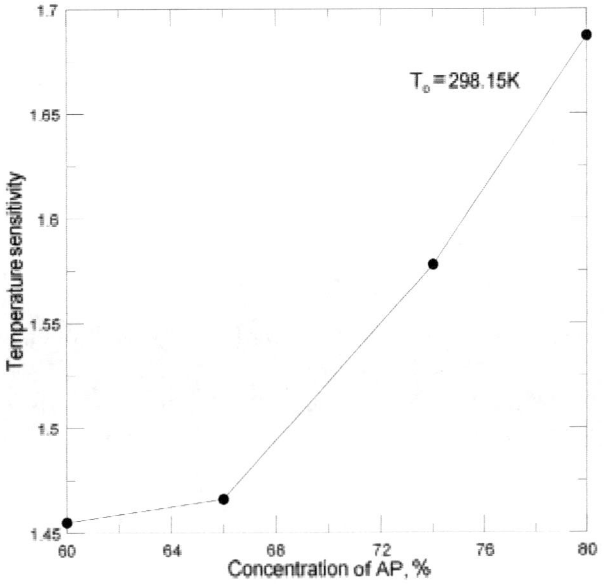

Figure 4. Effect of AP concentration on temperature sensitivity

5. Conclusion

The relationship between the temperature sensitivity in burning rate rule and different propellant composition of composite propellant was experimentally investigated. The results obtain shows that, the temperature sensitivity increase with increasing oxidizer-fuel ratio.

6. Acknowledgment

The authors gratefully acknowledge the support for this work through University Technology Malaysia and Malaysia Space Agency (ANGKASA) under vote no. 68730 for funding this project.

7. References

[1] M. V. M. a. V. S. Bozic, *"Small Motor Measurement method for Determining Burning Rate of Solid Rocket Propellants,"* presented at the 34th AIAA/ASME/SAE/ASEE Joint Propulsion Conference & Exhibit, Cleveland, OH, 1998.

[2] J. P. Renie and J. R. Osborn, *"Temperature and Pressure Sensitivity of Aluminized Propellants,"* presented at the AIAA/SAE/ASME 16th Joint Propulsion Conference, Hartford, Connecticut, 1980.

[3] G. P. Sutton, *Rocket Propulsion Element Sixth Edition*: John Willy & Sons, MIT, 1992.

[4] R. Mamat, *"Ciri-Ciri Propelan Berasaskan Kalium,"* Faculty of Mechanical Engineering, Universiti Teknologi Malaysia, Johor Bharu, 2002.

[5] B. J. M. a. S. Gordon, *"Computer Program For Calculation of Complex Chemical Equilibrium Compositions and Applications,"* NASA1996.

[6] K. K. S. Krishnan and Rajesh, *"Erosive Burning of Ammonium Perchlorate/Hydroxyl-Terminated Polybutadiene Propellants under Supersonic Crossflows,"* presented at the 37th AIAA/ASME/SAE/ASEE Joint Propulsion Committee Conference and Exhibit, Salt Palace - Salt Lake City, Utah, 2001.

[7] M. T. a. C. N. S. Hayakawa, *"An Effect of Oxidizer Particle Size on Combustion Stability in Composite Propellants,"* in 36th AIAA/ASME/SAE/ASEE Joint Propulsion Conference & Exhibit, Huntsville, AL, 2000.

[8] T. G. a. M. Shimizu, *"Pressure Exponent of Controllable Solid Rocket Propellants,"* presented at the AIAA/SAE 8th Joint Propulsion Specialist Conference, New Orleans, Louisiana, 1972.

[9] T. S. Matthew Stephens, Rodolphe Carro, Alexander LePage, David Reid, Sudipta Seal, and Eric Petersen, *"New Additives for Modifying the Burn Rate of Composite Solid Propellants,"* presented at the 42nd AIAA/ASME/SAE/ASEE Joint Propulsion Conference & Exhibit, Sacramento, California, 2006.

[10] M. S. Rodolphe Carro, Jason Arvanetes, Andreiev Powell, and Eric Petersen, *"High-Pressure Testing of Composite Solid Propellant Mixtures: Burner Facility Characterization,"* in 41st AIAA/ASME/SAE/ASEE Joint Propulsion Conference & Exhibit, Tucson, Arizona, 2005.

[11] P. L. S. J. A. Steinz, and M. Surmerfield, *"The Burning Mechanism Of Ammonium Perchlorate-Based Composite Solid Propellants,"* Princeton University Department Of Aerospace And Mechanical Sciences1969.

[12] R. S. Fry, *"Solid Propellant Test Motor Scaling,"* Chemical Propulsion Information Agency. Columbia, The John Hopkins University2001.

[13] K. V. A. K. Jayaraman, S.R. Chakravarthy and R. Sarathi, *"Effect of nano-aluminium in plateau-burning and catalyzed composite solid propellant combustion,"* Combustion and Flame vol. 156, pp. 1662-1673, 2009.

[14] R. E. H. a. J. R. Osborn, *"The Effect of Aluminum Concentration on Motor Temperature Sensitivity,"* in AIAA 26th Aerospace Sciences Meeting, Reno, Nevada, 1988.

[15] M. Q. B. a. J. C. Mullen, *"Burning-rate behavior in aluminized wide-distribution AP composite propellants,"* Combustion, Explosion, and Shock Waves, vol. Volume 47 pp. 200-208, 2011.

[16] M. K. a. H. Tsuzuki, *"Burning Characteristics of Composite Propellant Using Ammonium Perchlorate Modified by Ethylene Glycol,"* in 45th AIAA/ASME/SAE/ASEE Joint Propulsion Conference & Exhibit, Denver, Colorado, 2009

Power Plant

Thamir K. Ibrahim[1] and M M Rahman[1,2]
[1]Faculty of Mechanical Engineering
Universiti Malaysia Pahang
26600 Pekan, Pahang, Malaysia
Phone: +6094242246; Fax: +6094242202
[2]Automotive Engineering Centre
Universiti Malaysia Pahang
26600 Pekan, Pahang, Malaysia

E-mail: mustafizur@ump.edu.my

Abstract. In this paper, the parametric study of a two shafts gas turbine cycle model of the power plant was proposed. The power output, compression work, specific fuel consumption and thermal efficiency are evaluated with respect to the cycle temperature and compression ratio for a typical set of operating conditions. Two shafts gas turbine cycle with realistic parameters is modeled. The computational model was developed utilizing the MATLAB codes. Turbine work found to be decreases as ambient temperature increases as well as the thermal efficiency decreases. It can be seen that the thermal efficiency and power output increases linearly with increases of compression ratio while decreases of ambient temperature. The power of the simulated two shafts gas turbine reach to 135MW, which is higher than the simple gas-turbine cycle (Baiji gas turbine power plant, power < 131MW). The specific fuel consumption increases with increases of ambient temperature as well as the lower turbine inlet temperature. Even though at the lower turbine inlet temperature is decrement the thermal efficiency dramatically and the power output increases linearly with increases of compression ratio and decreases the ambient temperature.

1. Introduction

In a two-shaft gas turbine having an adjustable nozzle between the compressor high-pressure turbine and the load turbine, a control system is provided in which a biasing signal is added to the fuel temperature control increasing its base reference so that on an increase in called for speed or load the gas turbine is able to dynamically respond in an improved manner [1]. The steady-state temperature limits are exceeded transiently without damage to the gas turbine until the system returns to its new steady-state condition. Also provided for the two-shaft gas turbine is acceleration limiting control whereby the overall gas turbine is allowed to accelerate at the maximum allowable value of either of the two shafts. Further provision is made for a blow off valve control whereby excess energy from a regenerator is diverted from the turbines so that the speed and load control will remain in command, thus preventing over speed conditions and/or subsequent shutdown [2]. The gas turbine has experienced phenomenal progress and growth since its first successful development in the 1930s. The early gas turbine build in the 1940s and even 1950s had simple-cycle efficiencies of about 17 percent

[1,2]Correspondance author: M. Mustafizur Rahman
Email address: mustafizur@ump.edu.my

Published under licence by IOP Publishing Ltd

because of the low compressor and turbine efficiencies and low turbine inlet temperature due to metallurgical limitations of those times [3]. Therefore, gas-turbines found only limited use despite their versatility and their ability to burn a variety of fuel. The efforts to improve the cycle efficiency concentrated in three areas [4]:

i. Increasing the turbine inlet (or firing) temperatures
ii. Increasing the efficiencies of turbo machinery components.
iii. Adding modifications to the basic cycle.

Gas turbines manufactured by General Electric in the early 1990s had a pressure ratio of 13.5 and generated 135.7 MW of net power with a thermal efficiency of 33% in a simple cycle operation. A more recent gas turbine manufactured by General Electric used a turbine inlet temperature of 1425 ºC and produced up to 282 MW while achieving a thermal efficiency of 39.5% in the simple cycle mode [5]. Accordingly, gas turbines are rapidly becoming the choice for current and future power generation systems because they offer efficient fuel conversion, reduced cost of electricity, low installation and maintenance cost, can be put in service with minimum of delay time, occupy less room than other plants for the same capacity, and the ability to consume wide range of hydrocarbon fuels [6-7]. Their exhaust gases are relatively less pollutant to the environment, and can be used for preheating air before entering the combustion chamber, or for district heating as in combined heat and power plants [8].
Gas turbine performance can be qualified with respect to its efficiency, power output, specific fuel consumption and work ratio. There are several parameters that affect its performance, such as the compressor pressure ratio, the combustion inlet temperature and the turbine inlet temperature (TIT) [9]. Obviously, these parameters have been actually improved by various gas turbine manufactures, as it has been mentioned above. Meanwhile, operating parameters such as ambient temperature, altitude, humidity, inlet and exhaust losses also affect the performance of gas turbine units. Basically, gas turbines draw air directly from the atmosphere. Its performance is undoubtedly changed by anything that would affect the air density or mass flow rate of the air intake into the compressor. Consequently, a parametric study on the effect of operation conditions requires managing the operation conditions of the system. Thus, the aim of the present work is to develop a strategy to determine the performance of two shaft gas turbine power plant utilizing the effect of operating conditions.

2. Thermodynamic Modelling

The basic of a two shaft gas turbine is consists of a compressor where air is compressed adiabatically and directed to combustion chamber where the fuel is burned with the air, the maximum cycle temperature occurring at state 3 as a result from combustion. The exhaust gases that product from combustion process expanded adiabatically in the high pressure turbine, part from developed work in the high pressure turbine being used to drive the compressor, the remainder being delivered to drive the gas turbine generator. The additional turbine used to drive the gas turbine generator and can call power turbine, so that the gases can expand back to the pressure at the compressor inlet. The two shaft gas turbine illustrated in figure 1, where the compressor is keyed on the high pressure turbine shaft, whereas, the low pressure turbine is keyed on a second shaft that is mechanically separated from the first and activates the gas turbine user. The first assembly is called the gas generator and the second turbine is called the power turbine.

It is assumed that the compressor efficiency (η_c) and turbine efficiency (η_t). The ideal processes and actual processes are represented in dashed line and continuous line, respectively, on the *T-s* diagram as shown in figure 2. These parameters in terms of temperature are defined as [10].

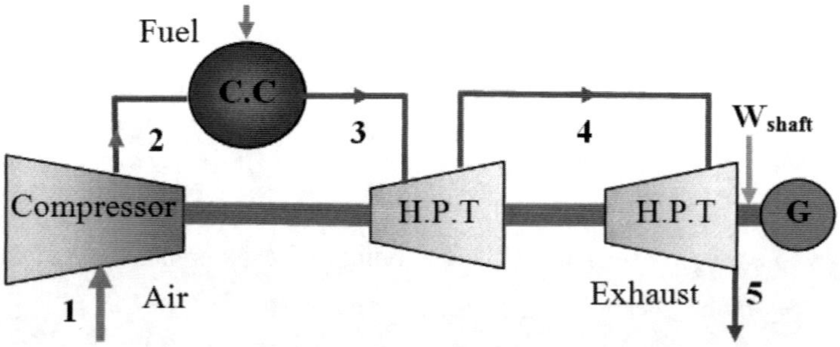

Figure 1. Schematic diagram of two shaft gas turbine

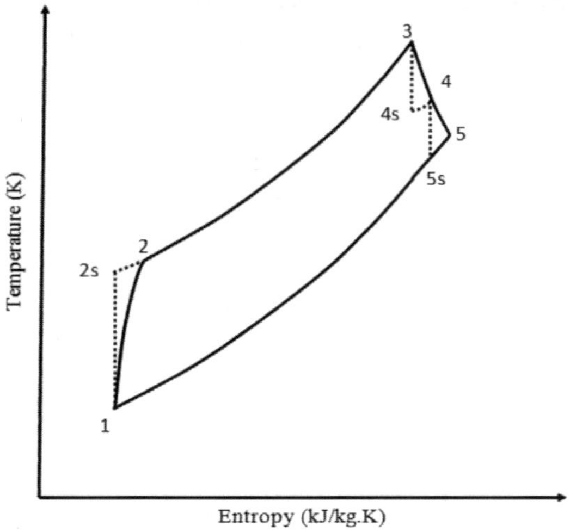

Figure 2. *T-s* diagram of two shaft gas turbine cycle.

$$\eta_c = \frac{T_{2s} - T_1}{T_2 - T_1} \tag{1}$$

$$\eta_{HPT} = \frac{T_3 - T_4}{T_3 - T_{4s}}, \qquad \eta_{LPT} = \frac{T_4 - T_5}{T_4 - T_{5s}} \tag{2}$$

$$\eta_{LPT} = \eta_{HPT} = \eta_t \tag{3}$$

where η_{HPT} and η_{LPT} represent the high and low pressure turbine efficiency respectively.
The final temperature of the compressor is calculated from Eq. (4) [11]:

$$T_2 = T_1\left(1 + \frac{r_p^{\frac{\gamma_a - 1}{\gamma_a}} - 1}{\eta_c}\right) \tag{4}$$

Therefore it can be simplified these relations by Eq. (5):

$$Rpa = \frac{r_p^{\frac{\gamma_a - 1}{\gamma_a}} - 1}{\eta_c}, \quad Rpg = \left(1 - \frac{1}{(r_p)^{\frac{\gamma_g - 1}{\gamma_g}}}\right) \tag{5}$$

where $\gamma_a = 1.4$ and $\gamma_g = 1.33$.

The work of the compressor (W_c) when blade cooling is not taken into account can be calculated as:

$$W_c = \frac{C_{pa} \times T_1 \left(r_p^{\frac{\gamma_a - 1}{\gamma_a}} - 1 \right)}{\eta_m \times \eta_c} = \frac{C_{pa} \times T_1 \times Rpa}{\eta_m} \tag{6}$$

where C_{pa} is the specific heat of air which can be fitted by Eq. (7) for the range of 200K<T<800K and η_m is the mechanical efficiency of the compressor and turbine [11]:

$$\begin{aligned} C_{pa} &= 1.0189 \times 10^3 - 0.13784 T_a + 1.9843 \times 10^{-4} T_a^2 \\ &+ 4.2399 \times 10^{-7} T_a^3 - 3.7632 \times 10^{-10} T_a^4 \end{aligned} \tag{7}$$

where $T_a = \frac{T_2 - T_1}{2}$ in Kelvin.

The specific heat of flue gas (C_{pg}) is given by Naradasu et al. [11]:

$$\begin{aligned} C_{pg} &= 1.8083 - 2.3127 \times 10^{-3} T + 4.045 \times 10^{-6} T^2 \\ &- 1.7363 \times 10^{-9} T^3 \end{aligned} \tag{8}$$

From the energy balance in the combustion chamber is expressed as:

$$\dot{m}_a C_{pa} T_2 + \dot{m}_f \times LHV + \dot{m}_f C_{pf} T_f = (\dot{m}_a + \dot{m}_f) C_{pg} \times TIT \tag{9}$$

where, $\dot{m}_f$ is fuel mass flow rate (kg/s), $\dot{m}_a$ is air mass flow rate (kg/s), LHV is low heating value, T_3=TIT = turbine inlet temperature, C_{pf} is specific heat of fuel and T_f is temperature of fuel. After manipulating Eq. (8), the fuel air ratio (f) is expressed as Eq. (10)

$$f = \frac{\dot{m}_f}{\dot{m}_a} = \frac{C_{pg} \times TIT - C_{pa} T_2}{LHV - C_{pg} \times TIT} \tag{10}$$

The exhaust gases temperature from high gas turbine is given by Eq. (11) [12].

$$\mathrm{T}_4 = TIT \left(1 - \eta_t \times \left(1 - \frac{1}{r_p^{\frac{\gamma_g - 1}{\gamma_g}}} \right) \right) = TIT (1 - \eta_t \times Rpg) \tag{11}$$

The exhaust gases temperature from low gas turbine is given by Eq. (12).

$$T_5 = T_4 \left[1 - \eta_t \left(1 - \frac{T_{5s}}{T_4} \right) \right] \tag{12}$$

where T_{5s} = adiabatic temperature at exit low pressure turbine, is then defined as Eq. (13).

$$T_{5s} = \frac{\left[T_3 - C_{pa} T_1 \left(\frac{r_p^{\frac{\gamma_a - 1}{\gamma_a}}}{\eta_c} \right) / C_{pg} \eta_m \right] \times \left[TIT / \left\{ TIT - C_{pa} T_1 \left(\frac{r_p^{\frac{\gamma_a - 1}{\gamma_a}}}{\eta_c} \right) / C_{pg} \eta_m \eta_t \right\} \right]}{r_p^{\frac{\gamma_a - 1}{\gamma_a}}} \tag{13}$$

The net work is expressed as in (14) [13]:

$$W_{\text{net}} = C_{pg} (T_4 - T_5) \tag{14}$$

In the combustion chamber, the heat supplied by the fuel is equal to the heat absorbed by air, Hence,

$$Q_{\text{add}} = C_{pg} (TIT - T_4) \tag{15}$$

The power output is,

$$\text{Power} = \dot{m}_g \times W_{net} \tag{16}$$

where $\dot{m}_g$ = exhaust gases mass flow rate.
The specific fuel consumption is expressed as (17):

$$\text{SFC} = \frac{3600}{\text{AFR} \times W_{\text{net}}} \tag{17}$$

Further the thermal efficiency of the cycle is then expressed as in (18).

$$\eta_{th} = \frac{W_{\text{add}}}{Q_{\text{net}}} \tag{18}$$

4. Results and Discussion

The cycle was modelled using the thermodynamic analysis for the simple gas turbine and two shaft gas turbine. The pressure losses are assumed negligent in this study. The effects of compression ratio across the compressor, turbine inlet temperature, ambient temperature and the air fuel ratio on the first-law efficiency and power are obtained by the energy-balance approach using MATLAB software.

4.1. Effect of Ambient Temperature

Figure 3 shows the effect of ambient temperature and air fuel ratio on thermal efficiency of two shaft gas turbine cycle. Turbine inlet temperature, compression ratio and the components efficiency are of 1425 K, 12 and 0.9 respectively. It can be seen that the thermal efficiency decreases with increases of ambient temperature while increases of air fuel ratio. The specific work of the compressors increases as the ambient temperature increases [10]. Thus the thermal efficiency for the two shaft gas turbine cycles is reduced. It can be noticed that the gain of thermal efficiency decrease of 3.6% with increases of air fuel ratio from (40-60). The density of air increases when ambient temperature decreases, which causes an increase in the air mass flow rate. The power and heat supplied to the combustion chamber increases with the gas mass flow rate increases. However, the power output increases is more than the increases in the heat supplied in the combustion chamber. The simulated thermal efficiency of a two shaft gas turbine is higher than the thermal efficiency of simple gas turbine cycle as shown in figure 4.

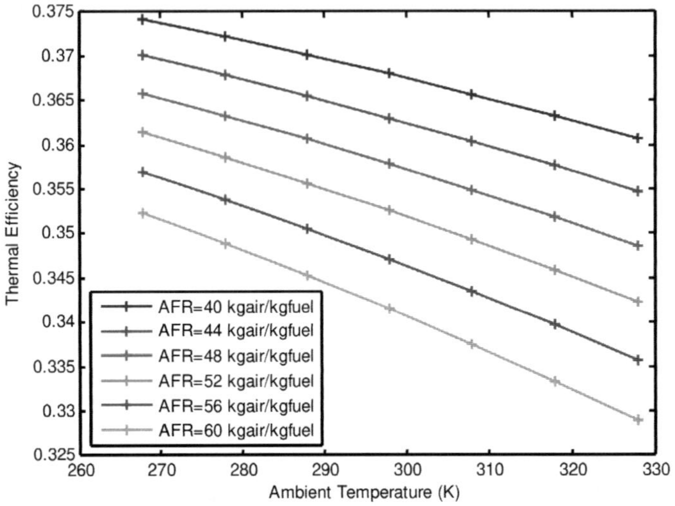

Figure 3. Effect of ambient temperature and air fuel ratio on thermal efficiency of two shaft gas turbine cycle.

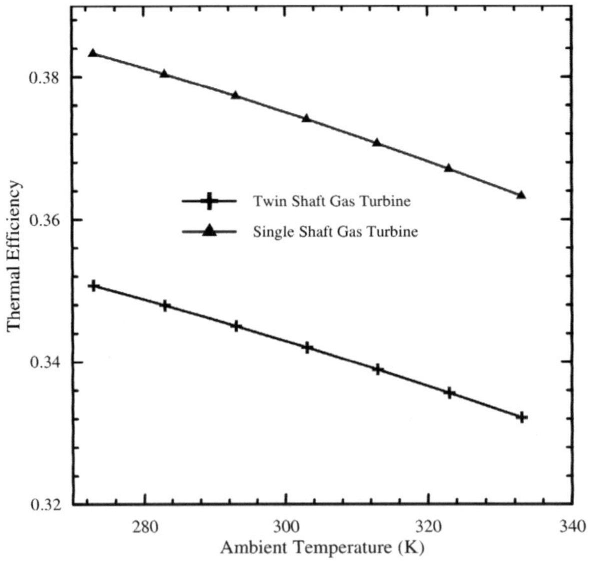

Figure 4. Effect of ambient temperature on thermal efficiency of two shaft and simple gas turbine cycle.

Figure 5 shows the variation of gas turbine power output with air fuel ratio as well as ambient temperature. It is noticed that the power output decreases linearly with increase of ambient temperature as well as increases the air fuel ratio. The simulated power output of a two shaft gas turbine is higher than the simple gas turbine cycle and practical data from Baiji gas turbine power plant as shown in figure 6. Figure 7 shows the effect of ambient temperature and air fuel ratio on specific fuel consumption. It can be seen that the specific fuel consumption increase linearly with increase of both ambient temperature and air to fuel ratio.

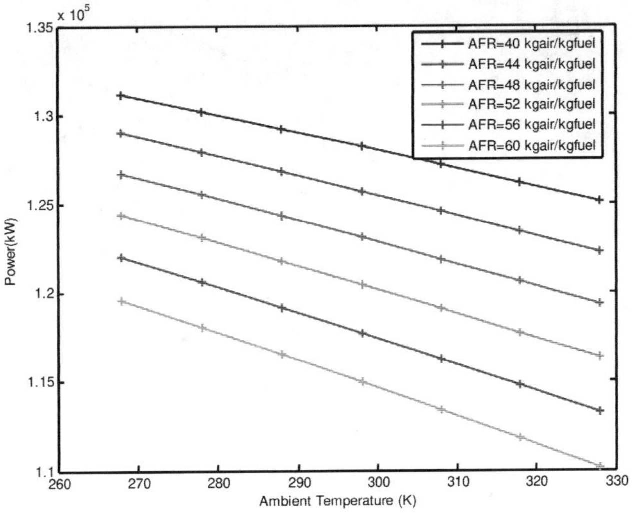

Figure 5. Effect of ambient temperature and air fuel ratio on power output of two shaft gas turbine cycle.

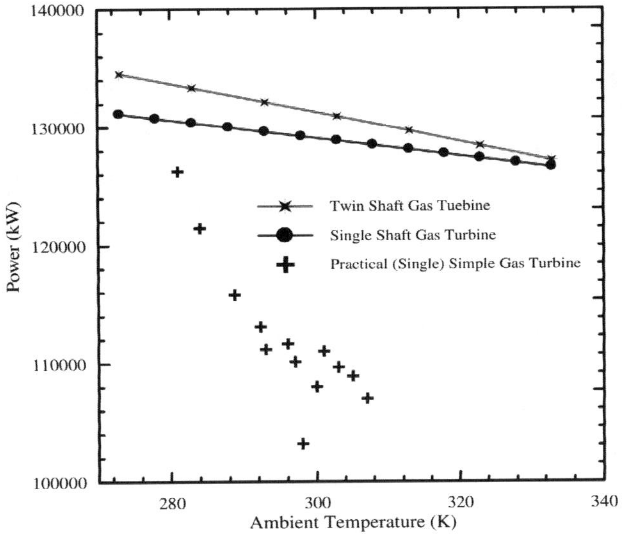

Figure 6. Effect of ambient temperature and air fuel ratio on power output of different gas turbine cycles.

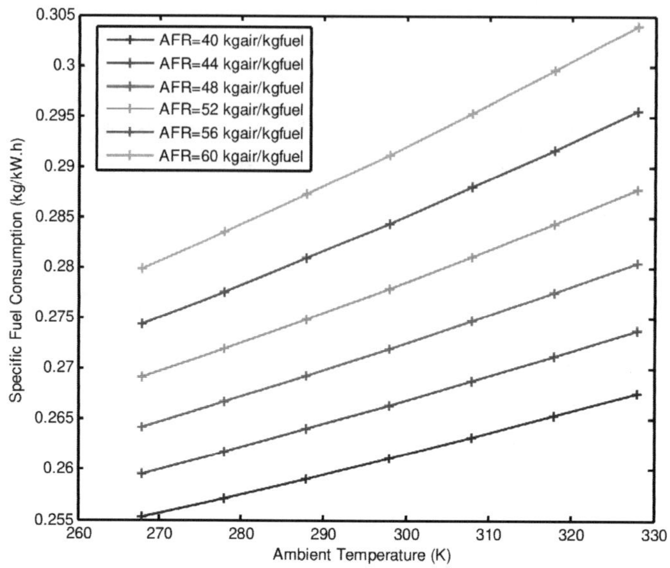

Figure 7. Effect of ambient temperature and air fuel ratio on specific fuel consumption of two shaft gas turbine cycle.

4.2. Effect of Compression Ratio

Figure 8 shows the variation of two shaft gas turbine cycle thermal efficiency with compression ratio and ambient temperature. The thermal efficiency increases with increases of compression ratio as well as lower ambient temperature. However, the variation of efficiency is insignificance at lower compression ratio. The thermal efficiency at high compression ratio decreases from 43.8% to 41% when the ambient temperature increases from 268 to 328K.

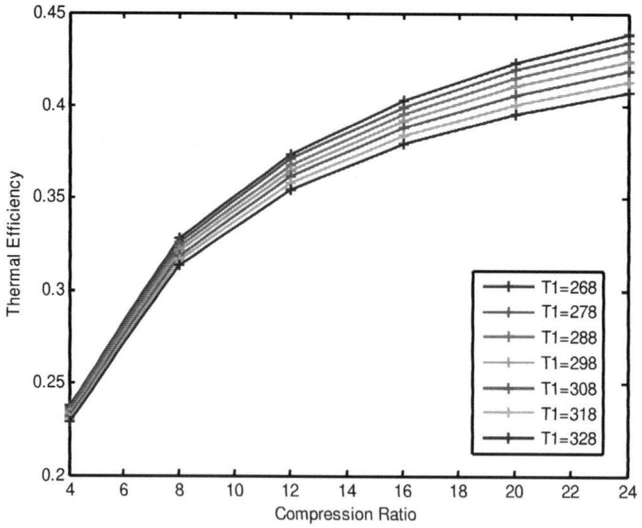

Figure 8. Effect of compression ratio and ambient temperature on thermal efficiency of two shaft gas turbine cycle.

The variation of two shaft gas turbine cycle power output with compression ratio and ambient temperature shown in figure 9. It can be seen that the power output insignificance at lower compression ratio and it significance at high compression ratio. The power output at high compression ratio decreases from 149MW to 137MW when the ambient temperature increases from 268 to 328K. Figure 10 shows effect of compression ratio and ambient temperature on specific fuel consumption of two shaft gas turbine cycle. The specific fuel consumption decreases from 0.4 to 0.22 kJ/kW.K when the compression ratio increases from 4 to 24. The effect of variation of SFC is more significance at higher ambient temperature than lower temperature.

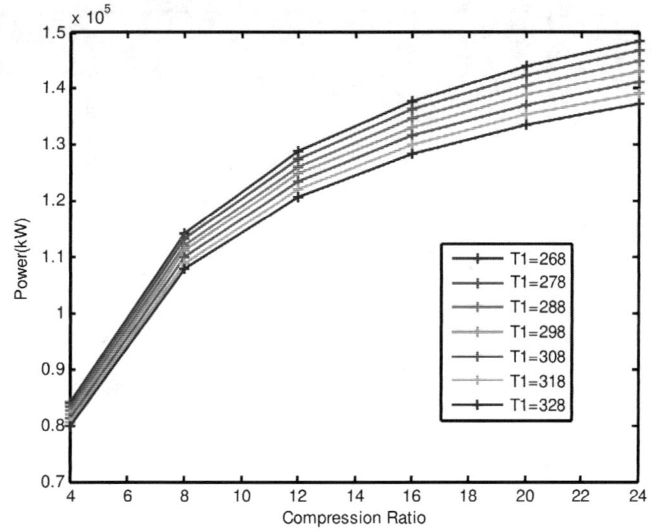

Figure 9. Effect of compression ratio and ambient temperature on power output of two shaft gas turbine cycle.

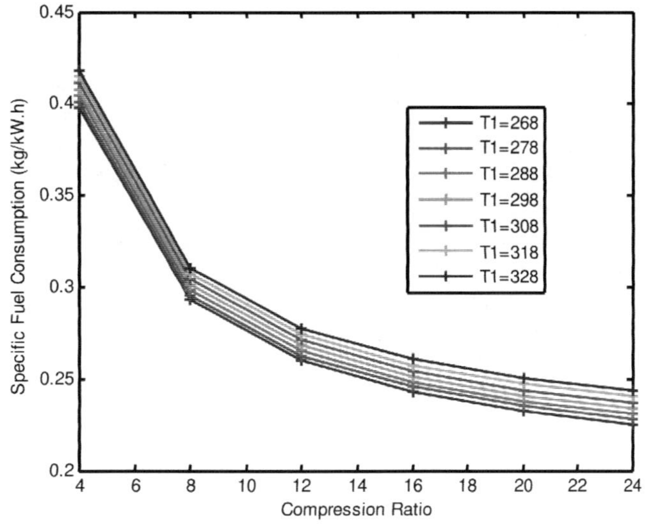

Figure 10. Effect of compression ratio and ambient temperature on specific fuel consumption of two shaft gas turbine cycle.

4.3. Effect of compressor and Turbine Isentropic Efficiency

Figure 11 shows the effect of the compressor isentropic efficiency on thermal efficiency for different ambient temperature. The thermal efficiency increases linearly with increase of the compressor isentropic efficiency. This means that the thermal losses reduced in compressor which leads to increase the power output. However, the variation of thermal efficiency is significance at lower compressor isentropic efficiency. The thermal efficiency decreases with increase of ambient temperature.

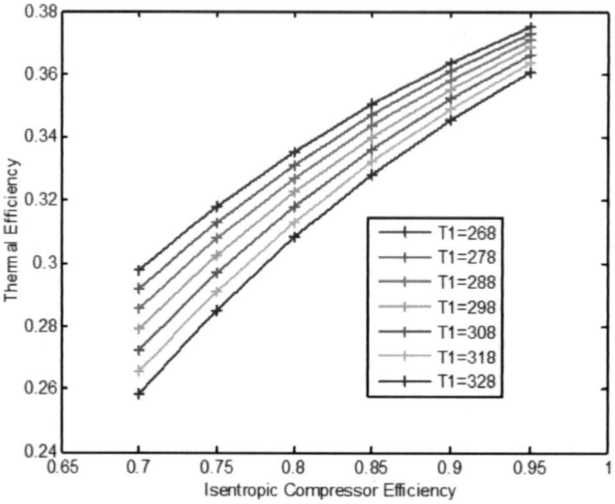

Figure 11. Effect of isentropic compressor efficiency and ambient temperature on thermal efficiency of two shaft gas turbine.

Figure 12 shows the effect of the turbine isentropic efficiency on thermal efficiency for different ambient temperature. The thermal efficiency increases linearly with increase of the turbine isentropic efficiency, however the variation thermal efficiency is more significant as lower isentropic efficiencies. The thermal efficiency decreases from 29.6% to 25.7% when the ambient temperature increases from 268K to 328K with effect of isentropic compressor efficiency, however the thermal efficiency decreases from 21.5% to 16.7% with effect of isentropic turbine efficiency for the same range of air to fuel ratio.

4.4. Effect of Turbine Inlet Temperature and exhaust temperature

Figure 13 represents the variation of exhaust temperature on thermal efficiency for several ambient temperature and air fuel ratio. It is noticed that the thermal efficiency increases with decrease of the ambient temperature and air fuel ratio while the thermal efficiency decreases with increase of the exhaust temperature. Figure 14 represents the variation of thermal efficiency against the turbine inlet temperature as well as compression ratio and air fuel ratio. The air fuel ratio varies of 40-52 (kgair/kgfuel) and compression ratio varies of 4-28. Thermal efficiency are strongly related with turbine inlet temperature, the efficiency increases when turbine inlet temperature increases and also decreases the air to fuel ratio caused increases of thermal efficiency and turbine inlet temperature.

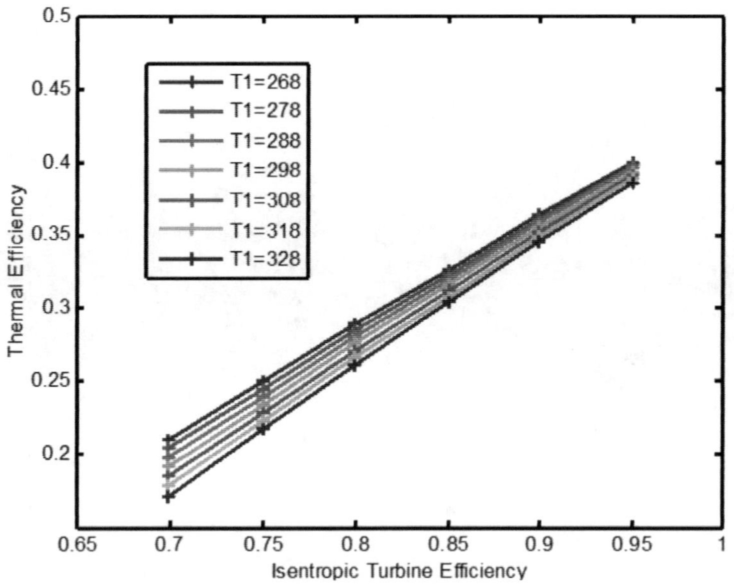

Figure 12. Effect of isentropic turbine efficiency and ambient temperature on thermal efficiency of two shaft gas turbine.

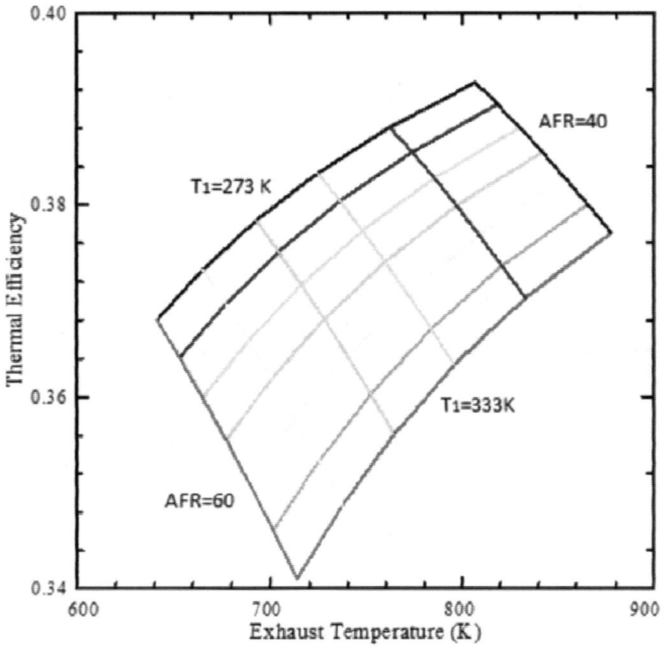

Figure 13. Effect of exhaust temperatures on thermal efficiency for ambient temperature and air to fuel ratio.

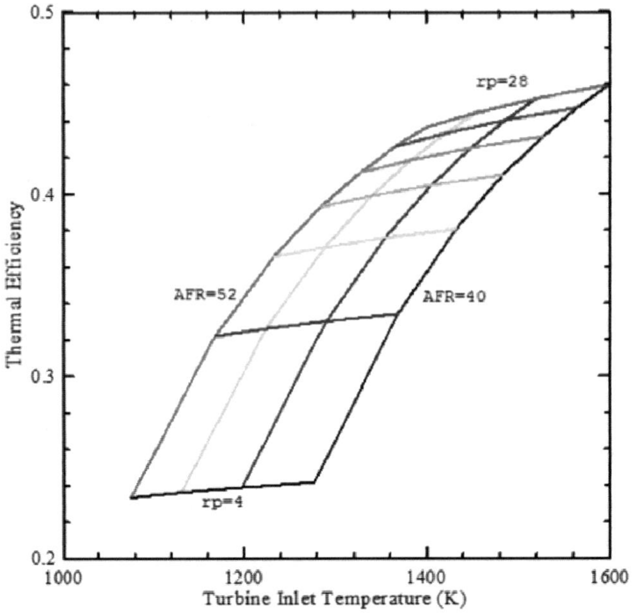

Figure 14. Variation of thermal efficiency against TIT, compression ratio and air fuel ratio.

5. Conclusions

The simulated modelling results shows that the influence of ambient temperature, compression ratio, air to fuel ratio, isentropic components efficiencies, turbine inlet temperature are significantly effect on performance of two shaft gas turbine power plant. The results were summarized as follows:

1. The compression ratios, ambient temperature, air fuel ratio as well as the isentropic efficiencies are strongly influence on the thermal efficiency of the two shaft gas turbine power plant.
2. The thermal efficiency and power output decreases linearly with increase of ambient temperature as well as the air to fuel ratio.
3. The peak efficiency, power and specific fuel consumption occur at when compression ratio increased in the gas turbine power plant.

The power of the simulated two shaft gas turbine is higher than the simple gas turbine cycle and real data from Baiji gas turbine power plant.

Acknowledgments

The authors would like to thank Universiti Malaysia Pahang and financial support under Doctoral Scholarship scheme.

References

[1] Boonnasa S, Namprakai P and Muangnapoh T 2006 *Energy* **31** 2036.
[2] Taniquchi H, Miyamae S 2000 *J. Propul. Power* **16** 557.
[3] Johnke T and Mast M 2002 Gas Turbine Power Boosters to Enhance Power Output. (Siemens Power for generation, Siemens Power J.)
[4] Mahmood F G, Mahdi D D 2009 *Appl. Energy* **86** 2750.
[5] Horlock J H 2003 *Advance Gas Turbine Cycles* (UK : Elsevier Sci. Ltd.)
[6] Ibrahim T K, Rahman M M and Abdalla A N 2011 *Int. J. Phys. Sci.* **6** 620.

[7] Rahman M M, Ibrahim T K, Taib M Y, Noor M M, Kadirgama K and Bakar R A 2010 *WASET*, **68** 94.

[8] Rahman M M, Ibrahim T K, Kadirgama K, Mamat R and Bakar R A 2011 *Adv. Mater. Res.*, **189-193** 3007.

[9] Saravanamuttoo H, Rogers G, Cohen H and Straznicky P 2009 *Gas Turbine Theory* (New York: Prentice Hall).

[10] Al-Sayed AF 2008 Aircraft Propulsion and Gas Turbine Engines (Taylor & Francis).

[11] Naradasu R K, Konijeti R K, Alluru V R 2007 *Therm. Sci.* **11** 143.

[12] Ibrahim T K, Rahman M M and Abdalla A N 2010 *Sci. Res. Essays* **5** 3760.

[13] Bathie W W 1996 Fundamentals of Gas Turbine (USA: John Wiley & Sons, INC.)

Determination of Burst Pressure of API Steel Pipes using Stress Modified Critical Strain Model

N. A. Alang[1], N. A. Razak[2] and A.S. Sulaiman[3]
Faculty of Mechanical Engineering, Universiti Malaysia Pahang,
26600 Pekan, Pahang.

E-mail: azuan@ump.edu.my[1], norhaida@ump.edu.my[2], syahrizan@ump.edu.my[3]

Abstract. This paper presents a technique which can be used to determine the burst pressure of defective steel pipes using non-linear finite element (FE) analysis. The technique uses stress modified critical strain (SMCS) failure criterion to study the effect of gouge defects on maximum working pressure of API X65 steel pipes. The procedures in determining the model parameters using 3-D, homogeneous isotropic elastic-plastic material model with large deformation finite element analyses from notched tensile bars were systematically discussed. The relationship between burst pressure and gouge depth was proposed. The burst pressure estimated then was compared to experimental data from the literature for validation showing overall good agreements.

1. Introduction

Various approaches to predict ductile fracture have been proposed in the past. Fracture mechanics, micro-mechanical and cohesive zone models are the main methods that emerged to describe ductile fracture of metals. A micro-mechanical model for ductile fracture, incorporating void nucleation, growth and coalescence, for instance, the Gurson–Tvergaard–Needleman (GTN) model, stress modified critical strain [1], void growth [2], and continuum damage model [3] have been widely used over several decades. As many researchers [1-11] have already published a number of papers using these methods, applicability and validity of these methods have been well discussed. However, few issues need to be resolved in practical application of these methods. For instance, GTN and continuum damage model (CDM) consist of relatively high number of parameters compare to SMCS and void growth model (VGM). Determination and validation of these parameters are not an easy task, and therefore often not reliable.

As recognized by Hancock et al [3], the SMCS model is quite simple since the critical plastic strain as a function of stress triaxiality can be directly calculated and it is not for VGM model where the triaxiality and plastic strain history have to explicitly integrate. Due to its simplicity and accuracy, SMCS model is frequently preferred by researchers to predict the ductile failure of the materials. Mathematically, SMCS model is evaluated by the equation (1):

$$\varepsilon_f = \beta \exp\left(-\frac{3\sigma_m}{2\sigma_e}\right) \quad (1)$$

where σ_m/σ_e represents stress state triaxiality, ε_f is the true fracture strain and β is a material constant.

[1]Correspondence author: N. A. Alang
Email address: azuan@ump.edu.my

Published under licence by IOP Publishing Ltd

Kanvinde et al [11] compares both VGM and SMCS models in predicting the ductile failure of engineering structural made from structural steels and concludes that both models can be applied accurately to the entire spectrum of structural steels in predicting the failure. Chang et al [8] developed SMCS failure criterion for API X65 steel pipes; widely used for gas pipelines using both smooth and notched tensile bar specimens then applying proposed model to predict the burst pressure of defective pipes using FE analysis and compare the results with experimental results. He reported that the errors are relatively small as less as 5%. However, the study is only limited to constant notch/defect depths and pipe diameters. As the depth of the defects on pipelines increase consistently over the time that mainly due to severity of corrosion [12], the study on its effect is significantly important. In this paper, the local failure criterion namely stress modified critical strain model is applied to predict the burst pressure of defective pipe of API X65 steel. The model was implementing in commercial FE analysis code, MSC PATRAN/MARC. As an extension of Chang et al [8] works, the effect of notch depth on burst pressure for API X65 steel pipes were investigated.

2. Finite Element Analysis

2.1. Materials

The material used in this study was API X65 steel, widely used for gas pipelines. The mechanical properties of the material are tabulated in table 1 [8].

Table 1. Mechanical properties of API X65 steel at room temperature.

Young Modulus, E (GPa)	Poisson Ratio, ν	Yield Strength, σ_y (MPa)	Tensile Strength, σ_u (MPa)
210.7	0.3	464.5	563.8

2.2. FE modelling

The test specimens based on Chang et al [8] were used as a model in FE analysis. Detailed dimensions of the test specimens are illustrated in figure 1, showing a round tensile bar with a circumferential notch at the center. In this paper, three different models with notch radius R=1.5, 3 and 6 mm were used. For efficient computation, symmetric conditions are applied. Therefore, all the specimens are modelled in one-four considering symmetry condition. Figure 2 shows the one-four tensile bars with notch radius of 6mm modelled in FE analysis.

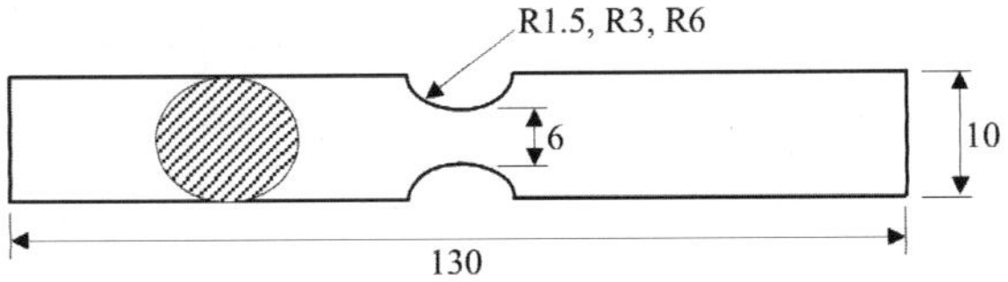

Figure 1. Notched round tensile bars.

Eight node solid elements with reduced integration scheme are applied by using the commercialize MSC PATRAN/MARC software. Based on authors' experiences, there would be little difference of final results by selecting the full integration element and reduced integration element in simulation. The true stress-plastic strain data from Chang et al [10], shown in figure 3, were used as an input in the FE analyses. Materials were modelled as elastic-plastic isotropic homogeneous with large strain/deformation are considered in all cases. Translational constraints are, respectively, imposed on the nodes in the three symmetry plane. The deformation boundary condition was applied to the end of the FE model, and the resulting tensile load was determined from nodal reaction forces.

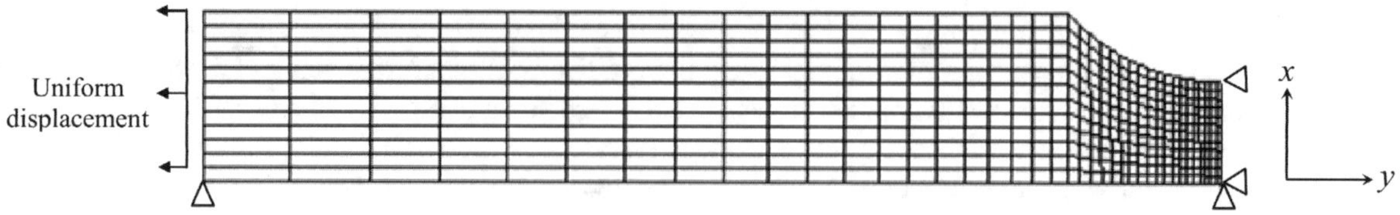

Figure 2. Detailed meshes in 2D views for tensile bars with notch radius of 6 mm.

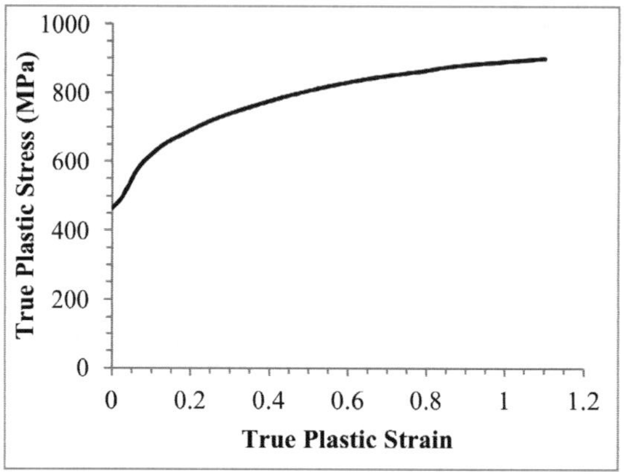

Figure 3. True plastic stress-strain data employed in FE analysis.

2.3. Comparison between FE results and experimental results

Figure 4 shows comparison between experimental engineering stress–strain data from notched tensile bars for notch radius, R=3, and 6 mm with the FE results. It is clearly shown that the FE analysis is able to simulate tensile deformation behaviour even after necking with excellent agreement. However, it cannot simulate the fracture point of tensile test specimens. Therefore, the experimental stress strain data have been compared to simulation results in order to detect the fracture strain of the material. This point was determined at the location where the simulation data divert from one in experiment results.

Figure 5 shows the variation of stress triaxiality and equivalent strain at the minimum section of the bar for three different notch radiuses. Both values are taken at the failure initiation points that have been identified from the previous section (see figure 4). The stress triaxiality, T is defined by the ratio of hydrostatic stress, σ_m and equivalent stress, σ_e and is expressed by:

$$T = \frac{\sigma_m}{\sigma_e} = \frac{\sigma_1+\sigma_2+\sigma_3}{3\sigma_e} \tag{2}$$

The equivalent stress is given by:

$$\sigma_e = \frac{1}{\sqrt{2}}[(\sigma_1-\sigma_2)^2+(\sigma_3-\sigma_1)^2+(\sigma_2-\sigma_3)^2]^{\frac{1}{2}} \tag{3}$$

On the other hand, the equivalent strain σ_e is given by:

$$\varepsilon_e = \frac{\sqrt{2}}{3}[(\varepsilon_1-\varepsilon_2)^2+(\varepsilon_3-\varepsilon_1)^2+(\varepsilon_2-\varepsilon_3)^2]^{\frac{1}{2}} \tag{4}$$

where the σ_1, σ_2, σ_3 and ε_1, ε_2, ε_3 are the principle stresses and principle strain respectively.

It is clearly shown that the stress triaxiality is higher at center of the tensile bar compared to the free surface. Conversely, as seen in figure 5b, the equivalent plastic strain for tensile bar with R=1.5 and 6 mm are higher at free surface than at the bar centre, and the distribution is considerably different for the two notch sizes. However, for tensile bar with R=6 mm, the maximum equivalent plastic strain occurred at the center of the bar. The fact that the effective plastic strain is maximum at the bar surface, whereas fracture initiates at the centre of the bar, emphasizes the significance of stress triaxiality in evaluating fracture initiation.

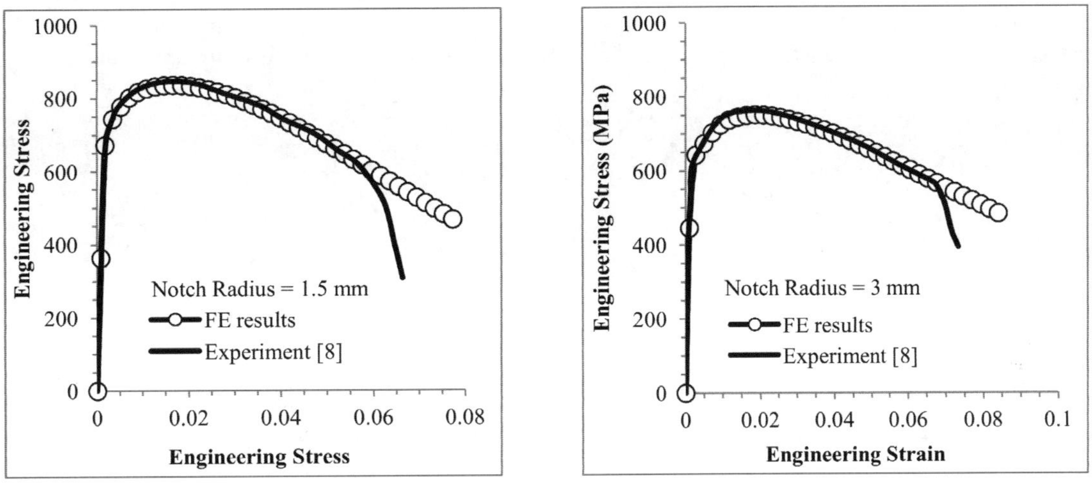

Figure 4. Comparison between experimental and FE results of engineering stress–strain data: (a) notched bar with the 1.5 mm notch radius and, (b) notched bar with the 3 mm notch radius.

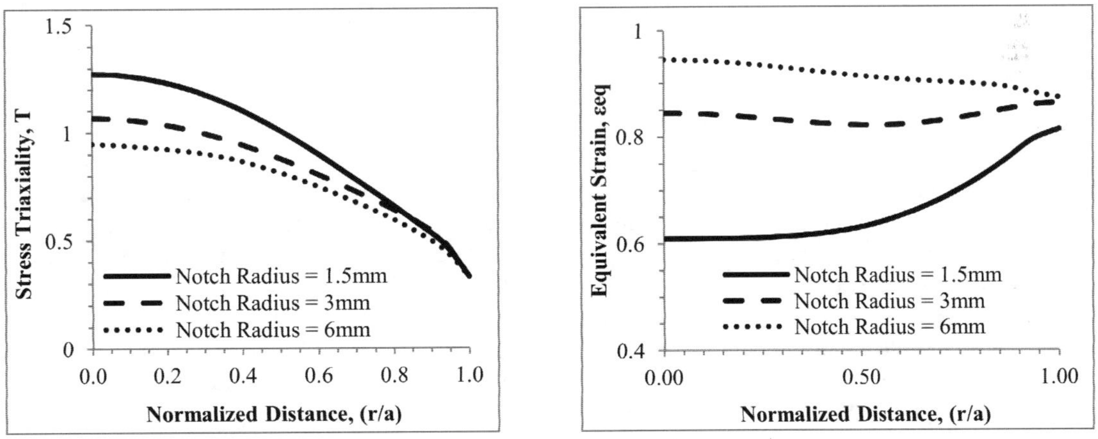

Figure 5. (a) Stress triaxiality profile, (b) equivalent strain profile over minimum cross section.

Figure 6 summarizes the resulting equivalent strain to fracture or true fracture strain, ε_f, as a function of the stress triaxiality for the API steel, considered in the present work. Based on the figure, it is noted that the true fracture strain decreases sharply with increasing of stress triaxiality. The true fracture strain can be represented as an exponentially dependent on the stress triaxiality [1]. Therefore, the following parameters for stress modified critical strain model are proposed:

$$\varepsilon_f = 4.0\exp\left(-1.5\frac{\sigma_m}{\sigma_e}\right) \tag{6}$$

To predict ductile failure of defective pipes made of API X65 steels (considered in present work) using the present approaches, the proposed equation, equation (6) should be combined with detailed elastic-plastic FE analyses from which local stresses and strains are determined. For instance, from the FE analysis, stress and strain information can be monitored as a function of load. Over the loading history, the stress triaxiality and equivalent strain were calculated through equations (3) and (5). Then, the equivalent strain to fracture or true fracture strain is estimated from equation (6). When the equivalent strain from the FE analysis equals the fracture strain, the failure is assumed to occur.

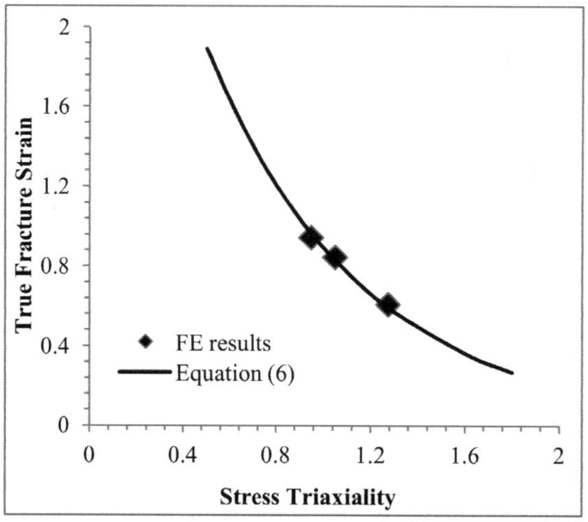

Figure 6. True fracture strain as a function of stress triaxiality

3. Validation

For validation, the burst pressure estimated from FE analysis using proposed model is compared with experimental data from full scale pipe tests [8]. The defective pipe with two different gouge length, l=100mm and 200mm were analyzed. The detailed dimensions of the pipe studied in this paper are according to Chang et al [8]. The pipe has the outside diameter, Do = 762mm, the thickness of t = 17.5mm, and the total length of L =2,300mm. The gouge is characterized by the 45 degree V-notch with the circular notch radius of 2mm. The depth of the gouge is d = 8.75mm which is 50% of the pipe thickness (d/t = 0.5). The pipes with a gouge defect were modelled using MSC PATRAN FE software. A typical finite element mesh is shown in figure 7, with enlargement on the defective region. Internal pressure was applied to the inner surface of the pipe, fixed at one end of the pipe to simulate the closed cap condition and the symmetrical condition was also applied. The results from FE analysis have been compared to experimental data as shown in table 3. The error between these two methods is relatively small therefore increasing the confident on FE analysis.

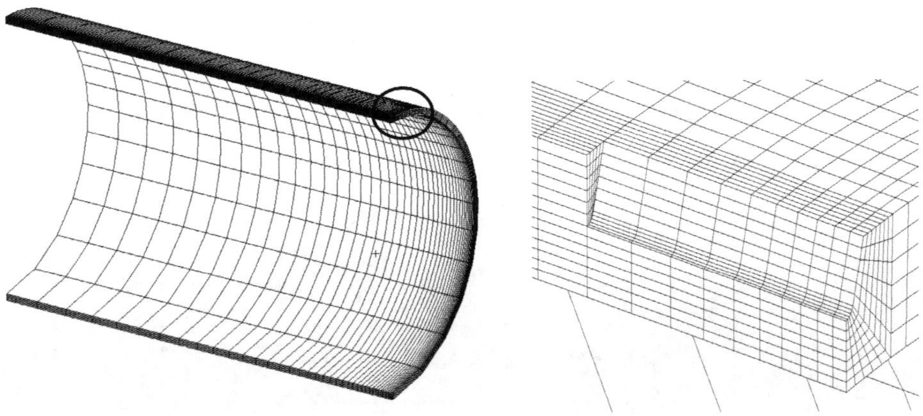

Figure 7. Detailed mesh of the defective pipe used in FE analysis.

Table 3. Comparison of burst pressure between experimental data and FE results.

	Burst Pressure (MPa)	
Gouge Defect	Experiment [8]	FE results Eq. (6)
100mm	24.68	25.8 (4.54%)
200mm	22.48	24.2 (7.65%)

4. Effect of gouge depth on burst pressure

Figure 8 shows schematic illustration of pipe with a gouge on its outer surface. For parametric study, the pipe with different gouge depth is simulated and the burst pressures for each case were predicted. Four gouge depth are chosen for analysis; d = 4.725, 6.563, 8.75 and 13.125 mm.

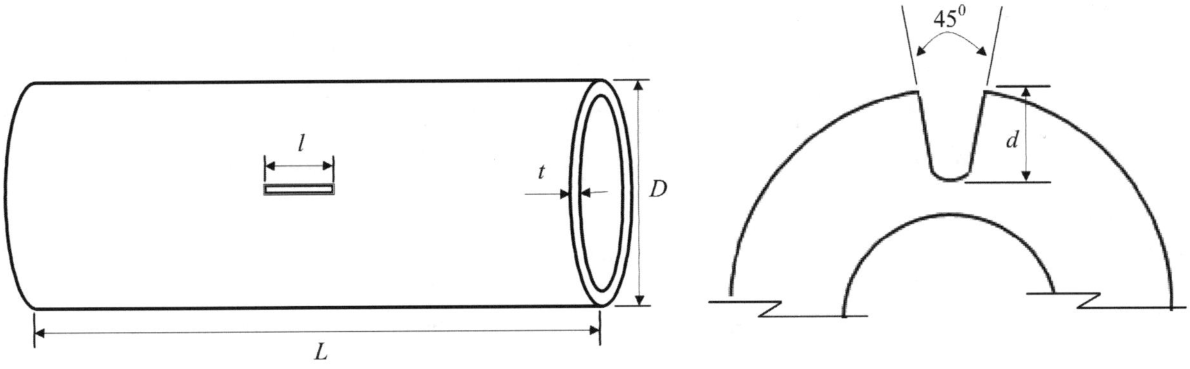

Figure 8. Schematic of pipes with gouge defect.

Shown in figure 9 is a plot of burst pressure versus gouge depth of the pipes in a range of 4.73-13.1 mm. The relationship between two parameters in this range of gouge depth can be represented by linear regression demonstrate a strong correlation. Based on the graph, the burst pressure is strongly dependent on the gouge depth. The burst pressure of the pipe decreases with increasing of gouge depth. Figure 10 shows the variation of pressure as a function of radial displacement taken from the critical location of defective pipes.

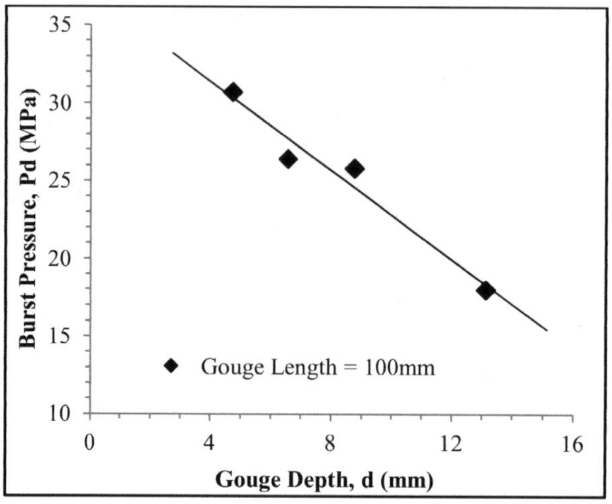

Figure 9. Relationship between burst pressure and gouge depth.

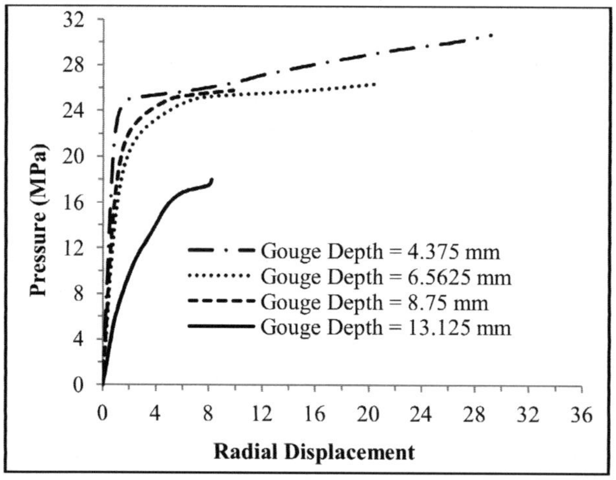

Figure 10. Predicted burst pressure of steel pipes with various gouge depth.

5. Conclusion

In this paper, the burst pressure of defective pipes was estimated using nonlinear FE analysis by implementing the SMCS model. Step by step procedures in determining the model parameters were systematically discussed in this paper. Determination of the parameters of stress-modified critical strain model for given material is quite simple and reliable, as explain in this paper. Using this model, the burst pressure of pipe with different gouge length was predicted and the parametric study of the gouge was also carried out. For validation, API X65 steel (considered in present study) pipes with two different gouge lengths are simulated using FE analysis with the proposed model and the predicted burst pressures are compared with experimental data from literature showing overall good agreements. This study indicates that the burst pressure decreases with increasing of gouge length. Based on parametric study, the relationship between burst pressure and gouge depth for API X65 can be represented by linear regression. In addition, the maximum radial displacement increases with decreasing of gouge depth.

References

[1] Hancock J.W and Mackenzie A.C 1976 On the mechanisms of ductile failure in high-strength steels subjected to multi-axial stress-states *J. Mech. Phys. Solids.* **24** 147-169

[2] Rice J.R and Tracey J 1969 On the ductile enlargement of voids in triaxial stress fields *J. Mech. Phys. Solids.* **17** 201-217

[3] Mackenzie A.C, Hancock J.W and Brown D.K On the influence of state of stress on ductile failure initiation in high strength steels, , *Int J Fract Mech.*

[4] Lemaitre J 1985 A continuous damage mechanics model for ductile fracture *J of Eng Mat and Tech.* **107** 83-89

[5] Bonora N, Gentile D, Pirondi A and Newaz G 2005 Ductile damage evolution under triaxial state of stress: theory and experiments, *Int J of Plasticity.* **21** 981–1007

[6] Zhoua Z, Kuwamura H, Nishida A 2011 Effect of micro voids on stress triaxiality-plastic strain states of notched steels *Procedia Eng.* **10** 1433–1439

[7] Bao Y and Wierzbicki T 2004 On fracture locus in the equivalent strain and strain triaxiality space *Int J of Mech Sciences.* **46** 81-98

[8] Oh C.K, Kim Y.J, Park J.M, Baek J.H and Kim W.S 2007 Development of stress-modified fracture strain for ductile failure of API X65 steel *Int J Fract.* **143** 119–33

[9] Oh C.K, Kim Y.K, Baek J.H and Kim W.S 2011 A finite element ductile failure simulation method using stress-modified fracture strain model *Int J Fract.* **78** 124–137

[10] Oh C.K, Kim Y.J, Baek J.H and Kim W.S 2007 Ductile failure analysis of API X65 pipes with notch-type defects using a local fracture criterion *Int J of Pressure Vessels and Piping.* **84** 512–525

[11] Kanvinde A.M and Deierlein G.G 2005 The void growth model and the stress modified critical strain model to predict ductile fracture in structural steels.

[12] Netto T.A, Ferraz U.S and Estefen S.F 2005 The effect of corrosion defects on the burst pressure of pipelines, *J Constructional Steel Research.* **61** 1185–1204

1st International Conference on Mechanical Engineering Research 2011 (ICMER2011) IOP Publishing
IOP Conf. Series: Materials Science and Engineering **36** (2012) 012026 doi:10.1088/1757-899X/36/1/012026

Microcontroller Based Proportional Derivative Plus Conditional Integral Controller for Electro-Mechanical Dual Acting Pulley Continuously Variable Transmission Ratio Control

A Budianto[1], K B Tawi[1], M Hussein[1], B Supriyo[1], S Ariyono[2], M S Che Kob[1], Mohd Ezlamy Zulkifli[1], Khairuldean A K[1], Aishah Daraoh[1]
[1] Faculty of Mechanical Engineering, Universiti Teknologi Malaysia, 81310, Skudai, Johor, Malaysia
[2] Faculty of Mechanical Engineering, Universiti Malaysia Pahang, 26600, Pekan, Pahang, Malaysia

Email: aries.budianto26@yahoo.com

Abstract. Electro-Mechanical Dual Acting Pulley (EMDAP) Continuously Variable Transmission (CVT) is a transmission utilized by electro-mechanical actuated system. It has a potential to reduce energy consumption because it only needs power during changing CVT ratio and no power is needed to maintain CVT ratio due to self lock mechanism design. This paper proposed simple proportional derivative plus conditional integral (PDCI) controller to control EMDAP CVT ratio which can be simply implemented on a microcontroller. This proposed controller used Astrom-Hagglund method and Ziegler-Nichols formula to tune PDCI gain. The Proportional Derivative controller is directly activated from the start but Integral controller is only activated when the error value reaches error value setting point. Simulation using Matlab/Simulink software was conducted to evaluate PDCI system performance. The simulation results showed PDCI controller has ability to perform maximum overshoot 0.1%, 0.001 steady state error and 0.5s settling time. For clamping condition, settling time is about 11.46s during changing ratio from 2.0 to 0.7, while for release condition, settling time is about 8.33s during changing ratio from 0.7 to 2.0.

1. Introduction

Continuously Variable Transmission (CVT) has a potential in saving fuel consumption because its wide variable ratio coverage enables the engine to run at either its fuel-efficient operating point or its maximum under various load condition [1]. Common CVT which widely used in the market is hydraulically actuated type. It needs continuous power to supply force to maintain the desired ratio and preventing gross belt slip. Thus, it can reduce CVT efficiency [2]. In order to increase CVT efficiency, Electro-Mechanical Dual Acting Pulley (EMDAP) CVT, which designed and developed by Drivetrain Research Group (DRG) Universiti Teknologi Malaysia (UTM), was introduced. EMDAP CVT has a potential to reduce energy consumption because it only needs power during changing CVT ratio and no power is needed to maintain CVT ratio due to self lock mechanism design. It utilizes two DC motors as actuator, power screw mechanism (PSM) and two moveable sheaves on both primary and secondary pulleys. Two moveable sheaves move identically to keep the belt align in the center thus, eliminate belt misalignment [3].

[1]Correspondence author: A. Budianto
Email address: aries.budianto26@yahoo.com

Published under licence by IOP Publishing Ltd

A good controller is also important in order to achieve high efficiency of EMDAP CVT. By controlling two DC motors precisely and accurately, desired CVT ratio can be achieved. More than 90% industrial controller used today is Proportional, Integral and Derivative (PID) controller. PID controller is easily tuned and combined to obtain the optimal performance. Its parameters can be set independently. Thus, PID becomes the simplest and most efficient controller [4]. The success reason of standard PID controllers is the tuning rules [5]. Ziegler–Nichols method is the most famous method in tuning PID parameters. This method has good load disturbance attenuation but shows unsatisfactory performance, with a large overshoot and long settling time [6].

Recent research in controlling EMDAP CVT ratio used PD controller then fine tuned by fuzzy controller. The integral gain was not used since it caused a big overshoot [7]. Since the fuzzy controller needs complex algorithms, which is consist of fuzzification, inference, rules and defuzzification processes, it needs a bigger memory size and longer processing time when it is implemented on a microcontroller. Thus, this paper proposes a simpler proportional derivative controller then fine tuned by conditional integral controller to control EMDAP CVT ratio so that it can be simply implemented on a microcontroller. In this paper, a simulation of proportional derivative plus conditional integral (PDCI) controller was conducted using Matlab/Simulink software to evaluate PDCI controller performances, including maximum overshoot, steady state error and settling time.

2. System modeling

Basically, EMDAP CVT consists of Metal pushing V-belt (MPVB), primary and secondary pulley, power screw, helical gear, pinion gear, gear reducer and DC motor. EMDAP CVT structure is shown in figure 1.

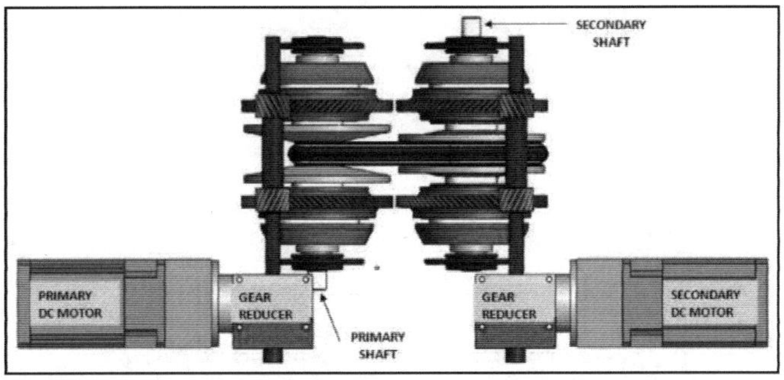

Figure 1. EMDAP CVT structure.

MPVB is used to transfer torque from primary pulley to secondary pulley. It also enables changing pulley contact radius by shifting up to increase the contact radius or by shifting down to decrease it. This system also utilizes dual acting pulley on both primary and secondary shafts. Each pulley consists of two moveable sheaves. These moveable sheaves are driven by axial movement of power screws which supply tangential force for clamping or releasing the belt. Helical gear is used to transfer torque from the pinion to the power screw, while the pinion gear receives the torque from the DC motor through gear reducer. Utilization of two moveable sheaves pulley enables to eliminate belt misalignment. Since the power screw utilizes square thread, it provides self locking mechanism which enables the system to maintain constant ratio without consuming energy. Therefore, this system only consumes energy during changing ratio [8].

2.1. DC motor modeling

DC motor transfer function model is created by determining DC motor block diagram and its parameters. DC motor block diagram is determined based on its basic equations, whereas DC motor parameters are identified based on datasheet. Equivalent model between electrical and mechanical relationship is shown in figure 2 [9].

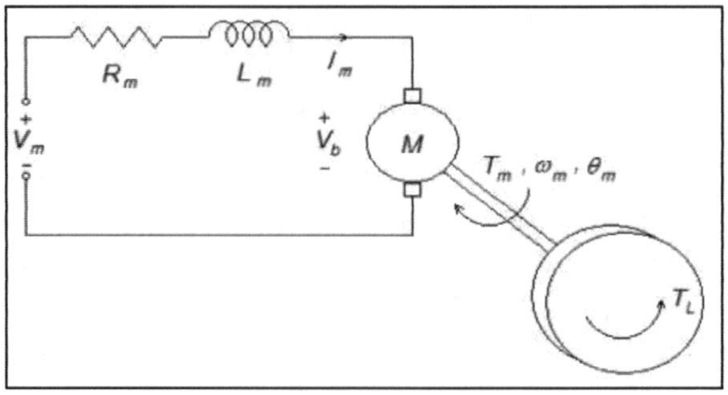

Figure 2. Equivalent DC motor model.

Transfer function of DC motor model is described as:

$$\omega_m = \left(\frac{(V_m - K_e\,\omega_m)}{(s\,L_m + R_m)} K_T - T_L\right)\frac{1}{(s\,J + B)} \tag{1}$$

Where:

ω_m = Motor Angular Speed (rad/s)
V_m = Motor Voltage (V)
V_b = Back EMF Voltage (V)
I_m = Motor Current (A)

The DC motor transfer function model is created based on equation (1) as shown in figure 3.

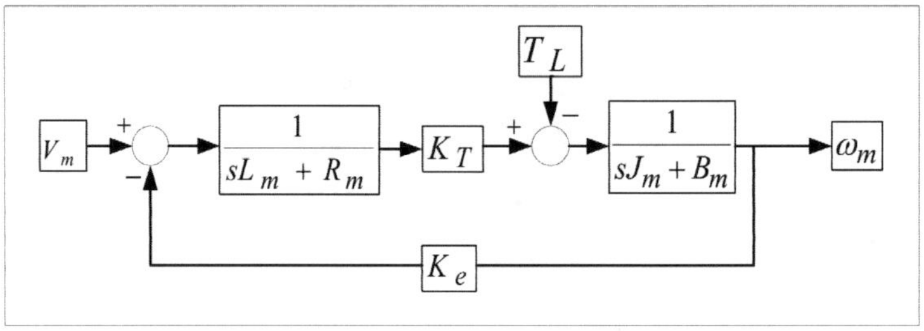

Figure 3. DC motor transfer function model.

Its parameters are shown in table 1.

Table 1. DC motor parameters values.

Parameters	Symbol	Value	Unit
Inductance	L_m	0.0228	H
Resistance	R_m	0.327	Ω
Back EMF constant	K_e	0.0535	V.s / rad
Torque Constant	K_T	0.0535	Nm / A
Rotor Inertia	J_m	7×10^{-5}	Kgm^2
Friction Coefficient	B	0.00166	Nm.s / rad

2.2. Mechanical modeling

Mechanical model consists of PSM and gear reducers models as shown in figure 4. Torque of power screw depends on clamping or release condition as represented by clamping or release equation. A condition selector is used to decide whether it is clamping or releasing. Then, the torque of power screw is transferred through interaction of helical and pinion gears. Both gears act as gear reducer. Next, torque of pinion gear is transferred to motor gear reducer. This torque of gear reducer acts as load torque for the DC motor.

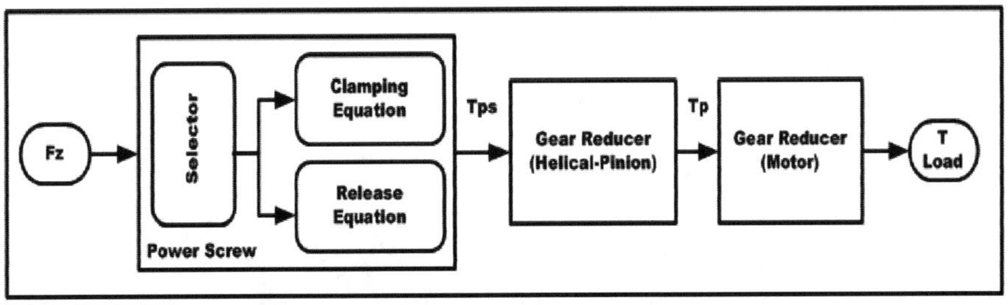

Figure 4. Mechanical model.

EMDAP CVT system utilizes PSM to convert rotational to axial movement. PSM schematic diagram is shown in figure 5. This design enables to convert 1 Nm torque of DC motor to approximately 20 kN of clamping force on MPVB [8]. Torque from DC motor is transferred to the pinion gear through gear reducer in order to produce bigger torque. Then, this torque is transferred to the helical gear and causes the helical gear to rotate. The helical gear rotation, either clockwise or counter clockwise, results the power screw movement, either inward or outward. As power screw moves, pulley sheaves also move in the same direction. The movements of pulley sheaves cause the metal pushing V-belt (MPVB) to shift up or down. The condition when the MPVB shifts up is called clamping condition. Otherwise, it is called release condition.

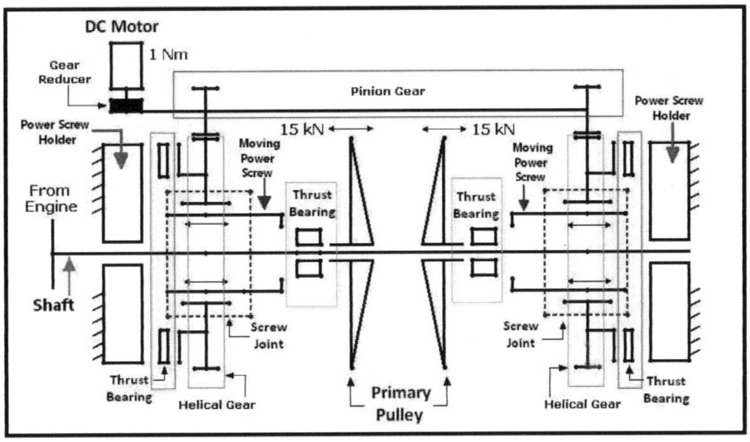

Figure 5. Power screw mechanism schematic diagram.

Torque of power screw for clamping condition is defined as:

$$T_{ps_C} = F_z \frac{d_p}{2} \frac{(\mu\pi d_p + L)}{(\pi d_p - \mu L)} \tag{2}$$

Whereas torque of power screw for release condition is defined as:

$$T_{ps_R} = F_z \frac{d_p}{2} \frac{(\mu\pi d_p - L)}{(\pi d_p + \mu L)} \tag{3}$$

Where:

F_z	=	Force for clamping or release MPVB	(N)
T_{ps_C}	=	Torque of power screw for clamping MPVB	(Nm)
T_{ps_R}	=	Torque of power screw for releasing MPVB	(Nm)

Dimensions of mechanical system are listed in table 2.

Table 2. Mechanical system dimensions.

Parameters	Symbol	Value	Unit
Pitch length	L	0.002	m
Pitch diameter of inner screw thread	d_p	0.0885	m
Friction coefficient	μ	0.15	
Motor Gear reducer ratio	GR_m	30:1	
Helical gear teeth	t_H	60	
Pinion gear teeth	t_P	14	

Combination of DC motor and mechanical models yields EMDAP CVT model as shown in figure 6. This model represents primary or secondary pulley system. Mechanical system acts as a load to the DC motor. The input of DC motor is motor voltage, whereas the output is motor speed.

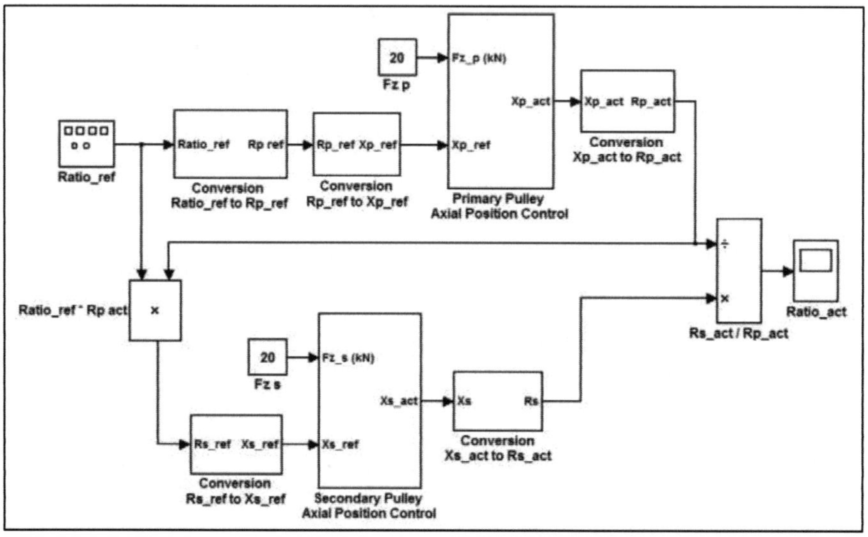

Figure 6. EMDAP CVT model.

2.3. CVT Ratio Variator Modeling

CVT variator consists of a primary pulley, a secondary pulley and a MPVB. CVT variator geometry diagram is shown in figure 7. By assuming that belt has a fixed length, does not slip and moves at perfect circles with primary radius, R_p, and secondary radius, R_s, the linear velocities of both pulleys and belt will be the same.

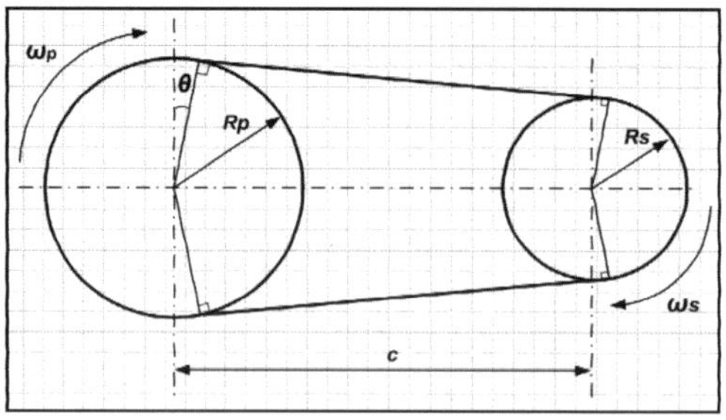

Figure 7. CVT variator geometry diagram.

Since the primary pulley is the driver whereas secondary pulley is the driven, speed and transmission ratios are the same and defined as:

$$r_s = \frac{\omega_p}{\omega_s} \quad (4)$$

$$r_{cvt} = \frac{R_s}{R_p} \quad (5)$$

Where:

r_{cvt} = CVT ratio
r_s = speed ratio
ω_p = angular speed of the primary pulley (rpm)
ω_s = angular speed of the secondary pulley (rpm)
R_s = secondary pulley radius (mm)
R_p = primary pulley radius (mm)

Geometric relationship between belt length and pulley radii is defined as:

$$R_p = R_s + c.sin\,(\theta) \quad (6)$$

$$L = (\pi + 2.\theta).R_p + (\pi - 2.\theta)\,R_s + 2.c.\cos\,(\theta) \quad (7)$$

Where:

L = belt length (mm)
c = pulley center distance (mm)
θ = wrapped angle on the primary pulley (rad)

Since the pulley driver is the primary side, radius of primary pulley becomes reference in changing CVT ratio. By substituting equations (5) and (6) to equation (7), primary radius is calculated as:

$$R_p = \frac{L-\left(\pi-2.\arcsin\{\frac{(1-r_{cvt}).R_p}{c}\}\right).r_{cvt}.R_p-2.c.\cos\,(\arcsin\{\frac{(1-r_{cvt}).R_p}{c}\}}{\left(\pi+2.\arcsin\{\frac{(1-r_{cvt}).R_p}{c}\}\right)} \quad (8)$$

According to EMDAP CVT design, the belt length, L, is 645.68 mm while the center distance between primary and secondary pulley, c, is 165 mm. By varying CVT ratio from 0.7 up to 2.0, relationship among R_p, R_s and CVT ratio is plotted as shown in figure 8.

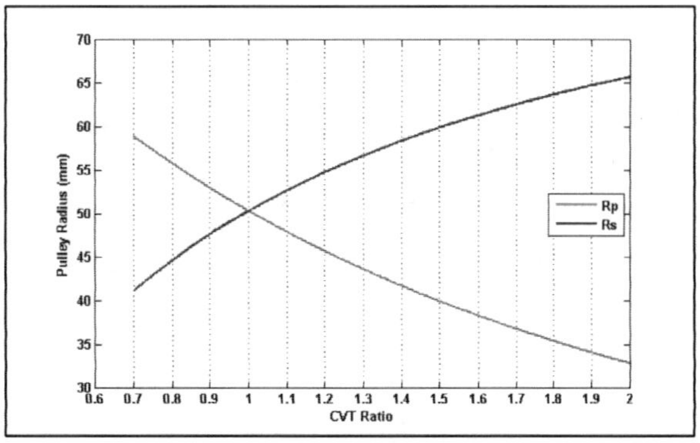

Figure 8. Relationship among CVT ratio, R_p and R_s.

2.4. Relationship between pulley radius and axial position

According to equation (5), CVT ratio is determined by both primary and secondary pulley radii. Pulley radius is calculated by considering pulley shaft diameter, R_{shaft}, sliding distance between pulley and shaft, *d,* contact point distance between belt and pulley sheave, *h*, pulley sheave angle, *α*, also effective pulley radius, *dR,* as described in figure 9. Since *x* is pulley axial position, while R_{shaft} (17.5mm), *d* (0.5mm), *α* (11o) and *h* (5.3mm) are constant, pulley radius is defined as:

$$R = R_{shaft} + d + h + (\frac{x}{\tan(\alpha)}) \quad (9)$$

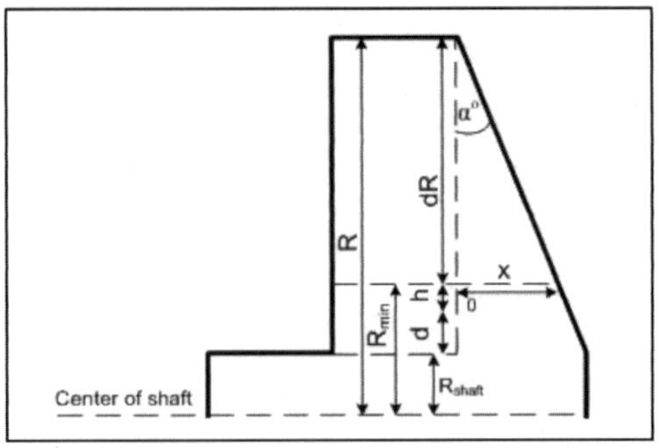

Figure 9. Relationship between pulley radius and axial position.

2.5. CVT Ratio Control Modeling

According to equation (9), changing pulley axial position means changing pulley radius, whereas according to equation (5), changing pulley radius means changing CVT ratio. Therefore, changing CVT ratio can be conducted by changing pulley axial position. Based on this condition, controlling CVT ratio can also be conducted by controlling pulley axial position. Pulley axial position controller is created by completing EMDAP CVT model with PID or relay feedback controller model as shown in figure 10.

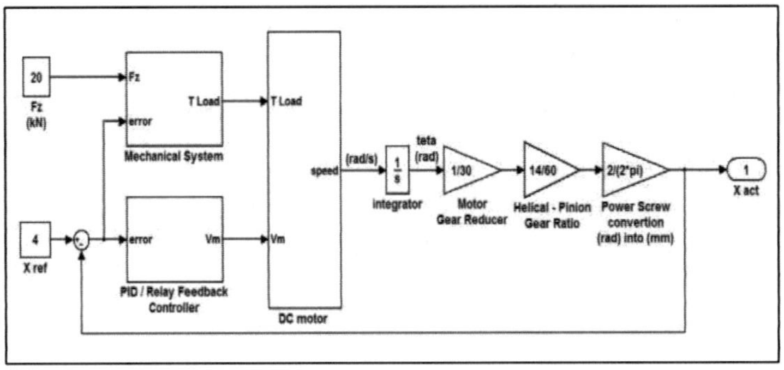

Figure 10. Pulley axial position control model.

In order to simulate EMDAP CVT ratio controller, CVT variator, axial position controllers for both primary and secondary pulleys, radius to axial position converter, and axial to radius converter models are combined as shown in figure 11.

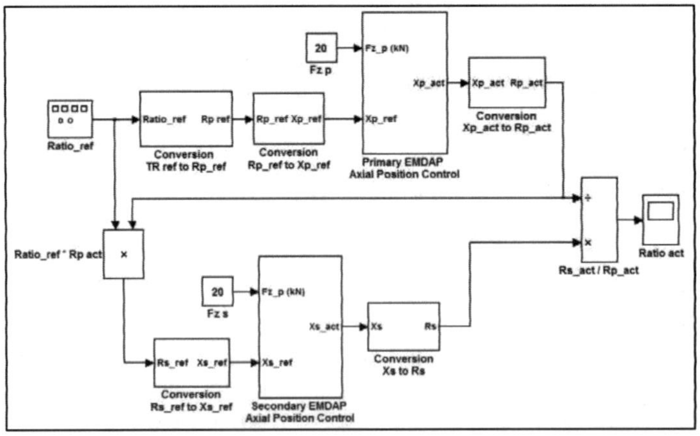

Figure 11. EMDAP CVT ratio controller model.

Primary pulley acts as reference for changing EMDAP CVT ratio. Once the CVT ratio input is set, it is converted to reference R_p and then, it is converted again to reference x_p as desired value for controlling primary pulley axial position. The result of pulley axial position control is actual x_p. Then, it is converted to actual R_p. It is multiplied with CVT ratio input to yield reference R_s. Then, it is converted to reference x_s as desired value for controlling secondary pulley axial position. The result of pulley axial position control is actual x_s. Then, it is converted to actual R_s. By dividing actual R_s with actual R_p, the actual CVT ratio is obtained.

3. Control method

3.1. Basic PID Controller

PID controller is a famous control method which has been widely used in control process. A PID controller involves three terms: the proportional gain, K_p, integral gain, K_i, and derivative gain, K_d. Transfer function of a PID controller has the general form as:

$$G_C(s) = K_p + \frac{K_i}{s} + K_d.s \tag{10}$$

Integral and derivative gains are also expressed as:

$$K_i = \frac{K_p}{T_i} \tag{11}$$

$$K_d = K_p.T_d \tag{12}$$

A common tuning method for determining PID parameters is Ziegler-Nichols frequency response method. It is conducted by generating oscillation response of system output. To generate oscillation, Astrom-Hagglund relay feedback method gives an advantage of less tuning time than manual tuning. A typical relay feedback signal is shown in figure 12.

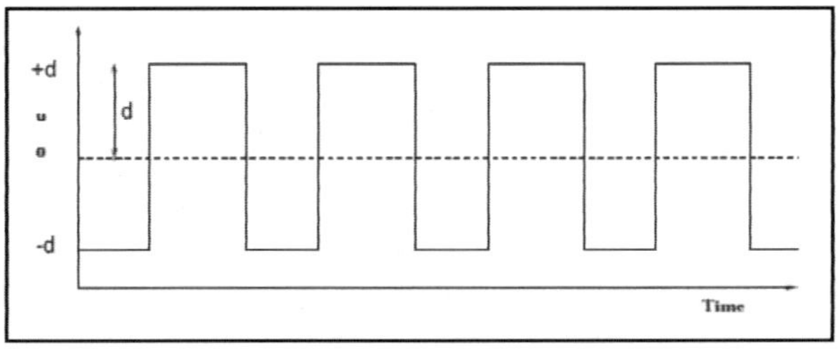

Figure 12. Relay feedback signal.

Once the oscillation generated, ultimate period, T_u, and oscillation amplitude, A, can be measured as shown in figure 13.

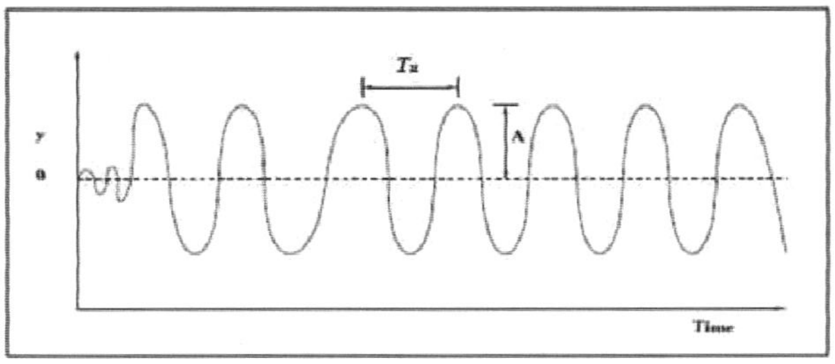

Figure 13. System output oscillation response.

The ultimate gain, K_u, can be calculated as:

$$K_u = \frac{4d}{\pi A} \tag{13}$$

Then, PID parameters can be calculated using Ziegler-Nichols formula as shown in table 3.

Table 3. Ziegler-Nichols formula frequency response.

Controller	K	T_i	T_d
P	$0.5K_u$		
PI	$0.4K_u$	$0.8T_u$	
PID	$0.6K_u$	$0.5T_u$	$0.125T_u$

3.2. PDCI Controller

This paper proposes Proportional Derivative plus conditional Integral (PDCI) controller for controlling primary and secondary pulley axial position. PDCI controller is used to minimize the overshoot, steady state error and settling time for clamping and release conditions. PDCI controller scheme is shown in figure 14.

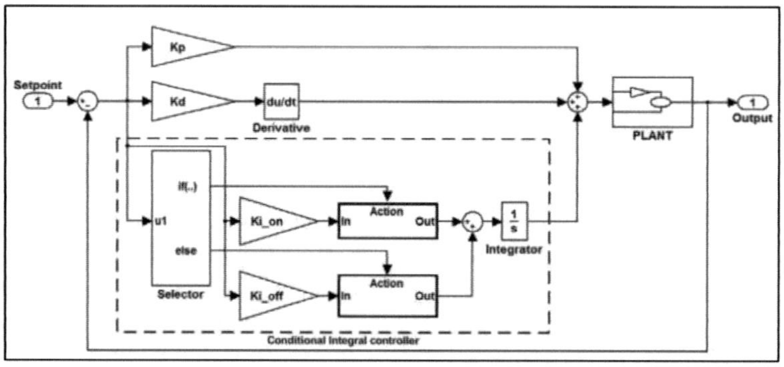

Figure 14. PDCI controller scheme.

Basically, PDCI controller consists of PD controller plus conditional integral controller. At first, PD controller is activated to control pulley axial position. When steady state error occurs, conditional integral controller is activated. This method is used to avoid a huge overshoot, steady state error and long settle time when integral controller is activated from the start. Conditional integral controller is activated based on error value setting using if-else selector. If the error meets the condition, the “if action block” is activated, the “else action block” is deactivated and the integral controller gain is set to Ki_on, which its value will be determined based on PID parameters tuning using Ziegler-Nichols and Astrom-Hagglund relay feedback methods. Thus, the output of integral controller equals to error value multiplied by integral controller gain. Otherwise, if the error doesn’t meet the condition, the “else action block” is activated, the “if action block” is deactivated and the integral controller gain is set to Ki_off, which its value is 0. Thus, the integral controller output equals to zero, which means the integral controller is not used and only PD controller is activated for controlling the pulley axial position.

4. Results and discussion

In this simulation, DC motor parameters are shown in table 1 and mechanical parameters are shown in table 2. Astrom-Hagglund relay feedback controller is used to tune PID parameters. Since the DC motor operates at ±24V, the amplitude of relay feedback controller is set to ±24V. Desired axial position is set to 4 mm, which is the center of pulley axial position, while the clamping force, F_z, is set to 0, which means there is no clamping force.

According to the system output response, the values of T_u, A and d are 0.14 s, 0.0315 and 24, respectively. Based on equation (13), the value of K_u is 970.58. Then, by using Ziegler-Nichols formula as shown in table 3, PID parameters values are obtained as follow: K_p= 582.05, K_i= 8315.03, K_d = 10.18. These parameters value then, implemented to basic PID controller scheme. After all PID parameters are set, simulation is run and the result is shown in figure 15.

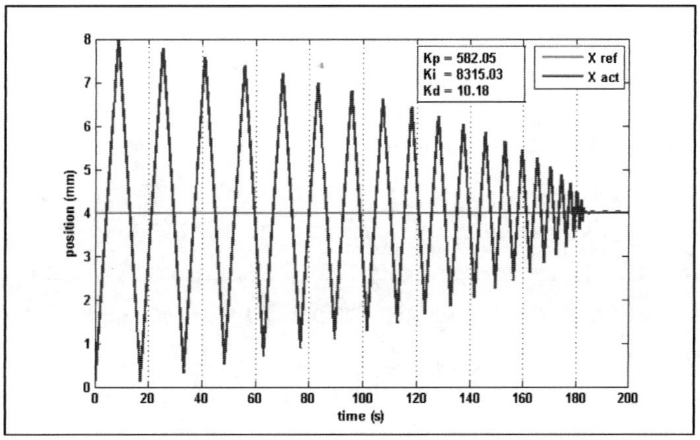

Figure 15. PID output response.

System output response shows a huge overshoot up to 100%, long settling time up to 183s and small error steady state. According to individual effect of increasing PID parameters [10], the factor which causes big overshoot and long settling time is the huge integral controller gain. Manual tuning by setting the same Kp and Kd and decreasing Ki, the system output response becomes better as shown in figure 16.

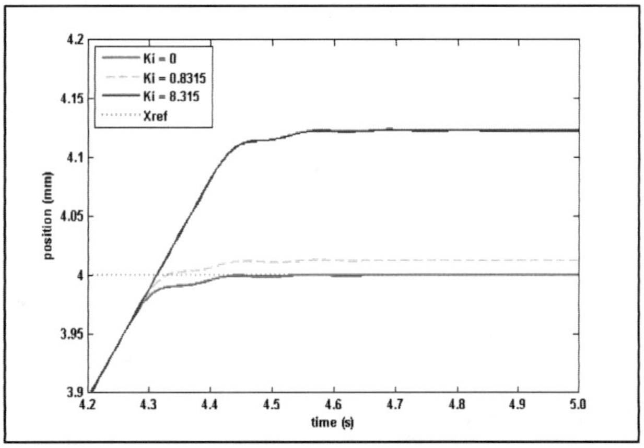

Figure 16. PID output response with manual tuning of Ki.

The smaller the integral gain, the better the system output response. No overshoot, quick settling time and no error steady state conditions can be achieved by setting integral gain to zero, which means it only uses PD controller.

Then, the PD controller is tested by applying clamping force, Fz, to the system, which is set to 5 kN, 10 kN, 15 kN and 20 kN. PD controller output response is shown in figure 17.

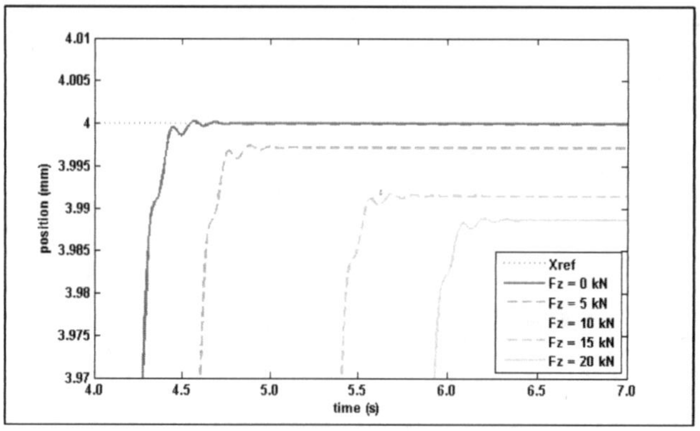

Figure 17. PD Output Response with Various F_z

Figure 17 shows that the bigger the clamping force, F_z, the longer the settling time and the bigger the steady state error although there is no overshoot in any clamping force. To eliminate this steady state error, integral controller should be conditionally activated in order to prevent a huge overshoot and long settling time. Thus, PDCI controller is designed to solve this problem. Parameters of PDCI controller are the same with PID controller which has been tuned using Ziegler-Nichols and Astrom-Hagglund methods. The only difference between PID and PDCI controller is the activation of integral controller.

Conditional integral controller is activated based on error value setting using if-else selector. If the error meets the condition, the "if action block" is activated, the "else action block" is deactivated and the integral controller gain is set to 8315.03, Otherwise, if the error doesn't meet the condition, the "else action block" is activated, the "if action block" is deactivated and the integral controller gain is set to 0. With the setpoint 4 mm and load 20 kN, the maximum steady state error shown in Figure 17 is 0.011. By using this steady state error value, the "if condition" is set to "$-0.011 \leq$ error ≤ 0.011". Since there is still steady state error, the "if condition" is set to several error values and the responses are shown in Figure 18 while its values are shown in table 4.

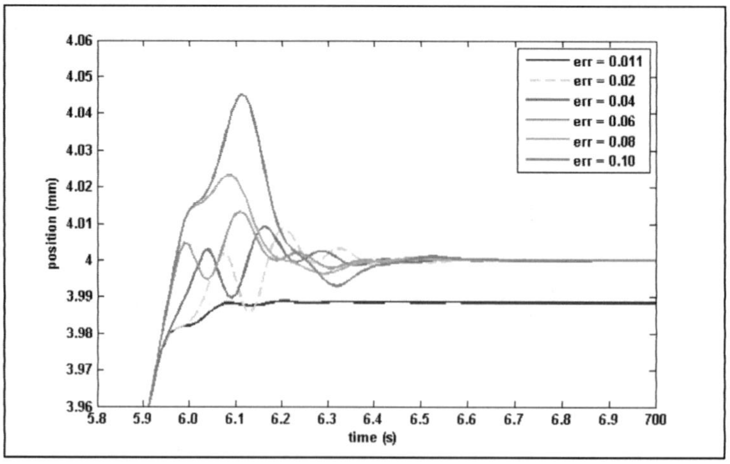

Figure 18. PDCI output response with various error setting.

Table 4. Maximum overshoot and time settling values.

Error Setting	Max Overshoot (%)	Settling Time (s)	Steady state error (mm)
0.011	0	6.1	0.011
0.02	0.200	6.4	0
0.04	0.225	6.4	0
0.06	0.325	6.4	0
0.08	0.575	6.4	0
0.10	1.125	6.4	0

Error setting 0.02 is chosen because it has zero steady state error, 6.4s settling time and 0.2% maximum overshoot. Then, PDCI controller is tested by applying various clamping force. The results are shown in figure 19. All responses show zero steady state error and maximum overshoot less than 0.25%.

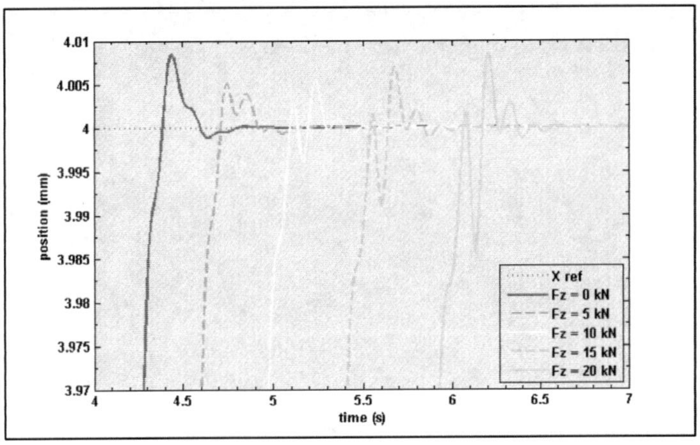

Figure 19. PDCI output response with various F_z.

By using PDCI controller for controlling both primary and secondary pulley axial position, EMDAP CVT ratio control is conducted based Matlab/Simulink model as shown in figure 11. By applying square wave input signal, which represents CVT ratio changing from 0.7 to 2.0, EMDAP CVT ratio controller response is shown in figure 20. For clamping condition, time taken to change CVT ratio from 0.7 to 2.0 is 11.46s and maximum overshoot is about 0.15% as shown in figure 21. For release condition, time taken to change CVT ratio from 2.0 to 0.7 is 8.33s and maximum overshoot is about 0.15% as shown in figure 22. Steady state error for clamping or release condition is less than 0.001.

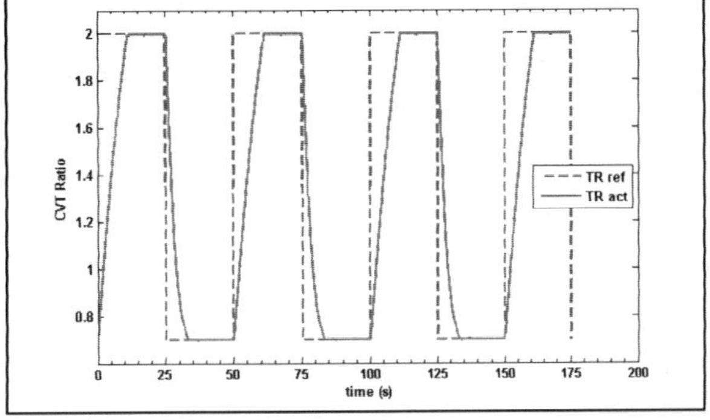

Figure 20. Output response of EMDAP CVT ratio control.

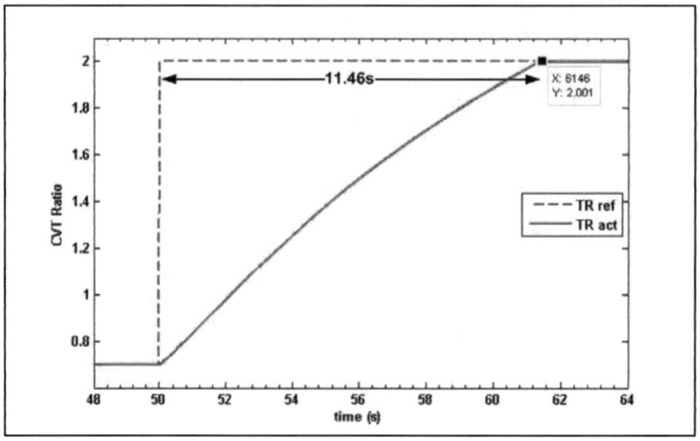

Figure 21. Output response of EMDAP CVT ratio control for clamping condition

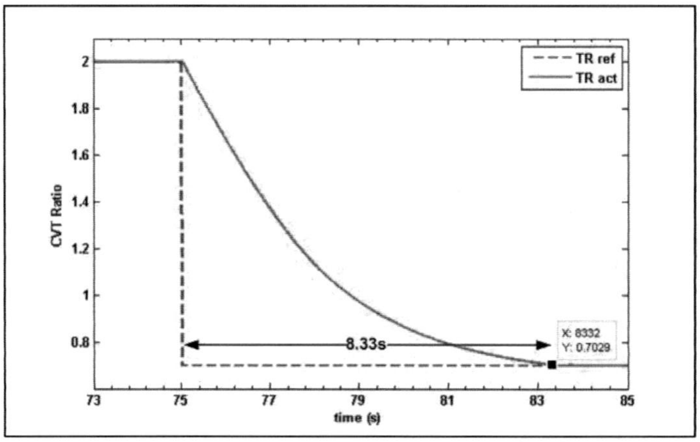

Figure 22. Output response of EMDAP CVT ratio control for release condition

5. Conclusion

Modification of PID control scheme, from basic PID to PDCI schemes, yields a good performance of PID control. By conditionally activating the integral controller, maximum overshoot is about 0.15% and transient response is about 0.5s. For clamping condition, settling time is about 11.46s, while for releasing condition, settling time is about 8.33s. Steady state error of CVT ratio is very small, less than 0.001. This PDCI controller scheme is possible to simplify the programming algorithm using C language on microcontroller for controlling EMDAP CVT ratio.

Acknowledgement

The authors would like to thank the Malaysian Ministry of Science, Technology and Innovation (MOSTI) and Universiti Teknologi Malaysia (UTM) for their continuous support in the research work. This work was financial supported in part by the Malaysia TechnoFund under Grant number Vot. 79908.

References

[1] Ariyono, S., K.B. Tawi, H. Jamaluddin, M. Hussein and B. Supriyo. (2007). "Adaptive Neural Network Optimisation Control of ICE for Vehicle with Continuously Variable Transmission", *ICIAS 2007: International Conference on Intelligent & Advanced Systems, Vols 1-3, 257-262.*

[2] Akerhurst, S. and N.D. Vaughan. (1999). "An Investigation into the Loss Mechanisms Associated with an Automotive Metal V-belt CVT", *European Automotive Congress Vehicle Systems Technology for The Next Century, STA991407, Barcelona.*

[3] Supriyo, B., K.B. Tawi, H. Jamaluddin and S. Ariyono. (2006). "DC Motor Position Control for Pulley Axial Movement of an Electromechanical Dual Acting Pulley (EMDAP) CVT System", *1st Regional Conference on Vehicle Engineering & Technology, 3-5 July 2006, Kuala Lumpur, Malaysia.*

[4] Ang, K. H., Chong, G., Li, Y. (2005). "PID Control System Analysis, Design and Technology", *IEEE Transactions on Control Systems Technology 13(4): 559-576.*

[5] Padula, F and A. Visioli. (2011). "Tuning rules for optimal PID and fractional-order PID controllers." *Journal of Process Control 21(1): 69-81.*

[6] Huang, C. P., Y. T. Juang, Y. T. Chang. (2008). "Design of Fuzzy PID Controllers Using Modified Triangular Membership Functions." *Information Sciences 178(5): 1325-1333.*

[7] Supriyo, B., K.B. Tawi, H. Jamaluddin and S. Ariyono. (2006). "Ratio Control of an Electro-Mechanical Dual Acting Pulley Continuously Variable Transmission (EMDAP CVT) System Using PD-Fuzzy Logic Controller". *Regional Postgraduate Conference on Engineering and Science (RPCES'06). 26-27 July. Johor Bahru, Malaysia.*

[8] Mohd Ezlamy Zulkili., K.B. Tawi., M. Hussein. B. Supriyo. Aishah Daraoh. (2010)."Clamping Force Characteristic Study of Power Screw Mechanism in EMDAP CVT", *3rd International Graduate Conference on Engineering, Science and Humanities (IGCESH 2010).*

[9] Aung, W. P. (2007). "Analysis on Modeling and Simulink of DC Motor and its Driving System Used for Wheeled Mobile Robot", *Proceedings of World Academy of Science, Engineering and Technology, Vol 26, Parts 1 and 2, December 2007 26: 299-306.*

[10] Satish. R. Vaishnav1 and Zafar. J. Khan. (2010). "Performance Of Tuned PID Controller And A New Hybrid Fuzzy PD + I Controller". *UK: World Journal of Modelling and Simulation.*

IOP Conf. Series: Materials Science and Engineering **36** (2012) 012027 doi:10.1088/1757-899X/36/1/012027

Interaction Effect of Pressurized Lamination Pipe by using 2D Finite Element Analysis

N. Razak[1],A.Sulaiman, N.Alang
Universiti Malaysia Pahang, 26600 Pekan, Pahang,Malaysia

E-mail[1]: norhaida@ump.edu.my

Abstract. The availability of the multiple sites is often observed in pipeline and their interaction and coalescence may significantly affect their life. In this paper the finite element method has been used to study the effect of pressure defect on the stress field and their interaction in the interlaminar region. The effect of the crack size and the pressure defect will be investigated. The results are presented as the evolution of the stress field in the interlaminar region as a function of the pressure defect. It is observed that for two cracks with equal length, the pressure inside the lamination give greater value than those for unequal length

1. Introduction

There are many ways of practical application in which the pipe experience complicated loading. In general, cracks may be found on either the inner surface or the outer surface of a pipeline, which are called internal and external cracks or surface cracks, respectively. The surface cracks can be generated by different reasons, for instance the corrosion in the case of chemically corrosive fluid outside the pipe such as in presence of wet hydrogen sulphide. The hydrogen atoms attempted to attack the metal matrix in pipeline and join together to form hydrogen molecules. The accumulations of hydrogen molecules can cause the buildup of internal pressure at this site thus forming a larger crack.

Zacaria [1] proposed the pressure mechanism; where the incremental of the internal pressure in the lamination which leads to the interaction of the stress field as shown in Figure 1. Iino [2] reported that the formation of internal blister caused by the hydrogen precipitation at an inclusion interface. This followed by the formation of the blister crack array by cracking the region connecting the blister through the action of internal and the external pressure.

[1] University Malaysia Pahang,26600,Pekan Pahang Malaysia.

Published under licence by IOP Publishing Ltd

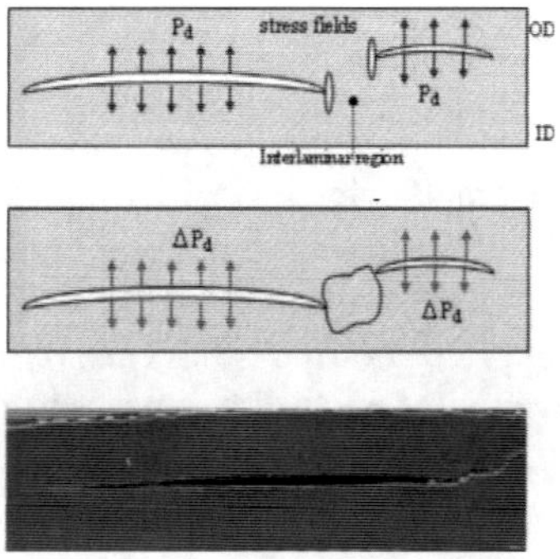

Figure 1: Schematic representation of the formation of a step wise cracks by pressure mechanism. (P_d = Pressure on defect)

This situation when there exist the pressure defect in the crack may increase the risk of unexpected and catastrophic failure during their nominal services. Furthermore the available design code may not reliable to assess these multiple cracks due to conservative method.

There are several methods for assessing the structural integrity of structural component that exist multiple cracks. These include body force method, boundary element method and. finite element method. Murakami [3] and Miyazaki [4] were among the first researcher who presented the stress intensity factor for coplanar surface cracks. Later, Jiang [5], Raju [6] and Moussa [7] employed the 3D finite element method to calculate the stress intensity factor for finite plate. Kirkhope [8] studied the stress intensity factor of coplanar cracks in cylindrical configuration. In this present paper, finite element method will be used to analyze the multiple non coplanar cracks in cylindrical configuration by considering the pressure on defect. The interaction effect in the interlaminar region will be investigated.

2. Material Properties

The material used for this paper was API X65. A true stress strain curve for this API X65 is shown in Figure 2 and the mechanical properties are tabulated in Table 1.

Table 1: Mechanical Properties of API X 65

	465 MPa
Ultimate strength	563 MPa
Elastic modulus	210 GPa
Poisson ratio	0.3

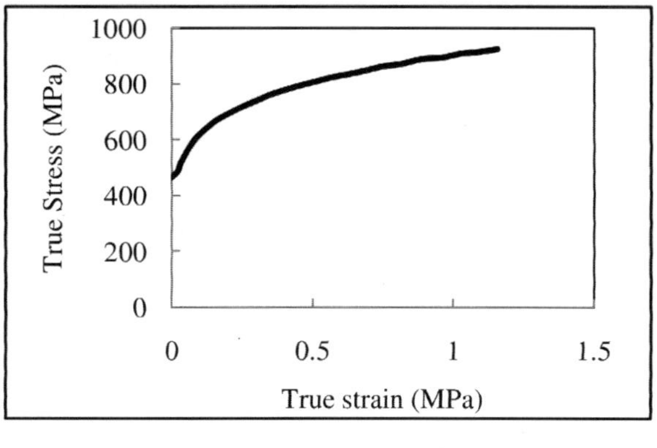

Figure 2: True stress strain curve for API X65

3. FE Modeling Conditions

In this study one quarter of the pipe was model due to an advantage of the symmetrical and to reduce the computation time. Two non-coplanar cracks was model by using MSc Marc. The pipe diameter was 610 mm and the wall thickness was 25.4 mm. The geometrical model was shown in figure 3 including the detail refinement mesh in the interlaminar region.

The boundary condition was applied both at the end of the pipe to allow free expansion. The operational pressure of 1.72 MPa was applied to the inner surface of the pipe. The load defect, P_d was pressurized until the interlaminar region become fully yielding. The dimensions of the pipe and the cracks are summarized in Table 2.The parameter varied in the model are shown in figure 5 and described as follows:

- Curvature of the defect on the right side,r_r
- Curvature of the defect on the left side,r_l
- Vertical separation,d_y
- Horizontal separation,d_x
- Pressure on defect,P_d
- Operational pressure,P_o

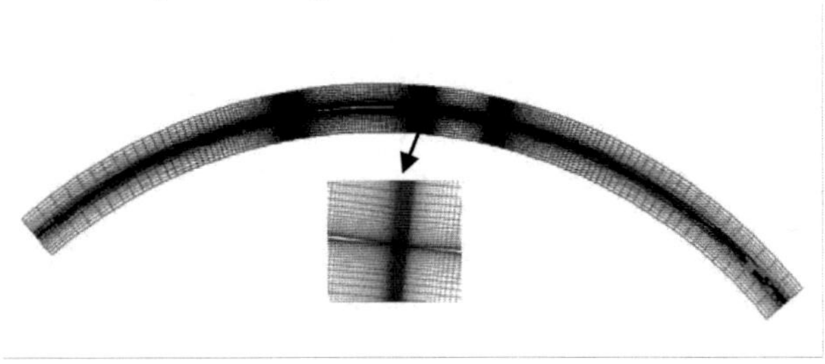

Figure 3: Geometrical modelling and mesh refinement in the interlaminar region

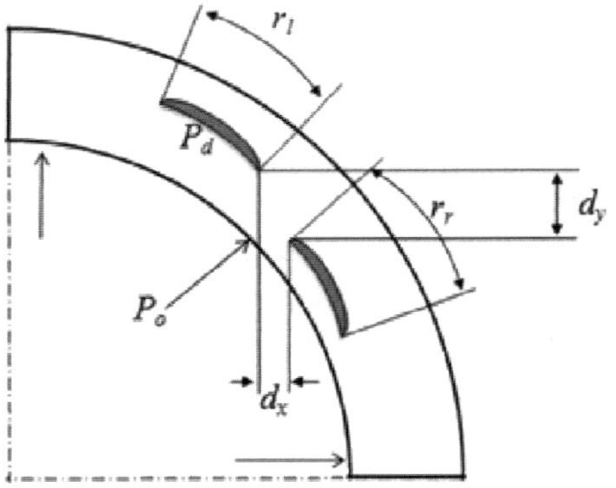

Figure 4: Variables of the model

Table 2: The crack geometry and parameter of the crack

Case number (CN)	r_l (mm)	r_r(mm)	d_x (mm)	d_y (mm)
1	6.35	6.35	0.7937	0.7937
2	38.1	19		
3		38.1		
4		63.5		
5	63.5	19		
6		38.1		
7		63.5		
8	89	19		
9		38.1		
10		63.5		

4. Results and Analysis

All the cases of study behaved in similar fashion. Due to that, the results of stress distribution for one case study are presented and described. The case described is CN 2, with dimensions given in Table 2.

4.1Effect of pressure defect and operational pressure along the crack path

The effect of zero pressure defects was shown in figure 5. It is observed that as the operational pressure increase the stress increases. The maximum stress occurs at the inner and outer crack tip and it was reduced at the midpoint. Figure 6 shows the similar fashion with the presence of pressure defect.

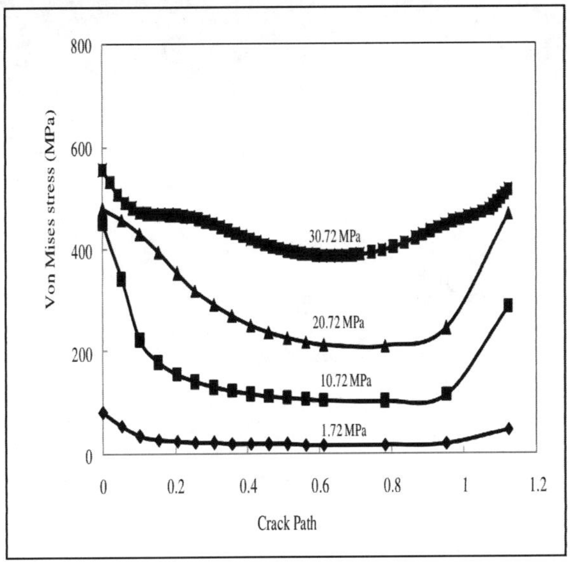

Figure 5: Stress distribution along the crack path without pressure defect in the interlaminar region

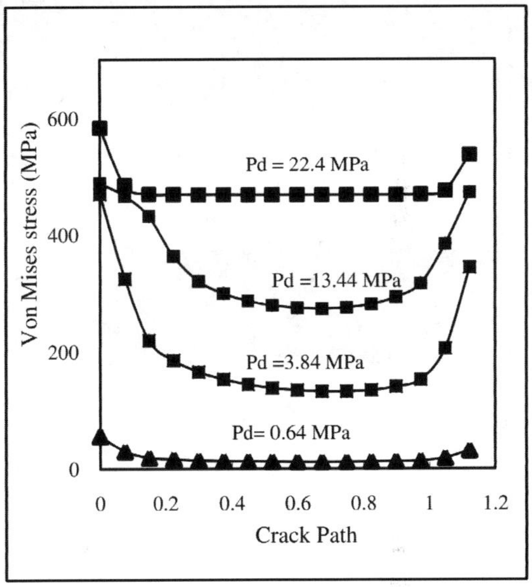

Figure 6: Stress distribution along the crack path with pressure defect in the interlaminar region .Operational pressure = 1.72 MPa

4.2 Stress field in the interlaminar region

The von Mises stress distribution for various pressure steps are shown in Figure 7(a-f). The maximum stress occurred at the internal tip, i.e in the interlaminar region, because this stress is higher than at the external tip. This indicates that there is an interaction of the stress field since the first pressure step. Figure 7(a) shows that at a pressure on defect of 3.82, MPa, there is a noticeable yielding in the at the internal crack tip. The von Mises stress increased when the pressure on the defect increased as shown in Figure 2(b and c) but not the entire interlaminar region.

As the pressure on defect increase to 23.04 MPa as shown in Figure 2(d), the von mises stress in the interlaminar region are above yield strength of the material. This region considered fully plastic. Finally, as shown in Figure 2(f), at the pressure on defect equal to 32 MPa, the von mises stress in the interlaminar region passes the ultimate strength of the material.

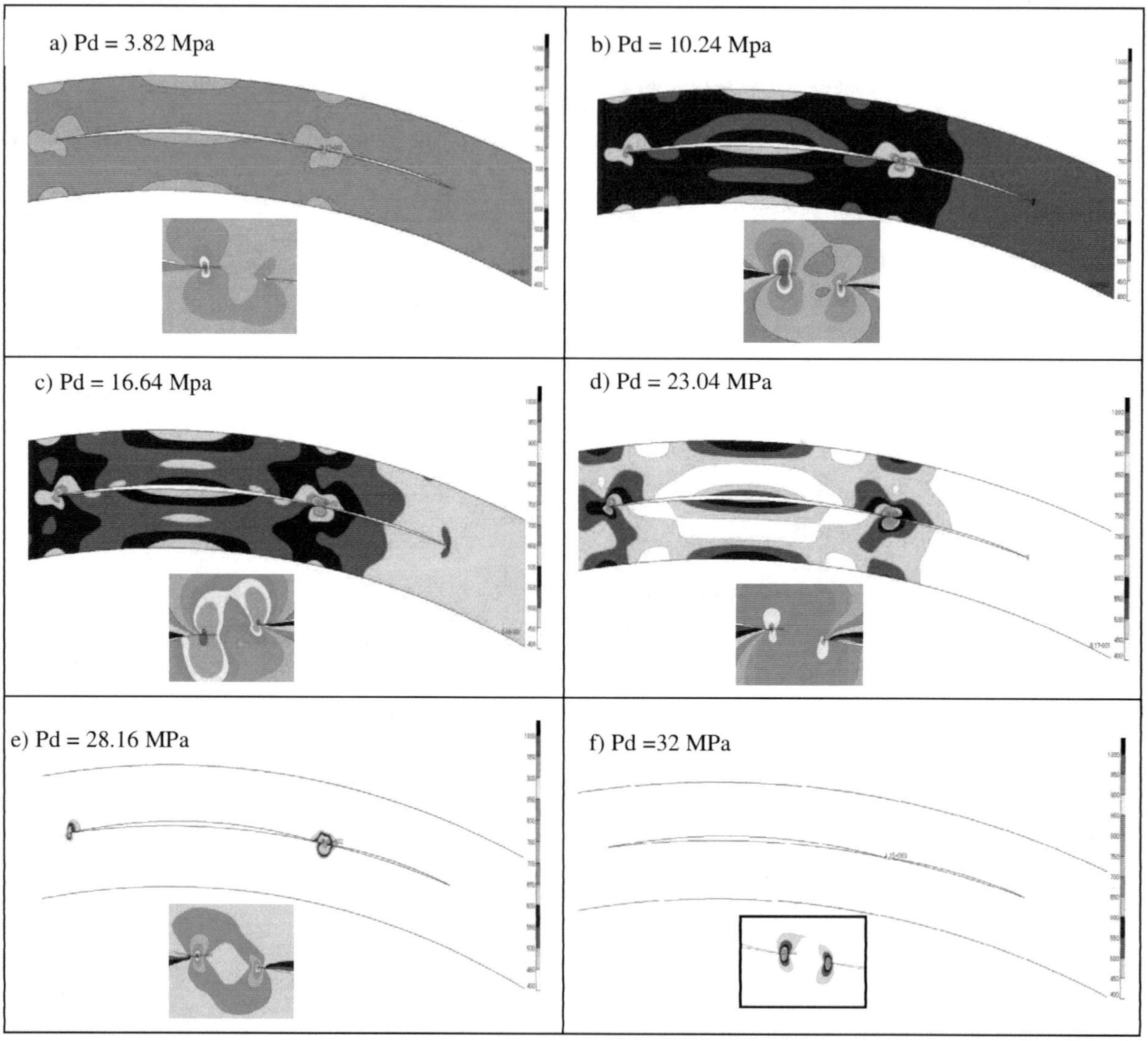

Figure 7: Von Mises stress distribution for various pressure steps

4.3 Variation of Pressure on defect and Von Mises stress

The maximum von mises stress were plotted to determine and analyze the mechanical behaviors of the material the in the region where the crack interact. By doing this, the transitions from linear to nonlinear behavior, i.e. the onset of the plastic deformation and the value of the critical lamination pressure can be observed. The critical lamination pressure is defines as the pressure in the interlaminar region when the von mises stress in the interlaminar region reach or surpass the ultimate strength of the material.

Figure 8 shows the maximum von Mises stress in the interlaminar region as a function of the pressure on defect for case 1 where the r_l is equal to r_r. In this case, the pressure that causes the interlaminar region reach or surpasses the ultimate strength of the material was 148 MPa. Figure 9 shows a reduction of pressure on defect almost 86%. For case 2, 3 and 4, the $r_l$was 38.1 mm and r_r increase from 19 to 63.5. It is shown that in case 3 where the radius of defect is similar, the pressure on defect slightly increases. The pressure on defect in case 4 is reduced to 14.72 MPa as the radius of crack become greater.

In case 7, it is obvious that the pressure on the defect was greater than in case 5 and 6. This is because the radius of defect on the right side becomes similar with the left side which is 63.5 mm. The pressure on defect decrease from 9.6 MPa to 7.4 MPa when the r_r increase. However for case 7 and 10, the pressure on defect reduces from 16.8 MPa to 9.6 Mpa. Figure 11 shows the variation of pressure on defect for ri equal to 6.35, 38.1 and 63.5. Obviously, the crack with size 6.53 mm, has greater pressure in defect as compared to the others. When the crack length increase from 38.1 mm to 63.5 mm, the pressure on defect reduce from 22.3 to 16.8 MPa. In reality, it difficult to have the crack with similar size.

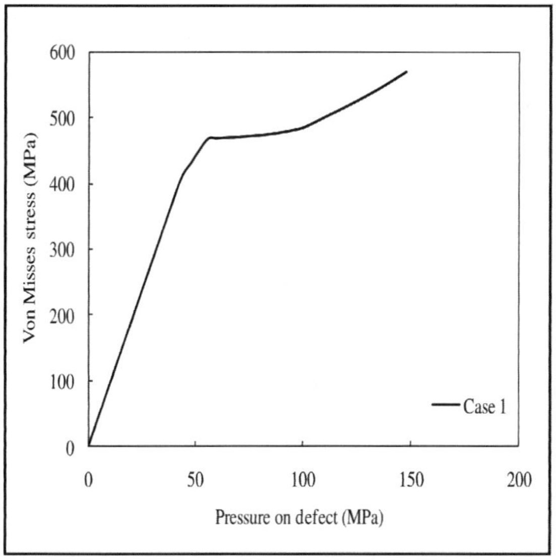

Figure 8: Maximum von Mises stress in the interlaminar region as a function of the pressure on the defect for Case 1, $r_l = 6.35$ mm

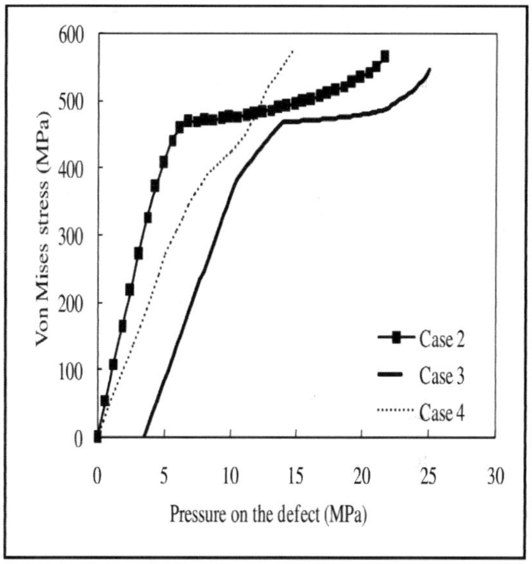

Figure 9: Maximum von Mises stress in the interlaminar region as a function of the pressure on the defect for r_l=38 mm

Figure 13 shows the maximum von mises stress in the interlaminar region for three equal crack length, 6.35 mm, 38.1 mm and 63.5 mm. The pressure defect for r_i equal to 6.35 mm, gives the higher value than larger cracks.

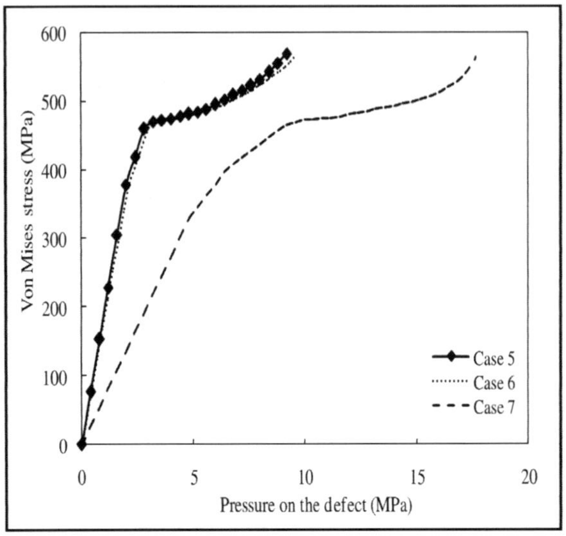

Figure 10. Maximum von Mises stress in the interlaminar region as a function of the pressure on the defect for r_l=63.5 mm

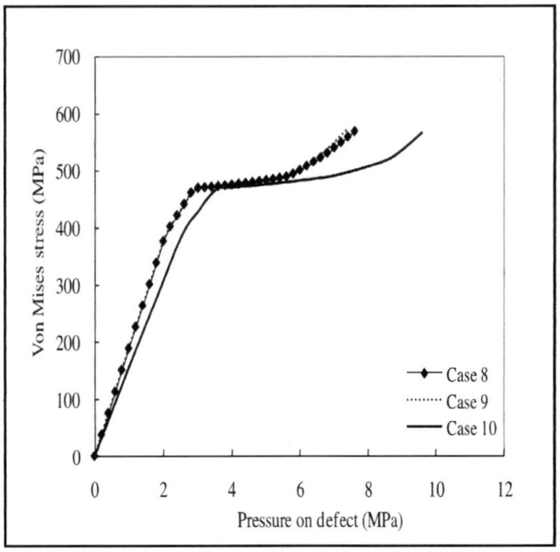

Figure 11. Maximum von Mises stress in the interlaminar region as a function of the pressure on the defect for r_l=89 mm

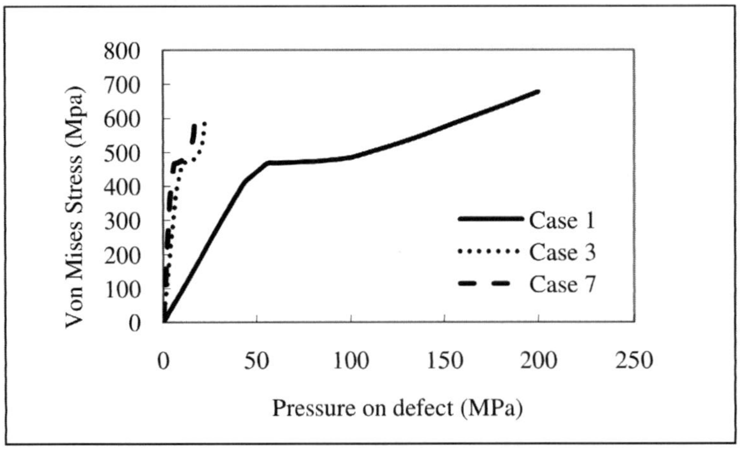

Figure 13. Maximum von Mises stress in the interlaminar region for three equal cracks.

5. Conclusion

The stresses at the internal crack tips are higher than at the external crack tip. This indicated that there is an interaction of the stress field since the first pressure step inside the lamination. The interaction of the stress field of two approaching non coplanar pressurized cracks will be higher and therefore will produce higher stresses as the cracks become similar size.

References

[1] Zacaria, M. Y. B., and Davies, T. T., 1993, "Formation and Analysis of Stack Cracks in Pipeline Steel," J. Mater. Sci., 28, pp. 322–328

[2] Makkio, Iino, 1978, "The Extension of Hydrogen Blister-Crack Array in LinePipe Steels," Metall. Trans. A, 9A, pp. 1581–1590.

[3] Murakami, Y. and S. Nemat-Nasser (1982). "Interacting dissimilar semi-elliptical surface flaws under tension and bending." Engineering Fracture Mechanics **16**(3): 373-386

[4] Miyazaki, N., H. Kaneko, et al. (1989). "Stress intensity factor analyses of interacting elliptical cracks using line-spring boundary element method." International Journal of Pressure Vessels and Piping **38**(1): 1-14.

[5] Jiang, Z. D., J. Petit, et al. (1991). "Stress intensity factors of two parallel 3d surface cracks." Engineering Fracture Mechanics **40**(2): 345-354.

[6] Raju, I. S. and J. C. Newman Jr (1979). "Stress-intensity factors for a wide range of semi-elliptical surface cracks in finite-thickness plates." Engineering Fracture Mechanics **11**(4): 817-829.

[7] Moussa, W. A., R. Bell, et al. (1999). "The interaction of two parallel non-coplanar identical surface cracks under tension and bending." International Journal of Pressure Vessels and Piping **76**(3): 135-145.

[8] Kirkhope, K. J., R. Bell, et al. (1990). "Stress intensity factor equations for single and multiple cracked pressurized thick-walled cylinders." International Journal of Pressure Vessels and Piping **41**(1): 103-111.

IOP Conf. Series: Materials Science and Engineering **36** (2012) 012028 doi:10.1088/1757-899X/36/1/012028

The Influence of impact on Composite Armour System Kevlar-29/polyester-Al_2O_3

A A Ramadhan, A R Abu Talib[1], A S Mohd Rafie and R Zahari

Department of Aerospace Engineering, Universiti Putra Malaysia, 43400 Selangor, Malaysia

E-mail[1]: abrahim@eng.upm.edu.my

Abstract. .An experimental investigation of high velocity impact responses of composite laminated plates using a helium gas gun has been presented in this paper. The aim of this study was to develop the novel composite structure that meets the specific requirements of ballistic resistance which used for body protections, vehicles and other applications. Thus the high velocity impact tests were performed on composite Kevlar-29 fiber/polyester resin with alumina powder (Al_2O_3). The impact test was conducted by using a cylindrical steel projectile of 7.62mm diameter at a velocity range of 160-400 m/s. The results (shown in this work) are in terms of varying plate thickness and the amount of energy absorbed by the laminated plates meanwhile we obtained that the 12mm thickness of composite plate suitable for impact loading up to 200m/s impact velocity. Therefore this composite structure (it is used to reduce the amount of Kevlar) considered most economical armoure products. We used the ANSYS AUTODYN 3D- v.12 software for our simulations. The results have been obtained a4.1% maximum errors with experimental work of energy absorption.

NOMENCLATURE

ρ	Density (g/cm3)
ν	Poisson's ratio
E_1	Longitudinal Young's Modulus (GN/m2)
E_2	Transverse Young's Modulus (GN/m2)
V_f	Volume fraction of fiber (N/m)
V_i	Initial velocity (m/s)
V_r	Residual velocity (m/s)
E_{Abs}	Energy Absorption by the specimen (J)
m_p	Mass of bullet (Kg)
V_b	Ballistic limit velocity (m/s)

1. Introduction

Human's protection has been an important issue since the beginning of creation. Throughout the recorded history, there have been various types of material that are utilized as protection garments from injury, such as in battles and other dangerous situations.[1].

The study of energy absorption behavior of metallic plates subjected to projectile impact requires attention due to the wide range of applications and complex nature of the problem. Extensive research

Published under licence by IOP Publishing Ltd

has been carried out in the past [2, 3] to understand various features of this complex phenomenon both from the point of view of the projectile as well as the target plate by using deferent impact conditions and varying target configuration.

R. Barauskas and A. Abraitiene et al [4]. used finite element model of the ballistic test against the multi-layer textiles package structure and they developed in LS-DYNA. The bullet has been considered as a deformable body in contact with the fabric package represented by an interwoven yarn structure. The energy absorption characteristics of thin single as well as layered in-contact aluminum target plates of different thicknesses impacted by blunt, ogive and hemispherical nosed steel projectiles. Influence of various parameters affecting the energy absorption capacity of the target, such as kinetic energy of the impact- ing projectile and its nose shape, configuration of the target plate, its mode of failure as well as thickness of the target plate was studied. The experimental findings were also verified with the numerical sim- ulations carried out using finite element code ABAQUS. Energy absorption capacity of the target plates obtained from the numerical analysis was found in good agreement with the experimental observations.[5]. Furthermore, the impact response of a non-woven symmetrical CFRP laminate has been compared to that of a woven laminate over an impact-energy regime of 92 J to 459 J. At lower impact- energies there were strong indications that the non-woven laminate out-performed the woven laminate whereas at the higher impact-energies the ballistic performance was seen to be approximately the same.[6], while the compression strengths of the polyester resins were found by M. Piggott and B. Harris [7]. Designing ballistic barriers of Kevlar 29 plates, based solely on physical models, would require a large number of experimental data, which are time consuming and costly to generate. Recent advances towards understanding damage mechanisms of laminated composites, (e.g. [8–10]) offer the possibility of avoiding many of the experimental tests by using computational simulation, bearing in mind that the numerical results should be used with precaution and be always certified by experimental tests.

The novel sandwich plate (Kevlar29/epoxy-Al_2O_3) laminated plate fabricated in this research can be contributed in many applications like Vehicle, body Armour, aerospace and military applications. The high velocity impact response of fiber (4mm, 8mm and 12mm) thicknesses of this composite material presented in this work by using cylindrical bullets type 4340 steel projectiles. Effects of projectile velocity, thickness of plates on the ballistic impact and energy absorption behavior of the targets are investigated experimentally and numerically simulation. The simulations are carried out using ANSYS AUTODYN 3D- v.12 software.

2. Experimental methodology

2.1. Materials used , properties and fabricating the specimens

The materials used in this work as follow as

- Woven fiber Kevlar-29
- Polyester.
- Alumina (Al_2O_3) powder.

The mechanical properties of these materials have been specified in table 1 as bellow

The volume fraction of fiber V_f (fiber volume/total volume) in composites is very often significantly less than 100%, and it was tested by Amotz and Schwartz [11] as shown in table 2, so the volume fraction was selected 55% in this work, it was a optimum value compared with others volume fraction for fabricating the specimens.

Table 1 The mechanical properties of fiber Kevlar-29 as shown in table 1 bellow

Properties	Material		
	Kevlar-29 fiber	Polyester (6061-T6)	Alumina a resin
Density (g/cm3)	1.4	2.7	1.12
Ultimate Tensile strength(MPa)	58.9	310	
Tensile Yield Strength (MPa)	2758	276	4.84
Modulus of Elasticity (GPa)	62	68.9	42
Poisson's Ratio	0.44	0.33	
Fatigue Strength (MPa)		96.5	
Shear Strength (MPa)		207	

The advancement in designing a new composite constructor is embodied in the composite compounds arrangement and the tactical ways that are depended on the assembled layers. Meanwhile, the target geometry is a sensitive point in this investigation, and the adjacent panels have the responsibility to increase energy absorption by the targets.

The target preparation processes include composite preparation and geometry of the target. Hence, the composite was constructed from the Kevlar29 and polyester resin mixed with alumina powder, since the target area was 10 × 10 cm, as illustrated in Figure.1

Figure 1. Kevlar29/polyester-Al_2O_3 laminated plate

Three thicknesses (4, 8 and 12 mm) of specimens were fabricated by using Hand lay-up method of all the specimens. Woven roving Kevlar-29 fibre is wound manually in the open mold of glass, which was coated with wax to easily remove the specimen later, and polyester resin mixed with alumina powder by mixer until homogenous occurred the ratio of polyester to alumina powder is 2:1, then brushed over and into the woven roving Kevlar-29 fibre , The time of solidification is about 72 hours to provide optimum blend. The tensile test as shown in figure 2, compression test also achieved in this work to obtain the basic mechanical properties to obtain the result of stress strain curve as shown in figure 3 of tensile and compression test for this laminated plate which it used later in simulation work, but the main objective in this work was fabricated novel sandwich structure and tested it under high impact loading.

Figure 2.Specimens under tensile test.

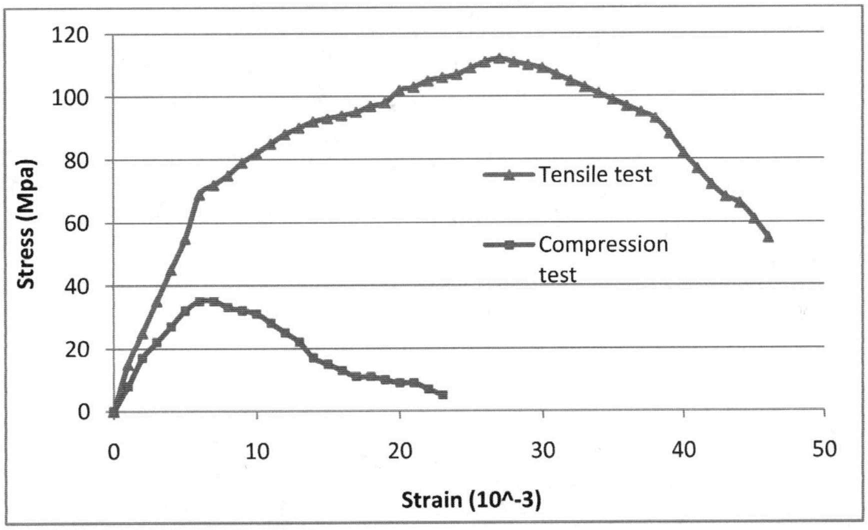

Figure 3. Stress-strain curve for tensile and compression test loading of Kevlar29/polyester with Al_2O_3 laminated plate.

The volume fraction used in all specimens fabricated is 55% and the total specimens were tested in this work 27 laminated plates; for compression and tensile test 6 and for high velocity impact test 21 specimens were subjected to cylindrical bullet with different initial velocity impact as shown in table 1.

Table 1. The volume fraction and number of specimens used in experimental work.

Thickness (4mm)	No. of Kevlar-29 layers	Volume fraction (V_f)	Initial Velocity impact m/s	No. of specimens
4	6	55%	160, 200,240,280,320,360,400	7
8	14	55%	160, 200,240,280,320,360,400	7
12	22	55%	160, 200,240,280,320,360,400	7

2.2. Gas Gun and Bullet selection

The instrumented impact test equipment used in this study was a gas gun impact tester M. T. H. Sultan [12] shown in Figure 4. The general features of the testing apparatus are shown in Figure 5 has been designed in order to launch the projectile. The main components of the gas gun are the 2200 Psi pressure tank, the purpose-built "ring section for compressed gas, the 4 m long smooth barrel and (60, 60 and 40 cm) dimensions of box chamber to hold the specimen inside this box which it has (60 x 60 cm) framing window to observe the behavior of target and bullet by photography high speed camera (200,000 fps) for this work.

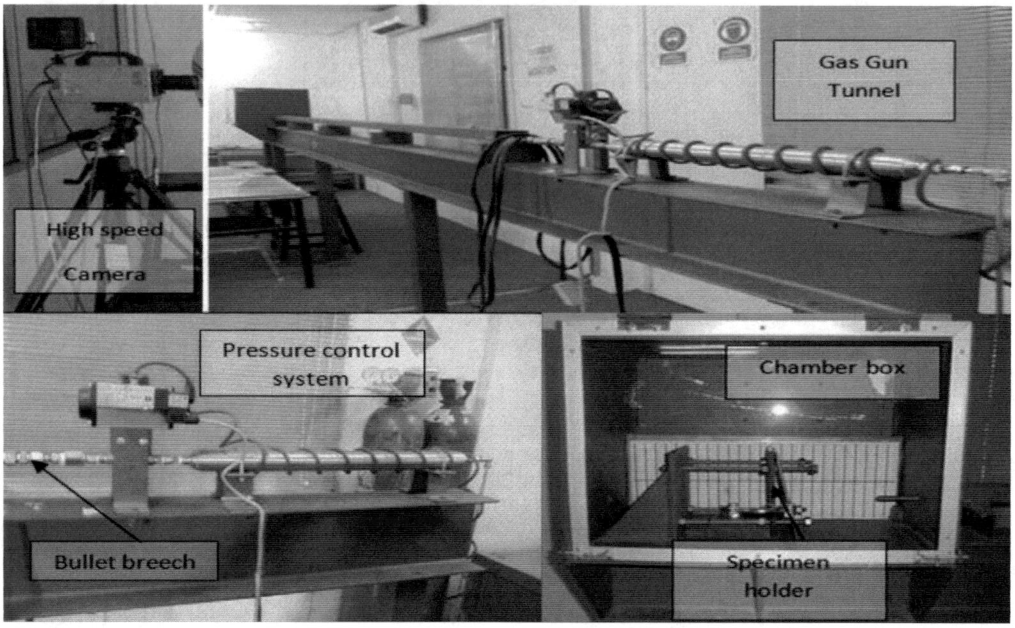

Figure 4. Gas gun tunnel to test the specimens.

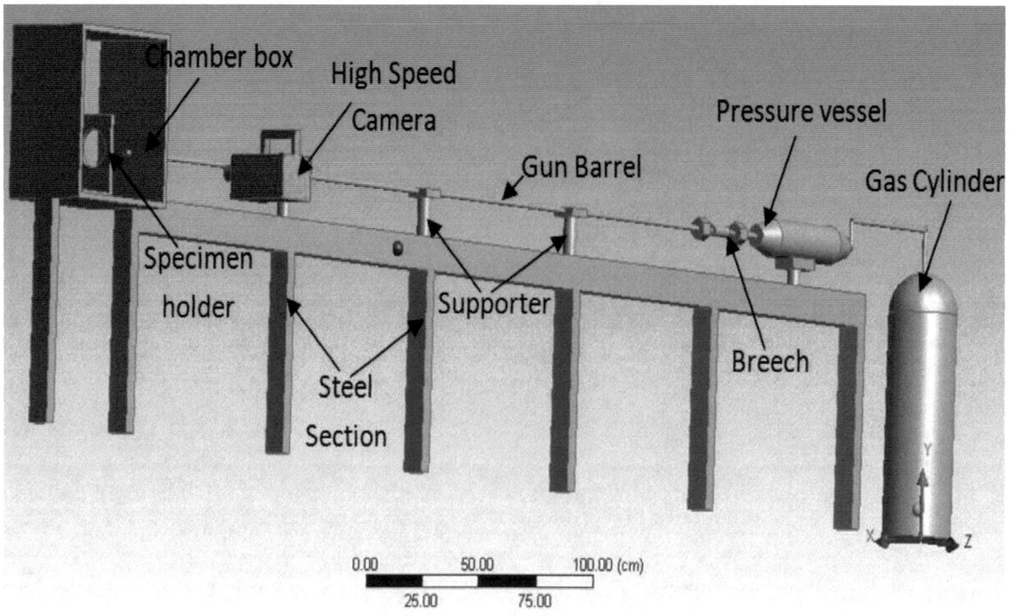

Figure 5. Gas gun tunnel (instrument impact tester).

The specimen mounted by (25 x 25cm) two frames to prevent the movement during the impact test as shown in figure 6.

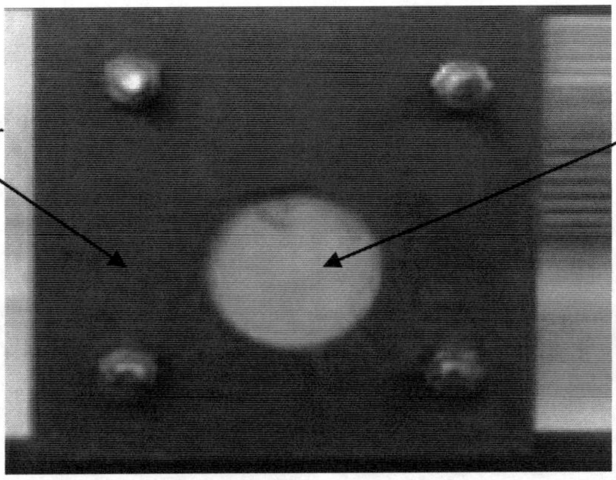

Figure 6.The target mounted with holding frame.

The gas gun is capable of launching three (cylindrical, spherical and nose tip) but in this work used cylindrical shape of 7.62mm diameter steel projectile and 5g of 4340 steel bullet (as shown in figure 7) package to a range of velocity of 180-400 m/s when helium is used as a propellant gas.

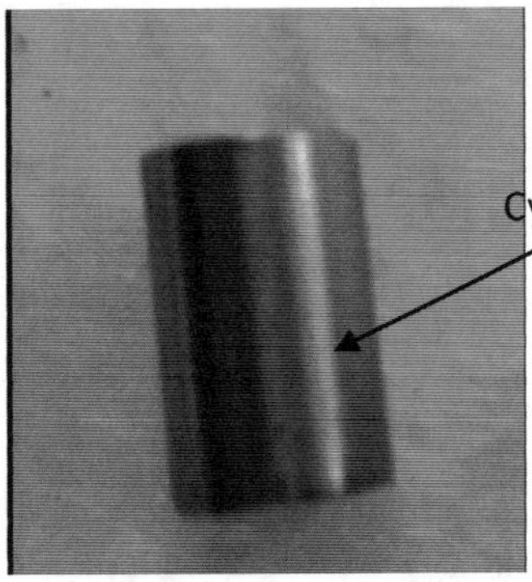

Figure 7. Cylindrical steel 4340 bullet.

2.3. Impact Testing

The impact testing was Installed the composite of woven Kevlar-29/polyester with alumina (Al_2O_3) laminated plates of square shape (100 x 100 mm) with different (4, 8 and 12 mm) thicknesses dimensions inside the specimen holder and focused the high speed camera with suitable resolution (200,000 fps) and adjusted the lights to get optimum conditions before test. However, both the aligning of the system and the luminance during testing are crucial parameters in order to have good, reliable images. As indicated above, the high-speed camera system is used for measuring the velocity of projectiles. This is made possible by advanced image processing of the digital pictures which it's connected with P.C. to calculate the time of bullet before impact and after penetration if accrued of specimens as shown in figure 8 an example of 280 m/s velocity impact. The most important feature is the possibility to obtain the initial and residual velocity by the following equations:

$$v_i = \frac{\Delta S_1}{\Delta t_1} \quad (1)$$

and

$$V_r = \frac{\Delta S_2}{\Delta t_2} \qquad (2)$$

Where

V_i and V_r initial and residual velocity (m/s) respectively

S_1 and S_2 displacement before and after penetration (m) respectively and t_1, t_2 time before and after penetration (s) respectively.

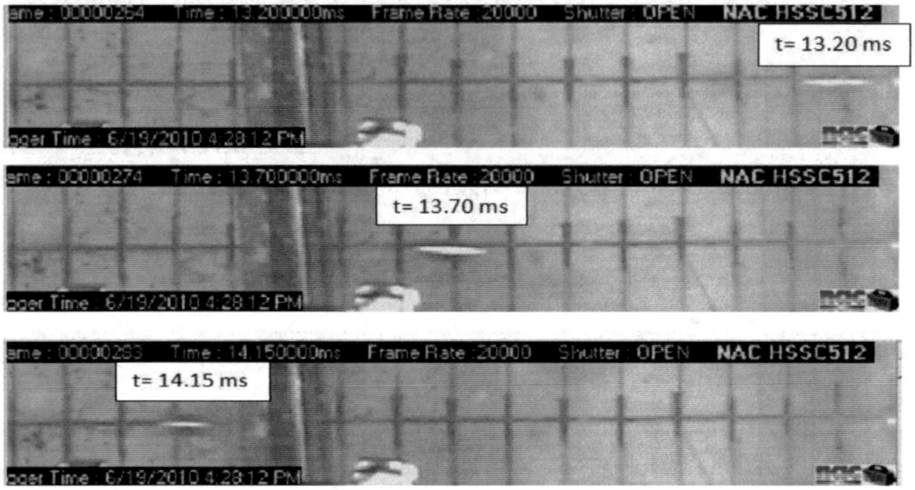

Figure 8. Frames record history by high velocity impact during impact test of specimen

The total energy absorption by the specimen according to the following equation :

$$E_{Abs} = \frac{1}{2} m_p \left(V_i^2 - V_r^2\right) \quad \text{or} \quad E_{Abs} = \frac{1}{2} m_p V_b^2 \qquad (3)$$

$$V_b = \sqrt{V_i^2 - V_r^2}$$

Where

E_{Abs}= Energy Absorption by the specimen (J), m_p = mass of bullet (Kg) and V_b= ballistic velocity (m/s).

3. Description of the numerical model

3.1 Material modeling

The main properties of fiber Kevlar29/polyester -Alumina (Al_2O_3) and Steel 4340 were used in this simulation at room temperature condition shown in tables 2 and 3.

Table 2. Kevlar29/polyester-Alumina(Al_2O_3) (orthotropic)

E_{11} (Kpa)	E_{22} (Kpa)	E_{33} (Kpa	Poisson's Ratio XY	Poisson's Ratio YZ	Poisson's Ratio XZ	Shear Modulus XY Pa	Shear Modulus YZ Pa	Shear Modulus XZ Pa
1.15e+7	1.15e+7	1.15e+6	0.08	0.7	0.075	1.e-004	1.e-004	1.e-004

The Johnson-Cook failure model has been used for projectile material, deformations were observed in the cylindrical steel 4340 bullet as shown in table 3.

Table 3. Steel 4340 impactor (Johnson Cook Strength)

Initial Yield Stress Pa	Density Kg/m^3	Bulk Modulus (Pa)	Strain Rate Constant	Thermal Softening Exponent	Melting Temperature C	Reference Strain Rate (/sec)
7.92e+008	7830	1.59e+011	1.4e-002	1.03	1519.8	1

3.2 Finite element mesh and Autodyn simulation

The finite element model by Lagrange model using ANSYS v12.1, where a sandwich plate is idealized using the same conditions in experimental work like fixed ended of plate thickness of plates mechanical properties, initial velocity of bullet and the properties of bullet as shown in figures 9 and 10-a. Explicit mesh element of 18551 nodes and 15266 elements for 12mm, 13349 nodes and 10266 elements for 8mm, 7803 nodes 5000 elements for 5mm target thickness, while for the bullet 344 nodes and 266 elements was used to simulate the target and the impactor by Academic Autodyn 3D version 12.1 as illustrated in figures 10,11 and 12 that show also the damage view occurring during different time . Each layer of the plate was modeled using two solid elements through the thickness, and the minimal element size was 0.2 x 0.2 mm at the plate, which was the point of impact

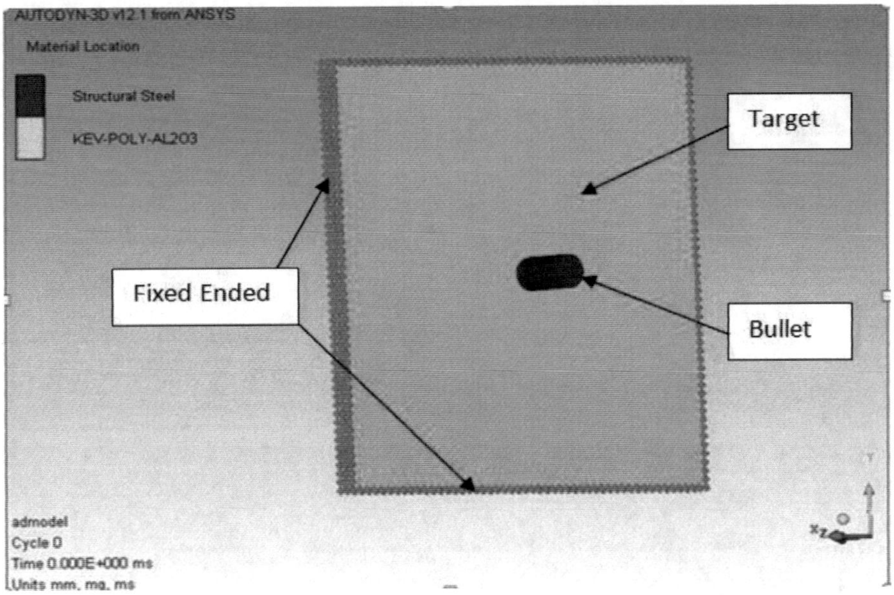

Figure 9. 4mm target of Kevlar29/polyester and Al_2O_3 fixed ended subjected to cylindrical steel bullet at time 0 ms and cycle 0.

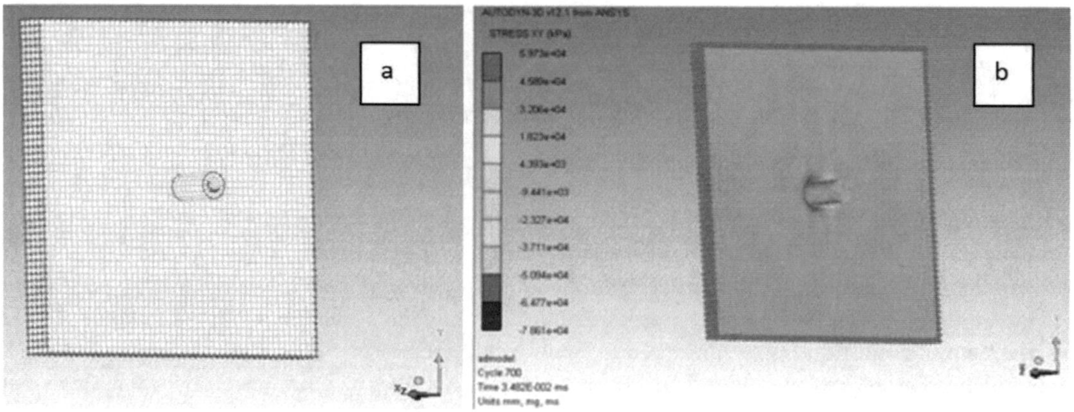

Figure 10. 4mm target of Kevlar29/polyester and Al_2O_3 subjected to 200m/s cylindrical steel bullet a- explicit mesh and b-stress damage occurred at time 3.4 ms.

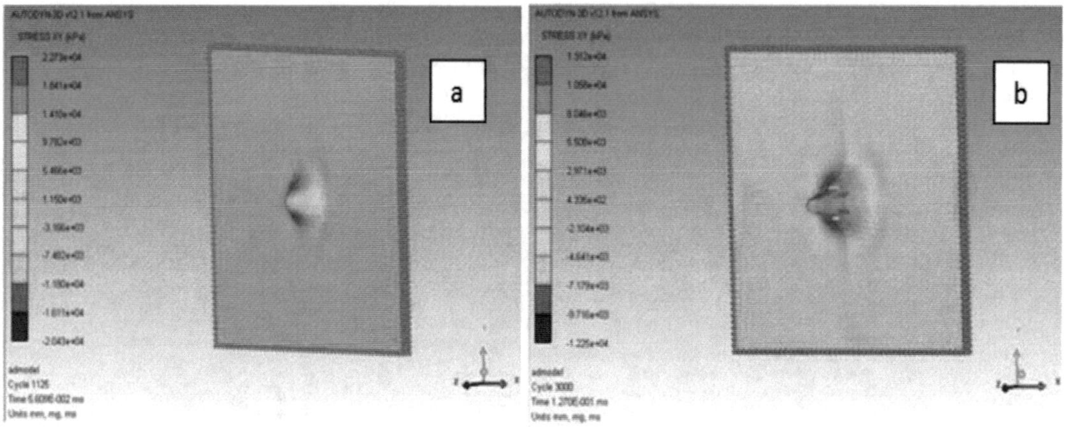

Figure 11. Back view stress damage of 4mm target of Kevlar29/polyester and Al_2O_3 subjected to 200m/s cylindrical steel bullet (a) at time 5.6ms and (b) at time 12.2ms.

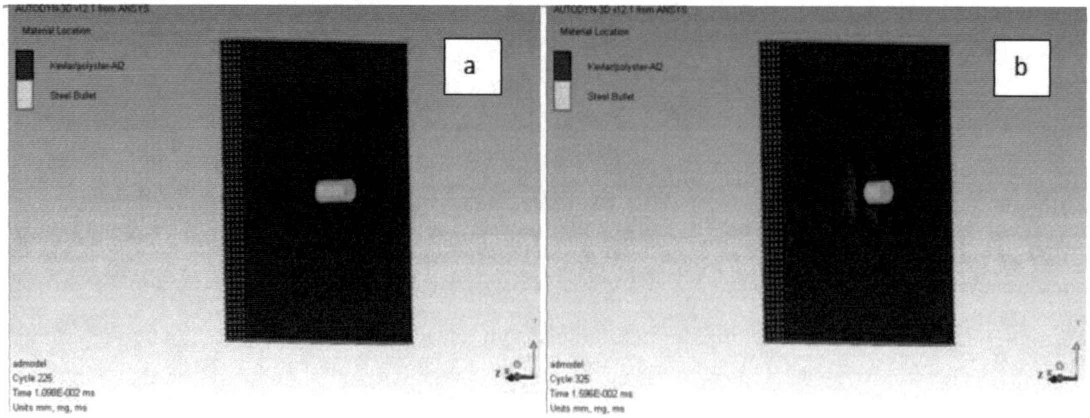

Figure 12. The damage of 8mm target of Kevlar29/polyester and Al_2O_3 subjected to 400m/s cylindrical steel bullet (a) at time 109ms and (b) at time 159ms.

4. Results and discussion

The predicted damages by Autodyn-3D v12.1 software of the composite Kevlar29/polyester-Al_2O_3 laminated plates perforating listed in table 4 with different (4, 8 and 12mm) thicknesses at range of (160 - 400 m/s) strike velocities of steel 4340 bullet, an example of (100 x 100 x 4mm) target under 240 m/s shown in figures 13 at different history cycles, this figure shows the stress and strain distribution which it results by the damage occurred in the specimen, hence the simulation was done as the same conditions in experimental work so as to obtain good comparison.

To comparison the damage occurred during this simulation with experimental test Figure14 shows the specimen tested of 4mm thickness of laminated plate and, while figure 15 for impactor steel bullet.

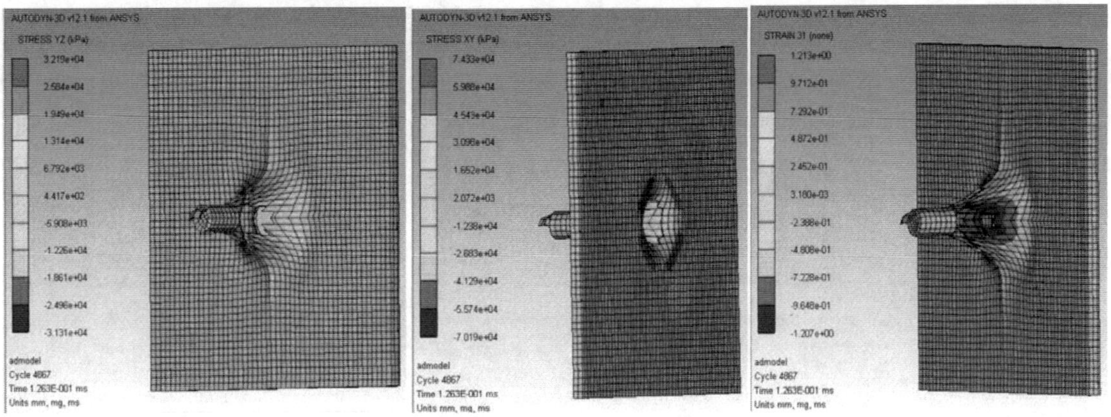

Figure 13. The stress and strain of (100 x 100 x 4 mm) Kevlar29/polyester and Al2O3 subjected to 240 m/s Steel 4340 at different times.

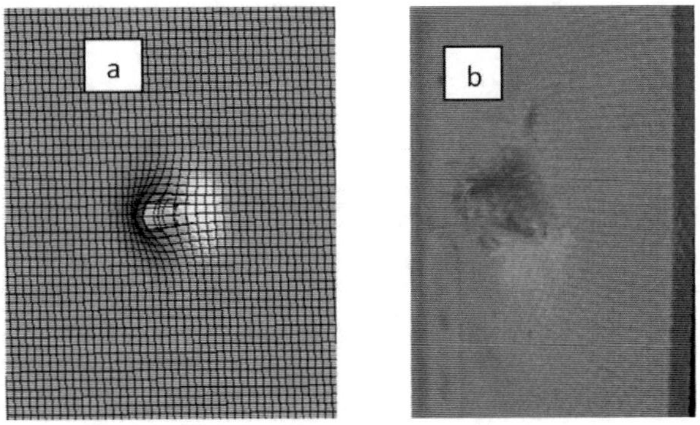

Figure 14. Comparison of the perforated damage in the fabric target in the experimental and theoretical analyses: (a) simulated and (b) observed

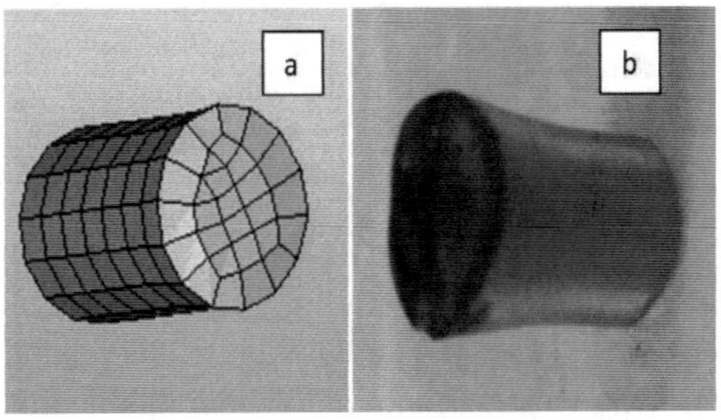

Figure 15. The damage occurred of Steel4340 bullet at 400m/s velocity on 12mm thickness of target (a) simulated and (b) observed

The initial velocity versus residual velocity of (4, 8 and 12mm) laminated plated thicknesses subjected to rang of impact velocity (180-400 m/s) steel bullet was presented in figure16 which it shows decreasing the residual velocities with increasing thickness of targets until no penetration occurred via 12mm thickness of laminated plate especially at initial velocity less than 200 m/s.

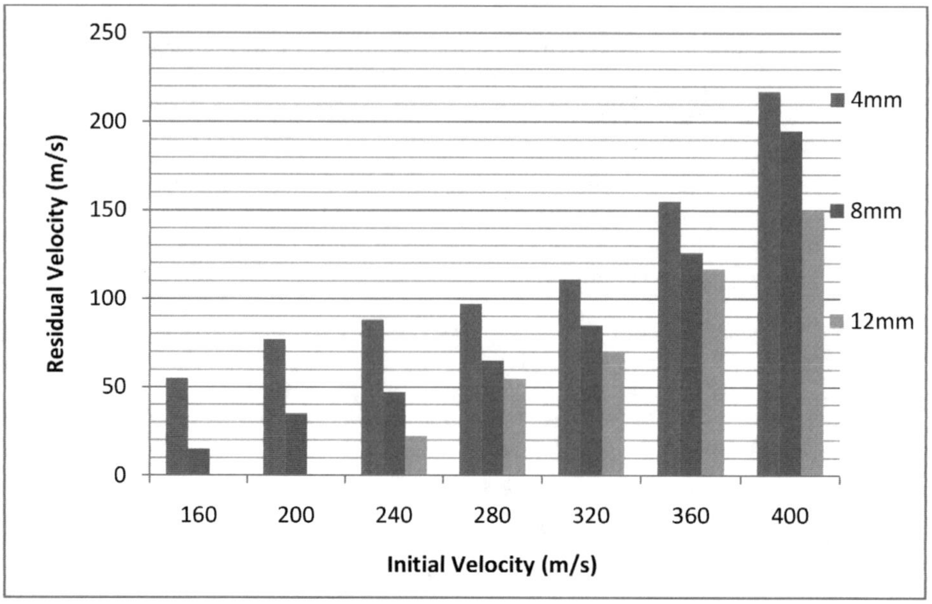

Figure 16. Simulation of time versus velocity of 4mm target (Front, Middle and back stacking sequence Al plate) plate subjected to 240 m/s steel bullet.

Figures 17 and 18 presented the experimental and simulation results respectively of initial velocity versus energy absorption of (4, 8 and 12mm) thicknesses of kevlar29/polyester with alumina laminated plates. The numerical simulation by using Autodyn-3D v12.1 obtained a good approximation to the experimental results.

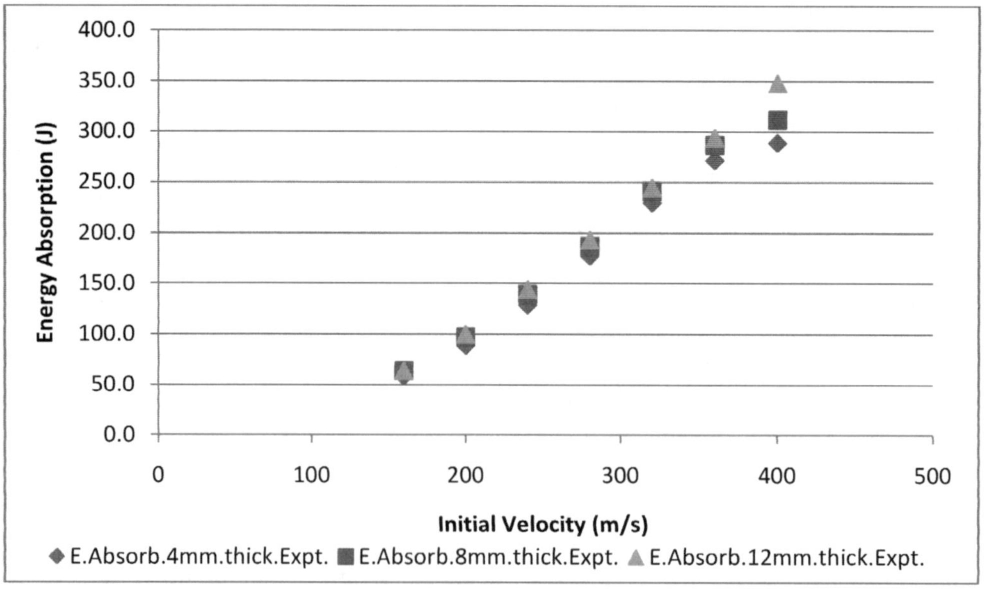

Figure 17. Experimental work of initial velocity versus energy absorption of 4mm, 8mm and 12mm targets laminated plate subjected to range (160-400 m/s) steel bullet .

The velocity history during impact also dependent on the thickness of plate as shown in figure 19, that shows an example of 200m/s of initial velocity for different thickness of kevlar29/polyester with alumina laminated plates, the time of drop slop increase with the thickness of plate.

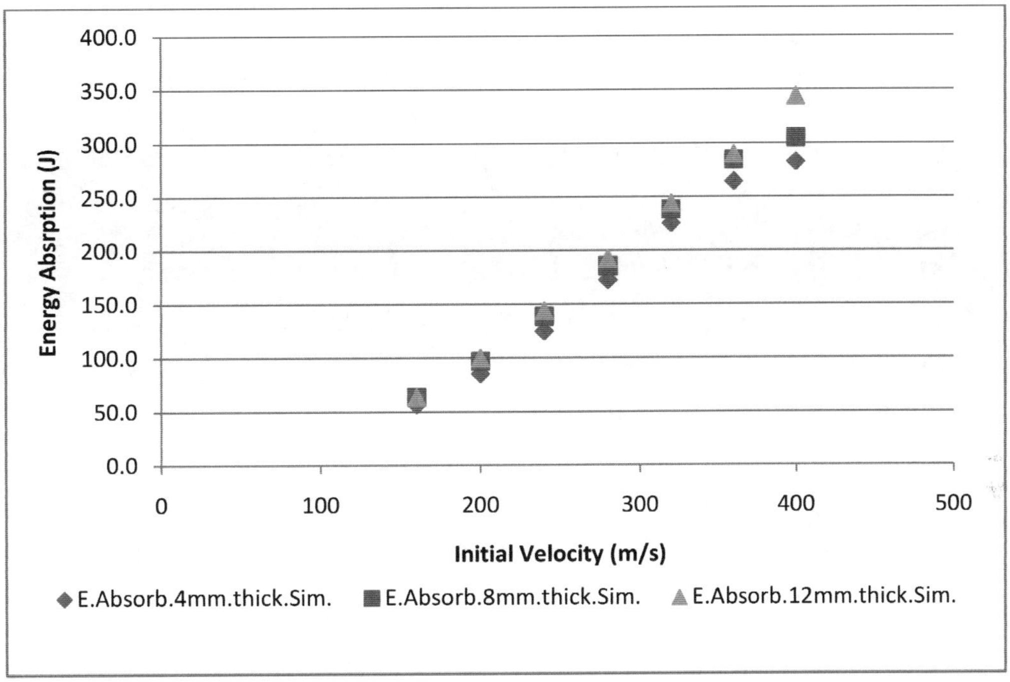

Figure 18. Simulation of initial velocity versus energy absorption of 4mm, 8mm and 12mm targets laminated plate subjected to range (160-400 m/s) steel bullet.

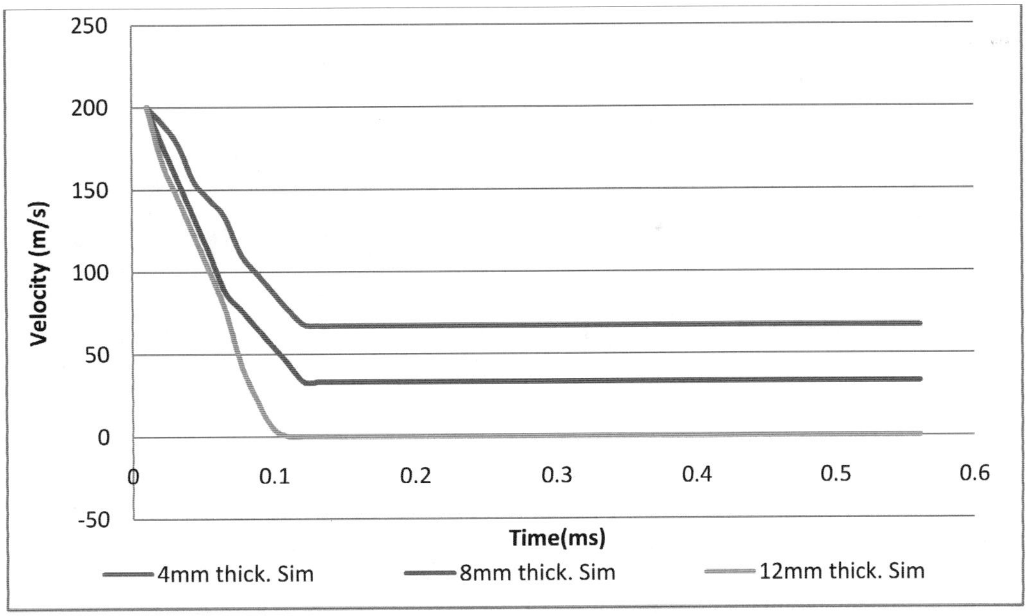

Figure 19. Simulation of initial velocity versus energy absorption of 4mm, 8mm and 12mm targets laminated plate subjected to range (160-400 m/s) steel bullet .

The energy absorption was studied in this work to obtain the optimum results value and it were found the maximum relative error of energy absorption between simulation and experimental values is 4.1% for overall results obtained with different thickness and initial velocity of composite laminated plates as shown in table 4.

Table 4. Experimental and simulation results of residual velocity and energy absorption of 4, 8 and 12mm thickness of Kevlar/polyester-Al_2O_3 laminated plate at different initial velocity.

Thick (mm)	V_i (m/s)	V_r (m/s) Simulation	V_r (m/s) Experimental	Error (%)	E. Absorption Simulation (J)	E. Absorption. Experimental (J)	Error (%)
4	160	55	48	12.7	56.4	58.2	3.1
4	200	77	68	11.7	85.2	88.4	3.7
4	240	88	79	10.2	124.6	129.9	4.1
4	280	97	88	9.3	172.5	176.6	2.4
4	320	111	103	7.2	225.2	229.5	1.9
4	360	155	145	6.5	263.9	271.4	2.8
4	400	217	211	2.8	282.3	288.7	2.2
8	160	12	11	8.3	63.6	63.7	0.1
8	200	35	33	5.7	96.9	97.3	0.3
8	240	47	44	6.4	138.5	139.2	0.5
8	280	65	61	6.2	185.4	186.7	0.7
8	320	85	77	9.4	237.9	241.2	1.3
8	360	126	122	3.2	284.3	286.8	0.9
8	400	195	188	3.6	304.9	311.6	2.2
12	160	0	0	0.0	64.0	64.0	0.0
12	200	0	0	0.0	100.0	100.0	0.0
12	240	8	7	12.5	143.8	143.9	0.0
12	280	37	33	10.8	192.6	193.3	0.4
12	320	70	66	5.7	243.8	245.1	0.6
12	360	117	109	6.8	289.8	294.3	1.5
12	400	150	144	4.0	343.8	348.2	1.3

From above table it has seen clear affect the thickness of target plate on the behavior of residual velocity which it leads to affect on the energy absorption with different velocity.

5. Conclusions

As a summary of this work, the following conclusions could be found:

- The energy absorption of the composites were increased as the initial velocity increases as shown in table 8 and the numerical simulation was developed and combined with academic 3D-AUTODYN v.12.1software by explicit mesh to calculate time vs. velocity curves and energy absorption of kevlar29/polyester with Al_2O_3 composite laminated plates. The impact energy absorptions of numerical simulation compared with experimental work which it was calculated by using equation (3). The good agreements of the comparisons with maximum errors were 4.1%.

- The increasing of thickness of plate affected on the behavior of energy absorption and ballistic limit for the projectile as shown in table 4 that shows the structure obtained of 12mm thickness was optimum structure to resist the impact loading under 200m/s impact velocity as armor application.

Acknowledgement
The authors would like to show their appreciation to Universiti Putra Malaysia for supporting the research activity.

References

[1] Z S Radif, A Ali and K Abdan 2010 Development of a Green Combat Armour from Rame-Kevlar-Polyester Composite vol. 19, no., pp. 339 - 348, 2011.
[2] Marom I and Bonder SR 1979 Projectile perforation of multi-layered beams. int. Journal of Mechanical Sciences, 21:489
[3] Gupta NK, Iqbal MA, and Sekhon GS 2006 Experimental and numerical studies on the behavior of thin aluminum plates subjected to impact by blunt- and hemispherical- nosed projectiles. International Journal of Impact Engineering, 32:1921
[4] R Barauskas and A Abraitiene 2007 Computational analysis of impact of a bullet against the multilayer fabrics in LS-DYNA," International Journal of Impact Engineering, vol. 34, no. 7, pp. 1286-1305
[5] M A Iqbal and N K Gupta 2008 Energy absorption characteristics of aluminum plates subjected to projectile impact," Latin American Journal of Solids and Structures, vol. 5, no. July, pp. 259-287
[6] C Defence 2009 A study on the energy dissipation of several different CFRP-based targets completely penetrated by a high velocity projectile PJ Hazell* and G Appleby-Thomas," Composite Structures, vol. 91, pp. 103-109
[7] M. Piggott and B Harris 1980 Compression strength of carbon, glass and Kevlar-49 fibre reinforced polyester resins, Journal of Materials Science, vol. 15, no. 10, pp. 2523–2538, 1980.
[8] Choi H, Downs R, Chang F 1991 A new approach toward understanding damage mechanisms and mechanics of laminated composites due to low-velocity impact: Part I—experiments. J Comp Mater ;25:992–1011.
[9] Choi H, Downs R, Chang F 1991 A new approach toward understanding damage mechanisms and mechanics of laminated composites due to low-velocity impact: Part II—analysis. J Comp Mater ;25:1012–38.
[10] Choi H, Chang F 1992 A model for predicting damage in graphite/epoxy laminated composites resulting from low velocity point impact. J Comp Mater;26(14):2134–69.
[11] AMOTZ WEINBERG and PETER SCHWARTZ 1987 Effect of Fibre Volume Fraction on the Strength of Kevlar-29/Epoxy Strands, Journal of Materials Science Letter 6, 183-184
[12] M T H Sultan 2007 High Velocity Impact Analysis of Glass-Epoxy Laminated Plates, Ms Thesis, Aerospace Engineering, UPM University, Malaysia.

Correlations for thermal conductivity and viscosity of water based nanofluids

W H Azmi[1,3], K V Sharma[1], Rizalman Mamat[1], A B S Alias[1], Izan Izwan Misnon[2]

[1]Faculty of Mechanical Engineering, UMP, Pekan campus, Pahang, Malaysia
[2]Faculty of Industrial Science and Technology, UMP, Gambang campus, Pahang, Malaysia

E-mail[3]: wanazmi2010@gmail.com

Abstract: The thermo-physical properties of nanofluids such as thermal conductivity, viscosity, density and specific heat of nanofluids are required for the analysis of convection heat transfer coefficients. The density and specific heat of nanofluids can be estimated with the mixture relations in literature. Information regarding the properties at different volume concentration and temperature is required for the estimation of heat transfer coefficient. The two most fundamental properties which are, experimentally, determined, are viscosity and thermal conductivity. Investigators have been determining the properties of nanofluids at different temperatures and base liquids. The present work is an attempt to analyze the available data to develop a non-linear regression equation for the estimation of thermal conductivity and viscosity of water based nanofluids. In the present study, nanofluids are considered as a homogenous medium and the parameters influencing the thermo physical properties identified. Equations are developed for the analysis of thermo-physical properties of nanofluids as a function of parameters viz., material, concentration, temperature and particle size useful for designer. The opposing nature of thermal conductivity rise and viscosity decrease with temperature; dependence of nanofluid thermal conductivity on material properties alters the range of applicability of nanofluids for heat transfer applications. The thermal conductivity and viscosity of Al_2O_3, ZnO and TiO_2 dispersed in water are measured to validate the proposed equations. The result shows that the equations are able to predict the thermal conductivity and viscosity of different types of nanofluids of different particle diameters closely.

1. Introduction

Experiments are undertaken to determine heat transfer coefficients and pressure drop for flow of viscous liquids in a tube under turbulent flow conditions. The study helped in fixing the dimensions for compactness of heat exchangers used in various process industries and thermal power plants. To achieve compactness and miniaturization of electronic equipment passive and active methods of augmentation have been suggested in heat transfer literature [1, 2] for a wide range of Prandtl numbers of the working medium. The initial studies relevant to the augmentation of heat transfer are mostly related to use of micro sized solid particles of various sizes in the base fluid. However, these techniques became impracticable and obsolete due severe practical problems arising due to erosion

Published under licence by IOP Publishing Ltd

and high pumping power requirements in spite of the fact that a certain degree of heat transfer augmentation is achieved. Further, agglomeration and resettlement of particles posed severe maintenance problem. In contrast, recent studies with the reduction of particle size to nano level proved effective in achieving heat transfer augmentation without any substantial increase in pumping power requirements. Results from several investigations have been coming to light from different sources and typical reviews of such literature can be cited.

Experiments are conducted for the determination of viscosity and thermal conductivity of nanofluids. Choi [3] conducted experiments with Carbon Nano Tubes (CNT) in motor oil at 1.0% volume concentration. The thermal conductivity enhancement obtained is up to 60% of the base liquid value. Various investigators [4-17] determined the properties of nanofluids dispersed with Cu and metal oxides such as Al_2O_3, TiO_2, CuO, SiO_2, etc in different base fluids, concentrations and temperatures. All these investigators observed greater values of thermal conductivity when compared to base fluid, water. The nanofluid thermal conductivity increased and viscosity decreased with temperature. Besides, the techniques employed for dispersion of particle suspension with surfactant and the pH, affect the values of thermal conductivity and the thermal stability of the nanofluid. The experimental values obtained by these investigators deviate from the theoretical models of Maxwell [18] and Hamilton and Crosser [19].

The primary consideration of the article is to review the salient physical properties of nanofluids and establish generalized correlations for different nano-base fluid combination for particle sizes ranging between 20 to 170 nm. It is fairly established from the regression analysis that the thermal conductivities and absolute viscosities of nanofluids are dependent on volumetric concentrations and temperature of the medium. The influence of particle size hither to not considered might also be an influencing factor.

2. Evaluation of thermo physical properties of nanofluid

2.1. Density, ρ_{nf}

Applying the principle of mass conservation of the two species in a finite control volume of the nano fluid, the nanofluid density ρ_{nf} can be obtained from the relation

$$\rho_{nf} = \varphi \rho_p + (1-\varphi)\rho_w \tag{1}$$

where φ is the volumetric fraction of nano particles in the base fluid.

2.2. Specific heat, C_{nf}

The thermal conservation of energy of the two species in a finite control volume will yield the overall specific heat C_{nf} of the nano fluid as

$$C_{nf} = \frac{(1-\phi)(\rho C)_w + \phi(\rho C)_p}{(1-\phi)\rho_w + \phi\rho_p} \tag{2}$$

2.3. Thermal conductivity, k_{nf}

The thermal conductivity of various nanofluids is determined experimentally by many [5-10]. The experiments are conducted mostly with spherical shape particles having diameters in the range of 20 – 170 nm, temperature of 20 – 70°C and volume concentration of less than 4.0% with Al_2O_3, Cu, CuO, SiC, TiO_2, ZrO_2 particles dispersed in water. The thermal conductivity data of metal and metal oxide nanofluids available in the literature are used in the development of regression equations. Pak and Cho [20], Williams et al. [13], Lee et al. [4], Murshed et al. [8], Das et al. [7], Chon and Kihm [21], Mintsa et al [22], Beck et al. [23], Avsec [24], Duangthongsuk and Wongwises [25], Sundar et al. [26] and Hong et al. [27] conducted experiments for the determination of thermal conductivity of nanofluids. The value of thermal conductivity depends primarily on the choice of material, concentration, and temperature and particle size. The influence of material on nanofluid thermal conductivity data is affected through the thermal diffusivity ratio of particle to water (α_{p}/α_{w}). Experimental values consisting of 252 data points are used in the regression analysis, to develop a correlation for the determination of thermal conductivity k_{nf}, given by

$$k_{nf} = k_w 0.8938\left(1+\frac{\phi}{100}\right)^{1.37}\left(1+\frac{T_{nf}}{70}\right)^{0.2777}\left(1+\frac{d_p}{150}\right)^{-0.0336}\left(\frac{\alpha_p}{\alpha_w}\right)^{0.01737} \tag{3}$$

where ϕ is volumetric concentration in %, temperature T_{nf} in °C and particle diameter d_p in nm.

2.4. Viscosity of nanofluid, μ_{nf}

Nguyen et al. [28] conducted experiments for the determination of viscosity of Al_2O_3 and CuO nanofluids in water at different concentrations and particle sizes under ambient conditions. The authors have stated the viscosity of Al_2O_3 nanofluid with particle size of 36 and 47nm and CuO of 29nm size predicted close values for volume concentration less than 4% and deviated at higher concentrations. Hence the experimental viscosity data of Nguyen et al. [28], Hwang et al. [29], Wang et al. [30], Zeinali Heris et al. [31], Nguyen et al. [32], Lee et al. [33], Pak and Cho [20], He et al. [34], Duangthongsuk and Wongwises [35] and Lee et al. [36] for volume concentration less than 4% consisting of 233 data points are subjected to regression and obtained the following correlation

$$\mu_{nf} = \mu_w\left(1+\frac{\phi}{100}\right)^{11.3}\left(1+\frac{T_{nf}}{70}\right)^{-0.038}\left(1+\frac{d_p}{170}\right)^{-0.061} \tag{4}$$

The properties of water are estimated with the aid of equations listed in equations (5) – (8) having a maximum deviation of 3.5 % applicable in the range $0 \le T_w \le 100$ where T_w is the temperature of water in °C.

$$\rho_w = 1000\times\left[1.0-\frac{(T_w-4.0)^2}{119000+1365\times T_w-4\times(T_w)^2}\right] \tag{5}$$

$$k_w = 0.56112+0.00193\times T_w-2.60152749e-6\times(T_w)^2-6.08803e-8\times(T_w)^3 \tag{6}$$

$$\mu_w = 0.00169-4.25263e-5\times T_w+4.9255e-7\times(T_w)^2-2.09935e-9\times(T_w)^3 \tag{7}$$

$$C_w = 4217.629-3.20888\times T_w+0.09503\times(T_w)^2-0.00132\times(T_w)^3+9.415e-6\times(T_w)^4-2.5479e-8\times(T_w)^5 \tag{8}$$

3. Thermal conductivity and viscosity measurement

The thermal conductivity of nanofluid was measured using KD2 Pro thermal property analyzer (Decagon Devices, Inc., USA). The viscosity of the nanofluid was measured using Brookfield LVDV-III Ultra Rheometer. Al_2O_3/water, ZnO/water and TiO2/water nanofluids having particle diameters of 50nm, 100nm and 150nm, respectively and up to 6% volume concentrations were used for thermal conductivity and viscosity measurements. The nanofluids are prepared by dilution process using mechanical homogenizer for uniform dispersion and stability. The measurements are taken at the room temperature.

4. Results and discussion

The salient observations from the regression modelling and experimental analysis on thermal conductivity and viscosity of nanofluids with different materials, particle sizes, temperatures and concentration can be listed as:

- The thermal conductivity and viscosity increases with volume concentration
- The thermal conductivity increases whereas viscosity decreases with temperature
- The thermal conductivity and viscosity increases as particle size decreases

The correlation for nanofluid thermal conductivity in equation (3) is validated with data taken from different sources and shown in the legend of figure 1 obtained with average deviation of 2.73%, standard deviation of 3.55% and maximum deviation of 10.96%. Equation (4) is validated with the data taken from various sources shown in the legend of figure 2 with $C_1 = 1.4$ for SiC and $C_1 = 1.0$ for other water based nanofluids. The data could be correlated with an average deviation of 2.89%, standard deviation of 3.8% with a maximum deviation of 12.96%. It can be observed that the present equation can predict the experimental data of water based nanofluids with a single equation.

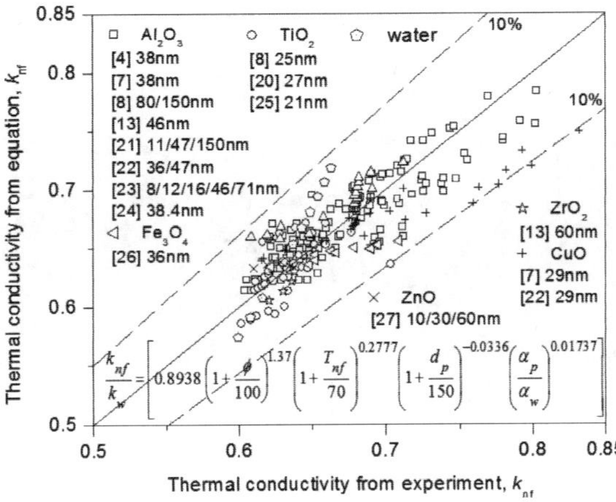

Figure 1. Validation of equation (3) using data from literature

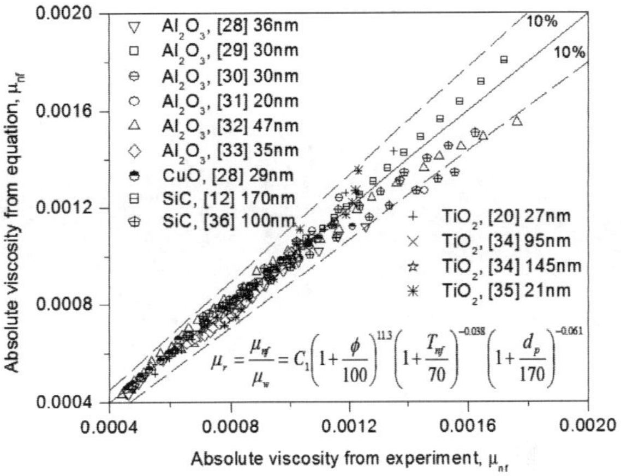

Figure 2. Validation of equation (4) using data from literature

The present thermal conductivity equation is compared with other equations from literature shown as in figure 3. It can be observed that the proposed equation is in satisfactory agreement with the equation by Corcione [37], Vajjha and Das [38] and Prasher et al. [39].

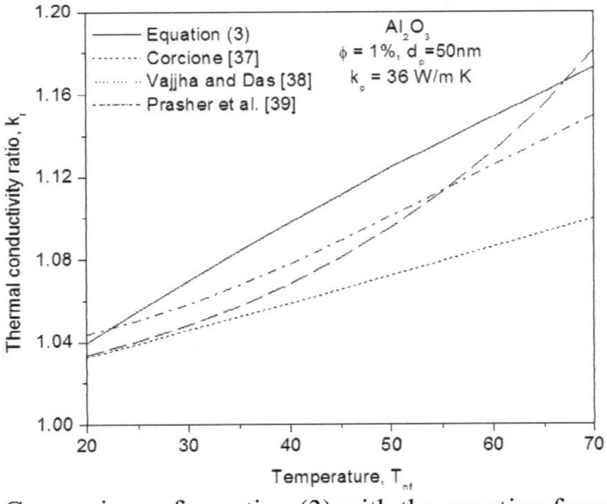

Figure 3. Comparison of equation (3) with the equation from literature

Comparison of the experimental data of measured thermal conductivity of Al_2O_3/water and ZnO/water nanofluids with values evaluated with equation (3) shown in figure 4. The figure shows the measured data of 50nm Al_2O_3/water nanofluid with a maximum deviation of 1.0%. Also the experimental data of ZnO/water nanofluids with 100nm particle diameter gives an average deviation of 0.9%. A satisfactory agreement between the two can be observed from the figure indicating the validity of the equations proposed for thermal conductivity estimation.

Comparison of the measured viscosity of water based nanofluids with the proposed correlation equation (4) is shown as figure 5. The figure shows Al_2O_3/water, ZnO/water and TiO_2/water nanofluids for particle diameters of 50nm, 100nm and 150nm, respectively. The measured data shows the effect of different materials, particle size and volume concentration to the values of viscosity at ambient temperature of 21.5°C. The viscosity estimated using the proposed equation is in a good agreement with the measured data for different nanofluids tested.

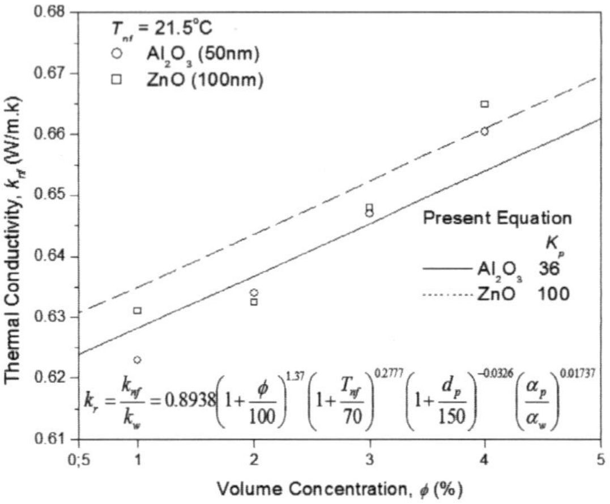

Figure 4. Comparison of measured thermal conductivity with proposed correlation

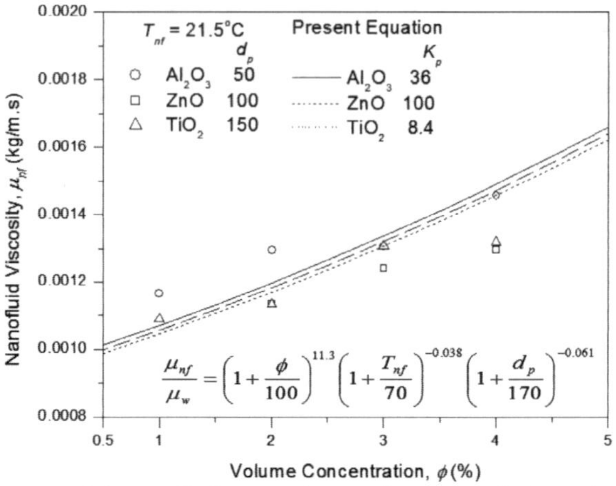

Figure 5. Comparison of measured viscosity with proposed correlation

5. Conclusion

The thermal conductivity and viscosity of metal and metal oxide nanofluids dispersed in water can be estimated using equations (3) and (4) respectively valid for volume concentration less than 4.0%. The proposed correlations predict good agreement with the experimental results. The material concentration, temperature and particle size are be considered simultaneously for evaluating the nanofluid thermal conductivity.

Acknowledgments

The financial support by Universiti Malaysia Pahang under RDU100349 and RDU0903100 and the technical support by Mecomb Malaysia Sdn Bhd are gratefully acknowledged.

References

[1] Ahuja A S 1975 *Journal of Applied Physics* **46** 3408
[2] Bergles A E 1985 in Handbook of heat transfer applications (W. M. Rohsenow, J. P. Hartnett, and E. N. Ganic, eds.), McGraw-Hill, New York, p. 31.
[3] Choi U S 1995 in Developments and Applications of Non-Newtonian Flows, Vol. FED-vol. 231/MD-Vol. 66 (D. A. Siginer and H. P. Wang, eds.), American Society of Mechanical Engineers (ASME), New York, p. 99.
[4] Lee S, Choi S U S, Li S, and Eastman J A 1999 *Journal of Heat Transfer* **121** 280
[5] Eastman J A, Choi S U S, Li S, Yu W, and Thompson L J 2001 *Applied Physics Letters* **78** 718
[6] Xie H, Wang J, Xi T, Liu Y, Ai F, and Wu Q 2002 *Journal of Applied Physics* **91** 4568
[7] Kumar Das S, Putra N, Thiesen P, and Roetzel W 2003 *Journal of Heat Transfer* **125** 567
[8] Murshed S M S, Leong K C, and Yang C 2005 *International Journal of Thermal Sciences* **44** 367
[9] Hong T-K, Yang H-S, and Choi C J 2005 *Journal of Applied Physics* **97** 064311
[10] Shima P D, Philip J, and Raj B 2009 *Applied Physics Letters* **94** 223101
[11] Duangthongsuk W and Wongwises S 2008 *International Communications in Heat and Mass Transfer* **35** 1320

[12] Yu W, France D M, Smith D S, Singh D, Timofeeva E V, and Routbort J L 2009 *International Journal of Heat and Mass Transfer* **52** 3606
[13] Williams W, Buongiorno J, and Hu L-W 2008 *Journal of Heat Transfer* **130** 042412
[14] Maïga S E B, Nguyen C T, Galanis N, Roy G, Maré T, and Coqueux M 2006 *International Journal of Numerical Methods for Heat and Fluid Flow* **16** 275
[15] Wen D and Ding Y 2004 *International Journal of Heat and Mass Transfer* **47** 5181
[16] Xuan Y and Li Q 2003 *Journal of Heat Transfer* **125** 151
[17] Fotukian S M and Nasr Esfahany M 2010 *International Communications in Heat and Mass Transfer* **37** 214
[18] Maxwell J C, A Treatise on Electricity and Magnetism, Oxford University Press, Cambridge, U.K., 1904.
[19] Hamilton R L and Crosser O K 1962 *I & EC Fundamentals* **1** 187
[20] Pak B C and Cho Y I 1998 *Experimental Heat Transfer* **11** 151
[21] Chon C H and Kihm K D 2005 *Journal of Heat Transfer* **127** 810
[22] Mintsa H A, Roy G, Nguyen C T, and Doucet D 2009 *International Journal of Thermal Sciences* **48** 363
[23] Beck M, Yuan Y, Warrier P, and Teja A 2009 *Journal of Nanoparticle Research* **11** 1129
[24] Avsec J 2008 *International Journal of Heat and Mass Transfer* **51** 4589
[25] Duangthongsuk W and Wongwises S 2009 *Experimental Thermal and Fluid Science* **33** 706
[26] Sundar L S, Naik M T, Sharma K V, Singh M K, and Siva Reddy T C 2011 *Experimental Thermal and Fluid Science* **37** 65
[27] Hong J, Kim S H, and Kim D 2007 *Journal of Physics* **59** 301
[28] Nguyen C T, Desgranges F, Roy G, Galanis N, Maré T, Boucher S, and Angue Mintsa H 2007 *International Journal of Heat and Fluid Flow* **28** 1492
[29] Hwang K S, Jang S P, and Choi S U S 2009 *International Journal of Heat and Mass Transfer* **52** 193
[30] Wang X, Xu X, and Choi S U S 1999 *Journal Of Thermophysics And Heat Transfer* **13** 474
[31] Zeinali Heris S, Etemad S G, and Nasr Esfahany M 2006 *International Communications in Heat and Mass Transfer* **33** 529
[32] Nguyen C T, Desgranges F, Galanis N, Roy G, Maré T, Boucher S, and Angue Mintsa H 2008 *International Journal of Thermal Sciences* **47** 103
[33] Lee J-H, Hwang K S, Jang S P, Lee B H, Kim J H, Choi S U S, and Choi C J 2008 *International Journal of Heat and Mass Transfer* **51** 2651
[34] He Y, Jin Y, Chen H, Ding Y, Cang D, and Lu H 2007 *International Journal of Heat and Mass Transfer* **50** 2272
[35] Duangthongsuk W and Wongwises S 2010 *International Journal of Heat and Mass Transfer* **53** 334
[36] Lee S W, Park S D, Kang S, Bang I C, and Kim J H 2011 *International Journal of Heat and Mass Transfer* **54** 433
[37] Corcione M 2011 *Energy Conversion and Management* **52** 789
[38] Vajjha R S and Das D K 2009 *International Journal of Heat and Mass Transfer* **52** 4675
[39] Prasher R, Bhattacharya P, and Phelan P E 2006 *Journal of Heat Transfer* **128** 588

The failure mode of natural silk epoxy triggered composite tubes

R A Eshkour[1], A K Ariffin, R Zulkifli, A B Sulong and C H Azhari

Department of Mechanical and Materials Engineering University Kebangsaan Malaysia

E-mail: sorena.a2569@gmail.com

Abstract In this study the quasi static compression test over natural silk epoxy triggered composite tubes has been carried out, the natural silk epoxy composite tubes consist of 24 layer of woven natural silk as reinforcement and thermoset epoxy resin as matrix which both of them i e natural silk and epoxy have excellent mechanical properties More over the natural silk have better moisture resistance in comparison with other natural reinforcements, the length of tubes are 50, 80 and 120 mm The natural silk epoxy composite tubes are associated with an external trigger which includes 4 steel pieces welded on downside flat plate fixture The hand lay up fabrication method has been used to make the natural silk epoxy composite tubes Instron universal testing machine with 250 KN load capacity has been employed to accomplish this investigation The failure modes of natural silk epoxy triggered composite tubes has been investigated by representative photographs which has been taken by a high resolution camera(12 2 Mp) during the quasi static compression test, from the photographs is observed the failure modes is progressive local buckling

1. Introduction

with development of societies, transportation vehicles also developed and the number of them growth day by day in the societies, in last decades safety and economical issues of them cause intensive amount of research in this field to design and develop such structures that be able to provide more functional and economic benefits such as enhanced strength, weight reduction, and lower fuel consumption[1-6]

Researchers' attention was addressed to use of new materials as an alternative to conventional metals like steel and aluminum Composite material with their special properties and flexibility in design make them attractive materials, and performed investigation on them showed if a composite structure be designed properly, it should be safer than a metallic structure and occupants will experience less force[7]

In the meanwhile with increasing environmental consciousness in societies, strict governments' legislations and anxiousness about finishing of fossil fuel with this consumption rate in 5-6 forward

[1] Corresponding author: Department of Mechanical & Materials Engineering, Faculty of Engineering, Universiti kebangsaan Malaysia 43600 Bangi, UKM, Malaysia.
Tel: ; +60 12 6796716
E-mail: sorena.a2569@gmail.com

Published under licence by IOP Publishing Ltd

decades open a new approach in the area of material science. To tackle on this problem, different types of fully biodegradable composites are being developed recently, as substitutions for non-biodegradable petroleum-based plastics, and even metallic components [8]

Recently, biodegradable materials have continued attracting worldwide attentions. Within the period of 2005 and 2009, the global market for the demand of biodegradable polymers was doubled in size In 2009, among all countries in the world, Europe had the largest growth in the market of biodegradable polymers in the range of 5–10% as compared with 2008 and this growth has kept continuing The forecast on the total consumption of biodegradable polymers is to grow at an average annual rate of nearly 13% from 2009 to 2014 in North America, Europe and Asia, which are accounted as the major global markets for materials consumption [8]

Biodegradable composites are generally believed as one of the key materials in coming centuries Their properties can be tailor-made to satisfy various product requirements for specific applications [8]

The combination of strength, stiffness and comparatively large elongation to failure, together with processability under environmentally friendly conditions, makes silk attractive as a potential reinforcing fiber for tough composite materials [8-10]

In this work, the failure mode of natural silk epoxy triggered composite tubes is investigated.

2. Experiment

Quasi static compression test of silk/epoxy square tubes were accomplished by a fully equipped and automated Instron MTS 810 universal testing machine with 250 KN loading capacity. All tests are performed at constant speed of test crosshead equals to 10 mm/min. Representative photographs has been taken by a high resolution camera(12 2 Mp) during the quasi static compression test.

The internal dimension of tubes are 80(mm)×80(mm)×length(mm) with 5(mm) radius curvature in the corners and the thickness is same for all tubes due to consist of 24 layer silk woven fabric as reinforcement and epoxy resin as matrix, their length, l is different 50 mm, 80 mm, 120 mm The hand lay up fabrication method has been used to make the natural silk epoxy composite tubes. A schematic sketch of tubes is shown in Figure 1

IOP Conf. Series: Materials Science and Engineering **36** (2012) 012030 doi:10.1088/1757-899X/36/1/012030

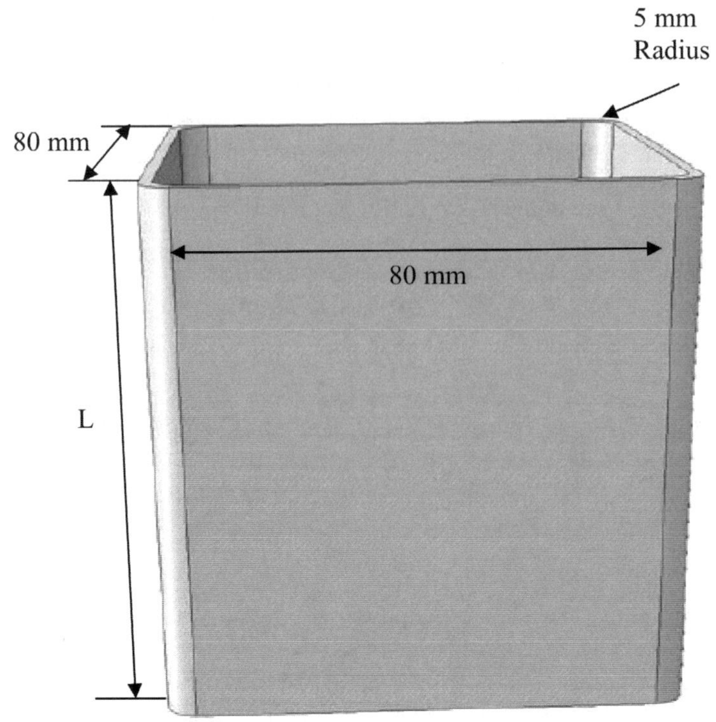

Figure 1 The schematic sketch of specimen

As described in[4] collapse triggering mechanism create an area of locally concentrated stress at one end of the tube from which failure commence and then propagates There are tree type of triggering collapse mechanism 1-no trigger 2- a direct modification to one end of the specimen 3- an external metallic trigger In type 3 no need to modification on specimen so it will take less time for machinery and modification and it would be re-usable

One flat plate is used as crosshead of the press at the top and another plate at the bottom that four rectangular shape pieces which are used as external metallic trigger are welded on it is employed as down side fixture. The pictures of these plates are depicted in figure 2

a b

Figure 2 a)fixture used at the top and b) metallic external trigger plate used at the bottom

3. Results

IOP Conf. Series: Materials Science and Engineering **36** (2012) 012030 doi:10.1088/1757-899X/36/1/012030

In the figure 3 the representative quasi static compression test photos of 50 mm triggered woven natural silk epoxy composite tube have been shown

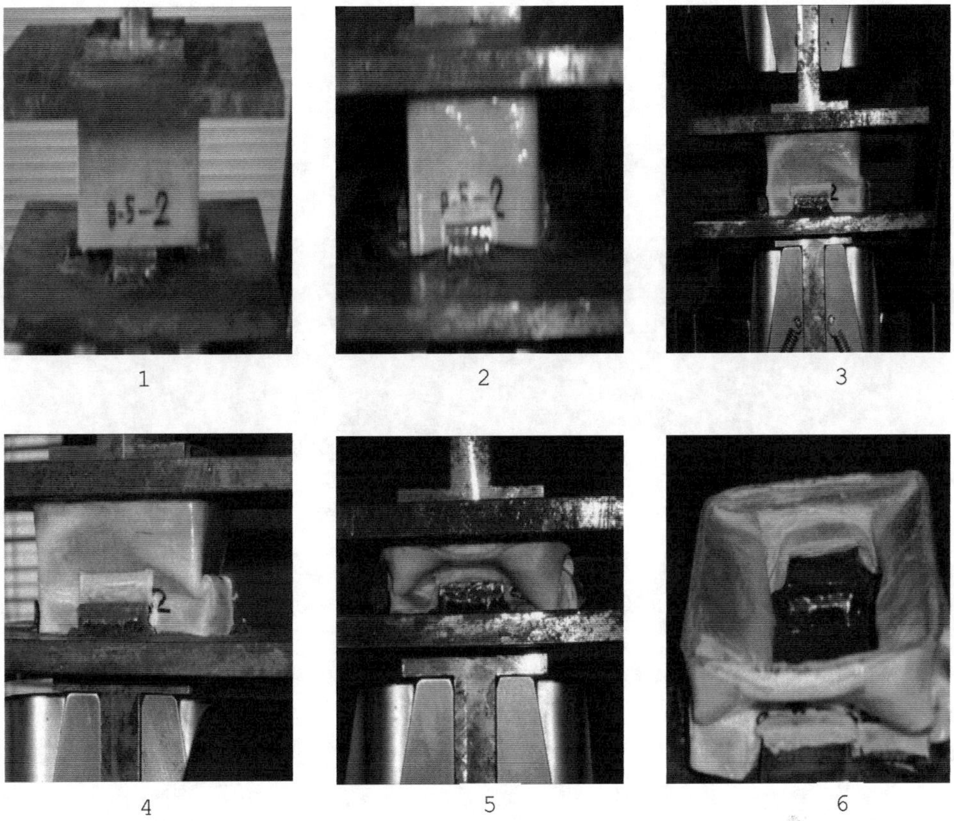

Figure 3 quasi static axial compression test of silk/epoxy tube with 50mm length

In the figure 4 the representative quasi static compression test photos of 80 mm triggered woven natural silk epoxy composite tube have been shown

Figure 4 quasi static axial compression test of silk/epoxy tube with 80mm length

In the figure 5 the representative quasi static compression test photos of 120 mm triggered woven natural silk epoxy composite tube have been shown

Figure 5 Quasi static axial compression test of silk/epoxy tube with 120mm length

In figures 3-5 it has been observed that 1- the failure start from the intersection location of tubes with triggers. After the trigger indented completely inside the tube wall and tube meet the down side plate, the progressive local buckling failure commenced. As the failure of tube proceeds due to the concentration of stress at the corners, rapture is another phenomenon that is observed especially at the top side of tubes. 2- Triggered woven natural silk epoxy composite tubes in contrary to brittle composite tubes deform in a plastic manner like metallic tubes and shown ductile failure. This phenomenon is attributed to high elongation characteristic of silk

4. References

[1] Li Y, Lin Z, Jiang A and Chen G, 2003. Materials & Design, **24**(3) 177-182.
[2] Mamalis A G, Manolakos D E, Ioannidis M B and Papapostolou D P, 2004. Composite Structures, **63**(3-4) 347-360.
[3] Mahdi E, Hamouda A S M, Mokhtar A S and Majid D L, 2005. Composite Structures, **67**(2) 175-187.
[4] Pitarresi G, Carruthers J J, Robinson A M, Torre G, Kenny J M, Ingleton S, Velecela O and Found M S, 2007. Composite Structures, **78**(1) 34-44.
[5] Olabi A G, Morris E and Hashmi M S J, 2007. Thin-Walled Structures, **45**(7-8) 706-726.
[6] Melo J D D, Silva A L S and Villena J E N, 2008. Composite Structures, **82**(4) 622-628.
[7] Zhang X, Cheng G, You Z and Zhang H, 2007. Thin-Walled Structures, **45**(9) 737-746.
[8] Ho M-p, Lau K-t, Wang H and Bhattacharyya D, 2011. Composites Part B: Engineering, **42**(2) 117-122.
[9] Craven J P, Cripps R and Viney C, 2000. Composites Part A: Applied Science and Manufacturing, **31**(7) 653-660.
[10] Shubhra Q T H, Alam A K M M and Beg M D H, 2011. Materials Letters, **65**(2) 333-336.

Fatigue Feature Classification for Automotive Strain Data

M.F.M. Yunoh[1,3], S. Abdullah[1], Z.M. Nopiah[1], M.Z. Nuawi[1] and N. Ismail[2]

[1]Dept. of Mechanical and Material, Faculty of Engineering and Built Enviroment, UniversitiKebangsaan Malaysia, 43600 UKM, Bangi Selangor, Malaysia.
[2]Faculty of Mechanical Engineering, Universiti Malaysia Pahang, 2660 Pekan, Pahang, Malaysia.

E-mail: shahrum@eng.ukm.my

Abstract. Fatigue strain signal were analysed using data segmentation and data clustering. For data segmentation, value of fatigue damage and global statistical signal analysis such as kurtosis was obtained using specific software. Data clustering were carried out using K-Mean clustering approaches. The objective function was calculated in order to determine the best numbers of groups. This method is used to calculate the average distance of each data in the group from its centroid. Finally, the fatigue failure indexes of metallic components were generated from the best number of group that has been acquired. Based on four data collect from two different roads which are D1, D2, the index value generated is not the same for all of data because due to K-Mean clustering, the best group is different for each of the data used. The maximum indexes generated are different for two types of road and namely the index 4 for D1 and index 5 for D2. Due to the road surface condition, higher distributions of the best groups give higher values of index and reflect to higher fatigue damage experienced by the system.

1. Introduction

Many engineering failures especially in mechanical engineering field involve a great loss in term of life and property. Phenomenon of fatigue failure is major factors contribute to the failure of products, component and structures in this field. Basically fatigue failure often occurs when repeated stresses imposed on the structural components within a certain time. More than 50% of mechanical failures lead by fatigue failure and mostly it unpredictable [1].

Material fatigue is a one of the most safety issues for structures subject to the cyclic loads and the cause of failure in a majority cases. In the automotive industry, most of the components are subjected in service to fatigue loading which may result in failures [2]. Most industries are susceptible to fatigue failure due to repetitive loading used every day. Nowadays, there are many techniques and tools developed to detect and prevent fatigue failure. Those techniques need to do the laboratory test, data collection, specific tools and software in order to do fatigue assessment [3].

Fatigue feature classification is a new analysis in fatigue related field and not been studied widely. Fatigue feature classification should be studied in order to produce a new standard and enable a design to the same reference level. When referring to some related studies. Mao et al. [4] had introduced a safety index of fatigue failure for ship structure details. In their research, the expected fatigue damage and its coefficient are estimated using Gussian and Bayesian process in order to

[1] Corresponding author. E-mail: shahrum@eng.ukm.my.

Published under licence by IOP Publishing Ltd

develop the index. In addition, fatigue life under Gaussian load was also performed in order to compute a fatigue risk of a component and the safety index [5].

In this paper, the main focus is developing a new fatigue feature classification using the data editing, segmentation and clustering approach to predict fatigue life of suspension car system names as fatigue failure indexes. Fatigue failure indexes are very important in component design in order to predict components life and to prevent sudden breakdown or failure [6]. Thus the developing a new fatigue failure indexes for automotive component will give a great advantage to automotive industry. Besides that, it will reduce maintenance cost, replacement a time saving.

2. Literature Background

2.1 Global Signal Statistical Parameters

In fatigue research area, the signals consist of a measurement of cyclic loads, i.e. force, strain, and stress against time. Time series typically consists of a set of observations of a variable were taken at equally spaced intervals of time. Global signal statistical parameters are frequently used to classify random signals and monitor the pattern of analyzed signals. For a signal with a numbers of data point n in a sampled sequence, the mean is given by: x

$$\bar{x} = \frac{1}{n}\sum_{j=1}^{n} x_j \tag{1}$$

In the fatigue signals, the calculation of the root-mean-square (r.m.s.) and the kurtosis are very important in order to retain a certain amount of the signal amplitude range characteristics. Both values are defined respectively as [7]:

$$r.m.s = \left\{\frac{1}{n}\sum_{j=1}^{n} x_j^{\,2}\right\}^{1/2} \tag{2}$$

$$K = \frac{1}{n(r.m.s)^4}\sum_{j-1}^{n}\left(x_j - \bar{x}\right)^4 \tag{3}$$

wherexj is the amplitude of signal.

2.2 Total Fatigue Damage

Three major approaches have widely been used toanalyse fatigue damage or fatigue life, namely thestress-life approach (*S-N*), the strain-life approach(ε-*N*) and the linear elastic fracture mechanics(*LEFM*) [8]. However, the strain-life approach (ε-*N*)is used for the analysis as the case study was relatedto low cycle fatigue, which is a suitable approach toanalyse random data collected from automotive components. ThePalmgren-Miner linear cumulative damaging rule normallyassociated with the established strain-life fatigue damaging models, i.e. the Coffin-Manson, the Morrow, and the Smith-Watson-Topper (SWT). Nevertheless, the Coffin-Manson relationship only considers the damaging calculation at zero mean stress. However, in real situation, some of the realistic service situations involve nonzero mean stresses. For example, in a case of the loading being predominantly compressive, particularly for wholly compressive cycles, the Morrow mean stress correction effect provides more realistic life estimates and seems to work reasonably well for steels. The strain-life model is mathematically defined as the following expression [9].

$$\varepsilon_a = \frac{\sigma'_f}{E}(2N_f)^b + \varepsilon'_f(2N_f)^c \tag{4}$$

where εa is the true strain amplitude, σ'f is the fatigue strength coefficient, E is the material modulus of elasticity, σm is the mean stress, Nf is the numbers of cycle to failure for a particular stress range and mean, b is the fatigue strength exponent, ε'f is the fatigue ductility coefficient, and c is the fatigue ductility exponent. The fatigue damage caused by each cycle of repeated loading is calculated by reference to material life curves, such as S-N or ε-N curves. The fatigue damage D for one cycle and the total fatigue damage ΣD caused by cycles are expressed respectively as [10]:

$$D = \left(\frac{1}{N_f}\right) \tag{5}$$

$$\Sigma D = \sum\left(\frac{N_i}{N_f}\right) \tag{6}$$

where*Ni* is the numbers of cycle within a particular stress range and mean.

2.3 Feature Classification

To extracting the information from the data columns, it is important for researcher to find the methods that can divide the data into specific parts and analysed each part in optimum [11]. Segmentation divided the non stationary times series segments with the constant statistics value. Therefore, segmentation technique can identify accurately and quickly the location changes in statistical signal [12]. Segmentation enables to divide the time series data and organised into several sections that have the same criteria [13]. Recently, many methods have been designed to segment the data as accurately as possible, but most of the methods have an own name and application. Thus, the segmentation algorithms in time series can be divided into three main groups which is sliding windows, top-down and bottom-up [14].

Clustering is a process a division of data into the same groups whereas the data in the groups will have the similar character and behaviour. There are two category of the clustering which is partitioning and hierarchical. For partitioning clustering, generally result in a set of M clusters, each object belonging to one cluster and each cluster may be represented by a centroid or a cluster.K-means is a fast and popular method to perform clustering because it easier to manage the huge of data [15]. The basic intuition behind K-means is the continuous reassignment of objects into different clusters so that the within-cluster distance is minimized. It uses an iterative algorithm divided in two phases to minimize the sum of point-to-centroid distances, over all K clusters[14]. Objective functions for K-means define as following expression:

$$J_h = \sum_{j=1}^{k} \sum_{i}^{c} \left\| x_i^{(j)} - c_j \right\|^2 \tag{7}$$

Where $\left\| x_i^{(j)} - c_j \right\|^2$is distances between point $x_i^{(j)}$ and mean of point c_j.

3. Methodology

Four data obtained from the suspension and lower arm is collected during the test at highway and country road. The highway and country road is then referred as D1 and D2 in this paper. The suspension and lower arm is attached with strain gauge in order to collect the strain signal during the test at D1 and D2 road. The data is recorded for 60 second with sampling frequency 500Hz gives 30000 discrete data[9]. This sampling frequency is chosen to be at 500Hz as the value was enough to record the data because of the strain gauge sensitivity in recording the changes of the strain response that exit during the test.

Figure 1shows the sequence method for data collection toward car suspension system. The strain gauge is attached at high stress region of the suspension and lower arm. The high stress region can be determined using finite element analysis (FEA). Strain gauge then connected to the data acquisition system for the purposed of strain signal data collection. The strain data is collected in the time history domain. The strain signal data after that is analysed using global statistical analysis in order to find the statistical parameter i.e kurtosis, root mean square (r.m.s)and fatigue damage that use for data clustering.

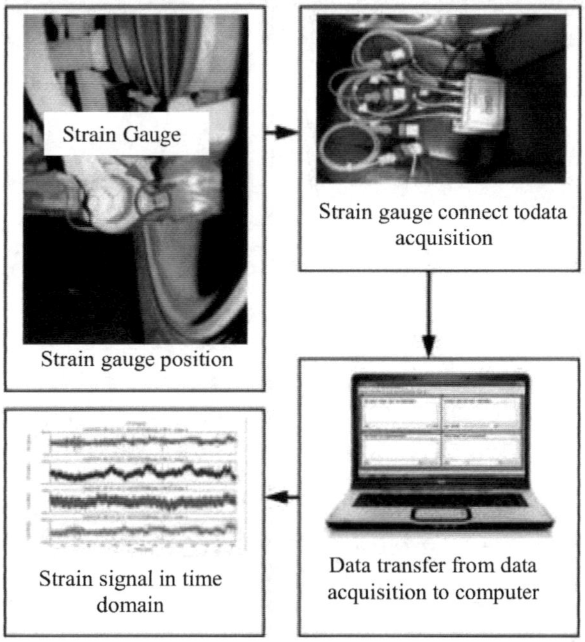

Figure 1. Sequence method for data collection toward car suspension system

Data segmentation and clustering is done by using Matlabsoftware. Data segmentation one of the processes divides data into certain section or segment. Segmentation method can be grouped into three categories such as sliding windows, top-down and bottom-up [14]. Segmentation using bottom-up method is the best approach segmentation for time history signal. Clustering data is analysed using K-mean approach as a clustering algorithm. Cluster with certain centroid is formed with the different colour for every object. The clustering is analysed based on the changing of fatigue damage. The best cluster is determined based on function objective value through Euclidean distance calculation. Euclidean distance gave the distance value for every cluster to centroid. The classification later is generated from the clustering data. The fatigue failure indexes are generated based on classification result.

4. Results and Discussion

Figure 2 (a) shows the data distribution with two group point for D1 road. The positions of the centroid are very close between each other and concentratewith the increasing of kurtosis. It show that the K-means clustering susceptible in initial centroid because the intake of random centroid point. Roughly, it can been seen more than two groups could be formed from the distribution of the data because there is a point that has far distances away from it centroid

The formation of the clusters can clearly see for three point centroid as shows in Figure 2 (b). There was a little difference in the distribution according to the groups of the fatigue damage in red and yellow. The four centroid point of data distribution as shows in Figure 2 (c) more organized than Figure 2 (d). Each of the centroid has a data distribution with uniform distances. The five centroid point of data distribution shows a little visible difference compared to the four centroid as the data seemingly too close between each group with low fatigue damage. Number of different group used to see the density distribution of the data in group. The data with the lower number of centroid have a high density distribution and the objective function also high.

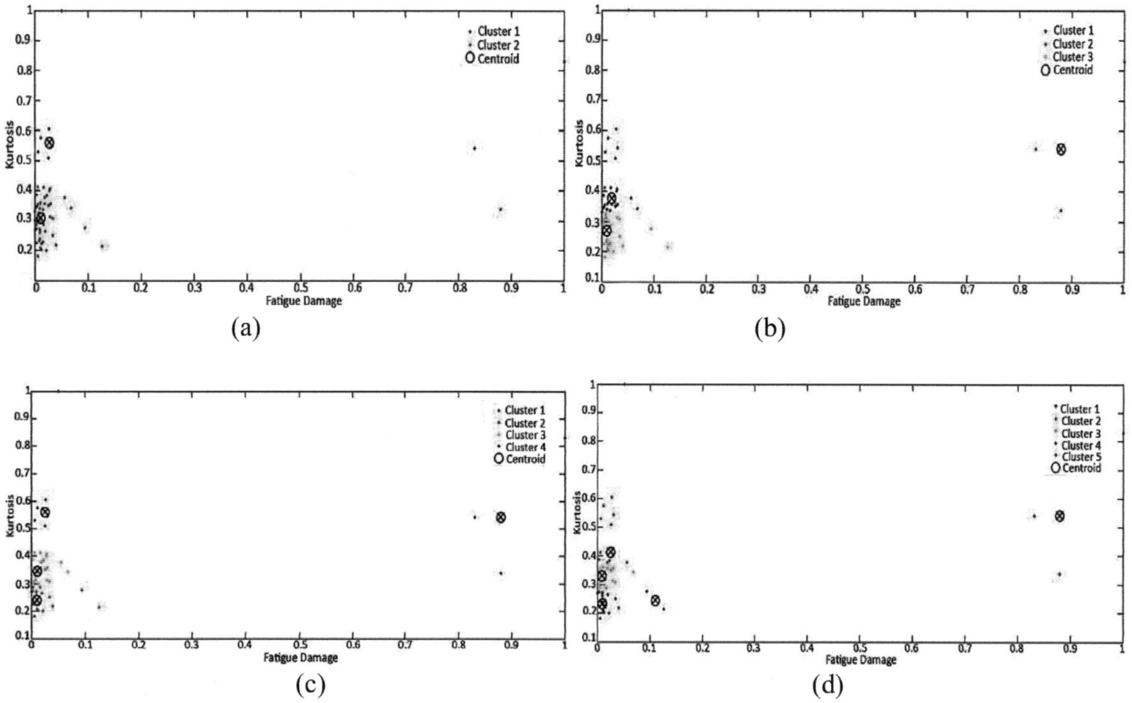

Figure 2. Kurtosis distributions of fatigue damage for D1 road condition; (a) two centroid, (b) three centroid, (c) four centroid, (d) five centroid

The clustering for the D2 road is start from two groups to five groups. Figure 3 (a) shows that the two centroid points is far between each other. However, both of the groups have a clear boundary between each other. This situation happen because of the data distribution is not too dense with each other because the surface D2 road always changing. Figure 3 (b) shows the group's formation seen more clearly for three points centroid. For the each group, there are a clear separation of data distribution and all of the data linear with the fatigue damage value. It also has a high kurtosis value with the ratio 1:1 after normalised in the data distribution and it namely outlier data

The distribution data for four and five centroid point as shown in figure 3 (c) and (d) is more organised compare to the two and three cnteroid point. Each of the centroid have a distribution data with a uniform distances. The boundary of the groups is clearly showed and the kurtosis ratio 1:1 still retained as a outlier data. However, the data with five groups is chosen as a best group because it have a small objective function.

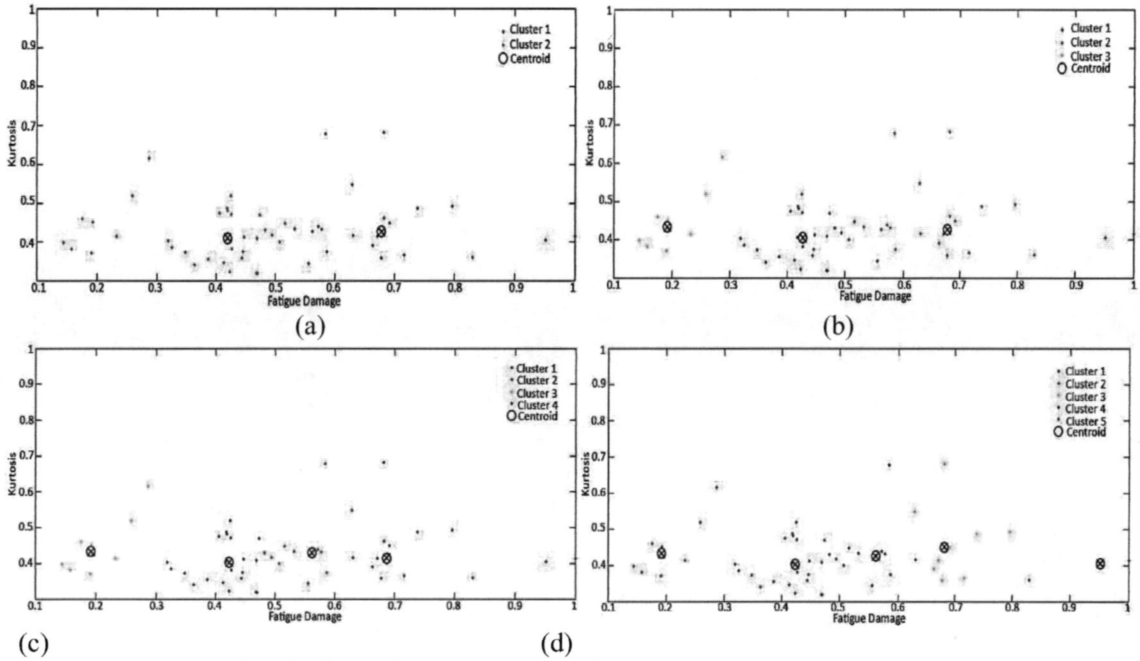

Figure 3. Kurtosis distributions of fatigue damage for D2 road condition; (a) two centroid, (b) three centroid, (c) four centroid, (d) five centroid

The maximum objective function value for D1 road occurs at data with two centroid as shown in Table 1. The objective function value is decrease with the increasing of centroid number but only for the data with five centroid has less objective function value than the data with four centroid. It show that the data with five group not necessary to generated because it only scattered the data getting away from the centroid. Therefore, the four clusters is enough for the D1 road. For the D2 road, the maximum objective function value also occurs at data with two centroid as shown in Table 1. The objective function value for five clusters is the best for D2 road. The increase to more than five centroid may cause the objective function value decreasing, but the clusters distribution after that not have many difference with the five clusters data.

Table 1. Average objective function value for D1 and D2 road.

	Objective Function value for D1 Road	Objective Function value for D2 Road
C=2	6.89	8.48
C=3	4.96	7.91
C=4	4.23	6.52
C=5	4.33	5.84

Fatigue failure index is generated using the data with four centroid as shown in Figure 4 for D1 road. The index from 1 to 4 is given for each group. The indexes that have a high value represent the group data that have high fatigue damage. Thus, it give a high probability of fatigue failure occurs. Index 1 to 3 is classification as an index with low fatigue damage. Index 4 is classified as index with high fatigue damage.

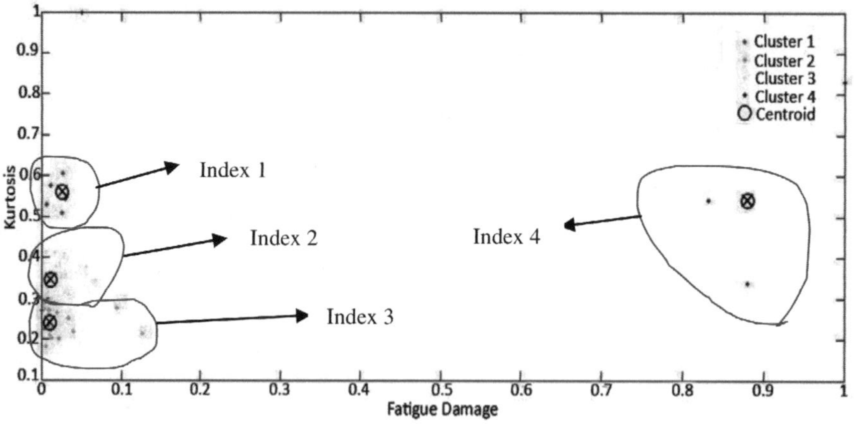

Figure 4. Selection of group to generate the index for D1 road

For D2 road, the fatigue failure index is generated using the data with five centroid as shown in Figure 5. Indexes from 1 to 5 are given for each group and have a high value represent the group data that have high fatigue damage. Group 1 with a green distribution point has a low fatigue damage value and it namely as an index 1. The second group with purple distribution data is namely as index 2. The third group with blue distribution data is namely as index 3. The second last group with the yellow distribution data start to show the high fatigue damage and it namely as index 4. For the last group with red distribution data is namely as index 5 shows the high fatigue damage. The outlier data is recognised for the data distribution at kurtosis ratio 0.7 and above. Although, the outlier data is not include in the group that given an index, this data cannot remove because it give the high amplitude for overall data.

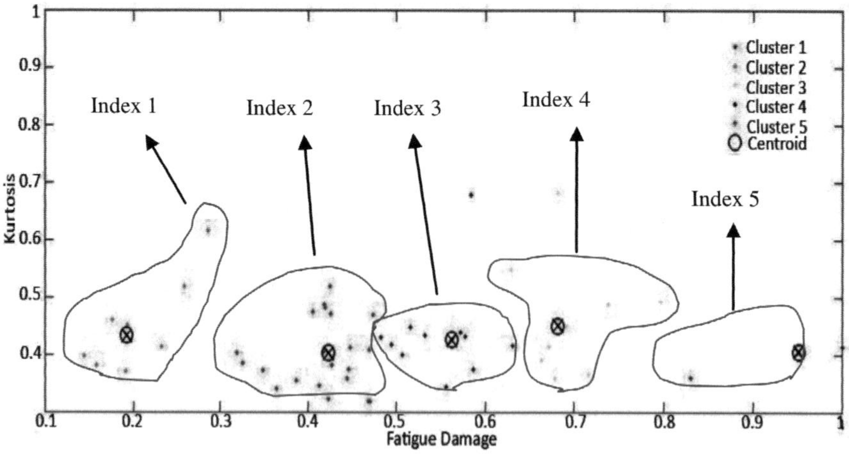

Figure 5. Selection of group to generate the index for D2 road

5. Conclusion

The fatigue failure index that generated using K-means clustering method is a new approach in fatigue study. Based on clustering process the data is cluster according to the required parameter and the best group can be determined. The objective function is calculated and the lower value of objective function show the K-means clustering is well organised data. From the results, it show that index 1 represent the low fatigue damage that occurs for the component. The increasing of the index number show more dominant of the effect of fatigue damage for the component. Index 4 for D1 road and index 5 for D2 road represent the high fatigue damage for the component. The high index number represent the high fatigue damage occurs for the system.

Acknowledgments
Authors would like to express their gratitude to Universiti Kebangsaan Malaysia (Research funding: UKM-KK-FRGS0188-2010 & UKM-OUP-NBT-28-135/2011) for supporting this research

6. References

[1] Stephens R I, Fatemi A, Stephens R R& Henry O F 2001 Metal Fatigue in Engineering. *John Wiley & Son, Inc: New York.*
[2] Fourlaris G, Ellwood R and Jones T B 2007 The reliability of the test results from simple test samples predicting the fatigue performance of automotive component *J. Material & Design***28(4)** 1198-1210
[3] Dong C L, Chang S H 2009 CAE (computer aided engineering) driven durability model verification for the automotive structure development*J. Finite Element in Analysis Design***45(5)** 324-332.
[4] Mao W, Rychlik I &Storhaug G 2008 Safety Index of Fatigue Failure for Ship Structure Details Department of Mathematical Science Chalmers University of Technology, University of Gothenburg.
[5] Bengston A And Rychlik I 2009 Uncertainty in fatigue life prediction of structures subject to Gaussian loads *J. Probabilistic Engineering Mechanics***24** 224-235
[6] Romaniv O N and Tkach A N 1987 A structural analysis of the kinetic fatigue failure curve of constructional steels *J. Materials Science***23(5)** 441-453
[7] Nuawi M Z, Abdullah S, Abdullah S, Arifin A and Haris S M 2009Matlab: a comprehensive reference for engineers, *Malaysia: McGraw-Hill Sdn. Bhd*
[8] Stephens R I, Dindinger P M and Gunger J E 1997 Fatigue damage editing for accelerated durability testing using strain range and SWT parameter criteria. *Int. J. of fatigue***19(8)** 599-606
[9] Oh C S 2001 Application of wavelet transform in fatigue history editing *Int J. of fatigue***23(3)** 241-250
[10] Abdullah S 2005 Wavelet Bump Extraction (WBE) for editing variable amplitude fatigue loadings *Ph.D, Thesis the University of Sheffield*
[11] Jeffrey D S 2007 Joint segmentation of multivariate astronomical time series: Bayesian sampling with a hierarchical model. *IEEE Trans on Signal Processing***55** 414-420
[12] Lopatka M, Laplanche C, Adam O, Motsch J F and Zarzycki J 2005 Non-stationary time-series segmentation based on the Schur prediction error analysis *IEEESP 13th Workshop on Statistical Signal Processing 2005*
[13] Lemire D 2007 A better alternative to piecewise linear time series segmentation Proc. of SIAM Data Mining 2007
[14] Keogh E, Selina C, David H and Pazzani M 2001 Segmenting time series: insight and recommendations *J. of IEEE Trans on Fuzzy Systems*
[15] Nopiah Z M, Abdullah S and Khairir M I 2008 Segmentation and scattering of fatigue time series data by kurtosis and root mean square *Proc. Int. Conf. on Signal Processing System* 64-68
[16] Alvarez F M, Troncoso A, Riquelme J C and Riquelme J M 2007 Partitioning-clustering technique applied to the electricity price time series *Proc. IDEAL'07 8th Int. Conf. on Intelligent Data Eng. And Automated Learning*

Review on CNC-Rapid Prototyping

M Nafis O Z, Nafrizuan M Y, Munira M A, Kartina J
Faculty of Manufacturing Engineering,
Universiti Malaysia Pahang,
26600 Pekan, Pahang,

Email: nafis@ump.edu.my

Abstract. This article reviewed developments of Computerized Numerical Control (CNC) technology in rapid prototyping process. Rapid prototyping (RP) can be classified into three major groups; subtractive, additive and virtual. CNC rapid prototyping is grouped under the subtractive category which involves material removal from the workpiece that is larger than the final part. Richard Wysk established the use of CNC machines for rapid prototyping using sets of 2½-D tool paths from various orientations about a rotary axis to machine parts without refixturing. Since then, there are few developments on this process mainly aimed to optimized the operation and increase the process capabilities to stand equal with common additive type of RP. These developments include the integration between machining and deposition process (hybrid RP), adoption of RP to the conventional machine and optimization of the CNC rapid prototyping process based on controlled parameters. The article ended by concluding that the CNC rapid prototyping research area has a vast space for improvement as in the conventional machining processes. Further developments and findings will enhance the usage of this method and minimize the limitation of current approach in building a prototype.

1. Introduction

Rapid prototyping (RP) can be defined as an automatic fabrication process of three dimensional physical models directly from computer-aided design (CAD)[1]. In designing processes, generating solution is largely associated with the use of external representations such as sketch and physical model [2]. The physical model does not only serve as external information storages but also act as example, solution development, testing and communication. Current rapid prototyping techniques are capable of automatically constructing physical models with minimal planning time, the '3D printing' concept where there is no human intervention. Additive layer-based manufacturing approaches are commonly adopted to automatically produce complex shapes. The promise of minimal process engineering is a major factor that has driven the use of free- form rapid prototyping (RP) techniques. Unfortunately, many of these processes have been restricted to a small variety of materials with limited geometric accuracy. Therefore, a new approach is developed aimed to produce quality parts with little or no human intervention by combining the important attributes of CNC machining and the existing rapid prototyping technologies.

Published under licence by IOP Publishing Ltd

2. Hybrid Rapid Prototyping

RP has proven to be an effective tool to reduce the time and expense involved in the realization of a new product. However, the fields of application are currently saturated. The current need is to produce the finished parts of required quality in the shortest possible time. This led to the focus of improvement in production speed, accuracy, variety of materials and cost [3].

The current additive RP fabricated the part layer by layer until the final part produce. This layered approach generates the stair-step effect which limits the accuracy of final part. Another limitation concerns on range of materials used in the process. Initially, RP used for visualization and assemblability test and none of these applications impose structural requirements on the part.

CNC process becomes a new substantial solution to the limitations of RP process. Modern CNC capable to operate at high speed, high precision and high flexibility. Consequently, the effective combination of RP and CNC technologies could bring a new hybrid manufacturing system that completely meets the requirements of the next generation of production systems, especially for short run production [3]. Figure 1 indicates the convergence between RP that is appropriate to meet the rapid manufacturing goals, and CNC that produces high accuracy final part.

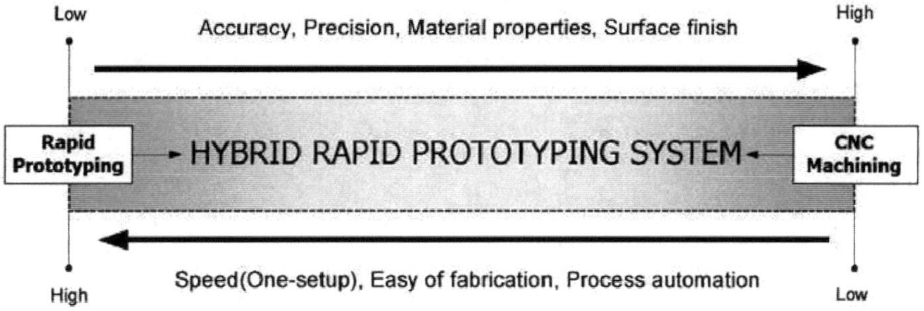

Figure 1. Convergence between RP & CNC [3]

The interesting features of these Hybrid-RP systems is that the integration of advantages and capabilities between the conventional RP, i.e. rapid process planning & independence geometry and CNC machining, i.e. accuracy, precision and good surface finish. Thus, some hybrid systems have ever been applied to the real manufacturing process such as mold making and tooling [4]

In addition, to assure the operation of the hybrid system, a test bench system, called ECLIPSE-RP has been developed used to verify the proposed approach [5]. The main features of the test bench system can be summarized into six core concepts. First, the ECLIPSE-RP system is a also a hybrid-RP system performing both deposition and machining. Second, ECLIPSE-RP builds a model by stacking thick layers with full three dimensional shapes. Third, each sheet is inverted to allow two machining setups in a building cycle. Fourth, a machining feature concept is adopted to maximize process efficiency. Fifth, ECLIPSE-RP uses a specially developed machining center based on a 6-axis parallel mechanism as a cutting station. Finally, this hybrid system imports the shape information through a physical file which carries exact geometry. Hence, the shape error does not occur and high accuracy model can be produced.

As the hybrid-RP system performs both deposition and machining, this technology is expected to become a new cost effective manufacturing technique that will produce completely finish quality parts in a very short time.

3. Development of CNC-Rapid Prototyping

Additive layer-based manufacturing approaches are commonly adopted to automatically produce complex shapes that would be difficult or impossible to create using conventional methods. These models are used as prototypes to evaluate form and function but RP is increasingly being used for production parts. Many different RP processes are available including Stereolithography, Fused Deposition Modelling and Selective Layer Sintering.

Some exploration into the use of CNC machines for rapid prototyping has been published. Layer-based robot machining for rapid prototyping have been developed using machined layers that are laminated during the process. The process is demonstrated using laminated slabs of plastic, machined as individual layers upon gluing to previous layers [6].

A hybrid approach using both deposition and machining called shape deposition manufacturing (SDM) continues to be developed [7]. For each layer, both support and build material is deposited and machined in a combined additive and subtractive process. This Shape Deposition Manufacturing (SDM) process integrates material layer deposition with CNC material removal processes. Recently, a hybrid layered manufacturing process is develop further which combines the features of subtractive and additive manufacturing to build near-net shape metallic prototypes using arc weld deposition [8].

Reference Free Part Encapsulation (RFPE) as the use of CNC machines for rapid prototyping has also been explored [9]. The approach was discussed recently in conjunction with high-speed machining (HisRP) [10]. RFPE in combination with feature-based CAD/CAM was proposed as an RP system [11]. Another approach is to use CNC machining for prototyping dies, an area called rapid tooling [12]. One approach to rapid tooling uses machined metal laminates stacked to form dies [13][14].

Many of these methods utilize CNC machining but do not address the fundamental problems of automating a fully subtractive rapid machining approach. Methods have been developed to cover all aspects of process planning for rapid machining, including toolpath planning, choosing tool geometries, calculating setup orientations and a concept for a universal approach to fixturing.

With regard to toolpath planning, the presented method borrows from layer-based RP methods. The general idea is to machine the visible surfaces of the part from each of a plurality of orientations. From each orientation, some but not all of the part surfaces will be visible. Only parts whose entire external surface is visible can be completely machined with this methodology. In some ways, this limits what can be done when comparing the methodology described herein to traditional RP techniques, but in no way reduces the flexibility when compared to traditional CNC machining [15].

The goal is to machine the part from enough orientations such that, after all toolpaths are complete, all surfaces have been fully machined from at least one orientation. For each orientation, there is not a particular plan for a set of feature machining operations; rather, each orientation is machined using simple 2½D layer-based toolpaths. This is very similar to the existing rapid prototyping systems; however, in this case, one is limited to removing only visible layers from each orientation instead of creating and stacking all of the true cross sections of the part from just one orientation.

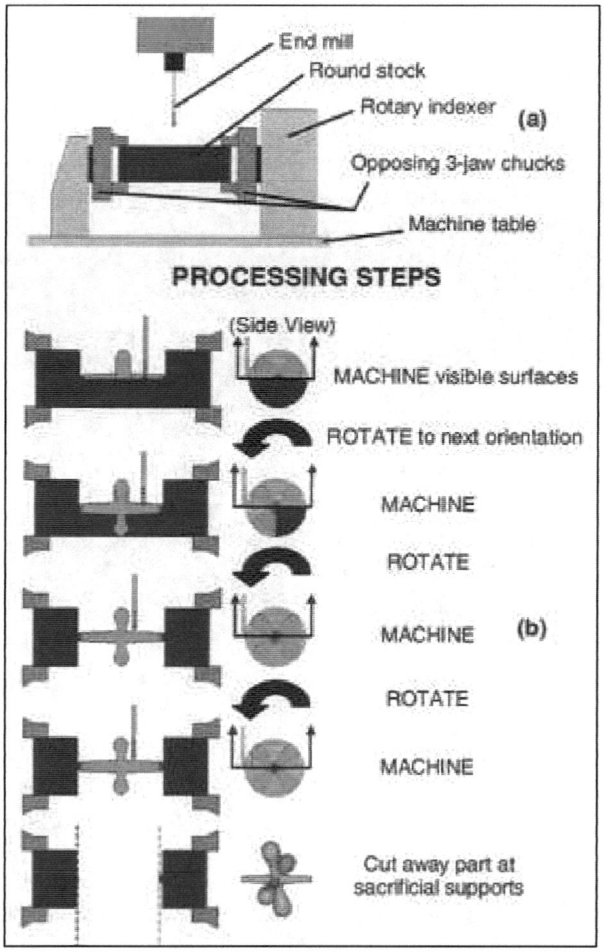

Figure 2. CNC-Rapid Prototyping processing steps[15]

Designing a fixture scheme for CNC machining is a difficult task that requires a significant amount of work from a highly skilled technician. In general, fixturing or workholding serves three primary functions: location, clamping, and support [16]. Currently, the approach to fixturing for CNC-RP adopts from the general idea of sacrificial supports, which are used in existing RP systems. In this work, the general intention is retained, however, the requirements for the support structures are different. The goal is to have a fixture solution that is created in process and is customized for each part. Specific to this work, the fixture supports need to allow the part to be rotated about the axis while providing access to as much of the part surface as possible [15].

The 'staircase' effect as seen on common RP method is still created on part surface contours. However, due to the capability of machining to make very shallow depths of cut, rapid machining able to produce very thin layer thickness. This reduction of layer thickness tends to increase the machining time but it does not necessarily do so proportionally because shallower depths of cut enable higher feed rates.

IOP Conf. Series: Materials Science and Engineering **36** (2012) 012032 doi:10.1088/1757-899X/36/1/012032

4. Optimization of CNC-Rapid Prototyping Process

The basic concept of CNC-RP involves machining all visible surfaces of a part from a particular orientation using a 3-axis vertical machining centre. Visible surfaces are those surfaces that can be 'seen' when looking down the axis of the cutting tool – the z-axis. Not all surfaces will be visible in a particular orientation and therefore a number of orientations are normally needed in order to machine all part surfaces but without refixturing. This reorientation is achieved by the use of two opposing fourth axis indexers. This removal of visible cross sections of the part using simple 2½ D layer-based tool paths is similar to a roughing process in traditional machining operations (Balasubranmaniam, 1999).

Figure 3. Setup for CNC-RP [16]

This process is feature free and successfully eliminates feature recognition and feature based process planning [18], as there is no need to plan the manufacture of each feature independently. Since only 2½ dimensional tool paths are to be machined and no feature information is required, a generic approach using a small diameter flat-end mill cutter is feasible. However, there are some draw-backs in using a small diameter tool. Avoiding tool deflection or breakage requires relatively low feed rates and depths of cut are limited by the required precision of machined surfaces and so material removal rates are low and machining times are consequently high.

The objective is however that overall efficiency is improved through the removal or very significant reduction of process planning and setup efforts [15]. Tool selection is very important to avoid collisions. The tool length must greater than the distance to the furthest visible surface for each orientation [18] so that the tool holder does not collide with the stock material. The tool diameter is defined by the need to manufacture the smallest features of the part.

Comparison study has been carried out to estimate process planning times for different machining approaches and to identify the suitability of the tool types to be used in CNC rapid prototyping [19]. Surface finish of the machined part was also observed and analysed. The three machining approaches considered were (i) conventional roughing, semi-finishing, and finishing using Ø8 endmill and Ø8 ballnose cutters (ii) single Ø8 endmill cutter and (iii) single Ø8 ballnose cutter. The results clearly demonstrated that the Roughing, Semi-finishing and Finishing approach offers shorter machining time but required intensive labour and time in process planning. The finishing approach using a single cutter gave longer machining time but reduced process planning. The study also showed that a single endmill cutter approach was the best option because it offered reasonably good dimensional control, simpler tool path strategy and no tool change was needed. Table 1 indicates the comparison of three machining approaches for CNC-RP.

Table 1. Comparison data on three machining approaches in CNC-RP

	Roughing/Finishing		Finishing		Finishing	
Material	Aluminium		Aluminium		Aluminium	
Tool	Endmill & Ballnose		Endmill		Ballnose	
Cutting Speed (m/min)	150-300		150-300		150-300	
Number of operations	8		4		4	
Type of operations	Rough Follow		Rest Milling		Rest Milling	
	Rest Milling					
	Countour Area					
Estimated Process Planning Time (minutes)	**Steps**	**Time**	**Steps**	**Time**	**Steps**	**Time**
	Operations selection	5	Operations selection	2	Operations selection	2
	Identify geometries	12	Identify geometries	4	Identify geometries	4
	Tool setup	6	Tool setup	3	Tool setup	3
	Path settings	25	Path settings	8	Path settings	8
	Post processing	6	Post processing	2	Post processing	2
	Program editing	8	Program editing	4	Program editing	4
	Machining setup	20	Machining setup	20	Machining setup	20
	Tool changes	2	Tool changes	0	Tool changes	0
	Total	84	**Total**	43	**Total**	43
Depth of Cut (mm)	Roughing	1	Finishing	0.1	Finishing	0.1
	Finishing	0.1				
Stepover	**Method**	**%**	**Method**	**%**	**Method**	**Height**
	Tool diameter (Endmill)	20	Tool diameter (Endmill)	20	Scallop (Ballnose)	0.1
	Method	**Height**				
	Scallop (Ballnose)	0.1				
Number of Tool Change	6		0		0	
Machining Time (hours)	4.57		37.73		23.95	
Total Time (hours)	**5.97**		**38.45**		**24.67**	

5. Conclusion

CNC Rapid Prototyping has been described in this paper as the significant solution to the limitation of RP process where it is able to meet the rapid manufacturing goals and also to produce high accuracy final part. CNC Rapid Prototyping has been highlighted as the new cost effective manufacturing technique that will maximize the process efficiency in terms of quality and also time of completion.

There are several approaches have been discussed in a number of researches on the development of CNC Rapid Prototyping where various RP processes were integrated with the use of CNC machine. However, the effectiveness of this techniques are still depending on a few factors which are the different types of machining approach that will give different effect on the machining time and process planning, the tools selection to avoid collision and also the highly skilled technician or operator to run this process since designing the fixture scheme for CNC machining is a difficult task.

Overall, there are still a lot of rooms for improvement in this CNC rapid prototyping research area. Future research needs to be done to enhance the usage of this method and minimize the limitation of current approach in building a prototype. Another interesting extension of this research would be the application of this method in other areas of the manufacturing decision making process such as material or tooling selection.

References

[1] Cooper K G 2001 *What is Rapid Prototyping? In Rapid Prototyping Technology Selection and Application*, ed. L.L.Faulkner, 1-10. Marcel Dekker, New York.

[2] Anne Ro¨mer, Guido Weißhahn and Winfried Hacker 2001 *Effort-saving product representations in design-results of a questionnaire survey* (aTechnische Universita¨t Dresden, Germany)

[3] Junghoon H, Kunwoo L, Zhu-hu and Jongwon K 2002 Hybrid rapid prototyping system using machining and deposition, *Computer-Aided Design* **34** 741-745

[4] Jeng J Y and Lin M C 2001 Mold fabrication and modification using hybrid processes of selective laser cladding and milling, *J Mater Process Technology*; **110(1)** :98-103

[5] Hur J H et al. 2000 Development of hybrid rapid prototyping system - hardware and software, *Proc of the Third World Congress on Intelligent Manufacturing Process and Systems*. Cambridge, Massachusetts, USA, Jun 28-30

[6] Chen Y H and Song Y 2001 The development of a layer based machining system. *Computer Aided Design* **33**, pp331-342

[7] Merz R, Prinz E B, Ramaswami, K Terk, M and Weiss L E 1994 Shape deposition manufacturing. *Proc of Solid Freeform Fabrication Syrup*. Austin, TX: Univ. of Texas at Austin, ppl-8

[8] Karunakaran K P, Suryakumar S, Pushpa V and Akula S 2009 Retrofitment of a CNC machine for Hybrid layered manufacturing. *The International Journal of Advanced Manufacturing Technology*. **45**:690-703

[9] Sarma S E and Wright E K 1997 Reference Free Part Encapsulation: a new universal fixturing concept. *Journal of Manufacturing Systems*. **l6**, pp35-47

[10] Shin B S, Yang D Y, Choi D S, Lee E S, Je T J and Whang K H 2002 Development of rapid manufacturing process by high-speed machining with automatic fixturing. *Journal of Materials Processing Technology*. **130**, pp363-371

[11] Choi D S, Lee S H, Shin B S, Whang K H, Yoon K K and Sarma S E 2001 A new rapid prototyping system using universal automated fixturing with feature-based CAD/CAM. *Journal of Materials Processing Technology*. **113**, pp285-290

[12] Radstok E. 1999 Rapid tooling. *Journal of Rapid Prototyping*. **5**, pp164-168

[13] Vouzelaud E A, Bagchi A and Sferro E E 1992 Adaptive laminated machining for prototyping of dies and molds *Proc of 3rd Solid Freeform Fabrication Syrnp*., pp291-300

[14] Walczyk D E and Hardt D E 1998 Rapid tooling for sheet metal forming using profiled edge laminations-design principles and demonstration. *Journal of Mfg. Science attd Engg., Trans. of ASME* **120**, pp746-754

[15] Frank M C, Wysk R A and Joshi S B 2004 Rapid Planning for CNC Milling - A New Approach for Rapid Prototyping. *Journal of Manufacturing Systems,* **23**:242-255

[16] Chang T C, Wysk R A and Wang H P 1998 *Computer-Aided Manufacturing* (Englewood Cliffs, NJ: Prentice-Hall)

[17] Balasubranmaniam M 1999 Tool selection and path planning for 3 axis rough cutting (Ph.D. thesis Department of Mechanical Engineering, Massachusetts Institute of Technology, Cambridge)

[18] Kamrani A K and Nasr E A 2006 Subtractive Rapid Prototyping: creating a completely automated process for rapid machining. Rapid Prototyping Theory and Practice (New York: Springer Science Business Media, New York) p 165-196

[19] Keat L L, Case K. 2010 Rapid Prototyping using Conventional 3-Axis Milling. *8th International Conference on Manufacturing Research ICMR*

CO_2 Laser Cutting of Glass Fiber Reinforce Polymer Composite

S Fatimah[1], M Ishak[1, 2] and S N Aqida[1, 2]
[1]Faculty of Mechanical Engineering,Universiti Malaysia Pahang, Malaysia.
[2]Automotive Engineering Center, University Malaysia Pahang, Malaysia

Email: fatimahmail86@gmail.com, mahadzir@ump.edu.my

Abstract. The lamination, matrix properties, fiber orientation, and relative volume fraction of matrix of polymer structure make this polymer hard to process. The cutting of polymer composite using CO_2 laser could involve in producing penetration energy in the process. Identification of the dominant factors that significantly affect the cut quality is important. The objective of this experiment is to evaluate the CO_2 spot size of beam cutting for Glass Fiber Reinforce Polymer Composite (GFRP). The focal length selected 9.5mm which gave smallest focus spot size according to the cutting requirements. The effect of the focal length on the cut quality was investigated by monitoring the surface profile and focus spot size. The beam parameter has great effect on both the focused spot size and surface quality.

1. Introduction

The laser industry has grown rapidly for the last 35 years where researchers all over the world have been investigating the use of laser radiation to process materials. Constant demand for laser machining by industries includes cutting, drilling, welding, thermal treatment and marking process [1-4]. A laser system consists of substance, energy source, and laser resonator. The output of the laser is the laser beam, which can be focused, directed, and controlled with great precision. Currently, laser beam cutting is a machining process frequently used to cut parts of different materials including composites[5]. Composite engineered products exhibit great structural integrity, higher dimensional stability and great flexibility in terms of shaping. Many interior automotive parts, electrical component and industrial applications suit with composite. In composite materials machining, by using conventional method like turning and milling, poor quality of cut surface obtained due to spalled fibres, fuzzing, and delamination [6-8]. Laser beam machining offers an ideal mean of fiber reinforced composites cutting, being a non-contact and virtually force-free manufacturing method.

Laser cutting is a thermal process in which a focused laser beam is used to melt and vaporized material in a localized area. A co-axial gas jet is used to eject the molten material from the cut and leave a clean edge. A particular characteristic of a laser cut found in previous work is the formation of striations on the cut edge [9]. This technology stimulates major interest due to the fact that it joins various advantages such as: non-contact process without tool wear, possibility of using a controlled atmosphere, high energy density, flexibility on beam delivering, easy automation, small heat affected zone, high speed and excellent edge quality [4, 10-11]. Laser cutting of composite materials has received much attention in the literature, there are exclusive work have been published on the laser cutting of this composite product [7],[12], [8]and [13]. J. Paulo Davim et al. [12] reported some experiment on CO_2 laser cutting quality of polymer and composite material. Though much works have been conducted, there are problems in laser beam machining of composites due to exothermic chemical reaction. The exothermic chemical reaction is influenced by several factors such as materials composition, density, moisture, thermal conductivity and internal bond strength [14-15]. The degree of surface roughness of the polymer matrix plays an important role since any surface irregularities may show through thin overlays and can reduce the final quality of the panel. In laser cutting of carbon fiber reinforced polymer (CFRP) composite, controlled parameters are laser power, repetition rate, pulse duration, cutting speed and pulse energy [7]. Other parameters include type of assisting gas and its flow rate, material type and thickness, and beam properties. To understand the quality of cut surface, the burn mark on materials surface resulted from the laser processing have been investigated[13]. The laser cutting on polymer composite excessive heat may produce sufficient damage on the surface layer based on different material composition and heat conduction[6-7]. In this paper the effect of laser spot size on burn mark profile of glass fiber reinforced epoxy composite was investigated.

2. Material and experiment set-up

A beam of finite diameter is focused by a lens to obtain a smaller beam spot, as shown in Figure 1. Where D is the entrance beam diameter, F is the focal length of focus beam and d_f is the diameter of the focus beam. The Gaussian beam focuses from a lens down to a waist and then expands from the focus point. Because the order of the beam mode has great effect on both the focused spot size and the depth of focus, the beam mode structure plays an important role in laser materials processing. Thus the focal length of the focus lens should be selected properly according to the cutting requirements.

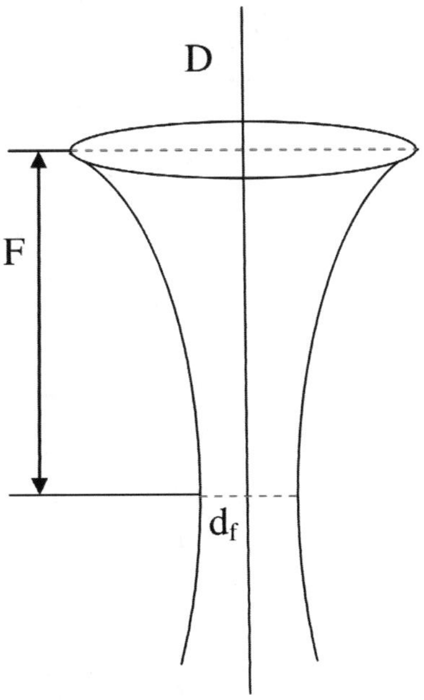

Figure 1.Schematic illustration of Gaussian beam.

In this study, glass fiber reinforced epoxy composites of 2.5 mm thickness were laser scanned. The fiber reinforcement material was E-glass fiber and the matrix material used was medium viscosity epoxy resin. The fabrication of the polymer composite was conducted at room temperature using curing polyamine hardener (K-6). The preliminary laser cutting experiment set-up of GFRP composite is shown in figure 2. The laser beam was focused on the sample surface using a Synrad CO_2 low power laser machining system with 30 W was used to produce profiles on the GFRP composites. Samples were mounted on two axes computer controlled table with integrated software.

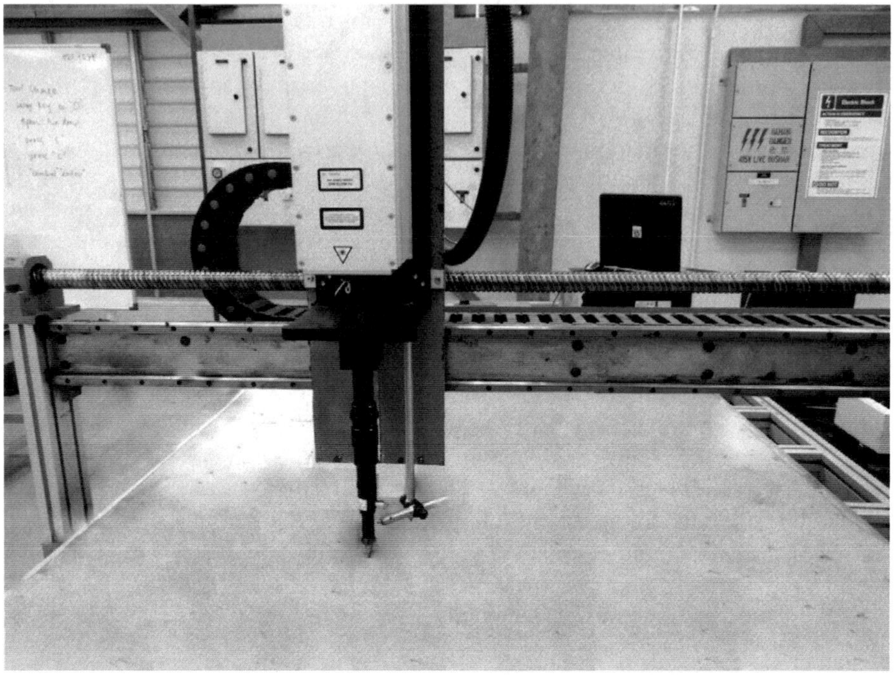

Figure 2. Experiment set-up of laser cutting.

The constant parameters of laser cutting of the composite were shown in the table 1. Parameter consider in the preliminary experiment was length of focal lens. The length was set-up differently for each test with range of between 6.0mm to 13.5mm from workpiece surface. The quality of CO_2 laser cutting was evaluated by standard process. The dimension of the burn and focus spot size was study by microscope, with low magnification. Power and the cutting speed were constant which are 10W and 2mm/s in order to investigate the interaction behavior laser beam and material. The 2.5mm glass fiber composite workpiece was set in flat surface position. The glass fiber composite was cut in rectangle shape with dimension of 10mm x 20mm.

Table 1. Parameters for laser cutting process.

Parameters	**Setting**
Laser wavelength	10.6 um
Power range	10W
Cutting speed	2mm/s
Cutting position	Flat surface

3. Result and Discussion

Laser cutting of GFRP composite is considered and the influence of laser focus setting on focus spot size and surface profile is investigated. The focus spot size, profile depth and material removal were evaluated and measured shown in the figure 3 and 4. The depth and removal rate of glass fiber reinforce polymer (GFRP) composite shows small changes with manipulating number of focal length. The fiber compound blocks the intensity of beam and reduced the depth and material removal.

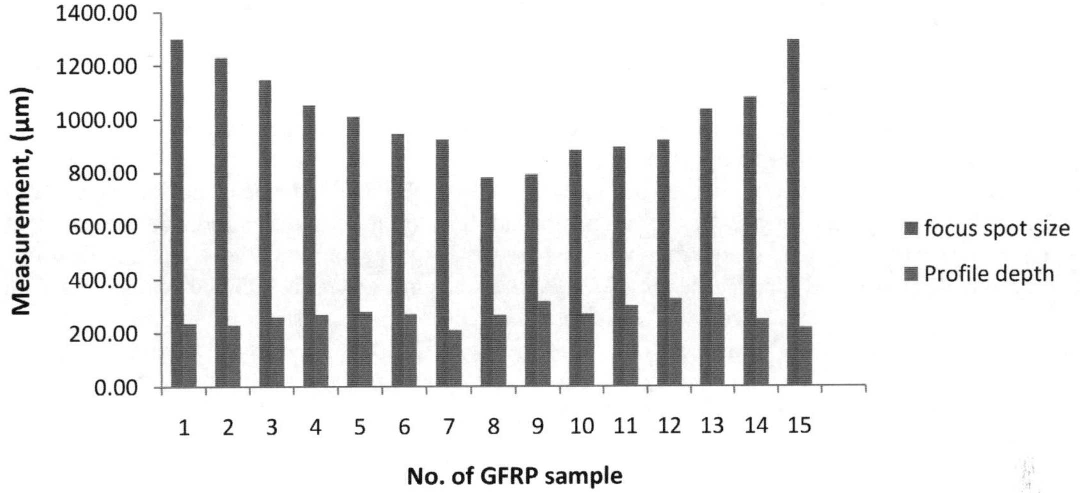

Figure 3. Result of focus spot size and profile depth of each glass fiber reinforced polymer (GFRP) composite sample.

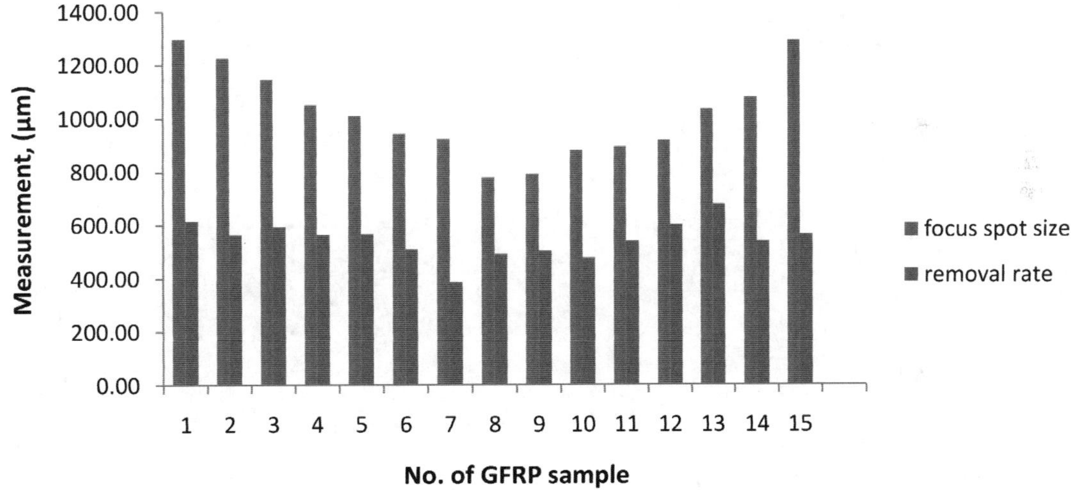

Figure 4. Result of focus spot size and material removing rate of each glass fiber reinforced polymer (GFRP) composite sample.

In the experiment, the sample eight shows the smaller value of focus spot size on the surface but the image of burn mark was shown clearly in the figure 4a. After continue with difference setting length the beam width value became larger. This is consequently related to the Gaussian beam focuses. The beam throughout lens down to a waist at 9.5mm length and then expands from the focus point. In this case, the nominal beam waist position of the focusing lens moves into the workpiece. This is due to that the beam divergence has significant influence the focus spot size when the nominal beam waist position is set at the workpiece, the beam power intensity distribution becomes more and the effective focus spot diameter becomes large as the beam penetrates into the workpiece. Therefore, molten material was removed at the surface and creates preferable value of depth in the figure3. But high intensity of beam with low cutting rate generates heat on the surface.

The different characteristic of glass fiber reinforced polymer (GFRP) composite and low heat conduction produced combustion during the process [6]. Thus, burn marks was appeared on the profile due to the high intensity of focus beam effect on the profile. Based on the surface profiles, burn marks have been seen decreasing due to expanding beam width. This is because of the material removed in the form of dross during matrix melting and accumulated fiber compound in the samples. In laser cutting, roles of gas assist were importance to eject the molten material and prevented flammability occurs during the process. Although, the cutting process does not occur, surface profile show that fiber was cut in the process can be seen in the figure 4b and 4c. No ejection and solidification on melting matrix give difficulties to cut more fiber and increase the cutting depth. Consequently, factor may cause the uncut and melted material is solidified back after cutting process. It is because this melted material has not been removed from the specimen. Therefore, gas assist plays an important role in laser cutting in order to make a material cut successfully [15], [16]. The gas helps much to eject the melt from the cutting zone and thus leaves a clean cut. It should be noted that the assisting gas under goes the exothermic reaction with the molten metal resulting in excessive energy in the cutting section.

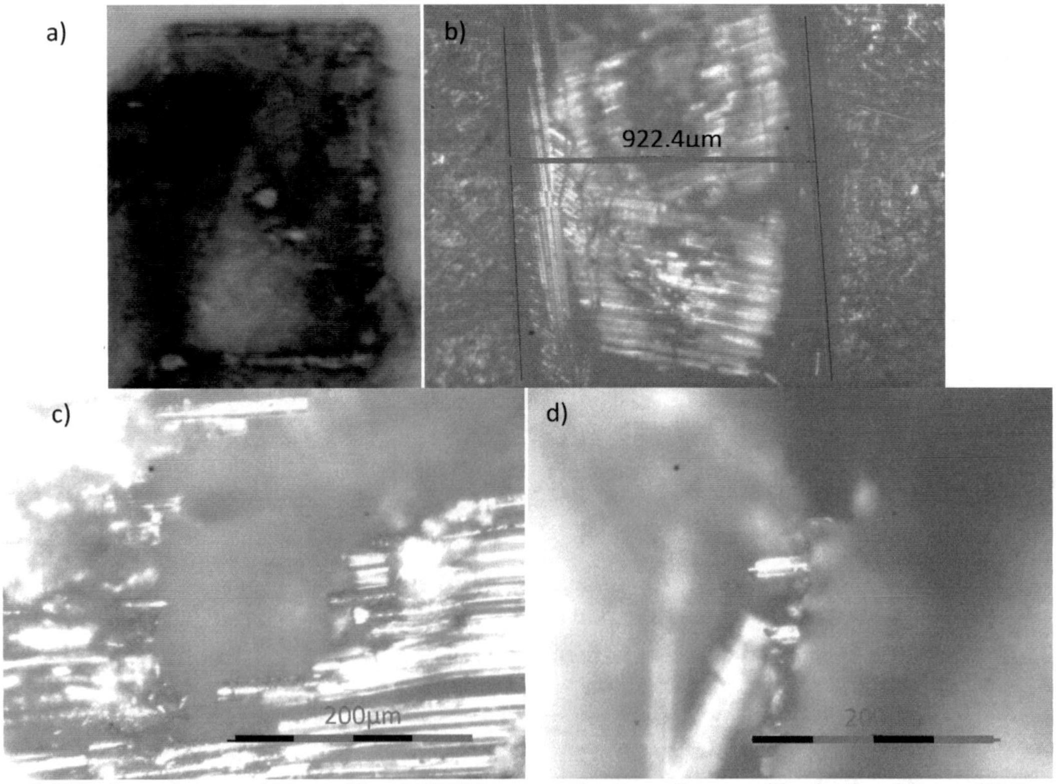

Figure 4. The focus beam width profile of GFRP composite a) surface profile the focal lens of 9.5 mm b)focus spot of the profile c) detail at the front surface and d) detail at the beside surface profile.

4. Conclusion

This preliminary experiment shows that the beam mode has great effect on the focused spot size, profile depth, material removal rate and surface quality. The focus beam parameter of laser cutting influence the behavior of surface, focus spot size, depth and removal rate in GFRP composite. This is due to that the beam divergence produce intensity energy distribution at the surface material. In this preliminary experiment was failed to cut the glass fiber composite but according to the profile fiber compound was cut with focal length of 9.5mm which gave smallest focus spot size. Therefore, other parameter such as power, cutting speed and pressure assist gas should give attention during cutting process of the polymer composite.

References

[1] B. Tirumala Rao, M.O.I.a.L.M.K., *A power ramped pulsed mode laser piercing technique for improved CO2 laser profile cutting.* Optics and Lasers in Engineering, 2009. **47**(11): p. 1108-1116.

[2] Neila Jebbari, M.M.J., Faycal Saadallaha, Annie Tarrats Saugnaca, Raouf Bennaceura andJean Paul Longuemard, *Thermal affected zone obtained in machining steel XC42 by high-power continuous CO2 laser.* Optics & Laser Technology, 2008. **40**(6): p. 864-873.

[3] Da-ping Wan, H.-b.L., Yu-ming Wangb, De-jin Hua and Zhen-xing Gui, *CO2 laser beam modulating for surface texturing machining.* Optics & Laser Technology, 2008. **40**(2): p. 309-314.

[4] A. Lamikiz, L.N.L.d.L., J.A. Sáncheza, D. del Pozo, J.M. Etayoa, and J.M. López, *CO2 laser cutting of advanced high strength steels (AHSS).* Applied Surface Science, 2005. **242**(3-4): p. 362-368.

[5] K. Li, D.W.Z., K. C. Yungb, H. L. W. Chana and C. L. Choy, *Study on ceramic/polymer composite fabricated by laser cutting.* Materials Chemistry and Physics, 2002. **75**(1-3): p. 147-150.

[6] A. Slocombe, L.L., *Laser ablation machining of metal/ polymer composite materials1.* Applied Surface Science, 2000. **154-155**: p. 617-621.

[7] Jose Mathew, G.L.G., N. Ramakrishnanc and N. K. Naik, *Parametric studies on pulsed Nd:YAG laser cutting of carbon fibre reinforced plastic composites.* Journal of Materials Processing Technology, 1999. **89-90**: p. 198-203.

[8] Z.L. Li, H.Y.Z., G.C. Lim, P.L. Chu and L. Li, *Study on UV laser machining quality of carbon fibre reinforced composites.* . Composites Part A: Applied Science and Manufacturing, 2010. **41**(10): p. 1403-1408.

[9] F.A. Al-Sulaiman, B.S.Y.a.M.A., *CO2 laser cutting of a carbon/carbon multi-lamelled plain-weave structure.* Journal of Materials Processing Technology, 2006. **173**(3): p. 345-351.

[10] Meijer, R.F.d.G.a.J., *Laser cutting of metal laminates: analysis and experimental validation* Journal of Materials Processing Technology, 2000. **103**(1): p. 23-28.

[11] F. Caiazzo, F.C., G. Daurelio and F. Memola Capece Minutolo, *Laser cutting of different polymeric plastics (PE, PP and PC) by a CO2 laser beam.* Journal of Materials Processing Technology, 2005. **159**(3): p. 279-285.

[12] J. Paulo Davim, N.B., Marta Conceicao and Carlos Oliveira, *Some experimental studies on CO2 laser cutting quality of polymeric materials.* Journal of Materials Processing Technology, 2008. **198**(1-3): p. 99-104.

[13] A. A Cennaa, P.M., *Analysis and prediction of laser cutting parameters of fibre reinforced plastics (FRP) composite materials.* International Journal of Machine Tools and Manufacture, 2002. **42**(1): p. 105-113.

[14] A Slocombe, J.C.a.L.L., *The effect of pigment addition in diode laser ablation machining of ceramic/ polymer composite material.* Applied Surface Science, 2000. **168**(1-4): p. 21-24.

[15] Hanadi G. Salem, M.S.M., Yehya Badr and Wafaa A. Abbas, *CW Nd:YAG laser cutting of ultra low carbon steel thin sheets using O2 assist gas.* Journal of Materials Processing Technology, 2008. **196**(1-3): p. 64-72.

[16] Chen., S.-L., *The effects of gas composition on the CO2 laser cutting of mild steel.* . Journal of Materials Processing Technology, 1998. **73**(1-3): p. 147-159.

Hull Surface Information Retrieval and Optimization of High Speed Planing Craft

A.F. Ayob[1], W.B. Wan Nik[1], T. Ray[2] and W.F. Smith[2]

[1]Maritime Technology Department, Faculty of Maritime Studies and Marine Science, Universiti Malaysia Terengganu, 21030 Kuala Terengganu, Terengganu, Malaysia
[2]School of Engineering and I.T., University of New South Wales, ACT 2600, Canberra, Australia

E-mail: ahmad.faisal@umt.edu.my, niksani@umt.edu..my

Abstract. Traditional approach on ship design involve the use of a method which takes a form that was earlier called the "general design diagram" and is now known as the "design spiral" – an iterative ship design process that allows for an increase in complexity and precision across the design cycle. Several advancements have been made towards the design spiral, however inefficient for handling complex simultaneous design changes, especially when later variable changes affect the ship's performance characteristics evaluated in earlier stages. Reviewed in this paper are several advancements in high speed planing craft design in preliminary design stage. An optimization framework for high speed planing craft is discussed which consist of surface information retrieval module, a suite of state-of-the-art optimization algorithms and standard naval architectural performance estimation tools. A summary of the implementation of the proposed hull surface information retrieval and several case studies are presented to demonstrate the capabilities of the framework.

1. Introduction

In conventional ship design, the process takes a form that was earlier called the general design diagram and is now known as the *design spiral* – an iterative ship design process that allows for an increase in complexity and precision across the design cycle. Significant contributions through the visualization and modelling of this process can be found in the work of Evans [1]. The ship design spiral combines synthesis (e.g. hull geometry and arrangement) and analysis (e.g. resistance, stability and seakeeping) in its sequential process. Refinements have been made to the design spiral, such as economic aspects by Buxton [2] and time Andrews [3]. The design spiral provides a *safe template* for producing a satisfactory hull for which the designs evolve over time. Although the Evans-Buxton-Andrews ship design spiral is able to capture the iterative practice and necessary elements in the ship design discipline, it is inefficient for handling complex simultaneous design changes, especially when later variable changes affect the ship's performance characteristics evaluated in earlier stages [4]. Mistree *et al.* [5] commented that, although the spiral approach may result in satisfactory designs, it does not promote the identification of superior solutions. The distinction here is between a design that is feasible and one that approaches optimality.

Published under licence by IOP Publishing Ltd

Prior to 1990, a significant amount of work conducted on ship design optimization was documented. Lyon and Mistree [6] and Smith and Woodhead [7] incorporated linear programming (LP) to solve ship design optimization that included structural problems. The use of non-linear programming was demonstrated by Moe and Gisvold [50], Moe and Lund [9], Nowacki *et al.* [10], Pal [11] and Lutkus *et al.* [12] using sequential unconstrained optimization technique (SUMT). Other optimization methods include Hooke and Jeeves [13] with penalty function [14], parametric study [15], simulated annealing [16, 17] and convergent random search technique [18]. The ship design optimization problems presented include structural optimization, transportation system optimization, construction cost and minimization of drag. Ship design optimization works before 1990 successfully demonstrated the use of empirical relations in which the full form surface geometry of the ship remained unknown, except for the principal parameters, form coefficients and performance characteristics information obtained following optimization.

In the early 1990s, the development of ship design optimization methods that combined analytical estimates, such as computational fluid dynamics (CFD), with full hull form geometry are observed. Motivation towards the inclusion of full hull geometry was driven by the advancements of CFD in the domain of ship design optimization. Janson [19] and Janson *et al.* [20] demonstrated calm water resistance minimization using standard Wigley and Series 60 hulls, where the offset data were treated as free variables. Although CFD requires the full geometry of a ship, using it within an optimization algorithm comes at a high computational cost and the final hull generated might be impractical due to substantial waviness [36]. However, utilizing full hull geometry in optimization is not limited to CFD as empirical relations can also be incorporated. Ganesan [22] proposed an optimization framework for a container ship using a hull blending technique following the work of Doctors [23]. Due to its extensive application, there are a significant number of optimization frameworks for the design of displacement craft such as the works of Ray [17] and Ganesan [22] on container ships, Smith [24] on destroyer frigates, Peacock [25] on Offshore Patrol Vessel (OPV) and Gammon [26] and Majumder *et al.* [27] on fishing vessels. The application of genetic algorithms for fast displacement craft resistance minimization with a first principle parameters technique and Mitchell thin ship theory has been demonstrated by Day and Doctors [28].

Although high speed planing vessels are popular as commercial and recreational craft, and widely used for life-saving, coastal patrol and surveillance missions, much information about them has been kept proprietary. As a result, there are only a small number of case studies available in the literature. Some earlier research includes the work of Almeter [29], in which resistance optimization of the Soviet's BK and MBK planing hulls was performed using Resistance Prediction and Optimization (RPO) software via a database of varied planing hulls. Jons *et al.* [30] presented an integrated CAD process that employed a comprehensive design geometry library (DGL) to satisfy the design goals required by the customer. Although using a database of various hulls is helpful, given a parent hull as a starting point, a comprehensive database is not necessarily important if the parametric transformation method driven by an optimization algorithm is applied. Bearing in mind such a paradigm, a high speed planing craft optimization framework was proposed by Ayob *et al.* [31, 32, 33] that include the use of Savitsky's [34] equation.

In this paper, a state-of-the-art optimization framework for the design of high speed planing craft is utilized to solve calm water resistance minimization problem. A proposal to retrieve surface information of high speed planing craft hull with hard chines/non-differentiable edges using B-spline method is presented. Finally the capability to represent other hull shape of DTMB 5415 is presented followed by a summary of a planing craft optimization case study.

2. Optimization framework components

The optimization framework utilized in this paper consists of three applications namely Matlab, Microsoft® Excel and Maxsurf. Maxsurf Automation Library built upon Microsoft® COM interface is used as a medium of communication (inter-process) between applications. Presented in Figure 1 is a flowchart of the optimization framework. The inter-process communication is initialized with the selection of principal dimensions (*L*, *B*, *T*) by the optimizer module in Matlab The optimization module consist of a library of optimization algorithm, namely Non-dominated Sorting Genetic Algorithm (NSGA-II) [41], Infeasibility Driven Evolutionary Algorithm (IDEA) [39] and Surrogate-Assisted Evolutionary Algorithm (SA-EA) [40]. Parametric transformation is invoked to generate a candidate hull followed by evaluation of the hydrostatics and calm water resistance of the candidate hull in Maxsurf using the methods of Savitsky [34]. Finally the seakeeping performance is evaluated using the Savitsky and Koelbel [36] method. This completes one workflow loop. The details of the optimization framework can be found from the work of Ayob *et al.* [31, 32, 33].

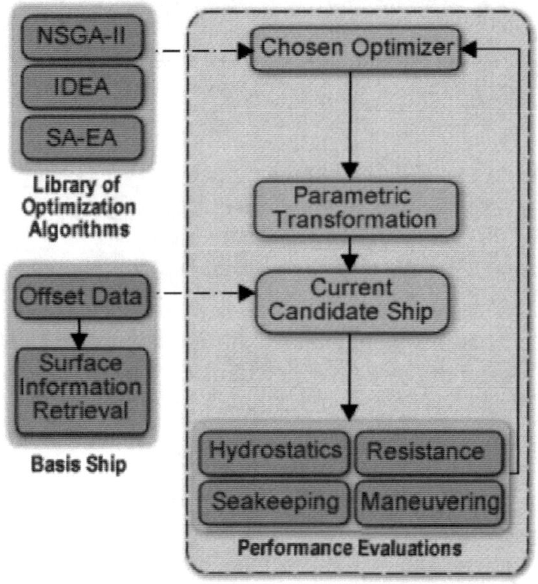

Figure 1. Flowchart of the optimization framework

2.1 Hull form representation and surface information retrieval

2.1.1 B-spline curve and surface

As shown in Figure 1, the geometry tools consist of a *surface information retrieval* and a *parametric transformation* module. The surface information retrieval module is employed to generate B-spline representation of the hull while the parametric transformation module changes the shape of the hull based on principal dimensions given by the optimizer. The formulation of surface information retrieval module is based on the inverse B-spline method [35]. A set of known surface (offset) data is used to determine the defining polygon net for a B-spline surface that best interpolates the data (Section 2.1.2). This method is further expanded to yield a representation of a hard chine form that normally represents a planing craft, discussed in Section 2.1.3. Detail explanations which include mathematical definition and proof of B-spline method can be found from the work of Rogers *et al.* [37], and Piegl and Tiller [35].

2.1.2 One-surface hull fitting approach

A one-surface hull form geometry representation approach is the simplest embodiment of a three-dimensional hull form representation method. It is assumed that a hull can be approximated using a set of continuous data points located on its surface. This means that the method eliminates several characteristics of the hull, such as knuckles (hard chines) or other forms of C^0 discontinuities. The topology of the hull (e.g. keel, stations, deck) is assumed to be correct when the resulting surface matches the shape desired by the ship designer. Complementary works in the literature that discuss one-surface hull geometry representation are Rogers *et al.* [37], Bole [38] and Harries [21].

A one-surface hull approach is best for a simple hull form (e.g. no bulbous bow, no hard chine, etc.). A fair hull surface can be generated using a minimal number of control points, which prevent unwanted wiggles and undulations. Shown in Figure 2 are the steps taken to fit a one-surface hull using a simple yacht. The original definition of the yacht consisted of 1282 three-dimensional data points of the x, y and z coordinates. The three-dimensional fitting error, ε between two 3D data points, Q and S, can be measured using Euclidean distance method that yields,

$$\varepsilon = \sqrt{\left(Q_x - S_x\right)^2 + \left(Q_y - S_y\right)^2 + \left(Q_z - S_z\right)^2}$$

1. The original coordinate data that defined the parent hull was arranged station-wise, from aft to forward (Figure 2(a)). The stations are numbered for clarity (in this case, stern to bow).
2. For each station, the data points were fitted to obtain the station-wise control points. Shown in Figure 2(b) are the fitted/extrapolated lines together with the corresponding control points. A good fitting error was obtained with a mean error, $\bar{\varepsilon} = 0.34 \times 10^{-2}$.
3. The station-wise control points were then fitted to obtain the control points in the longitudinal direction. Shown in Figure 2(c) are the longitudinal control points obtained with respect to the station-wise control points. The number of longitudinal control points was independent of the number of station-wise control points. It could be increased providing the desired number of longitudinal control points was less than the number of stations.
4. Finally, the control polygon net that defines the whole hull surface was obtained, as shown in Figure 2(d). In this example, 5 x 4 control points were employed to approximate the whole surface of the yacht.

2.1.3 Multi-surface hull approach

In real-world design, hull form geometry may contain several complex features, such as surface discontinuities (e.g. a planing craft's hard chine) and a bulbous bow. For planing craft, a one-surface fitting approach was attempted by Rogers *et al.* [37]. Their implementation involved interactive manipulation of the control vertices to produce knuckles/hard chines of the planing craft. Although it is possible to manually relocate the control points to represent a hard chine, there are two problems: firstly, the initial approximations of the hull are poorly faired, especially on the hard edges; and secondly, allowing the designer to interactively modify the location of the control points interferes with the fairness of the B-spline surfaces.

The work reported in this paper proposes the splitting of surface data points at the edge of the surface for the design of a planing craft; for example, a ship designer identifies the knuckle positions and feeds the coordinate data points to a module that defines the control points of the surface data that fit the parent ship. The steps taken to approximate a three-surface hull are demonstrated using the basic planing craft shown in Figure . The original planing craft example was defined by 6066 discrete x, y and z data points.

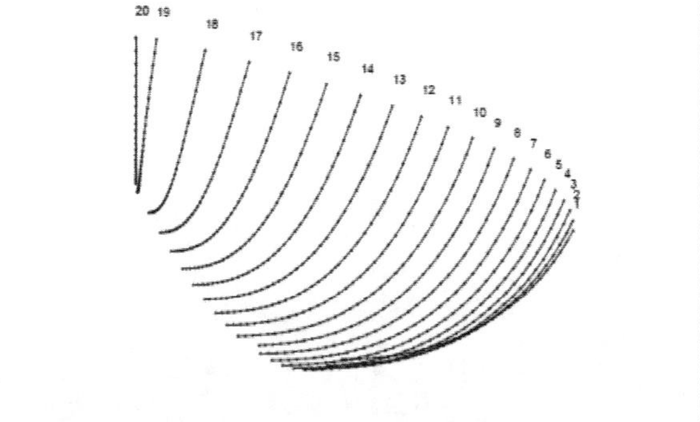

(a) Data points arranged in numbered stations offsets.

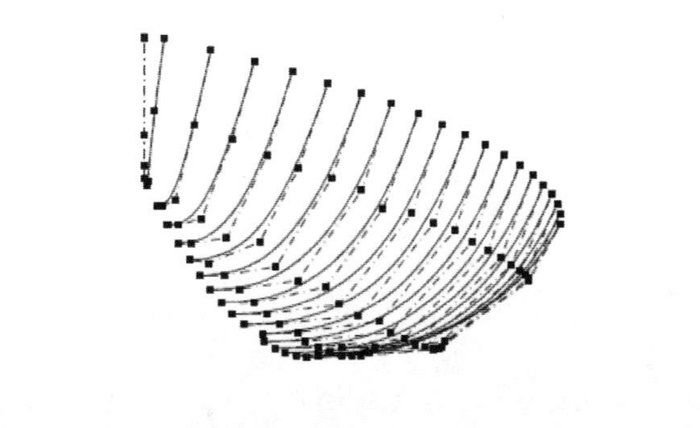

(b) Fitting station-directional data points ($\bar{\varepsilon} = 0.34 \times 10^{-2}$).

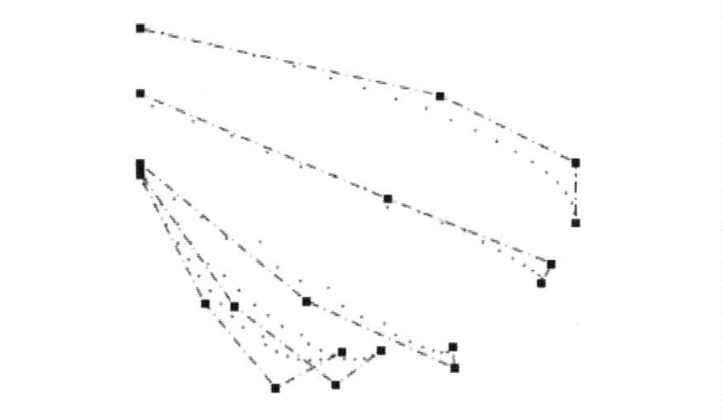

(c) Fitting longitudinal direction through control points of station-directional control points ($\bar{\varepsilon} = 0.15$)

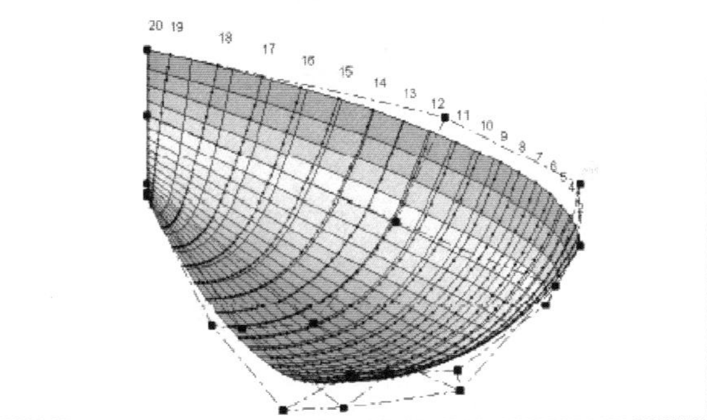

(d) Surface defined from B-spline fitting process

Figure 2: Example of one-surface hull fitting

1. The coordinates of the parent hull were partitioned into three segments and, for clarity, labelled 'top offset data', 'middle offset data' and 'bottom offset data' respectively (Figure 3(a)). Each set of offset data had common points representing the hard edges/discontinuities on the planing hull form surface.
2. The three segments of the offset data were fitted station-wise independently, as shown in Figure 3(b). The top, middle and bottom offset data are fitted using 4 control points respectively, which were sufficient to cater for the minimal curvature of such planing craft forms. The mean fitting errors, $\bar{\varepsilon}$, for the top, middle and bottom segments were 0.92 × 10−2, 0.78 × 10−9 and 0.95 × 10−8.
3. Longitudinal control points were fitted using the station-wise control points (Figure 3(c)). This second fitting process produced a set of global control points that was independent of the number of control points in the station-wise direction.
4. Shown in Figure 3(d) is the final form of the parent planing vessel. The three surfaces, each with a 4 × 4 control polygon net with appropriate matching of the edges on the hull form, are illustrated.

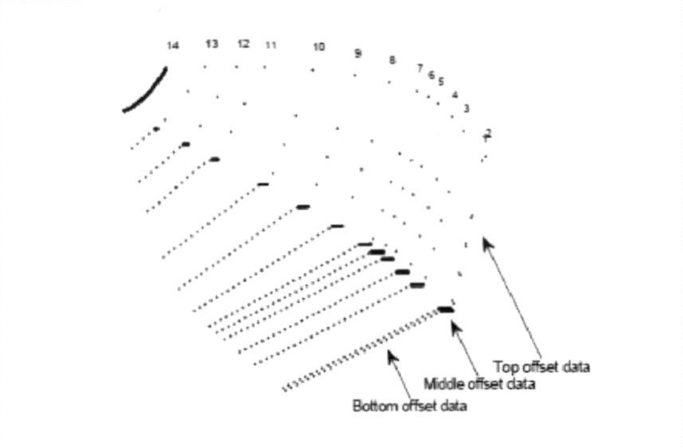

(a) Data points arranged in numbered stations offsets.

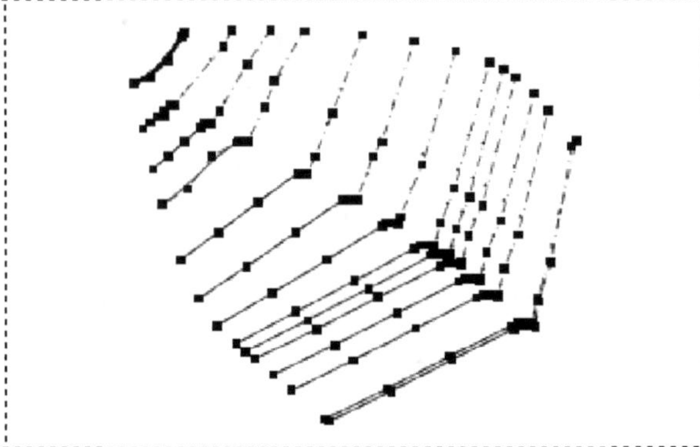

(b) Fitting station-directional data points ($\bar{\varepsilon}_{top} = 0.92\times10^{-2}$, $\bar{\varepsilon}_{middle} = 0.78\times10^{-9}$, $\bar{\varepsilon}_{bottom} = 0.95\times10^{-8}$).

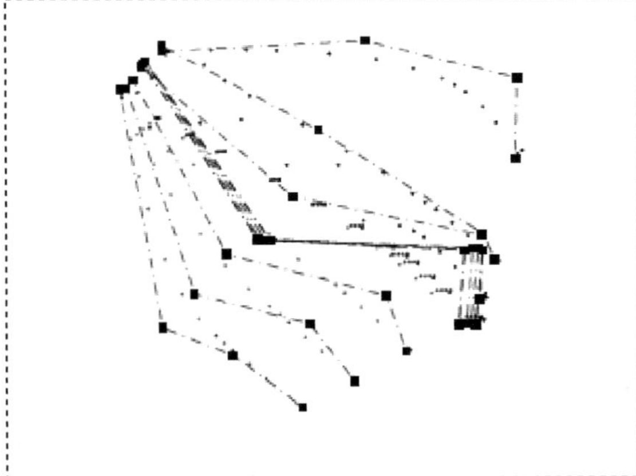

(c) Longitudinal control points and corresponding B-spline ($\bar{\varepsilon}_{top} = 0.67$, $\bar{\varepsilon}_{middle} = 0.67$, $\bar{\varepsilon}_{bottom} = 0.48$)

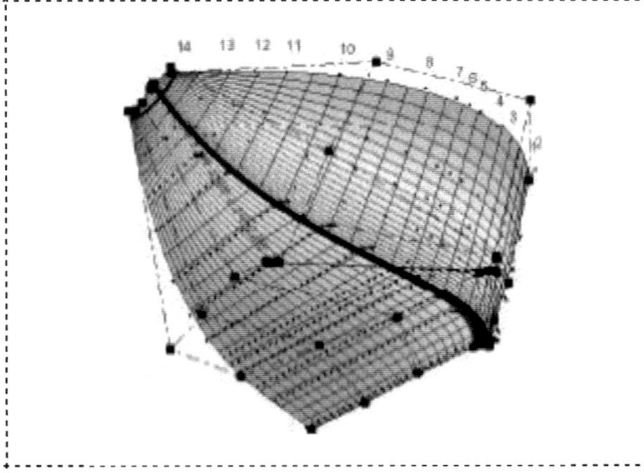

(d) Surface defined from longitudinal net

Figure 3: Example of one-surface hull fitting

3. Implementation and case studies

3.1 Shape retrieval of other hull shapes

A US Navy Combatant, DTMB 5415, represented using three B-spline surfaces segmented at the bow, middle body and transom is shown in Figure 4. The three polygon nets at these surfaces consisted of 8 × 10, 8 × 7 and 8 × 5 control points respectively. The original parent hull was defined by 1323 data points. The respective error values for each segment are reported in Table 1.

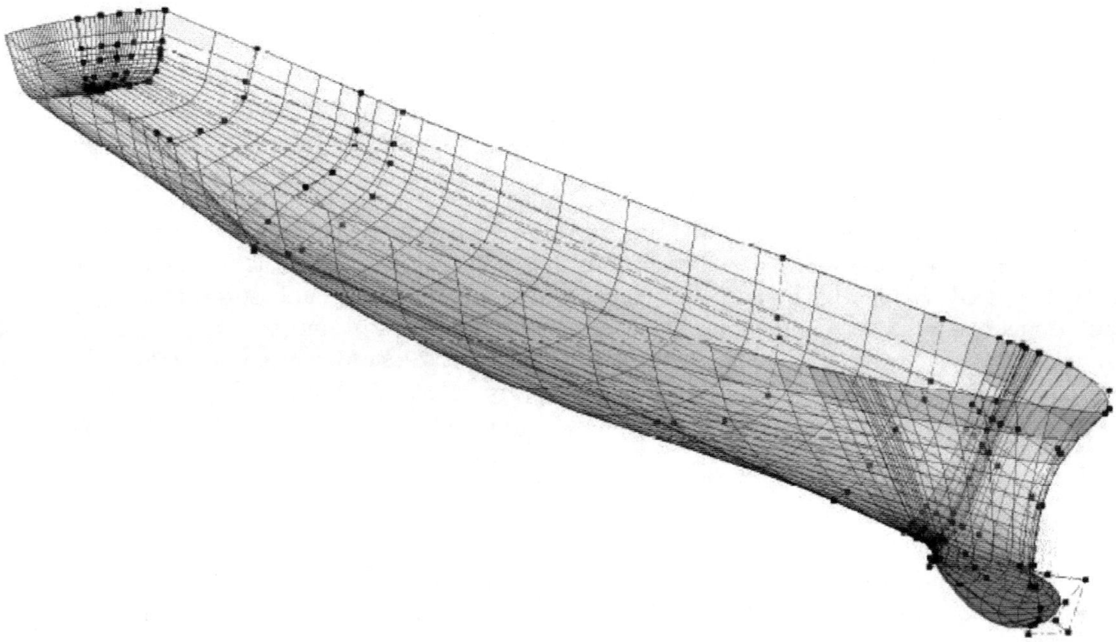

Figure 4. US Navy Combatant DTMB 5415 craft

Table 1. US Navy DTMB 5415 fitting error

Segment	Error Description	Error Value
Bow	$\bar{\varepsilon}_{station}$	4.5×10^{-3}
	$\bar{\varepsilon}_{longitudinal}$	4.3×10^{-3}
Middle body	$\bar{\varepsilon}_{station}$	4.6×10^{-8}
	$\bar{\varepsilon}_{longitudinal}$	2.3×10^{-3}
Transom	$\bar{\varepsilon}_{station}$	5.4214×10^{-7}
	$\bar{\varepsilon}_{longitudinal}$	5.6330×10^{-7}

3.1 Minimization of calm water resistance

Success on the use of the proposed optimization framework for high speed planing craft has been demonstrated through several case studies. The case studies include the minimization of calm water resistance using the example of a 10m Indian Fast Craft (IFC) [31] and a typical coast guard patrol vessel [33] derived from the United Coast Guard WPB-110ft (USCG WPB-110ft) craft. The initial/basis candidate vessels retrieved using the multi-surface approach has been compared with the original craft in order to validate the mathematical model constructed prior to optimization. Due to the stochastic nature of the employed algorithms, thirty independent runs of each algorithm, NSGA-II, IDEA and SA-EA, were performed. The performance of the optimized candidate vessel is then compared with the initial craft in order to measure the savings obtained from the optimization exercise.

Summarizing from the case study of IFC, the optimization problem formulation has been set as a minimization of calm water resistance, subject to the constraint on the displacement and stability (metacentric height). The minimization of calm water resistance formulation has been executed using the library of optimization algorithms as summarized in Section 2. The calm water resistance values of all the optimized candidate designs were approximately 9.8kN, compared to 11.5kN for the basis ship, accounted for 15% reduction of calm water resistance when compared with the basis hull. The lengths, beams and drafts of the best designs (presented to two decimal places) were approximately at around 11.00m, 3.04m and 0.60m for the optimized candidate design, and 10.04m, 2.86m and 0.7m for the basis ship, respectively.

In the case of study of the USCG WPB-110ft, similar objective function, constraints and optimization algorithms have been used. Significant reductions in calm water resistance values have been observed when compared with the basis ship. The lengths, beams and drafts of the best designs (presented to two decimal places) were approximately at around 35.92m, 7.28m and 1.50m for the optimized candidate design, and 32.93m, 7.52m and 1.60m for the basis ship, respectively. The calm water resistance values of the best candidate designs are reported at around 122.37kN, compared to 135.93kN for the basis ship. This reflects about 10% reduction in calm water resistance when compared with the basis ship.

4. Conclusions

Demonstrated in this paper is the application of an optimization framework for optimum design of high speed planing craft. The application of one-surface and multi-surface fitting has been demonstrated using several examples which include a simple yacht, a planing craft and a DTMB 5415 destroyer hull. The proposal of splitting the surface prior to fitting the offset data is capable to maintain the non-differential edges of the hull surface. It has been furthermore supported by small fitting error obtained from the surface information retrieval process.

Two case studies have been summarized, which concerning on the minimization of calm water resistance of two candidate hulls, namely IFC and USCG WPB-110ft craft. The case studies reported significant reduction of calm water resistance of 15% and 10% respectively for IFC and USCG WPB-110ft craft. The utilization of the optimization framework for high speed planing craft shows a promise to assist high speed craft designer to provide optimum design of high speed planing craft in concept and preliminary design stage.

5. References

[1] J. H. Evans. Basic Design Concepts. *Naval Engineers Journal*, **71**:671–678, 1959.
[2] I.L. Buxton. Engineering Economics Applied to Ship Design. *Transactions of the Royal Institution of Naval Architects (RINA)*, **114**:409–428, 1972.
[3] D. Andrews. Creative Ship Design. *Transactions of the Royal Institution of Naval Architects (RINA)*, **123**:447–458, 1981.
[4] A. J. Keane, W. G. Price, and R. D. Schachter. Optimization Techniques in Ship Concept Design. *Transactions of the Society of Naval Architects and Marine Engineers (SNAME)*, **133**:123–140, 1991.
[5] F. Mistree, W. F. Smith, B. A. Bras, J.K. Allen, and D. Muster. Decision-Based Design: A Contemporary Paradigm in Ship Design. *Transactions of the Society of Naval Architects and Marine Engineers (SNAME)*, **98**:565597, 1990.
[6] T. D. Lyon and F. Mistree. A Computer Based Method for Preliminary Design of Ships. *Journal of Ship Research*, **29**:251–269, 1985.
[7] C. Smith and R. G. Woodhead. A Design Scheme for Ship Structures. *Transactions of the Royal Institution of Naval Architects (RINA)*, **116**:47–66, 1974.
[8] J. Moe and K. M. Gisvold. *Optimization and Automated Design of Structures*. Technical report, Division of Ship Structures, University of Trondheim, 1968.

[9] J. Moe and S. Lund. Cost and Weight Minimization of Structures with Special Emphasis on Longitudinal Strength Members of Tankers. *Transaction of the Royal Institution of Naval Architects (RINA)*, **110**:43–70, 1968.
[10] H. Nowacki, F. Brusis, and P. M. Swift. Tanker Preliminary Design - An Optimization Problem with Constraints. *Transactions of the Society of Naval Architects and Marine Engineers (SNAME)*, **78**:357–390, 1970.
[11] P.K. Pal. *Optimum Design of Trawlers*. PhD thesis, Indian Institure of Technology, Kharagpur, India, 1981.
[12] A. J. Lutkus, G. G. Piche, and K. E. Wagner. *Sequential Optimization Technique for Trawler Design*. Technical report, College of Engineering, University of Michingan, 1970.
[13] R. Hooke and T. A. Jeeves. Direct Search Solution of Numerical and Statistical Problems. *Journal of ACM*, **8**:212–229, 1961.
[14] A. Z. Fiacco and G. P. McCormick. *Non-linear Programming: Sequential Unconstrained Minimization Technique*. Wiley, 1968.
[15] J. H. Evans and D. Khoushy. Optimized Design of Midship Section Structure. *Transaction of the Society of Naval Architects and Marine Engineers (SNAME)*, **71**:144–191, 1975.
[16] A. H. Day. The Optimization of Aerodynamic Lift Distribution for a Heeled Yacht in a Wind Gradient. *Transaction of the Royal Institution of Naval Architects (RINA)*, **134**:91–108, 1991.
[17] T. Ray. *Preliminary Ship Design in the Multiple Criteria Decision Making Framework*. PhD thesis, Indian Institute of Technology, Kharagpur, 1995.
[18] R. Leopold. *Mathematical Optimization Methods Applied to Ship Design*. Technical report, Department of Naval Architecture and Marine Engineering, Massachusetts Institute of Technology, 1965.
[19] C. E. Janson. *Shape Optimization of a Simple Hull Fom with Respect to Total Resistance as Computed by SHIPFLOW system*. Technical report, FLOWTECH Int. AB., Gothenburg, Sweden, 1992.
[20] C. E. Janson, K. J. Kim, L. Larsson, D. Holm, and B. Esping. Optimization of the Series 60 Hull from a Resistance Point of View. *In Prooceedings of the CFD Workshop* Tokyo 1994, 1994.
[21] S. Harries. *Parametric Design and Hydrodynamic Optimization of Ship Hull Forms*. PhD thesis, Institute of Naval Architecture and Ocean Engineering, Technische Universitt Berlin, 1998.
[22] V. Ganesan. *A Model for Multidisciplinary Design Optimization of Containership*. PhD thesis, Virginia Polytechnic Institute and State University, Blacksburg, Virginia, 1999.
[23] L. J Doctors. A Versatile Hull-Generator Program. *In Proceedings of the Twenty-First Century Shipping Symposium*, pages 140–158, University of New South Wales, Sydney, Australia, 1995.
[24] W. F. Smith. *The Modeling and Exploration of Ship Systems in the Early Stages of Decision-based Design*. PhD thesis, University of Houston, 1992.
[25] D. A. Peacock. *Decision-Based Hydrodynamic Design of Displacement Monohulls*. PhD thesis, School of Mechanical and Manufacturing Engineering, July 1998.
[26] M. A. Gammon. *Ship Hull Form Optimization by Evolutionary Algorithm*. PhD thesis, Institute for Graduate Studies in Science and Engineering, Yildiz Technical University, Istanbul, 2004.
[27] M. Majumder, A. Akinturk, and S. M. Calisal. Optimized Design of Small Craft. *Marine Technology*, **39**:67–76, 2002.
[28] A. H. Day and L. J. Doctors. Resistance Optimization of Displacement Vessels on the Basis of Principal Parameters. *Journal of Ship Research*, **41**:249–259, 1997.
[29] J. M. Almeter. Resistance Prediction and Optimization of Low Deadrise, Hard Chine, Stepless Planing Hulls. *In Proceedings of the Ship Technology and Research Symposium (STAR)*, 1995.
[30] O. P. Jons, J. Koelbel, and R. Sheldon. A New Generation of High-performance Planing Craft. *Naval Engineers Journal*, pages 234–247, 1985.
[31] A. F. Ayob, T. Ray, and W. Smith. An Optimization Framework for the Design of Planing Craft. *In Proceedings of the International Conference on Computer Applications in Shipbuilding 2009 (ICCAS09)*, pages 181–186, Shanghai, 2009.

[32] A. F. Ayob, T. Ray, and W. F. Smith. A Framework for Scenario-Based Hydrodynamic Design Optimization of Hard Chine Planing Craft. *In Proceedings of the 9th International Conference in Computer and IT Applications in the Maritime Industries (COMPIT'10).* Technische Universitat Hamburg-Harburg, 2010.

[33] A. F. Ayob, T. Ray, and W. F. Smith. Hydrodynamic Design Optimization of a Hard Chine Planing Craft for Coastal Surveillance. *In Proceedings of the Pacific 2010 International Maritime Conference (Pacific 2010)*, Sydney, Australia, 2010.

[34] D. Savitsky. Hydrodynamic Design of Planing Hull. *Marine Technology*, **1**:71–95, 1964.

[35] D. Savitsky and P. W. Brown. Procedures for Hydrodynamic Evaluation of Planing Hulls in Smooth and Rough Water. *Marine Technology*, **13**:381–400, 1976.

[34] D. Savitsky and J. G. Koelbel, Jr. *Seakeeping of Hard Chine Planing Hull.* Technical report, Society of Naval Architects and Marine Engineers, 1993.

[35] L. Piegl and W. Tiller. *The NURBS Book.* Springer-Verlag Berlin / Heidelberg / New York, 1997.

[36] D. F. Rogers and S. G. Satterfield. B-spline Surfaces for Ship Hull Design. *ACM SIGGRAPH Computer Graphics*, **14**(3):211–217, 1980.

[37] D. F. Rogers, S. Satterfield, and F. Rodriguez. Ship Hulls, B-spline Surfaces, and CAD/CAM. *IEEE Computer Graphics and Applications*, **3**:37–45, 1983.

[38] M. Bole. *Parametric Generation of Yacht Hulls*. Master's thesis, Naval Architecture and Small Craft Design, University of Strathclyde, 1997.

[39] H. K. Singh, A. Isaacs, T. Ray, and W. F. Smith. Infeasibility Driven Evolutionary Algorithm (IDEA) for Engineering Design Optimization. *In Proceedings of the 21st Australasian Joint Conference on Artificial Intelligence (AI'08)*, pages 104–115, Berlin, Heidelberg, 2008. Springer-Verlag.

[40] A. Isaacs, T. Ray, and W. F. Smith. An Evolutionary Algorithm with Spatially Distributed Surrogates for Multiobjective Optimization. *In Progress in Artificial Life*, volume 4828/2007, pages 257–268. Springer Berlin / Heidelberg, 2007.

[41] K. Deb, A. Pratap, S. Agarwal, and T. Meyarivan. A Fast and Elitist Multiobjective Genetic Algorithm: NSGA-II. *IEEE Transactions on Evolutionary Computation*, **6**:182–197, 2002.

DOE applied to study the effect of process parameters on

A Shayganpour[1], M H Idris[2], S Izman[1] and H Jafari[2,3]

[1]Manufacturing & Industrial Engineering Dept., Faculty of Mechanical Engineering, Universiti Teknologi Malaysia, 81310 UTM Skudai, Johor, Malaysia
[2]Materials Engineering Dept., Faculty of Mechanical Engineering, Universiti Teknologi Malaysia, 81310 skudai, Johor, Malaysia
[3]Materials Engineering Dept., Faculty of Mechanical Engineering, Shahid Rajaee Teacher Training University, 16785-136 Tehran, Iran

E-mail: Shayeganpour@gmail.com

Abstract. Lost foam casting as a relatively new manufacturing process is extensively employed to produce sound complicated castings. In this study, an experimental investigation on lost foam casting of an Al-Si-Cu aluminium cast alloy was conducted. The research was aimed in evaluating the effect of different pouring temperatures, slurry viscosities, vibration durations and sand grain sizes on eutectic silicon spacing of thin-wall castings. A stepped-pattern was used in the study and the focus of the investigations was at the thinnest 3 mm section. A full two-level factorial design experimental technique was used to plan the experiments and afterwards identify the significant factors affecting casting silicon spacing. The results showed that pouring temperature and its interaction with vibration time have pronounced effect on eutectic silicon phase size. Increasing pouring temperature coarsened the eutectic silicon spacing while the higher vibration time diminished coarsening effect. Moreover, no significant effects on silicon spacing were found with variation of sand size and slurry viscosity.

1. Introduction

Lost foam casting (LFC) is a relatively new process in commercial terms which is gaining an increased attentions because of its technical and environmental advantages [1,2]. No requirement of binders, elimination of some machining steps and capability of producing complex casting make this process to be a suitable manufacturing method for shortening the production time and reducing environmental loads which is a demand for automotive and aerospace industries. Aluminium-silicon-copper alloys are broadly used in automotive and aerospace industries due to their high possibilities of producing imperfection free castings and opportunity to replace forged steel or cast iron parts resulting in more fuel-efficient automobiles or aerospace [3]. Theses alloy can be employed effectively in order to manufacture engineering components by LFC technique. However, several disadvantages have been reported using this method [4,5]. Consequently, many considerations arise in processing to achieve the appropriate parameters setting for getting premium experiment results. LFC is believed more susceptible to any process variation compared to other casting methods due to a large number of

[1] To whom any correspondence should be addressed.

Published under licence by IOP Publishing Ltd

variables and complex interactions between them. Therefore, a well established and designed process parameters can contribute to approach high quality and sound aluminium alloy lost foam castings.

Many researchers have extensively used design of experiment (DOE) methods in order to ascertain significant factors and optimum parameters influencing different objective responses, and finally establish practical model that represent correlation between the significant factors. Taguchi method with $L_9(3^4)$ array was used by Vijian and Arunachalam [6] to optimize the surface finish LM6 aluminium alloy produced by squeezed casting. Similar method, but using L_9 orthogonal array, was taken into consideration by other researchers [7] in order to plan and analyse the experiments to investigate the optimization of tensile properties of Al-7%Si alloy castings manufactured by evaporative pattern casting (EPC). A five-factor, two-level, half-fractional factorial designs with four centre points was employed by Gunasegaram *et al.* [8] to identify critical factors influencing shrinkage porosity in A356.0 aluminum alloy castings produced by permanent mould casting. Verran *et al.* [9] used the same half-fractional design, but without centre points, for analysing and optimising the process parameters on the internal quality of SAE305 aluminium alloy die cast parts. A full-factorial design method was recently used by the authors [10] to assess the effect of sand grain size and vibration time, as the two of important process parameters, on the surface roughness of lost foam aluminium alloy casting. However, limited papers can be found which studied numerical experiments based on the DOE methods. On the other hand, a growing trend of researcher's interest to investigate the use of different DOE methods including Taguchi's method, fractional design, full-factorial and response surface methodology has been reported in casting industry over the past five years. Therefore, the recent published papers indicate increasing application of DOE methods in different aluminium alloys casting methods.

This paper aims to investigate the effects of four important parameters of lost foam casting, namely pouring temperature, slurry viscosity, vibration time and sand size, on eutectic silicon spacing of a thin wall Al-Si-Cu alloy casting, as a vital factor influences mechanical properties of the casting.

2. Experiment Details

An Ishikawa cause–effect diagram illustrated in Figure 1 was constructed to summarize the factors affecting the quality of LFC castings in terms of silicon spacing. As highlighted in the figure, only the process parameters including pouring temperature, slurry viscosity, vibration time and sand size were taken into account to reveal their influences on eutectic silicon phase fineness of a Al-Si-Cu alloy castings in this study. Table 1 shows the chemical composition of the alloy. A stepped polystyrene foam pattern having 20 kg/m^3 density with the size of 100 mm (width) × 260 mm (length) and different step thicknesses of 3, 6, 12, 18, 24 mm as shown in Figure 2 was designed, fabricated and used in the present research. The pattern was coated by dipping into slurry prepared using a mixture of zircon and colloidal silica for 20 s, then left to drip dry for up to 24 hours in a controlled environment. The slurry viscosities were measured at 20, 28 and 36 s using a Zahn cup (No. 5) in order to achieve the required coating thickness in the range of 0.3-0.5 mm.

Table 1. Chemical composition of the alloy used during the study

Si	Fe	Mg	Mn	Cu	Ti	Zn	Al
10.55	0.62	0.27	0.23	1.79	0.043	0.852	Bal.

Silica sands with the grain size of AFS 16-30, 40-60 and 60-100 were used for mould making during this experiment. A four-point clamping vibrating table was used to produce uniform distribution of the sands inside flask. A horizontal vibration of 50 Hz for durations of 30, 60 and 90 s was utilized for mould making. The alloy was melted in an induction furnace and poured at three different temperatures of 680, 720 and 760 °C. Samples with the size of 10 × 10 mm were cut from the 3 mm thickness section of the castings, mounted, ground and finally polished at their cross section.

Linear intercept method was used to measure the eutectic silicon spacing in microstructures using an image analyzer.

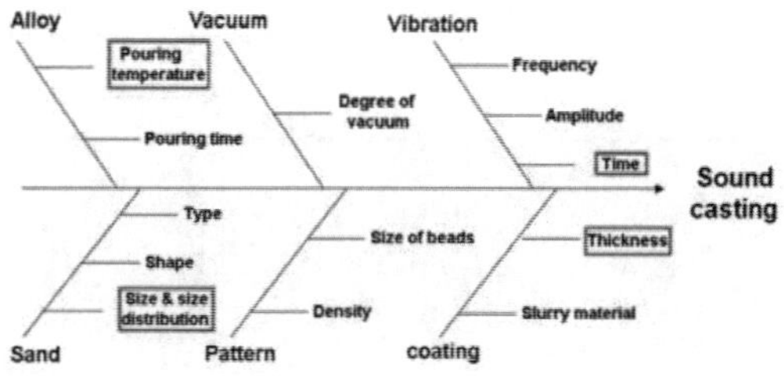

Figure 1. Ishikawa diagram summarizing factors affecting the LFC quality

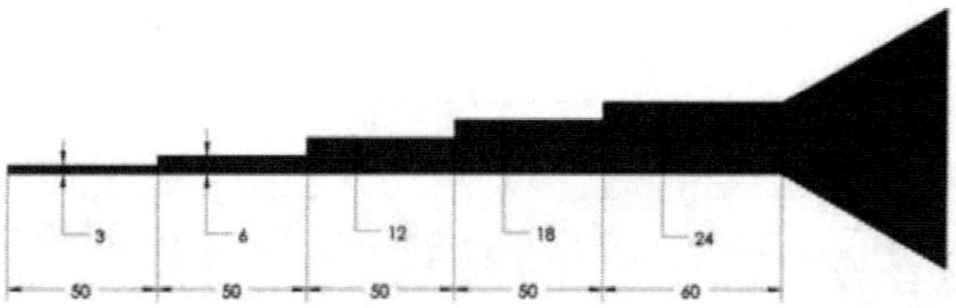

Figure 2. Pattern drawing and dimensions

The effect of selected process parameters on the silicon spacing of the castings were analyzed simultaneously using a statistical technique. A two-level full factorial test was used to plan and analyze the experimental trials. A linear mathematical model (Equation 1) was used to model the real interaction between the casting process parameters. It was used in order to ascertain the most significant factors in the process.

$$Y= X_0 + X_1A + X_2B + X_3C + X_4D + X_5AB + X_6AC + X_7AD + X_8BC + X_9BD + X_{10}CD + X_{11}ABC + X_{12}ABD + X_{13}ACD + X_{14} BCD+ X_{15}ABCD \quad (1)$$

where Y is silicon spacing response, X_0 is the overall mean effect, A, B, C, D represent the effects of experimented factors, X_i represents the regression coefficients and other terms represent the interaction factors. Table 2 summarizes the factors and levels varied in the experiment. A centre point was added in each factor for testing the presence of curvature. An optimum region is detected when the curvature is significant.

Table2. Levels of the independent variables

Independent Variables	Symbol	Levels		
		-1	0	1
Pouring temperature	A (°C)	680	720	760
Slurry viscosity	B (s)	20	28	36
Vibration time	C (s)	30	60	90
Sand grain size	D (AFS)	30	60	90

3. Results and discussion

Table 3 gives the results of silicon spacing measurements performed on the thinnest section of the produced castings. The number of cast samples was calculated to be sixteen together with other four more trials proposed by DOE which are considered as the centre points for each variable. The examination of ANOVA output (Table 4) reveals that the model used for the analysis is significant due to the F-value of model is 5.42. According to the DOE if the P-value of the source is less than 0.05, the contribution of the source is substantial; therefore, pouring temperature (Factor A) and the interaction of it with vibration time (Factor C) have more significant effect on eutectic silicon spacing compared to other parameters. It is worth noting that based on the F value in ANOVA result the pouring temperature shows much greater effect (more than three times) on silicon spacing compared to that of the interacted parameters (Factor AC). Equation 2 designates the response for eutectic silicon spacing:

$$\text{Silicon spacing} = 11.10 + 1.34A +0.43B -0.17C +0.36D -0.75AC + 0.47BC -0.38BD +0.58CD \quad (2)$$

Table 3. Experimental results

Sample no.	Run	Factor A: Pouring temperature (°C)	Factor B: Slurry viscosity (s)	Factor C: Vibration time (s)	Factor D: Sand size (AFS)	Response: Silicon spacing (μm)
1	1	680.00	20.00	30.00	30.00	9.21
2	2	760.00	20.00	30.00	30.00	12.96
3	16	680.00	36.00	30.00	30.00	10
4	8	760.00	36.00	30.00	30.00	13.8
5	12	680.00	20.00	90.00	30.00	7.61
6	15	760.00	20.00	90.00	30.00	9.95
7	11	680.00	36.00	90.00	30.00	9.99
8	10	760.00	36.00	90.00	30.00	12.43
9	9	680.00	20.00	30.00	90.00	9.28
10	19	760.00	20.00	30.00	90.00	13.8
11	17	680.00	36.00	30.00	90.00	8.24
12	5	760.00	36.00	30.00	90.00	12.88
13	4	680.00	20.00	90.00	90.00	11.22
14	6	760.00	20.00	90.00	90.00	11.34
15	18	680.00	36.00	90.00	90.00	12.5
16	7	760.00	36.00	90.00	90.00	12.38
17	13	720.00	28.00	60.00	60.00	7.87
18	14	720.00	28.00	60.00	60.00	10.06
19	20	720.00	28.00	60.00	60.00	12.1
20	3	720.00	28.00	60.00	60.00	9.35

Table 4. Analysis of variance table

Source	Sum of squares	d.f.	Mean square	F-value	Prob. > F	
Model	54.39	8	6.80	5.42	0.0078	significant
A	28.86	1	28.86	23	0.0007	
B	2.93	1	2.93	2.34	0.1573	
C	0.47	1	0.47	0.38	0.5531	
D	2.02	1	2.02	1.61	0.2329	
AC	8.90	1	8.90	7.09	0.0238	
BC	3.53	1	3.53	2.81	0.1246	
BD	2.35	1	2.35	1.87	0.2012	
CD	5.32	1	5.32	4.24	0.0664	
Curvature	5.04	1	5.04	4.01	0.0730	not significant
Residual	12.55	10	1.25			
Lack of fit	3.27	7	0.47	0.15	0.9812	not significant
Pure error	9.28	3	3.09			
Cor. total	71.97	19				

Figure 3 illustrates graphically the effect of pouring temperature on eutectic silicon spacing of thin section of the castings. It can be seen that the silicon spacing of the casting increases when the pouring temperature increases. Therefore, it is necessary to decrease the pouring temperature in order to develop finer eutectic silicon microstructure to improve mechanical properties of the castings. The result of interacted pouring temperature and vibration time on silicon spacing is shown in Figure 4. It can be observed that an increase in pouring temperature resulted in coarsening eutectic silicon phase of the castings. The result is in good agreement with that of reported by Ding et al. [11]. However, the degree of increment was considerably lower when higher vibration time was used (90 s). It is interesting to note that at lowest experimented pouring temperature (680 °C) an inverse behaviour in terms of eutectic silicon fineness was observed. In other words, finer silicon microstructure in aluminium alloy casting was developed at that temperature. It can be inferred that lower pouring temperature together with lower vibration time would be preferable in LFC of Al-Si-Cu alloy. In fact, the mentioned conditions can develop higher solidification rate which accompanies finer eutectic silicon structure and better mechanical properties.

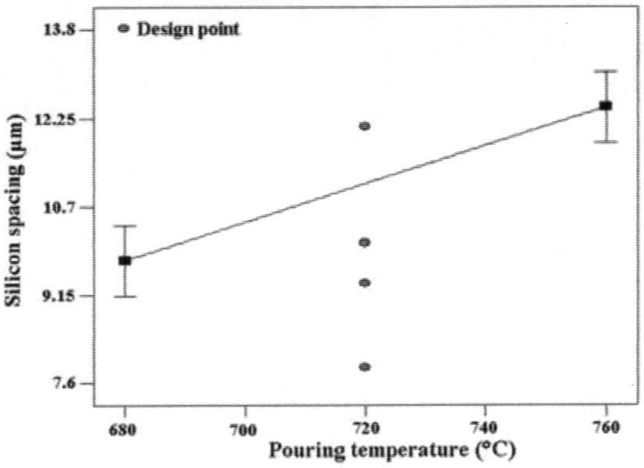

Figure 3. Effect of pouring temperature on silicon spacing

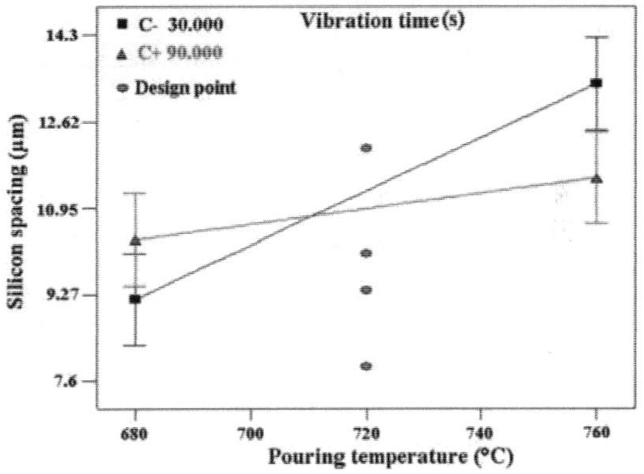

Figure 4. Interacted effect of pouring temperature and vibration time on silicon spacing

4. Conclusions

A two-level full-factorial design of experiment was effectively applied in order to plan and investigate the experimental results. It was found that pouring temperature acts as the most significant factor influencing eutectic silicon phase fineness compared to other process parameters. Decreasing pouring temperature produced finer eutectic silicon phase which is a demand for improving mechanical properties of aluminium alloy castings. An interaction between lower pouring temperature and vibration time was observed quite significant on producing Al-Si-Cu castings having smaller silicon spacing, although their effects was not as strong as pouring temperature solely.

Acknowledgement

The authors would like to acknowledge the Ministry of Higher Education of Malaysia for the financial support (Vote Number 78610) and faculty of Mechanical Engineering of Universiti Teknologi Malaysia for providing research facilities.

References

[1] Kang B, Kim Y, Kim K, Cho G, Choe K and Lee K 2007 *J. Mater. Sci. Technol.* **23** 838-832.
[2] Kumar S, Kumar P and Shan H S 2007 *J. Mater. Process. Technol.* **182** 615-623
[3] Griffiths W D, Davies P J 2008 *J. Mater Sci.* **43** 5441-5447
[4] Wall K F, Bhavnani S H, Overfelt R A, Sheldon D S and Williams K 2003 *Metall. Mater. Trans. B.* **34B** 843
[5] Molerus O 1996 *Powder Technol.* **88** 309-321
[6] Vijian P, Arunachalam V P 2006 *J. Mater. Process. Technol.* **180** 161-166
[7] Kalyan Kumar K V B S and Choudhury S K 2008 *J. Mater. Process. Technol.* **203** 95-101
[8] Gunasegaram D R, Farnsworth D J and Nguyen T T 2009 J. *Mater. Process. Technol.* **209** 1209-1219

[9] Verran, G O, Mendes R P K and Dalla Valentina L V O 2008 *J. Mater. Process. Technol.* **200** 120-125
[10] Izman S, Shayganpour A and Idris M H 2011 *Adv. Mat. Res.* **337** 422-425
[11] Ding H, Guo J, Jia J, Bi W and Chen L 2001 *Trans. Nonferr. Met. Soc. China* **11** 540-546

Modelling of Non-Premixed Turbulent Combustion of Hydrogen using Conditional Moment Closure Method

M.M.Noor[1,3], A.Aziz Hairuddin[1,4], Andrew P.Wandel[1] and T.F.Yusaf[2,3]
[1]Computational Engineering and Science Research Centre, Department of Mechanical and Mechatronic, Engineering University of Southern Queensland (USQ), Australia
[2]National Centre for Eng. in Agriculture (NCEA), USQ, Australia
[3]Faculty of Mechanical Engineering, Universiti Malaysia Pahang (UMP), Malaysia
[4]Faculty of Engineering, Universiti Putra Malaysia (UPM), Malaysia

E-mail: Muhamad.MatNoor@usq.edu.au / muhamad@ump.edu.my

Abstract. Most of the electricity generation and energy for transport is still generated by the conversion of chemical to mechanical energy by burning the fuels in the combustion chamber. Regulation for pollution and the demand for more fuel economy had driven worldwide researcher to focus on combustion efficiency. In order to reduce experimental cost, accurate modelling and simulation is very critical step. Taylor series expansion was utilised to reduce the error term for the discretization. FORTRAN code was used to execute the discretized partial differential equation. Hydrogen combustion was simulated using Conditional Moment Closure (CMC) model. Combustion of hydrogen with oxygen was successfully simulated and reported in this paper.

1. Introduction

Energy is a human basic need. Currently, combustion remains the main source of energy for transportation, industrial, electricity generation and other human activities. World energy demand for 2030 is estimated at about 18 billion tons of oil equivalent and about 80 percent will be fulfilled by oil, gas and coal [1,2]. Worldwide concern on fuel prices, environmental pollution and energy sustainability has led to an increased interest in energy efficiency improvement with lower emissions. This issue has driven the combustion community to focus on the research for experimental and modelling. Turbulent combustion is divided into two classes depending on when the fuel is mixed with the oxidizer: premixed and non-premixed. Premixed combustion is usually modelled using eddy break-up model, coherent flame, flamelet based on G-equation or linear eddy model. However, premixed combustion is not of interest of this paper. For non-premixed combustion, there are a few types of modelling methods for the mixing process: probability density function (PDF) based model [3-5], flamelet model [6-9], fast chemistry limit model [10], and mapping closure model [11,12]. Besides those models, there are a few other developments such as conditional moment closure (CMC) by Klimenko and Bilger [13-15] and multiple mapping conditioning (MMC) by Klimenko and Pope [16], with further developments [17-20] for the mixing process. The CMC model for non-premixed combustion will be investigated in this paper.

Published under licence by IOP Publishing Ltd

The concept of turbulent combustion modelling describes the transport of reactive scalars in conserved scalar spaces. The novelty of CMC model was the capability of the model to take into account the conditional averages in the combustion modelling. This is due to the chemical source term is not a function of unconditional averages, which are used in conventional modelling. Ternat et al [21] computed stable solutions using two finite difference methods, namely the Euler method and the Crank–Nicolson method, to advance the solution of the heat equation in time. Noor et al [22] use Taylor expansion to discretize the CMC model and finite difference method to solve the equation through explicit and implicit method. Clarke et al [23] used direct numerical simulation (DNS) results to model the parameters of CMC for combustion systems with droplets. This is a complicated process because of the interaction of the evaporating liquid with the gaseous phase; the usage of a spark to evaporate the fuel and ignite the mixture exacerbates this complexity in comparison with auto ignition.

A mixing model is required to close the molecular diffusion term in the PDF transport equation [3], which is the last term in equation (1) and contains the molecular diffusion flux vector (J_{ik}):

$$\bar{\rho}\frac{\partial \tilde{P}(\underline{\psi})}{\partial t} + \bar{\rho}\tilde{u}_m\frac{\partial \tilde{P}(\underline{\psi})}{\partial x_m} + \bar{\rho}\sum_{k=1}^{n}\frac{\partial}{\partial \psi_k}\left\{\dot{\omega}_k(\underline{\psi})\tilde{P}(\underline{\psi})\right\} = -\frac{\partial}{\partial x_m}\left\{\bar{\rho}\left\langle u_k'' \middle| \underline{\phi} = \underline{\psi}\right\rangle\right\}\tilde{P}(\underline{\psi})$$

$$+\sum_{k=1}^{n}\frac{\partial}{\partial \psi_k}\left\{\left\langle\frac{\partial J_{i,k}}{\partial x_i}\middle|\underline{\phi} = \underline{\psi}\right\rangle\tilde{P}(\underline{\psi})\right\} \quad (1)$$

Mixing plays a crucial role in the non-premixed combustion process; a number of mixing models have been developed. The Coalescence and Dispersion process was modelled by Curl in 1963 [24] and Levenspiel and Spielman in 1965 [25] and is often called "Curl's model". The governing equation for Curl's model is shown in equation (2). The left part of the equation is the particle composition before mixing. This particle then Coalescences with another particle and mixing occurs. The mixing process can be referred to at the middle part of equation (2) and then this mixed particle will be dispersed as shown in the last part of equation (2).

$$\begin{pmatrix}(\phi_{A1},\phi_{B1})_1\\(\phi_{A2},\phi_{B2})_2\end{pmatrix} \rightarrow \begin{pmatrix}\phi_A^*=(\phi_{A1}+\phi_{A2})/2\\ \phi_A^*=(\phi_{B1}+\phi_{B2})/2\end{pmatrix} \rightarrow \begin{pmatrix}(\phi_A^*,\phi_B^*)_1\\(\phi_A^*,\phi_B^*)_2\end{pmatrix} \quad (2)$$

Here ϕ_{A1} is the composition of species *A* for particle 1, ϕ_A^* is the new composition of species *A* and (ϕ_A, ϕ_B) is a particle which consists of species *A* and *B*. The Curl's model was modified by Janicka et. al [26] and Dopazo [27] in 1979. This model was called modified Curl's (MC) model (equations (3) and (4)) where β can take any value from 0 to 1 and can be a random variable. If $\beta = 0$, then no mixing occurs, whilst $\beta = 1$ reproduces Curl's model.

$$\phi_i^{MC}(t+\delta t) = (1-\beta t)\phi_i^{MC}(t) + \tfrac{1}{2}\beta\left(\phi_j^{MC}(t) + \phi_i^{MC}(t)\right) \quad (3)$$

$$\phi_j^{MC}(t+\delta t) = (1-\beta t)\phi_j^{MC}(t) + \tfrac{1}{2}\beta\left(\phi_j^{MC}(t) + \phi_i^{MC}(t)\right) \quad (4)$$

Here ϕ_i and ϕ_j is the composition of particles *i* and *j*, *t* is time and δt is the time interval.

The weakness of the MC model is that this mixing process does not enforce the requirement that only particles that are close to each other are allowed to mix and interact with each other. This issue was solved by the Euclidean Minimum Spanning Tree (EMST) model [28] whereby particles that mix are close together in composition space as shown in equations (5) and (6). In these equations, there are two new constants introduced where d is determined so that the desired amount of mixing is obtained and P_b is the position of particle in the EMST branch. Particles near to the centre will have higher P_b values.

$$\phi_i^{EMST}(t+\delta t) = \phi_i^{EMST}(t) + dP_b\delta t\left(\phi_j^{EMST}(t) + \phi_i^{EMST}(t)\right) \tag{5}$$

$$\phi_j^{EMST}(t+\delta t) = \phi_j^{EMST}(t) + dP_b\delta t\left(\phi_i^{EMST}(t) + \phi_j^{EMST}(t)\right) \tag{6}$$

Another mixing model is the interaction by exchange with the mean (IEM) model [29] where the composition of all particles in a cell are moved a small distance toward the mean composition using a characteristic mixing timescale. The IEM equation is shown in equation (7) where $\tilde{\phi}$ is the Favre mean-composition vector at the particle's location and τ_t is the turbulence time scale. The scalar mixing time scale τ_ϕ in equation (7) is often modelled as proportional to τ_t as in equation (8).

$$\frac{\partial\phi}{\partial t} = -\frac{\phi-\tilde{\phi}}{2\tau_\phi} \tag{7}$$

$$\tau_\phi \equiv \frac{\tau_t}{c_\phi} \tag{8}$$

The mixing models just described are either non-local [24-27] or over-local [28], thereby producing imperfect combustion modelling processes. Recently Wandel [30] has proposed a new mixing model which randomizes the particle interaction in a local manner. The proposed model is the Stochastic Particle Diffusion Length (SPDL) [30] model, which is based upon the practical localness of the random inter-particle distance [31].

Combustion processes are very complex especially because the chemical reactions between chemical species involved in burning fuel (gas, liquid or solid) are highly non-linear functions of temperature and species concentration. Significant errors are produced when a computational fluid dynamics (CFD) code only solves the Reynolds Averaged Navier-Stokes (RANS) model as written in equation (9), which involves averaging the source terms. All terms are averaged and the models work reasonably well when solving $\bar{\phi}$ for velocity but not for chemical species.

$$\frac{\partial\bar{\rho}\bar{\phi}}{\partial t} + \frac{\partial\bar{\rho}\bar{\phi}\,\bar{u}_i}{\partial x_i} = \frac{\partial}{\partial x_i}\left[D\frac{\partial\bar{\phi}}{\partial x_i}\right] + \bar{W} + \bar{T} \tag{9}$$

These models are inaccurate and produce errors by ignoring the correlation between the source term variables, which are generally significant, compared to the corresponding averages. These errors were overcome by the CMC model using the conditional averages method. It takes the average of the variables for a specified value of the mixture fraction and effectively averages over a smaller region of space. Compare to the IEM and EMST model, CMC is more accurate at the cost of more complex simulations that require more computational time. The chemical source term $\langle W|Z\rangle$ can be calculated using equation (10) and the reaction rate is obtained by using equation (11). The Arrhenius equation [18] in equation (12) is used to determine the reaction rate constant. The chemical source term, reaction rate and reaction rate constant can be written as:

$$W_I = \frac{w_I}{\rho}\sum_j v_{Ij}\,\omega_j \tag{10}$$

$$\omega_j = \left[k_{fj}\prod_{I=1}^{n_s}[X_I]^{v'_{Ij}} - k_{bj}\prod_{I=1}^{n_s}[X_I]^{v''_{Ij}}\right] \tag{11}$$

$$k = AT^\beta exp\left(-\frac{E_a}{RT}\right) \tag{12}$$

with the rate of progress variable ω_j the net strength of reaction j in the forward direction, w_I is the mean molecular weight, ρ is density, E_a is the activation energy, A and β are Arrhenius constants, R is the ideal gas constant, and T is temperature. The CMC equation for two-phase homogenous systems is shown in equation (13):

$$\frac{\partial\langle Y|Z\rangle}{\partial t} = \langle N|Z\rangle\frac{\partial^2\langle Y|Z\rangle}{\partial Z^2} + \langle W|Z\rangle + \langle S|Z\rangle \tag{13}$$

The combustion of hydrogen produces the most clean and very low emission in combustion. Hydrogen's low density giving a challenging medium for the storage (requires very high pressure tank). Table 1 shows the properties of natural gas, hydrogen and diesel.

Table 1 Diesel properties compared to hydrogen and natural gas [32-35]

Properties	Diesel	Hydrogen	Natural Gas
Main component	C and H	H Only	Methane (CH_4)
Auto-ignition Temperature (K)	553	858	923
Lower heating value (MJ/kg)	42.5	119.93	50
Density (kg/m^3)	833-881	0.08	692
Molecular weight (g/mol)	170	2.016	16.043
Flammability limits in air (vol %)	0.7-5	4-75	5-15
Flame velocity (m/s)	0.3	2.65-3.25	0.45
Specific gravity	0.83	0.091	0.55
Boiling point (K)	453-653	20.2	111.5
Cetane number	40-60	-	-
Octane number	30	130	120
CO_2 emissions (%)	13.4	0	9.5
Mass diffusivity in air (cm^2/s)	-	0.61	0.16
Min ignition energy (mJ)	-	0.02	0.28

Fuel blend with hydrogen additive will increase the molecular diffusion with the increase of hydrogen [36]. Recently Mardani et al. [36,37] and Wang et al. [38] investigated the effects of hydrogen addition and found that flameless combustion occurred more easily. Yu et al. [39] found that pure hydrogen could not reduce thermal NO_x emission in the flameless combustion regime. Hydrogen properties show a lot of advantage over fossil fuels. Hydrogen is produced mainly from fossil fuel resources and only 4% generated by electrolysis. In the future, when fossil fuel depleted, the raw material will be changed to water and biomass [40]. The important properties that serve best for combustion are: high auto-ignition temperature. The higher auto-ignition temperature allows higher compression ratio and produce higher engine thermal efficiencies; high diffusivity. More homogenous combustion due to more uniform air-fuel mixture and will disperse quickly if leaking; wide range of flammability. Fuel air mixing ratios from 4% to 74% and can be burned on a lean mixture with fuel economy and produce less NO_x. The disadvantage is the power produced will be significantly lower than fossil fuel; low ignition energy. To ignite, hydrogen only needs 10% of what gasoline needs. The disadvantage is possible early ignition causing knocking problems; low density leads to low energy density and larger storage; high flame velocity requires tighter ignition timing and burns at about 2.83 m/s compare to 0.34 m/s for gasoline (at stoichiometric and atmospheric pressure).

In this paper, Taylor series expansion method was utilised to discretise the CMC equation and FORTRAN code was used to run a simulation and produce results. The purpose of the study is to simulate the reaction of hydrogen and oxygen using CMC mixing model. Explicit and implicit method was used and the outcome from both methods will be discussed.

2. Taylor Expansion

Numerical method is very important to reduce the experimental cost. After simulation results are obtained and ready to be validated, then experimentation is used to confirm the findings and results from the simulation. There are many numerical methods can be utilized to solve partial differential equation (PDE). Implicit and explicit methods are the common methods. Implicit finite difference relations have been derived by many mathematicians and physicists with various methods [41-53]. Most of them claim that all the implicit formulas can be derived from a Taylor series expansion. The Taylor series expansion is a good basis for studying numerical methods since it provides a means to predict a function's value at one point in terms of the function's value and derivatives at another point. In particular, the theorem states that any smooth function can be approximated as a polynomial [53]. There are many different types of numerical differentiation formulations, depending on the number of points, direction of the formula and the required derivative order [54]. The Taylor expansion is a useful method to discretise partial differential equations to minimise and accurately predict the value of the error term. The expansions for y_1 and y_{-1} which are to the right and to the left of y_0 respectively are shown below up to the 7th-order derivatives as equations (14) and (15).

$$y_1 = y_0 + (\Delta x_1)\left.\frac{dy}{dx}\right|_{x_0} + \frac{(\Delta x_1)^2}{2!}\left.\frac{d^2y}{dx^2}\right|_{x_0} + \frac{(\Delta x_1)^3}{3!}\left.\frac{d^3y}{dx^3}\right|_{x_0} + \frac{(\Delta x_1)^4}{4!}\left.\frac{d^4y}{dx^4}\right|_{x_0} + \frac{(\Delta x_1)^5}{5!}\left.\frac{d^5y}{dx^5}\right|_{x_0} + \frac{(\Delta x_1)^6}{6!}\left.\frac{d^6y}{dx^6}\right|_{x_0} + \frac{(\Delta x_1)^7}{7!}\left.\frac{d^7y}{dx^7}\right|_{x_0} \quad (14)$$

$$y_{-1} = y_0 + (\Delta x_{-1})\left.\frac{dy}{dx}\right|_{x_0} + \frac{(\Delta x_{-1})^2}{2!}\left.\frac{d^2y}{dx^2}\right|_{x_0} + \frac{(\Delta x_{-1})^3}{3!}\left.\frac{d^3y}{dx^3}\right|_{x_0} + \frac{(\Delta x_{-1})^4}{4!}\left.\frac{d^4y}{dx^4}\right|_{x_0} + \frac{(\Delta x_{-1})^5}{5!}\left.\frac{d^5y}{dx^5}\right|_{x_0} + \frac{(\Delta x_{-1})^6}{6!}\left.\frac{d^6y}{dx^6}\right|_{x_0} + \frac{(\Delta x_{-1})^7}{7!}\left.\frac{d^7y}{dx^7}\right|_{x_0} \quad (15)$$

We assume Δ_i is constant, then y_1 and y_0 become

$$y_1 = y_0 + (\Delta x)\left.\frac{dy}{dx}\right|_{x_0} + \frac{(\Delta x)^2}{2!}\left.\frac{d^2y}{dx^2}\right|_{x_0} + \frac{(\Delta x)^3}{3!}\left.\frac{d^3y}{dx^3}\right|_{x_0} + \frac{(\Delta x)^4}{4!}\left.\frac{d^4y}{dx^4}\right|_{x_0} + \frac{(\Delta x)^5}{5!}\left.\frac{d^5y}{dx^5}\right|_{x_0} + \frac{(\Delta x)^6}{6!}\left.\frac{d^6y}{dx^6}\right|_{x_0} + \frac{(\Delta x)^7}{7!}\left.\frac{d^7y}{dx^7}\right|_{x_0} \quad (16)$$

$$y_{-1} = y_0 + (\Delta x)\left.\frac{dy}{dx}\right|_{x_0} + \frac{(\Delta x)^2}{2!}\left.\frac{d^2y}{dx^2}\right|_{x_0} + \frac{(\Delta x)^3}{3!}\left.\frac{d^3y}{dx^3}\right|_{x_0} + \frac{(\Delta x)^4}{4!}\left.\frac{d^4y}{dx^4}\right|_{x_0} + \frac{(\Delta x)^5}{5!}\left.\frac{d^5y}{dx^5}\right|_{x_0} + \frac{(\Delta x)^6}{6!}\left.\frac{d^6y}{dx^6}\right|_{x_0} + \frac{(\Delta x)^7}{7!}\left.\frac{d^7y}{dx^7}\right|_{x_0} \quad (17)$$

Rearrangement of equations (16) and (17) becomes the first order derivatives at x_0 as below (equations (18), (19) and (20)). Equation (16) is used to obtain the forward difference method (which calculates $\left.\frac{dy}{dx}\right|_{x_0}$ based on forward movement from y_0 to y_1):

$$\frac{dy}{dx} = \frac{(y_1 - y_0)}{\Delta x} - \frac{(\Delta x)}{2!}\frac{d^2y}{dx^2} - \frac{(\Delta x)^2}{3!}\frac{d^3y}{dx^3} - \frac{(\Delta x)^3}{4!}\frac{d^4y}{dx^4} - \frac{(\Delta x)^4}{5!}\frac{d^5y}{dx^5} - \frac{(\Delta x)^5}{6!}\frac{d^6y}{dx^6} - \frac{(\Delta x)^6}{7!}\frac{d^7y}{dx^7} \quad (18)$$

Equation (17) is used to obtain the backward difference method, which calculates $\left.\frac{dy}{dx}\right|_{x_0}$ based on backward movement from y_0 to y_{-1}.

$$\frac{dy}{dx} = \frac{(y_0 - y_{-1})}{\Delta x} - \frac{(\Delta x)}{2!}\frac{d^2y}{dx^2} - \frac{(\Delta x)^2}{3!}\frac{d^3y}{dx^3} - \frac{(\Delta x)^3}{4!}\frac{d^4y}{dx^4} - \frac{(\Delta x)^4}{5!}\frac{d^5y}{dx^5} - \frac{(\Delta x)^5}{6!}\frac{d^6y}{dx^6} - \frac{(\Delta x)^6}{7!}\frac{d^7y}{dx^7} \quad (19)$$

By taking the difference between equations (16) and (17), the central difference method is derived which calculates $\left.\frac{dy}{dx}\right|_{x_0}$ based on the domain between y_1 and y_{-1}

$$\frac{dy}{dx} = \frac{(y_1 - y_{-1})}{2\Delta x} - \frac{(\Delta x)^2}{3!}\frac{d^3y}{dx^3} - \frac{(\Delta x)^4}{5!}\frac{d^5y}{dx^5} - \frac{(\Delta x)^6}{7!}\frac{d^7y}{dx^7} \tag{20}$$

with leading error term of $\frac{(\Delta x)^2}{6!}\frac{d^3y}{dx^3}$. This equation (20) is a second-order accurate method: $O(\Delta x^2)$. Since the differences actually evaluate the derivative at the midpoint of the finite difference, equation (20) estimates the derivative at x_0, while equation (18) and (19) estimate the derivative either side of x_0. Using central difference derivative, we can obtain $\frac{d^2y}{dx^2}$ from equations (16) and (17), as shown in equation (21). The first term in equation (21) can be re-arranged as equation (22) to show that this is simply a central difference of the first derivative.

$$\frac{d^2y}{dx^2} = \frac{(y_1 - 2y_0 + y_{-1})}{(\Delta x)^2} - \frac{2!(\Delta x)^2}{4!}\frac{d^4y}{dx^4} - \frac{2!(\Delta x)^4}{6!}\frac{d^6y}{dx^6} \tag{21}$$

$$= \frac{1}{\Delta x}\left(\frac{y_1 - y_0}{\Delta x} - \frac{y_0 - y_{-1}}{\Delta x}\right) \tag{22}$$

The leading error term is $\frac{(\Delta x)^2}{12!}\frac{d^4y}{dx^4}$, so this is a second-order accurate method: $O(\Delta x^2)$. The first-order derivative using the fourth-order Taylor expansion scheme is below, for the difference between y_2 and y_{-2}:

$$\frac{dy}{dx} = \frac{(y_{-2} - 8y_{-1} + 8y_1 - y_2)}{12\Delta x} + \frac{4(\Delta x)^4}{5!}\frac{d^5y}{dx^5} \tag{23}$$

with leading error term of $\frac{(\Delta x)^4}{30}\frac{d^5y}{dx^5}$. This is a fourth-order accurate method: $O(\Delta x^4)$ and can be summarized as equation (24) to show that it is a weighted average of the "near" and "far" central differences:

$$\frac{dy}{dx} = \left(\frac{4}{3}\right)\frac{(y_1 - y_{-1})}{(2\Delta x)} - \left(\frac{1}{3}\right)\frac{(y_2 - y_{-2})}{(4\Delta x)} \tag{24}$$

The second-order derivative for the fourth-order Taylor expansion scheme is equation (25):

$$\frac{d^2y}{dx^2} = \frac{(-y_{-2} + 16_{y_{-1}} - 30_{y_0} + 16_{y_1} - y_2)}{12(\Delta x)^2} + \frac{8(\Delta x^4)}{6!}\frac{d^6y}{dx^6} \tag{25}$$

with leading error term of $\frac{(\Delta x^4)}{90}\frac{d^6y}{dx^6}$, so this is a fourth-order accurate method: $O(\Delta x^4)$. This can be summarized as equation (26):

$$\frac{d^2y}{dx^2} = \left(\frac{4}{3}\right)\frac{(y_{-1} - 2y_0 + y_1)}{(\Delta x)^2} - \left(\frac{1}{3}\right)\frac{(y_{-2} - 2y_0 + y_2)}{(2\Delta x)^2} \tag{26}$$

Interestingly, the weightings of the terms in equation (26) for the second-order derivative are identical to those in equation (24) for the first-order derivative. The third-order derivative for the second-order Taylor expansion scheme is shown as equation (27):

$$\frac{d^3y}{dx^3} = \frac{(-y_{-2} - 2y_{-1} - 2y_1 + y_2)}{2(\Delta x)^3} - \frac{(\Delta x)^2}{4}\frac{d^5y}{dx^5} \tag{27}$$

with leading error term of $\frac{(\Delta x)^2}{4}\frac{d^5y}{dx^5}$, so this is a second-order accurate method: $O(\Delta x^2)$.

Assuming that there is a uniform spacing of Δx, using notation that $y^{(k)} = \frac{d^k y}{dx^k}$, for the Taylor series expansion, central difference derivatives can be summarised as equation (28):

$$y^{(k)} = C\sum_{i=\frac{(n-1)}{2}}^{\frac{(n-1)}{2}} z_i y_i + ET \tag{28}$$

Forward difference derivatives can be written as equation (29):

$$y^{(k)} = C\sum_{i=0}^{(n-1)} z_i y_i + ET \tag{29}$$

Backward difference derivatives can be written as equation (30):

$$y^{(k)} = C\sum_{i=-(n-1)}^{0} z_i y_i + ET \tag{30}$$

where n is the number of points ($y_{-2}, y_{-1}, y_0, y_1, y_2$ is equal to five points), ET is the leading error term and z_i is the coefficient of y for each point i.

3. Numerical Method

The finite difference schemes, as agreed by most of the scientific community, were first used by Euler (1707-1783) [55] to find an approximate solution of a differential equation. It was invented prior to boundary element methods (BEM), finite element methods (FEM), spectral methods, and discontinuous spectral element methods [56]. FDM is still relevant and remain competitive as a discretisation method for use in many applications and can be used to solve problems with simple and complex geometry, such as fluid flows and gas reaction [57,58]. The Finite difference method (FDM) is a numerical method for approximating the solutions to partial differential equations by using finite difference equations to approximate derivatives based on the properties of Taylor expansions and on the straightforward application of the definition of derivatives [59]. The objectives are to transform the calculus problem to algebra as from a continuous equation to a discrete equation. The discretisation process is a mathematical process that divides the continuous physical domain into a discrete finite difference grid and then approximates each individual partial derivative in the partial differential equation. Using the Taylor expansion method, a partial differential equation was discretised in order to transform it to FORTRAN code. FORTRAN is a high level of programming language developed by team of IBM programmers led by John Backus in 1954. The name of FORTRAN was derived from the words "Formula Translation", started from 1957 when the first FORTRAN compiler was used. It has evolved through FORTRAN II, FORTRAN 66, until now FORTRAN 2008. FORTRAN 90 was used in this study. From the CMC equation, to study the discretisation and code it in FORTRAN, simplifications of the CMC equation were used: the conditional chemical source term ($\langle W|Z\rangle$) and conditional generation due to droplet evaporation term ($\langle S|Z\rangle$) were not considered. So the homogeneous and passive CMC is equation (31):

$$\frac{\partial\langle Y|Z\rangle}{\partial t} = \langle N|Z\rangle\frac{\partial^2\langle Y|Z\rangle}{\partial z^2} \tag{31}$$

In this equation, the conditional mass fraction quantity $\langle Y|Z\rangle$ can be considered as "Y is a function of Z" (written as y(Z)) and conditional scalar dissipation $\langle N|Z\rangle$ is expressed as "N is a function of Z" (written as N(Z)). After summarizing all the assumption, the CMC equation becomes equation (32):

$$\frac{dy}{dt} = N\frac{d^2y}{dz^2} \tag{32}$$

The Taylor expansion equations (18) and (21) can be expressed in this nomenclature as:

$$\frac{dy}{\partial t} = \frac{y(z,t+\Delta t)-y(z,t)}{(\Delta t)} \tag{33}$$

$$\frac{d^2y}{\partial z^2} = \frac{y(z+\Delta z,t)-2y(z,t)+y(z-\Delta z,t)}{(\Delta z)^2} \tag{34}$$

The final form of the CMC equation after discretization is represented by equation (35) for the explicit method and equation (36) for implicit.

$$YH(i,j+1) = \big((1-(2B)) * YH(i,j)\big) + \big(B * YH(i+1,j)\big) + \big(B * YH(i-1,j)\big) \tag{35}$$

$$(1+2B) * YH(i,j+1) - \big(B * YH(i+1,j+1)\big) - \big(B * YH(i-1,j+1)\big) = YH(i,j) \tag{36}$$

where $B = N\frac{\Delta t}{(\Delta z)^2}$, *YH* is the array in the computer code for the variable mass fraction of fuel, *i* is the index for mixture fraction and *j* is the index of the time step. Equations (35) and (36) were coded in FORTRAN to simulate the CMC modelling. Parts of the FORTRAN code for the explicit and implicit methods are listed in appendixes A and B.

Three parameters were used as input to the code: *dt* is time step size, *dz* is step size in mixture fraction space and *TT* is total time for the simulation. These three parameters will determine the accuracy and total time taken to run the code. The more steps taken, the longer it will take to complete the simulation. The result of the simulation must be checked to ensure its convergence meets expectations. These parameter must also comply with the Courant-Friedrichs-Lewy (CFL) condition [60,61] to ensure the stability of the solution and that the result acceptably reaches a converged solution. The stability of the solution is very important because an unstable condition will create large errors in the solution and wrong predictions of the result. The time-step must satisfy the condition shown in equation (37) otherwise the simulation will produce incorrect results.

$$CFL = N\frac{dt}{dz^2} \leq \frac{1}{2} \tag{37}$$

In order to achieve this condition, the time step must be small enough for the flow conditions.

4. Results and Discussions

In this study, the value for the conditional scalar dissipation *N* in was assumed to be the constant 0.5. Results from the simulation are plotted in Figures 1. The explicit method was found to converge faster than the implicit method, reaching the steady-state condition after 190 time steps whereas implicit required 247 time steps. ("Steady-state" is defined as no variation in 5 significant figures.) In addition, the time required for the computer to calculate one time step was significantly shorter for the explicit method. However, the explicit method reaches the steady-state too quickly due to greater errors in this method for the same time-step. The implicit method can have a bigger time-step (δt) for the same accuracy as the explicit method.

The first lines (from Y = 1.0 to 0.0 over the gap of Z = 0.1) are the initial conditions that oxidizer (left lines) is zero everywhere except at Z = 0 and fuel (right lines) is zero everywhere except at Z =1. The cell size used here is Z = 0.1, which is why the line drops at both edges for both methods. The explicit and implicit methods start to show changes for both air and fuel mass fraction immediately (after the first time step: the second lines in Figure 1). For the explicit method at mixture fraction equal to 0.2, mass fraction is still 0.00 whereas for the implicit method, for mixture fraction equal to 0.2, mass fraction values are between 0.0 and 0.2. This variation is because differences between adjacent cells in the explicit method can only march one cell at a time, while the difference influences all cells immediately in the implicit method. Because of this, the explicit method influences the adjacent cell too much, which results in the method reaching steady-state quicker. These mixing processes were repeated over many time steps. The mixing process will reach steady-state and equilibrium when both air and fuel completely mix with both reaching 0.5 at mixture fraction equal to 0.5 (Figure 1(b) and Figure 1(d)).

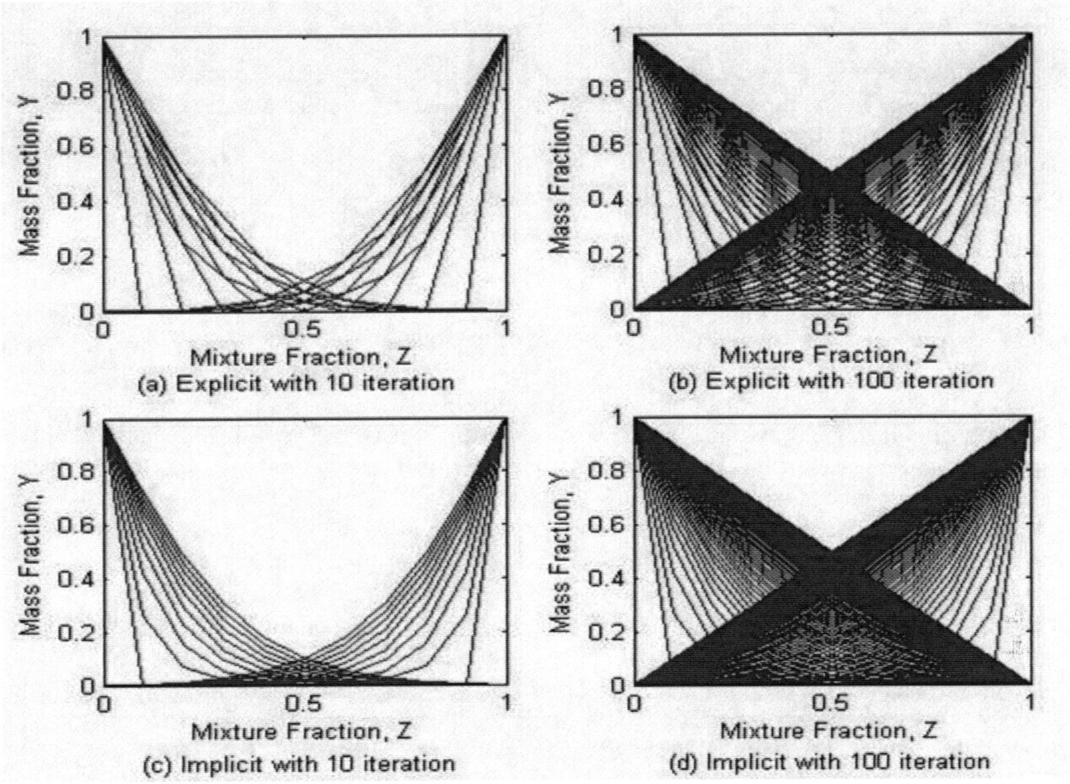

Figure 1..CMC mass fraction vs mixture fraction for iteration 10 and 100 for explicit and implicit method

The mixing behaviour between fuels and oxidisers in the previous discussion was studied without combustion process. Next step of this study is taking into account the combustion effect in addition to the mixing process, and after certain time, the mixing behaves differently compared to without combustion. A one-step chemical reaction of hydrogen react with oxygen was used to study the effect of combustion, which is represented in the following formula:

$$2H_2O + O_2 \leftrightarrow 2H_2O \qquad (38)$$

The combustion effect was studied in a closed system, with the same amount of initial volume. The change in mass fraction due to the chemical reaction is given by:

$$\frac{\partial Y_i}{\partial t} = \frac{w_I}{\rho} \sum_j v_{Ij}\, \omega_j \qquad (39)$$

where, Y is the mass fraction, I is the species, W is the molecular weight, is the density, v_{Ij} is the stoichiometric coefficient for species I in reaction j and ω_j is the reaction rate. The overall stoichiometric coefficient of a chemical reaction is:

$$v_{Ij} = v''_{Ij} - v'_{Ij} \tag{40}$$

where v''_{Ij} is the stoichiometric coefficient of product and v is the stoichiometric coefficient of reactant. The reaction rate is given by:

$$\omega_j = \left[k_{fj} \prod_{I=1}^{n_s}[X_I]^{v'_{Ij}} - k_{bj} \prod_{I=1}^{n_s}[X_I]^{v''_{Ij}}\right] \tag{41}$$

where, k is the reaction rate coefficient with subscripts f and b represent "forward" and "backward" reaction. X_I is the mole fraction of species I. Summarise and combining equation (20) and (21), the equation (19) in its full form is represented by:

$$\frac{\partial Y_I}{\partial t} = \frac{w_I}{\rho}\sum_j \left[\left(v''_{Ij} - v'_{Ij}\right)\left(k_{fj} \prod_{I=1}^{n_s}[X_I]^{v'_{Ij}} - k_{fj} \prod_{I=1}^{n_s}[X_I]^{v''_{Ij}}\right)\right] \tag{42}$$

where the equation (22) is used to determine the change in mass fraction of the species due to combustion. The result for the combustion effect by using explicit method is shown in Fig. 2. It shows that the fuel will consume the oxidiser if we allow time for the species to mix. The mixture is not completely mixed at 10 iterations, where it will reach steady state condition (completely mixed and chemical equilibrium) when the iterations is increased to 100.

For the results in Figs. 3 and 4, the step size in mixture fraction space (dz) for both explicit and implicit methods for combustion is 0.01 while the time step (dt) used is 0.0001. This is smaller (both dz and dt) compare to the simulation without combustion where *dz* was reduce from 0.1 to 0.01 and *dt* was reduce from 0.01 to 0.0001. The reason is to achieve more stable computation and obtain an appropriate resolution when the combustion is involved in the mixing process. It can be seen from Figs. 3 and 4 that the mass fraction and the mixture fraction are different compared to without combustion (Fig. 1), after the mixing process reaches steady state condition. The result is showing significant difference in mixture fraction between fuels and oxidisers, where fuels consume some amounts of oxidisers to reach the steady state condition. The mass fraction is not completely zero at the intersection line because the effect of backward reaction. Therefore, CMC modelling would be useful to study the mixing behaviour due to chemical reactions [12,14,15].

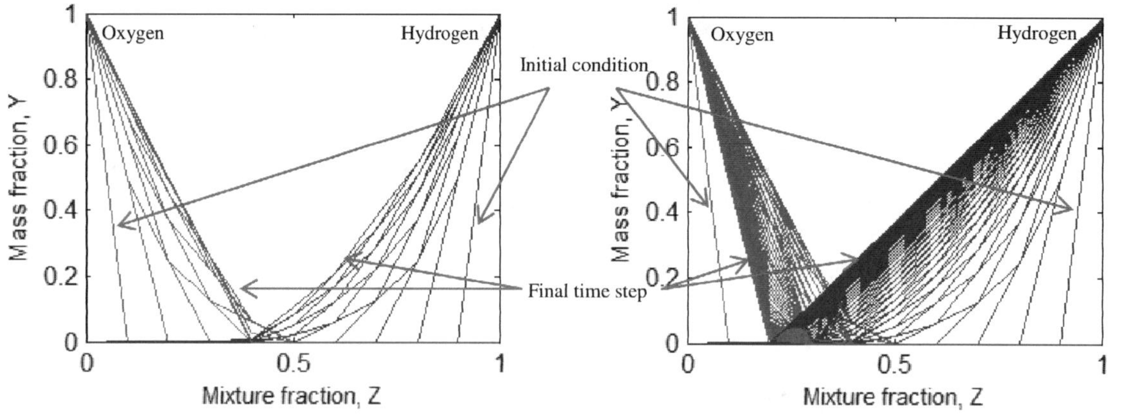

(a) (b)

Figure 2. Mixing behaviour between hydrogen and oxygen for (a) 10 iterations, and (b) 100 iterations

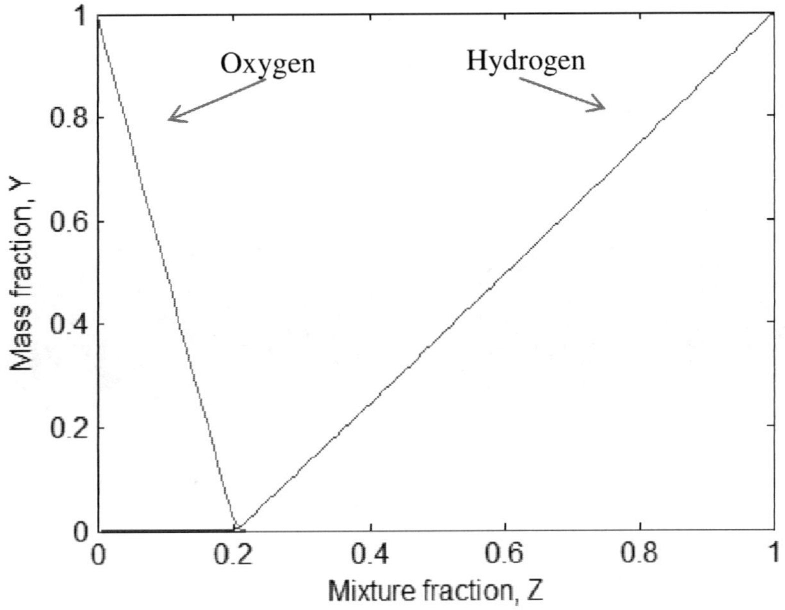

Figure 3. CMC simulation for mass fraction vs mixture fraction at final time step with hydrogen combustion using implicit method

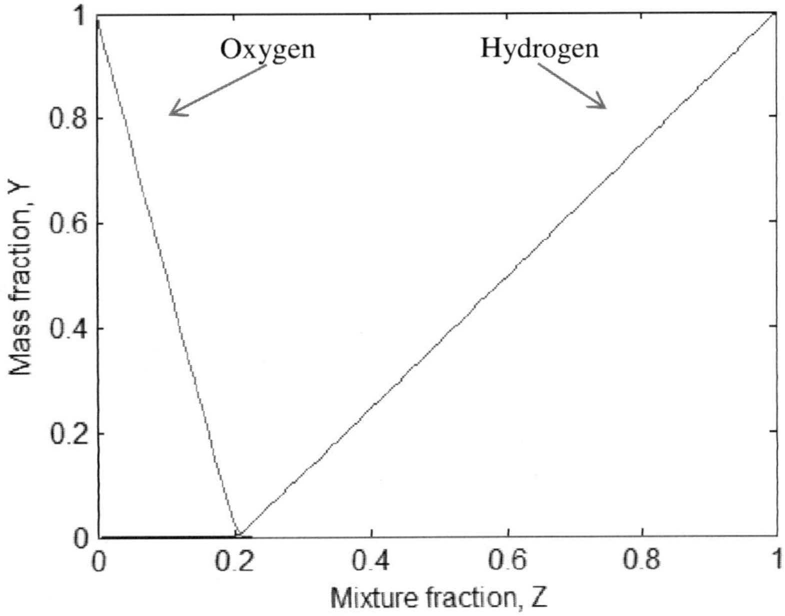

Figure 4. CMC simulation for mass fraction vs mixture fraction with hydrogen combustion at final time step using explicit method.

Explicit method is faster to converge than the implicit method, where the time taken for convergence for explicit method is 3867.75 s, while 4088.07 s for implicit method. The difference between the two methods at the final time step is shown in Fig. 5, where the error is small for $Z > 0.2$, which corresponds to richer compositions than the point where the mass fractions of oxygen and hydrogen intersect (Figs. 3 and 4). For the lower mixture fractions, the explicit results for both species are smaller than the implicit values, while the reverse is true where the mass fraction of hydrogen is significant. The error has its greatest magnitude near to where it changes sign on both sides of this change. A smaller time step might be required to reduce the error in this region. It seems that the explicit method is faster to converge at the expense of accuracy.

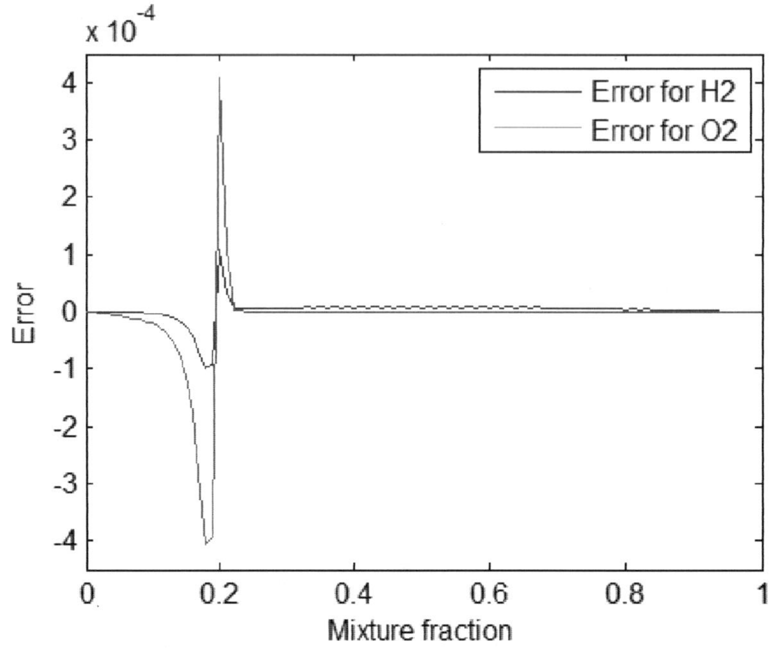

Figure 5. Result comparison for both oxygen and hydrogen between explicit and implicit methods at final time step. Error represents explicit – implicit.

5. Conclusions

The simulation process is very important to reduce high cost on the extensive experimental work. After simulation results are obtained and ready to be validated, then experimentation can take place to confirm the findings and results from the simulation. The Taylor expansion was utilized to discretize the partial differential equation for the CMC model. The modelling of CMC using explicit and implicit methods was successfully implemented using FORTRAN as the simulation software.

From the results, we conclude that the implicit method is more accurate for the same time step, whereas it is much easier to write the FORTRAN code for the explicit method and the computational time to calculate is much shorter for the same time step. When preparing to conduct simulations, the researcher needs to balance the requirements of time step size with the necessary accuracy and time required for the simulations to be run. Further work on modelling the combustion of hydrogen with oxygen using CMC was successfully carried out. From the results of the simulations, CMC modelling using FORTRAN code is very useful to analyse the mixing process and behaviour for the case of mixing coupled with chemical reactions.

Acknowledgments

The authors would like to thank University of Southern Queensland, Universiti Malaysia Pahang and Ministry of Higher Education, Malaysia for providing laboratory facilities and financial support.

References

[1] IEA (International Energy Agency) Organisation for Economic Co-Operation and Development, 2009 World Energy Outlook (WEO), Int. Energy Agency, IEA, Paris

[2] Maczulak A. 2010 ‘Renewable Energy, Sources and Methods’, Facts on File Inc., New York, USA

[3] Pope, SB 1985, 'PDF Methods for Turbulent Reactive Flows', Progress in Energy and Combustion Science, **11**, pp. 119-192.

[4] Pope, SB 2000, *Turbulent Flows*, Cambridge University Press.

[5] Haworth, DC 2010, 'Progress in Probability Density Function Methods for Turbulent Reacting Flows', Progress in Energy and Combustion Science **36(2)**, pp. 168-259.

[6] Peters, N 1980 'Local Quenching Due to Flame Stretch and Non Premixed Turbulent Combustion', Western States Section of the Combustion Inst., spring meeting, Paper WSS-80-4.

[7] Peters, N 1983, 'Local Quenching Due to Flame Stretch and Non Premixed Turbulent Combustion' Combustion Science Technology **30**, pp. 1-17.

[8] Peters, N 1984, 'Laminar Diffusion Flamelet Models in Non-Premixed Turbulent Combustion," Progress in Energy and Combustion Science, **10**, pp. 319-339.

[9] Peters, N 2000, Turbulent Combustion, Cambridge University Press.

[10] Bilger, RW 1980,"in: P.A. Libby, F.A. Williams (Eds.), Turbulent Reacting Flows, Springer-Verlag, Berlin, pp. 65-113.

[11] Chen, H, Chen, S, Kraichnan, RH 1989, 'Probability Distribution of a Stochastically Advected Scalar Field', Physical Review Letters, **63**, pp. 2657-2660.

[12] Pope SB 1991, 'Mapping Closures for Turbulent Mixing and Reaction 'Theoretical and Computational Fluid Dynamics **2(5–6)**, pp.255-270.

[13] Klimenko, AY 1990, 'Multi-component Diffusion of Various Admixtures in Turbulent Flow', Fluid Dynamics, **25(3),** pp. 327-334.

[14] Bilger, RW 1993, 'Conditional Moment Closure for Turbulent Reacting Flow', Physics of Fluids, **5**, pp. 436-444.

[15] Klimenko, AY, Bilger, RW 1999, 'Conditional Moment Closure for Turbulent Combustion', Progress in Energy and Combustion Science, **25**, pp. 595-687.

[16] Klimenko, AY and Pope, SB 2003, 'The Modelling of Turbulent Reactive Flows based on Multiple Mapping Conditioning', Physical of Fluids, **15**, pp. 1907-1925.

[17] Klimenko, AY 2004,'Matching the Conditional Variance as a Criterion for Selecting Parameters in the Simplest Multiple Mapping Conditioning Models', Physical of Fluids, **16**, pp. 4754-4757.

[18] Wandel, AP, 2005 'Development of Multiple Mapping Conditioning (MMC) for Application to Turbulent Combustion', PhD Thesis, University of Queensland, Australia.

[19] Wandel, AP and Klimenko, AY, 2005, 'Multiple Mapping Conditioning Applied to Homogeneous Turbulent Combustion', Physics of Fluids, **17**, p. 128105.

[20] Vogiatzaki, K, Kronenburg, A, Navarro-Martinez, S and Jones WP 2011 'Stochastic Multiple Mapping Conditioning for a Piloted, Turbulent Jet Diffusion Flame', Proceedings of the Combustion Institute **33**, pp. 1523-1531.

[21] 19 Ternat, F, Orellana, O and Daripa, P 2011 'Two Stable Methods with Numerical Experiments for Solving the Backward Heat Equation', Applied numerical Method, **61**, pp. 266-284.

[22] Noor MM, Hairuddin, AA, Wandel AP and Yusaf, TF 2011 'Implementation of Conditional Moment Closure using Taylor Expansion and Finite Different Method', Int. Conf. of Mechanical Eng. Research (ICMER2011), 5-7 Dec, Malaysia, Paper ID:2011-151

[23] Clarke, J, Wandel, AP and Mastorakos, E 2010 'Analysis of Data to Develop Models for Spray Combustion', Southern Region Engineering Conference (SREC), 11-12 Nov, Australia, Paper ID SREC2010-F2-1.

[24] Curl, RL 1963, 'Dispersed Phase Mixing: I. Theory and Effects in Simple Reactors', AIChE Journal, **9**, pp. 175-181.

[25] Levenspiel, O and Spielman, LA 1965, 'A Monte Carlo Treatment for Reacting and Coalescing Dispersed Phase Systems. Chemical Engineering Science, **20**, pp. 247-254.
[26] Janicka, J, Kolbe, W, Kollmann, W 1979, 'Closure of the Transport Equation for the Probability Density Function of Turbulent Scalar Fields', Journal of Non-Equilibrium Thermodynamics, 4, pp. 47-66.
[27] Dopazo, C 1979, 'Relaxation of Initial Probability Density Functions in the Turbulent Convection of Scalar Fields', Physics of Fluids, **4(22)**, pp. 20-30.
[28] Subramaniam, S and Pope, SB 1998, 'A Mixing Model for Turbulent Reactive Flows based on Euclidean Minimum Spanning Trees', Combustion and Flame, **115**, pp.487-514.
[29] Villermaux,J., Devillon, JC.,1972 'Representation de la Coalescence et de la Redispersion des Domaines de Segregation dans un uide par un Modµele d'Interaction Phenomenologique, Second International Symposium on Chemical Reaction Engineering, pp. 1-13.
[30] Wandel, AP, 2011 A Stochastic Micromixing Model based on the Turbulent Diffusion Length Scale, 2011 Australian Combustion Symposium (ACS2011), Paper ID: ACS2011-20.
[31] Noor, MM, Yusaf, TF and Wandel AP, 2011 Study of Random Particle Interactions for Analysis of Diffusion Lengths in Turbulent Combustion Modelling, 2011 Australian Combustion Symposium (ACS2011), Paper ID: ACS2011-36.
[32] Liu, C & Karim, GA 2008, 'A simulation of the combustion of hydrogen in HCCI engines using a 3D model with detailed chemical kinetics', Int. J Hydrogen Energy, 33(14), pp. 3863-3875
[33] Saravanan N and Nagarajan G 2010 'An experimental investigation on hydrogen fuel injection in intake port and manifold with different EGR rates', Energy and Env., 1(2), pp. 221-248.
[34] Saravanan, N, Nagarajan, G, Sanjay, G, Dhanasekaran, C and Kalaiselvan, KM 2008, 'Combustion analysis on a DI diesel engine with hydrogen in dual fuel mode', Fuel, vol. 87, no. 17-18, pp. 3591-3599.
[35] Verhelst, S and Wallner, T 2009, 'Hydrogen-fuelled internal combustion engines', Progress in Energy and Combustion Science, 35(6), pp. 490-527
[36] Mardani A, Tabejamaat S and Ghamari M. 2010 Numerical study of influence of molecular diffusion in the MILD combustion regime, Combust Theory Model, 14, pp. 747-774
[37] Mardani A and Tabejamaat S, 2010 Effect of hydrogen on hydrogenemethane turbulent non-premixed flame under MILD condition, Int. J Hydrog Energy, 35, pp. 11324-11331
[38] Wang F, Mi J, Li P and Zheng C. 2011 Diffusion flame of a CH4/H2 jet in hot low-oxygen coflow, Int. J. Hydrogen Energy, 36, pp. 9267-9277
[39] Yu Y, Wang G, Lin Q, Ma C and Xing X 2010 'Flameless combustion for hydrogen containing fuels', Int. J Hydrogen Energy, 35, pp. 2694-2697
[40] Hollinger T and Bose T 2008 Hydrogen Internal Combustion Engine, Chapter 7a, in L´eon (ed.), Hydrogen Technology, Springer-Verlag, Berlin Heidelberg
[41] Collatz, L 1966, 'The Numerical Treatment of Differential Equations', Springer Verlag, Berlin.
[42] Krause, E. 1971 'Mehrstellen Verfahren zur Integration der Grenzschicht gleichungen', DLR Mitteilungen, **71(13)**, pp. 109-140.
[43] Adam, Y 1975 'A Hermitian Finite Difference Method for the Solution of Parabolic Equations' Computational Mathematics Application, **1**, pp. 393–406.
[44] Rubin, SG and Graves, RA 1975 'Viscous Flow Solutions with a Cubic Spline Approximation' Computational Fluid, **3**, pp. 1-36.
[45] Hirsh, RS 1975, 'Higher Order Accurate Difference Solutions of Fluid Mechanics Problems by a Compact Differencing Scheme' Journal Computational Physics, **19**, pp. 20-109.
[46] Ciment, M and Leventhal, SH 1975, 'Higher Order Compact Implicit Schemes for the Wave Equation' Mathematics Computational, **29**, pp. 885-944.
[47] Adam, Y 1977 'Highly Accurate Compact Implicit Methods and Boundary Conditions', Journal Computational Physics, **24**, pp. 10-22.
[48] Leventhal, SH 1980, 'The Operator Compact Implicit Method for Reservoir Simulation' Journal Society of Petroleum Engineering, **20**, pp. 120-128.

[49] Peyret, R 1978 'A Hermitian Finite Difference Method for the Solution of the Navier–Stokes Equations', Proceedings First of the Conference on Numerical Methods in Laminar and Turbulent Flows, Pentech Press, Plymouth (UK) pp. 43-54.

[50] Peyret, R and Taylor, TD 1982, 'Computational Methods for Fluid Flow', Springer Verlag, New York.

[51] Lele, SK 1992, 'Compact Finite Differences with Spectral-like Resolution', Journal Computational Physics, **103(1)**, pp. 16-42.

[52] Rubin, SG and Khosla, PK 1977 'Polynomial Interpolation Method for Viscous Flow Calculations' Journal Computational Physics, **24**, pp. 217-246.

[53] Chapra CS and Canale RP 2006, 'Numerical Methods for Engineers', McGraw Hill, 5th Edition, Singapore.

[54] Griffiths DV and Smith IM 2006, 'Numerical Methods for Engineers' Chapman and Hall/CRC, 2nd Edition, New York.

[55] D'Acunto, B, 2004, Computational Methods for PDE in Mechanics, World Scientific Publishing Co. Italy.

[56] Kopriva DA and Kolias JH 1996, A Conservative Staggered-grid Chebyshev Multidomain Method for Compressible Flows, Journal of Computational Physics, **125**, pp. 244-261.

[57] Yeung, RW and Ananthakrishnan, P 1992, Oscillation of a Floating Body in a Viscous Fluid, Journal of Engineering Mathematics, **26**, pp. 211-230.

[58] Yeung, RW and Ananthakrishnan, P 1997, Viscosity and Surface-Tension Effects on Wave Generation by a Translating Body, Journal of Engineering Mathematics, **32**, pp. 257-280.

[59] Hirsh, C 2007, Numerical Computation of Internal and External Flows: Fundamentals of Computational Fluid Dynamics, Butterworth-Heinemann, 1, Burlington, USA.

[60] Courant, R, Friedrichs, K and Lewy, H 1928, Über Die Partiellen Differenzengleichungen Der Mathematischen Physik, Mathematische Annalen, **100(1)**, pp 32-74.

[61] Courant, R, Friedrichs, K and Lewy, H 1967, On the Partial Difference Equations of Mathematical Physics, English Translation from German 1928 Paper. IBM Journal, pp. 215-234.

Nomenclature

T	Temperature, K
D	Diffusivity
N	Scalar Dissipation Rate
V	Volume, m^3
W	Chemical Source Term
Z	Mixture Fraction (a Conserved Scalar)
P	Favre Joint PDF of Composition
B	Constant (Function of *dt* and *dz*)
E_a	Activation Energy
A, β	Constants
K, j	Grid Point Involved in Space Difference
L, i	Grid Point Involved in Time Difference
S_k	Reaction Rate for Species k
J_{ik}	Molecular Diffusion Flux Vector
RANS	Reynolds Averaged Navier Stokes
CMC	Conditional Moment Closure
IEM	Interaction by Exchange with the Mean
DNS	Direct Numerical Simulation
MMC	Multiple Mapping Conditioning
CFD	Computational Fluid Dynamics
CFL	Courant-Friedrichs-Lewy
PDF	Probability Density Function
EMST	Euclidean Minimum Spanning Tree
MC	Modified Curl's model
TT	Total time for the simulation
YA	Air Mass Fraction
YH	Fuel Mass Fraction
dt	Time step size
dz	Step size in mixture fraction space
u_i	Favre Mean Fluid Velocity Vector
ψ	Composition Space Vector
u_i''	Fluid Velocity Fluctuation Vector
$\dot{\omega}_i$	Net Formation Rate per Unit Volume
v	Kinematic Viscosity,
α	Thermal Diffusivity, m^2/s
ϕ	Particle Composition
k	Thermal Conductivity, W/mK
k_j	Arrhenius Reaction Rate Coefficient
ρ	Density or Mean Fluid Density, kg/m^3
R	Universal Gas Constant (8.31431 kJ $kmol^{-1}K^{-1}$)
$\bar{w}$	Molecular Weight of a Gas Mixture
$\langle \ \vert \ \rangle$	Ensemble Average
$\langle Y \vert Z \rangle$	Mass Fraction of Fuel
$\langle N \vert Z \rangle$	Conditional Scalar Dissipation
$\langle W \vert Z \rangle$	Conditional Chemical Source Term
$\langle S \vert Z \rangle$	Conditional Generation Due to Droplet Evaporation

Appendix A: Part of FORTRAN code for explicit method

```
YH(:,1)  = 0.0    Comment: Initial Condition
YH(L,1) = 1.0
YA(1,:)  = 1.0
YA(L,1) = 0.0
Comment: Main Calculation
Do j=1, K
Do i=2, L
YH(i,j+1) =((1-(2*B))*YH(i,j)) + (B*YH(i+1,j)) + (B*YH(i-1,j))
YA(i,j+1) =((1-(2*B))*YA(i,j)) + (B*YA(i+1,j)) + (B*YA(i-1,j))
End Do
YH(1,j+1)=0   Comment: Boundary Condition
YH(L,j+1)=1
YA(1,j+1)=1
YA(L,j+1)=0
End Do
```

Appendix B: Part of FORTRAN code for implicit method

```
A(1,1) = 0.0    Comment: Initial Condition
A(2,1) = 1.0
A(1,2) = 0.0
Do i=2, dz-1
A(3,i-1) = - B
A(2,i ) = 1.0 + 2.0 * B
A(1,i+1) = - B
End Do
A(3,dz-1) = 0.0
A(2,dz) = 1.0
A(3,dz) = 0.0
Comment: Factor the matrix
Call MatrixC (dz,a,b,FF)
Do j = 2, dt
Call YH (z_min,z_max,t_min,t(j),B(1))
B(2:dz-1) = YH(2:dz-1,j-1)
Call YA (z_min,z_max,t_min,t(j),B(dz))
WW = 0
Call MatrixD (dz,a,b,WW)
YH(1:dz,j) = B(1:dz)
End Do
Comment: Subroutine for matrix
Do i = 1, n-1
If (P(2,i) .eq. 0.0 ) then
info = i
Write (*, '(P)') 'MatrixD - error'
Return
End If
P(3,i) = P(3,i) / P(2,i)
P(2,i+1) = P(2,i+1) - P(3,i) * a(1,i+1)
End Do
```

Improving the performance of ultrasonic horn reactor for deactivating microorganisms in water

R A Al-juboori[1] and T F Yusaf[1,2]

[1] Faculty of Engineering and Surveying, University of Southern Queensland, Toowoomba 4350 Toowoomba, Australia
[2]National Centre for Engineering in Agriculture, University of Southern Queensland, Australia

E-mail[1]: RaedAhmed.Mahmood@usq.edu.au

Abstract. The research on enhancing the performance of ultrasonic reactor for the purpose of microorganisms' inactivation is still ongoing. In this work, covering the cavitation chamber bottom with a corrugated surface as a source for heterogeneous cavities has been proposed as a simple modification to improve ultrasonic deactivation for ultrasonic horn reactor. *Escherichia coli* ATCC 25922 was used as a model microorganism in this study. Before using the corrugated surface, the configuration of the cavitation chamber was optimized experimentally in regards to cavitation chamber diameter and the depth of ultrasonic probe tip in the suspension. The optimization of the aforementioned factors was conducted on a basis of using constant suspension volume of 50ml. The depth of the ultrasonic probe tip in the suspension was changed from 2-10mm with a step of 2mm in overall depth of the suspension of 2cm, while the diameter of the chamber was changed using five Pyrex beakers with different diameters. The study was carried out using three level of ultrasonic intensity; low (17.56), intermediate (21.49) and high (24.17) W/cm^2. The results of the optimization showed that increasing the diameter of cavitation chamber can decrease the log reduction of *E.coli* significantly. However, changing the depth of ultrasound probe in the suspension within the studied range was found to have only slight effect on the log reduction of *E.coli* in the order of approximately 0.5-$\log_{10}$. When using the corrugated surface with optimum chamber design, the results revealed that the corrugated surface can increase the log reduction of *E.coli* for the applied ultrasonic intensities. This effect was more discernable with low ultrasonic intensity than intermediate and high intensities.

1. Introduction

Ultrasonic reactors have various uses in different application including food processing, gene transfer, synthesis of biodiesel emulsifier preparation and water disinfection [1]. However the use of ultrasonic reactor for water disinfection is considered to be one of the valuable uses of ultrasound now days, due to the environmentally friendliness of this technique and its ability to deactivate and disintegrate the

Published under licence by IOP Publishing Ltd

clusters of the pathogenic microorganisms. The efficacy of ultrasound in deactivating microorganisms in the water lies in the simultaneous effects of bubble collapse which includes the generation of localized spots with high temperature and pressure, reactive free radicals and turbulences induced by the oscillations of bubbles [2]. The bubbles that are generated under the effect of acoustic cavitation can be categorised into two types; transient and stable bubbles. Transient bubbles last for one or few sound cycles and implode subsequently generating high temperature and pressure. Stable bubbles exist for many cycles during which the bubbles oscillate around its mean radius producing micro-streaming [2, 3]. Although the implosion of the stable bubbles is less violent than that of the transient bubbles, micro-streaming resulting from the oscillation of the stable bubbles was found to be producing shear stress that is sufficient to rupture the cell of microorganism [4, 5]. The bubbles generally develop and grow on the nucleation sites in the medium, as these sites are represented by the locations of the dissolved gases in the liquid body and the crevices and irregularities of the solid surfaces [6, 7]. In spite of the remarkable results that are achieved with using ultrasound alone or augmented with other techniques in deactivating microorganisms [8-13], the energy demand associated with using ultrasound for deactivating microorganism is considered to be high [14].

There are several types of ultrasonic reactors that are used for various applications; ultrasonic horn, ultrasonic bath, multiple transducer reactor and tubular reactor with two ends irradiated by transducers or transducer on one end and reflector on the other end [2]. Ultrasonic horn reactor is considered to be the most used reactor among the other types of ultrasonic reactors [2]. However the mechanical and chemical effects of ultrasound in this type of reactor is restricted to the axial and radial distances near to ultrasonic horn [15]. Another shortcoming of ultrasonic horn reactor is the shedding and the erosion of its irradiating surface when using it with high ultrasonic intensity [1]. To use ultrasonic horn reactor for deactivating microorganisms effectively, many factors have to be considered carefully including; the position of the ultrasonic horn in the suspension in regard to the axial and radial distances [2], and the depth of ultrasonic probe in the suspension. In addition to these factors, the efficiency of ultrasound reactor in the process of water disinfection can be improved by promoting cavitation intensity using different means such as the use of solid additives and combining ultrasound with other techniques such chlorination and ultraviolet [2] .

The present work was undertaken to examine the effect of covering the bottom of cavitation chamber by corrugated surface on the inactivation of *E.coli*. The corrugated surface was used in this study to serve as a source for the heterogeneous cavities. Prior to the use of corrugated surface, the configuration of cavitation chamber was optimized with regard to two parameters, the diameter of the chamber and the depth of ultrasonic probe in the suspension.

2. Materials and methods

2.1. Preparing the suspension of E. coli ATCC25922

E. coli was selected to be the sample microorganisms in this work owing to its widespread use as indicator microbial contamination in most of the water resources [16]. A 75 µl of saline (autoclaved to 121°C) was spread with 25 µl of *E. coli* ATCC 25922 stock culture (cooked meat medium) on an agar plate (DifcoTM Nutrient Agar) and then incubated at 37°C for 24 hours. Thereafter, the prepared plate was washed with 2ml of sterilised saline and small amount of the suspending fluid was diluted with 10 ml of sterilised saline to prepare the inoculum of *E. coli*. The concentration of *E. coli* in the prepared inoculum was estimated by measuring the optical density at 625 nm (OD625) of the inoculum using Hach DR/2000 direct reading spectrophotometer. Saline was used as a blank in the spectrophotometer measurements. When the optical density of the inoculum reaches between 0.08-0.13, this indicates that the concentration of *E. coli* in this inoculum is equivalent to approximately $1\text{-}2\times10^8$ (Colony Forming Unit) CFU/ml. One millilitre of the prepared inoculum was added to 200 ml of autoclaved distilled water to obtain suspension with concentration of *E. coli* about 1×10^6 CFU/ml.

2.2. Measurement log reduction of E. coli

Viable cell count technique was used in the present work to measure the reduction of *E. coli* after ultrasound treatment due to its reliability in measuring the CFU of the suspension after ultrasound treatment [17].The concentration of *E. coli* expressed in CFU/ml of the untreated suspension was measured by diluting 0.01 ml of untreated suspension with 10 ml of saline, then spreading 0.1 ml of this solution onto three plates of Difco^{TM} Nutrient Agar. The three plates were then incubated at 37°C for 18-24 hours. Subsequently, the colonies were counted using SUNTEX colony counter model 570 with electronic register. The number of colonies for the untreated suspension was obtained by taking the average of the colonies of these plates. While, the concentration of *E. coli* in the treated suspension was measured using an appropriate serial 5-fold dilutions of the treated suspension. The dilution series consisted of 4 centrifuge tubes (1.5 ml) filled with 0.9 ml of sterilised saline. A 0.1 ml of the treated suspension was added to the first plate of the series and another 0.1 ml of the treated suspension was added to the first centrifuge tube. After vortexing the mixture in the first centrifuge tube, a 0.1 ml of the mixture was added to the second tube and then the same procedure was repeated for the rest of the series. Following, a 0.1 ml of each tube was spread on the plates of Difco^{TM} Nutrient Agar and again the plates were incubated for 24 at 37°C. Log reduction of *E.coli* was obtained by subtracting log (CFU/ml) of the treated suspension from that of the untreated suspension. To ensure that there was no contamination taking place during ultrasound treatment, samples of the treated suspensions were cultured in MacCONKEY agar, as this agar displays the grown colonies of *E. coli* with a pink colour while other microorganisms are displayed in other colours. Additionally, triplication for the plates of the entire dilution series was applied to improve the accuracy of the viable cell count.

2.3. Ultrasound treatment

A commercial ultrasonic horn device (HielscherUIP500) with a 22 mm diameter probe tip (sonotrode, BS20d22 titanium) and a 55 kHz frequency was used to perform the experiments of ultrasound treatment as shown in figure1. The applied ultrasonic intensity can be controlled through regulating the amplitude of ultrasonic device. In this study, three levels of amplitude were applied; low 50% that gives 17.56W/cm^2, intermediate 75% gives 21.49W/cm^{2}and high 100% gives 24.17 W/cm^2.

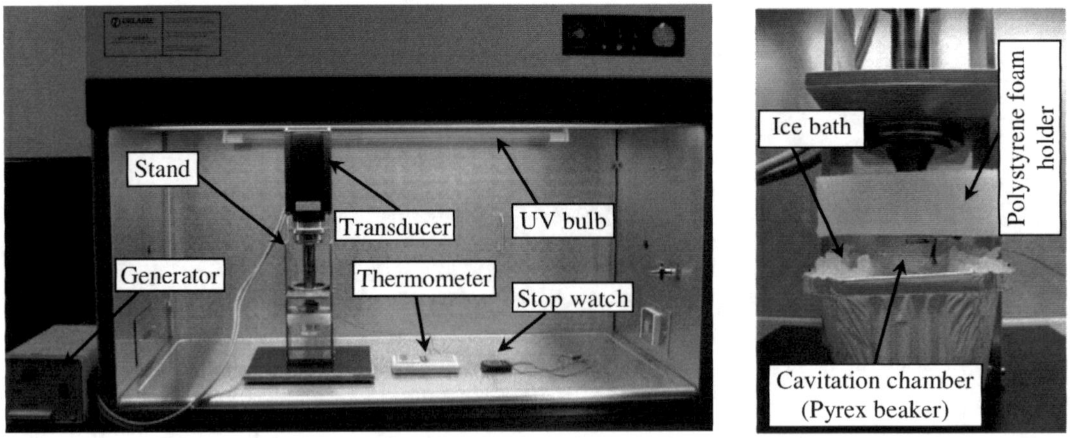

(a) (b)

Figure 1. Sonication setup (a) the entire setup inside the laminar flow (b) enlarged view of the beaker immersed in the ice bath

3. Optimizing the configuration of cavitation chamber

Prior using ultrasonic horn device for corrugated surface experiments, the configuration of cavitation chamber was optimized in regards to two parameters, the diameter of the chamber and the depth of ultrasonic horn tip in the suspension. It should be noted that cavitation chamber refers to the beaker that was used to carry the suspension. The optimization experiments were conducted with three ultrasonic intensities 17.56, 21.49 and 24.17 W/cm^2 and fixed treatment time of 4minutes.

3.1. Chamber diameter

3.1.1. Procedure

The optimum diameter of cavitation chamber was indicated simply using five Pyrex beakers with different diameters 3.6, 3.8, 4.5, 5, 6.5 cm. It should be noted that changing the diameter of the chamber that carried a constant volume of *E.coli* suspension (50ml) with maintaining the depth of the ultrasonic horn tip at 2mm required changing the vertical distance between the irradiating surface of the probe and the bottom of the chamber (*h*) which was maintained within the effective axial range of ultrasonic reactor between 0 to 4cm [15]. Each diameter has its corresponding value of *h* as shown in table 1.

Table1. Beakers' radius with their corresponding *h*.

Diameter (cm)	*h* (cm)
3.6	3.6
3.8	3.3
4.5	2
5	1.8
6.5	1.1

3.1.2. Results

The effect of chamber diameter on the reduction of *E.coli* for three ultrasonic intensities; 17.56, 21.49 and 24.17W/cm^2 for treatment time of 4 minutes is shown in figure 2. The values of *E. coli* log reduction presented in this figure and the following figures in this paper are expressed by mean value and the error bars represent the standard error of plate triplication.

The general trend of figure 2 suggests that increasing the chamber diameter within the effective axial distance of ultrasonic horn reactor can lead to a decrease in the log reduction of *E. coil*. The results in figure 2 showed that the maximum reduction of *E.coli* was achieved using the smallest diameter of 3.5cm. The maximum values of log reduction were 4.51, 5.05 and 5.45 for the intensities 17.56, 21.49 and 24.17W/cm^2 respectively. These results are in agreement with the results obtained by Gogate et al. [15], as the latter confirmed that the mechanical and the chemical effects of ultrasound for ultrasonic horn reactor occur effectively under the centre of the horn tip and started to decrease away from the horn centre up to a distance of 3 cm. Beyond this limit the effect of ultrasound almost vanished. It can be also noticed from figure 2 that increasing the diameter of the chamber has a significant effect on the intermediate intensity of 21.49 W/cm^2 as compared to the other intensities. The log reduction of 21.49 W/cm^2 with small diameters became closer to that of high ultrasonic intensity (24.17W/cm^2), while with larger diameters the log reduction became closer to that of low ultrasonic intensity (17.56W/cm^2).

From the results presented here, it can be said that the smaller the diameter of the beaker the better the results that can be obtained. However, the temperature rise of the suspension due to ultrasound treatment in the small beakers was significantly faster than that of the large beakers which can be a problem. It is worth to mention that temperature rise in a small diameter beaker can cause some problems including requirement for extensive cooling process and giving misleading results of ultrasound treatments. The increase in the temperature of the treated suspension can affect on the viability of microorganisms which makes it hard to identify the actual reduction of microorganisms caused by ultrasound. Condón et al.[18] suggested that the change in the outer membrane of the gram-negative microorganisms due to the thermal effect of ultrasound can be a possible reason for the synergistic inactivation. To avoid the rapid temperature rise in the treated suspension during ultrasound treatment, beaker with intermediate diameter of 4.5 cm was chosen to be used in the following ultrasound treatments.

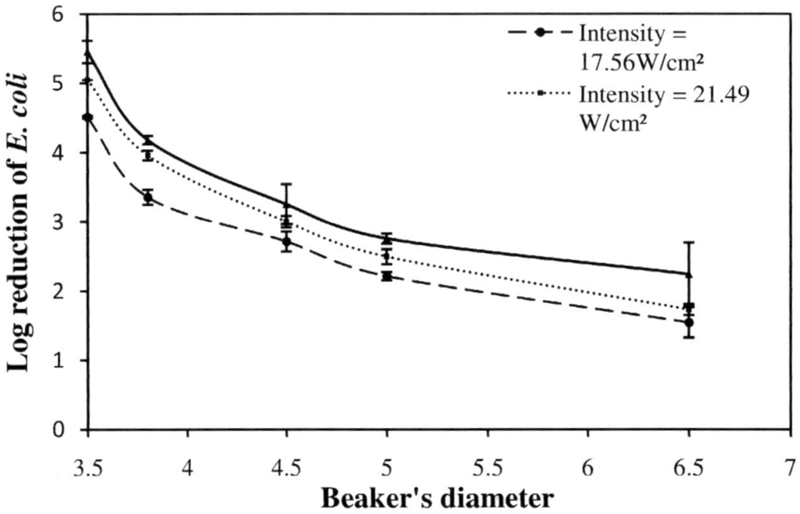

Figure2. Log reduction of *E.coli* with different chamber diameters for three ultrasonic intensities (●) 17.56, (■) 21.49 and (▲) 24.17 W/cm^2 and treatment time of 4 minutes.

3.2. Depth of ultrasonic horn tip

3.2.1. Procedure

The other parameter that is associated with the configuration of cavitation chamber is the depth of the ultrasonic horn tip. The effect of this parameter on the reduction of *E.coli* was investigated within the range of 2 to 10mm, as the ultrasonic horn was set at a fixed level while the depth of the horn tip in the suspension was changed by lifting the beaker up with a step of 2mm.

3.2.2. Results

The effect of changing the depth of the horn tip within the range of 2-10mm on the viability of *E.coli* for three ultrasonic intensities 17.56, 21.49 and 24.17W/cm^2 and fixed treatment time of 4minutes is shown in figure 3. The results presented in figure 3 show that changing the depth of probe from 2 to 10mm has a slight effect on the log reduction of *E.coli* of approximately 0.5-$\log_{10}$for all the applied ultrasonic intensities.

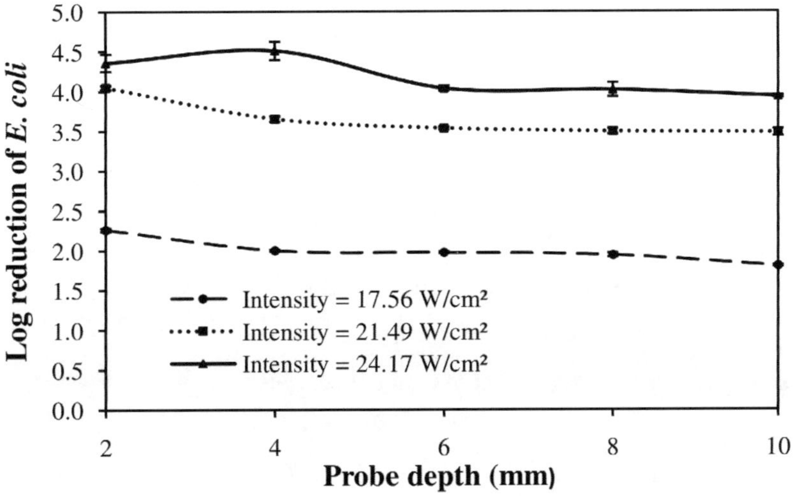

Figure 3. The change in log reduction of *E.coli* versus probe depth for three ultrasonic intensities; (●) 17.56, (■) 21.49 and (▲) 24.17 W/cm2 and treatment time of 4 minutes.

Log reduction of *E.coli* for the intensities 17.56 and 21.49W/cm^2 decreased with increasing the depth of the probe in the suspension from 2 to 10mm. This could be attributed to the increase in the lost volume of the suspension that is not subjected to the effect of ultrasound with increasing the depth of the probe as demonstrated in figure 4. However, log reduction of *E. coli* for the intensity 24.17W/cm^2 increased when changing the depth from 2 to 4mm and then decreased with increasing the depth. Subjecting a suspension with small volume (50ml) to ultrasonic intensity of 24.17W/cm^2 was found to be causing high level of turbulences in the suspension especially at the surface. When the depth of the ultrasonic probe is as small as 2mm, the turbulences at the suspension surface allow the air from the atmosphere to penetrate into the suspension and cause formation of large bubbles near to the irradiating surface as revealed in figure 5.

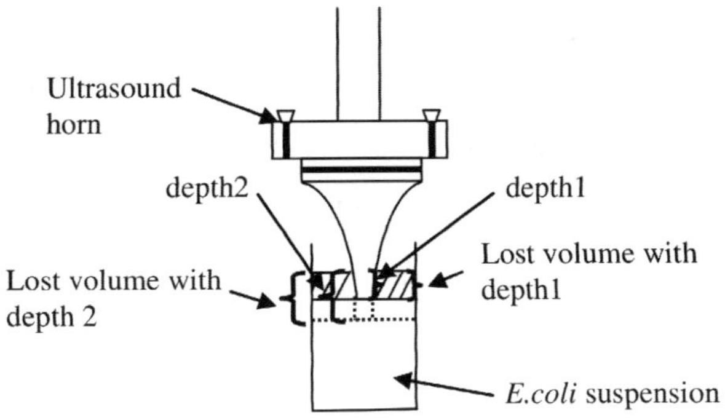

Figure4. Demonstration for the effect of probe depth on the lost volume of the suspension.

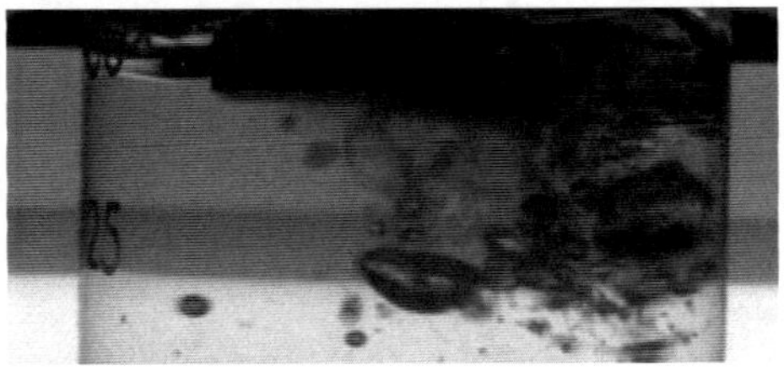

Figure5. Generation of large bubbles in the suspension with ultrasonic intensity of 24.17 W/cm^2.

The formation of large bubbles near to the ultrasound horn in this figure was captured using SONEY hybrid Handycam. The presence of the large bubbles in the affinity of the irradiating surface can hinder the transition of ultrasonic energy into the suspension. The increase in the log reduction at the depth of 4mm suggests that at this depth ultrasonic energy transmitted into the suspension body without obstacles (formation of large bubbles). This discussion is valid when ultrasonic horn reactor used to treat solution open to the atmosphere.

The outcome of the optimization for ultrasonic probe depth showed that the smaller the depth of the probe, the higher the reduction of microorganisms can be achieved for low and intermediate ultrasonic intensities, while in the case of high ultrasonic intensity the probe has to be immersed to a depth that is sufficient to prevent any air penetration from the atmosphere.

4. Experiments of immersing corrugated surface in *E.coli* suspension

4.1. Preparing the corrugated surface

The corrugated surface used in this study was constructed from pebbles with an average length of 10mm adhered on a circular plate with thickness of 1mm using commercial silicon.

4.2. Particle shedding

Irradiating *E. coli* suspension with the corrugated surface immersed in it can cause particle shedding from the corrugated surface. To detect the occurrence of particle shedding during ultrasound treatment, the corrugated surface was submerged in a 50mL of nano-pure water and subjected to ultrasound irradiation with three intensities 17.56, 21.49 and 24.17 W/cm^2 for 4 minutes. Thereafter, the treated water was filtered using flat-sheet RO membrane. The RO flat-sheet membrane was used due to its fine pores so as the particles with nano-scale size detained. Later, the filtrate particles were visualized using Motic Stereo-microscope. Photos of the eroded particles with the three ultrasonic intensities were captured using CC12 camera which is attached to the microscope as illustrated in Figure 6.

The photos were analysed using *AnalySIS* software. It is shown in figure 6 that the number of the eroded particles with ultrasonic intensity of 21.49W/cm^2 is higher than that of 17.56 and 24.17W/cm^2. This can be explained as increasing the ultrasonic intensity beyond 21.49W/cm^2 did not result in an increase in the mechanical effects of ultrasound. As 21.49 W/cm^2 can be considered as the optimum ultrasonic intensity for the used ultrasonic device in this study, and increasing the ultrasonic intensity more than its optimum level will generate a cloud of bubbles near to the irradiating surface that impairs the transition of ultrasonic energy into the body of the suspension [3, 19].

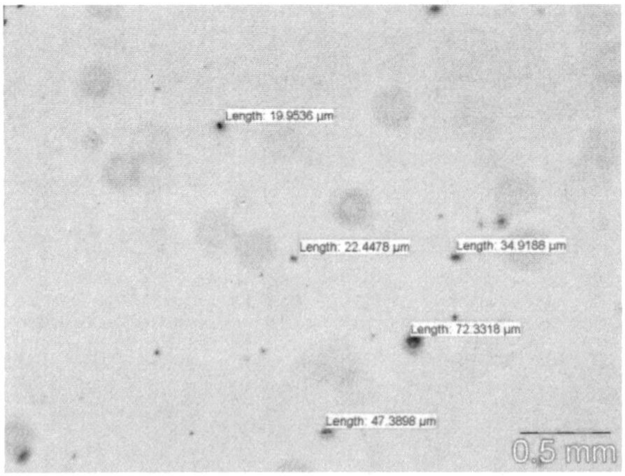

(a)

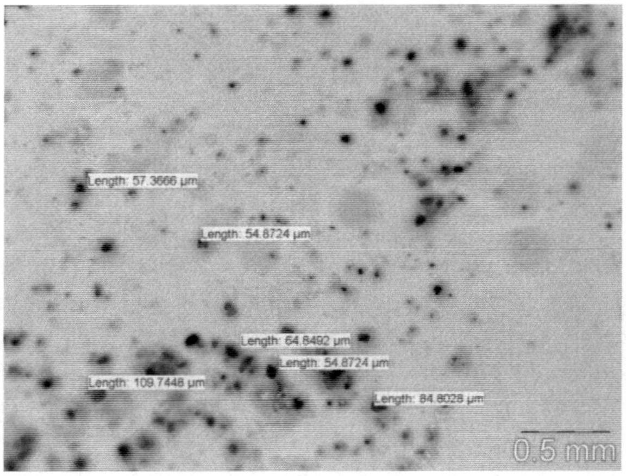

(b)

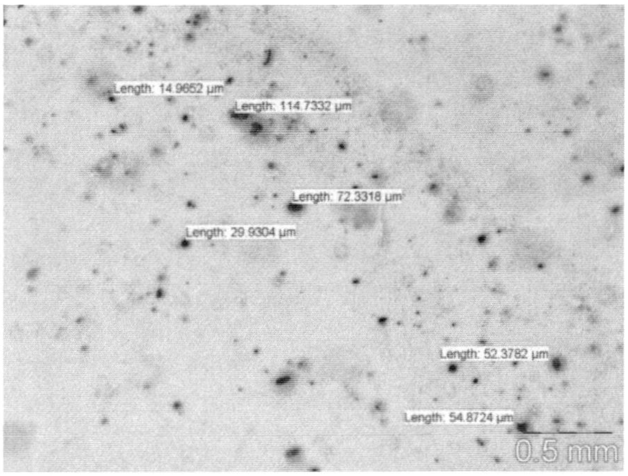

(c)

Figure6. Eroded particles from the corrugated surface with three different ultrasonic intensities (a) 17.56W/, (b) 21.49W/cm^2 and (c) 24.17W/cm^2.

4.3. Procedure for the corrugated surface experiments

The setup used to perform sonication experiments is a batch configuration as shown in figure 1, where the generator was placed outside the laminar flow while the transducer rested on an aluminium stand inside the laminar flow. The selected Pyrex beaker from the optimization was used to hold the suspension. The capacity of optimum Pyrex beaker is 80ml. before using the beaker for holding the suspension; the beaker was autoclaved for 15 minutes. The beaker was fitted between the two plates of the aluminium stand using a fitted holder made from foamed polystyrene (figure 1b). Treatment times of ultrasound treatments were measured using stop watch. The treatment time and the temperature of the treated suspension were measured using stop watch and *DiGi-Sense* thermometer type K.

Before performing the experiments of corrugated surface, the following precautions were adopted to avoid possible contamination.

1. UV light in the laminar flow was operated for one hour before starting the experiments to sterilise the air inside the laminar flow,
2. Ethanol 75% (volumetric percentage) was used to sterilise the steel bottom of the laminar flow,
3. The Ultrasound horn tip and the probe of thermometer were sterilised with ethanol, and
4. The corrugated surface was sterilised with ethanol 75% before immersing it in the suspension.

The corrugated surface was placed on the beaker's bottom before filling the beaker with *E. coli* suspension. After that, the beaker was filled with 50 ml of the suspension and placed in-between the plates of the aluminium stand (Figure 1 b). Next, the horn tip of the ultrasound device was immersed in two different depths in the suspension depending on the supplied ultrasonic intensity. The depth of the ultrasonic horn tip was 2mm in case of 17.56 and 21.49 W/cm^2, and 4mm in case of 24.17 W/cm^2.

The pressure of the suspension in sonication treatment was constant at the atmospheric pressure as the suspension was exposed to the atmosphere, while the temperature of the suspension was variable due to ultrasound treatment. The temperature was controlled using an ice bath set up below the suspension beaker to cool down the temperature and maintain it to be less 40°C. The generator of the ultrasonic device was operated at the targeted intensities for 5 minutes during which a 2 ml sample of the treated suspensions was withdrawn at 1, 2, 3, 4 and 5 minutes for analyses.

4.4. Investigating the effect of the eroded particles on the viability of E. coli

The filtrate particles from the corrugated surface with three ultrasonic intensities 17.56, 21.49 and 24.17W/cm^2 were seeded in 50mL of *E.coli* suspension and treated with their corresponding ultrasonic intensities for 3 minutes. The purpose of these experiments is to identify the effects of the eroded particles on the viability of *E.coli*. The results of viable cell count for these tests compared to the results of the treated suspension with and without corrugated surface are illustrated in table 2 in average values of viable cell count of three plates.

Table 2. Comparison between three different treatments; without corrugated surface, with corrugated surface and with the eroded particles suspended in *E.coli* suspension.

Ultrasonic intensity(W/cm^2)	Log reduction of *E.coli* (CFU/ml)		
	Without corrugated surface	With corrugated surface	With eroded particles
17.56	1.966527	2.685172	2.045417
21.49	3.104567	3.485542	3.205275
24.17	3.567778	3.71963	3.602783

Generally, the presence of the eroded particles in the suspension has a slight effect on *E.coli* reduction as compared to the overall effect of the corrugated surface. Submerging the corrugated surface in the suspension of *E.coli* led to increase the log reduction of *E.coli* by almost 0.72, 0.38 and 0.152, while seeding the eroded particles in the suspension resulted in an increase in the log reduction by approximately 0.08, 0.1 and 0.035 for the intensities 17.56, 21.49 and 24.17W/cm^2 respectively.

However, table 2 shows that the effect of eroded particles on *E.coli* viability is more sensible in ultrasonic intensity of 21.49W/cm^2 than the other two intensities (17.56 and 24.17 W/cm^2) and this can be ascribed to the high loading of eroded particles of the intensity 21.49 W/cm^2 as opposed to the other intensities as shown in figure 6. The eroded particles in the suspension act as beads in bead mill treatment, as increasing the loading of the beads in the suspension results in an increase in the disruption rate of the microorganisms [20].

4.5. Effect of corrugated surface on log reduction of E.coli

Figure 7 shows the log reduction of *E.coli* under the effect of three different ultrasonic intensities 17.56W/cm^2, 21.49W/cm^2 and 24.17W/cm^2 for two treatments; with and without corrugated surface.

The improvement in the reduction of *E.coli* due to immersing a corrugated surface in the suspension could be due to the combination of the synergistic effects of the surface heterogeneity that served as cavitation nuclei and the effect of the eroded particles from the corrugated surface. The effect of the eroded particles was identified in the previous section (4.4) as having slight effects on the viability of *E. coli* as compared to the overall effect of the corrugated surface. Therefore, the increase in the log reduction is believed to be due to the heterogeneity of the corrugated surface that contributed to the formation of large number of bubbles and thus increase the cavitational events. Ince, N.H. and R. Belen [21] found that the decrease in *E.coli* viability treated by ultrasound when seeding granules of activated carbon in *E.coli* suspension is attributed mainly to the cavitational effects induced by the formation of bubbles on the solid-liquid interface. Similarly, Dadjour, M.F., et al [22] used pellets of TiO_2 with the diameter of 2mm to enhance the deactivation of *E.coli* using ultrasound treatment, and the decrease in the *E.coli* viability with presence of TiO_2 was attributed to the catalytic effect of TiO_2 and the surface heterogeneity of the pellets that contributed to the formation of additional bubbles.

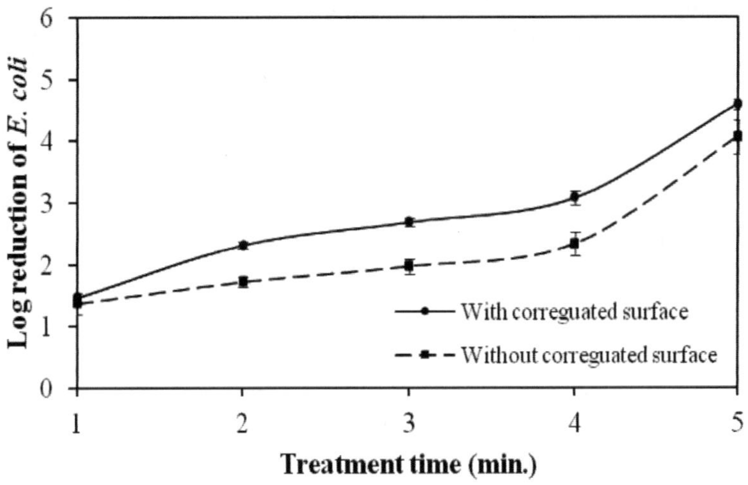

(a)

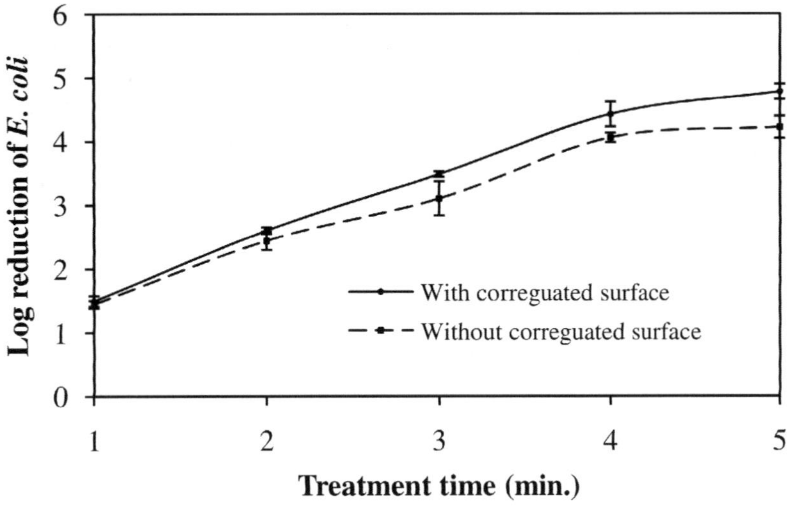

(b)

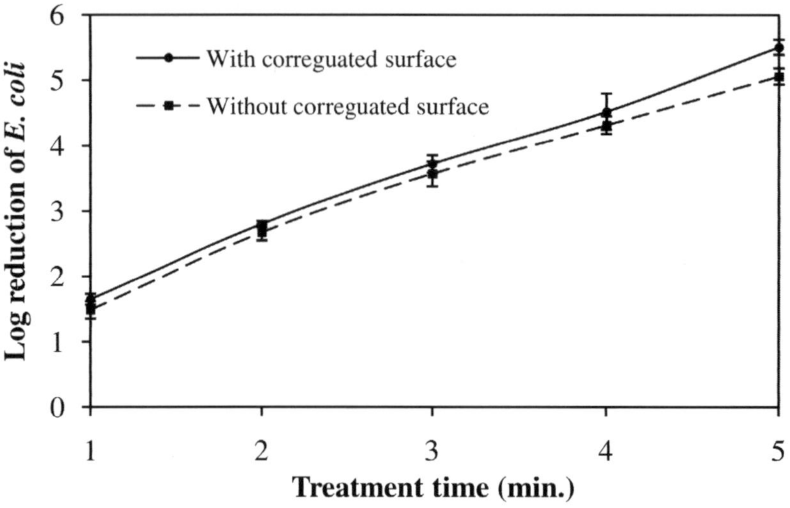

(c)

Figure7. Log reduction of E.coli for three ultrasonic intensities a) 17.56, b) 21.49 and c) 24.17W/cm^2 using corrugated surface (solid line) and without corrugated surface (dashed line).

From Figure 7, it can be said that the presence of corrugated surface in the suspension is more effective with low ultrasonic intensities than higher intensities. This could be explained as the irregularities of the corrugated surface acted as a source for the heterogeneous cavities which in turn helped the low ultrasonic intensity to form additional bubbles at negative pressure less than the required pressure to overcome the cavitation threshold of the water [23]. Similar observations were reported by many studies, for instance Broekman, S., et al [24] reported that the presence of micro-bubbles in the medium can reduce the cavitation threshold and help the low power ultrasound to generate cavitation in that medium. Likewise, Mahulkar, A.V., et al.[25] injected steam bubbles into ultrasonic bath to enhance the cavitational yield. They succeeded to increase the efficiency of acoustic cavitation in the non-degassed and degassed water by 4 and 16 times respectively. However, the case of higher ultrasonic intensities is different, as these intensities may have sufficient negative pressure to rupture the water and generate bubbles. Hence, the effect of the corrugated surface wasn't as pronounced as it was with low ultrasonic intensity.

Over all, immersing corrugated surface or constructing cavitation chamber with corrugated internal surface could increase the reduction of microorganisms for the same ultrasonic intensity. This technique could be even more feasible with low ultrasonic intensities than high intensities.

5. CONCLUSIONS

The effect of immersing corrugated surface in the cavitation chamber of ultrasonic horn reactor on the viability of *E.coli* was investigated in this work. Prior using corrugated surface, the configuration of the cavitation chamber was optimized in terms of chamber diameter and the depth of ultrasonic probe within the reported effective axial distance of ultrasonic horn reactor in the literature. The outcomes of the optimization are summarized in the following points

1. The general effect of the ultrasonic probe depth within the range of 2-10 mm in overall depth of the suspension of 2cm is considered to be slight in the order of 0.5-$\log_{10}$. However, increasing the depth of the horn tip in the suspension resulted in a decrease in the log reduction of *E.coli* for all ultrasonic intensities with exception for high ultrasonic intensity with small probe depth. With high

ultrasonic intensity of 24.17W/cm^2 the ultrasonic probe needs to be immersed to a sufficient depth to avoid the penetration of the atmospheric air into the suspension causing turbulences that affect negatively on the delivery of ultrasonic energy into the suspension.

2. Increasing the radius of the cavitation chamber within the effective axial distance of the ultrasonic horn reactor led to decrease in the log reduction of *E.coli* for different ultrasonic intensities.

The optimum configuration of cavitation chamber was used in the experiments of corrugated surface. Covering the bottom of the cavitation chamber with corrugated surface resulted in an increase in the log reduction of *E.coli* for different ultrasonic intensities. However, the effect of corrugated surface on the viability of *E. coli* appeared more clearly with low ultrasonic intensities than higher intensities. The heterogeneity of the corrugated surface seems to have the main role in increasing the log reduction of *E. coli* for the ultrasonic horn reactor.

From the results obtained in this study, it can be said that designing cavitation reactor with internal corrugated surface can decrease the required energy to obtain better killing of microorganisms especially for low ultrasonic power. This can be a possible solution to overcome the problem of high energy demand in ultrasound treatment.

ACKNOWLEDGEMENT

The authors would like to thank Mrs Morwenna Boddington and Mr Adrian Blokland for their technical support in the laboratory work.

References

[1] Gogate, P.R. and A.M. Kabadi, A review of applications of cavitation in biochemical engineering/biotechnology. Biochemical Engineering Journal, 2009. **44**(1): p. 60-72.

[2] Gogate, P.R., Application of cavitational reactors for water disinfection: Current status and path forward. Journal of Environmental Management, 2007. **85**(4): p. 801-815.

[3] Thompson, L.H. and L.K. Doraiswamy, Sonochemistry: Science and Engineering. Industrial & Engineering Chemistry Research, 1999. **38**(4): p. 1215-1249.

[4] Hua, I. and J.E. Thompson, Inactivation of Escherichia coli by sonication at discrete ultrasonic frequencies. Water Research, 2000. **34**(15): p. 3888-3893.

[5] Doulah, M.S., Mechanism of disintegration of biological cells in ultrasonic cavitation. Biotechnology and Bioengineering, 1977. **19**(5): p. 649-660.

[6] Young, F.R., Cavitation. 1999, London: Imperial College Press.

[7] Marschall, H.B., K.A. Mørch, A.P. Keller, and M. Kjeldsen, Cavitation inception by almost spherical solid particles in water. Physics of Fluids, 2003. **15**(2): p. 545.

[8] Phull, S.S., A.P. Newman, J.P. Lorimer, B. Pollet, and T.J. Mason, The development and evaluation of ultrasound in the biocidal treatment of water. Ultrasonics Sonochemistry, 1997. **4**(2): p. 157-164.

[9] Blume, T. and U. Neis, Improving chlorine disinfection of wastewater by ultrasound application. Water Science and Technology, 2005. **52(10-11)**: p. 139-44.

[10] Joyce, E., T.J. Mason, S.S. Phull, and J.P. Lorimer, The development and evaluation of electrolysis in conjunction with power ultrasound for the disinfection of bacterial suspensions. Ultrasonics Sonochemistry, 2003. **10**(4-5): p. 231-234.

[11] Blume, T. and U. Neis, Improved wastewater disinfection by ultrasonic pre-treatment. Ultrasonics Sonochemistry, 2004. **11**(5): p. 333-336.

[12] Lee, H., B. Zhou, W. Liang, H. Feng, and S.E. Martin, Inactivation of Escherichia coli cells with sonication, manosonication, thermosonication, and manothermosonication: Microbial responses and kinetics modeling. Journal of Food Engineering, 2009. **93**(3): p. 354-364.

[13] Ayyildiz, O., S. Sanik, and B. Ileri, Effect of ultrasonic pretreatment on chlorine dioxide disinfection efficiency. Ultrasonics Sonochemistry, 2011. **18**(2): p. 683-688.

[14] Hulsmans, A., K. Joris, N. Lambert, H. Rediers, P. Declerck, Y. Delaedt, F. Ollevier, and S. Liers, Evaluation of process parameters of ultrasonic treatment of bacterial suspensions in a pilot scale water disinfection system. Ultrasonics Sonochemistry, 2010. **17**(6): p. 1004-1009.

[15] Gogate, P.R., P.A. Tatake, P.M. Kanthale, and A.B. Pandit, Mapping of sonochemical reactors: Review, analysis, and experimental verification. Aiche Journal, 2002. **48**(7): p. 1542-1560.

[16] Szewzyk, U., R. Szewzyk, W.Manz, and K.H. Schleifer3, Microbiological safety of drinking water. Annu. Rev. Microbiol., 2000. **54**: p. 81–127.

[17] Mason, T.J., E. Joyce, S.S. Phull, and J.P. Lorimer, Potential uses of ultrasound in the biological decontamination of water. Ultrasonics Sonochemistry, 2003. **10**(6): p. 319-323.

[18] Condón, S., J. Raso, and R. Pagán, Microbial inactivation by ultrasound, in *Novel food processing technologies*, M.S. Tapia and M.P. Cano, Editors. 2005.

[19] Gogate, P.R. and A.B. Pandit, Engineering design method for cavitational reactors: I. Sonochemical reactors. Aiche Journal, 2000. **46**(2): p. 372-379.

[20] Tamer, I.M., M. Moo-Young, and Y. Chisti, Disruption of Alcaligenes latus for Recovery of Poly(β-hydroxybutyric acid): Comparison of High-Pressure Homogenization, Bead Milling, and Chemically Induced Lysis. Industrial & Engineering Chemistry Research, 1998. **37**(5): p. 1807-1814.

[21] Ince, N.H. and R. Belen, Aqueous Phase Disinfection with Power Ultrasound: Process Kinetics and Effect of Solid Catalysts. Environmental Science & Technology, 2001. **35**(9): p. 1885-1888.

[22] Dadjour, M.F., C. Ogino, S. Matsumura, and N. Shimizu, Kinetics of disinfection of Escherichia coli by catalytic ultrasonic irradiation with TiO2. Biochemical Engineering Journal, 2005. **25**(3): p. 243-248.

[23] Bremond, N. and et al., Cavitation on surfaces. Journal of Physics: Condensed Matter, 2005. **17**(45): p. S3603.

[24] Broekman, S., O. Pohlmann, E.S. Beardwood, and E.C. de Meulenaer, Ultrasonic treatment for microbiological control of water systems. Ultrasonics Sonochemistry, 2010. **17**(6): p. 1041-1048.

[25] Mahulkar, A.V., P.S. Bapat, A.B. Pandit, and F.M. Lewis, Steam bubble cavitation. Aiche Journal, 2008. **54**(7): p. 1711-1724.

IOP Conf. Series: Materials Science and Engineering **36** (2012) 012038 doi:10.1088/1757-899X/36/1/012038

Coefficient of restitution of sports balls: A normal drop test

Adli Haron* and K A Ismail
School of Manufacturing Engineering, Universiti Malaysia Perlis, 02600 Arau, Perlis, Malaysia.

*E-mail: mami1702@yahoo.com

Abstract. Dynamic behaviour of bodies during impact is investigated through impact experiment, the simplest being a normal drop test. Normally, a drop test impact experiment involves measurement of kinematic data; this includes measurement of incident and rebound velocity in order to calculate a coefficient of restitution (COR). A high speed video camera is employed for measuring the kinematic data where speed is calculated from displacement of the bodies. Alternatively, sensors can be employed to measure speeds, especially for a normal impact where there is no spin of the bodies. This paper compares experimental coefficients of restitution (COR) for various sports balls, namely golf, table tennis, hockey and cricket. The energy loss in term of measured COR and effects of target plate are discussed in relation to the material and construction of these sports balls.

Keyword: coefficient of restitution, sports balls, normal drop test, normal impact, energy loss

1. Introduction

Predicting the dynamics behaviours of sports balls on court surfaces during ball impacts can be extremely complicated. Dynamic behaviour of sports balls with interacted surfaces during ball impacts plays a major role in many popular sports, particularly, tennis, golf, and cricket and can have great effect on the style of play adopted by players, such as the changes seen in the game of tennis when moving from hard to clay surface, normally classified as fast to slow court respectively. In such sports as mentioned, ball –surface interaction can also affect the outcome of a match as well.

Recently, the dynamics behaviours of spherical sports balls during impacts have been investigated from theoretical, numerical and experimental viewpoints incorporated with some approximations [1-12]. According to these approaches of studies within the specific purpose noted, theoretical model or similarly mentioned as mathematical model used to be the main component of the analyses attempted. By considering to engineering practices and disciplines respectively, theoretical model is initiated the studies while either numerical or experimental studies as well are necessarily important to prove or verify the theoretical analyses. Thus, it is crucially important to determine the dynamics behaviours of sports balls experimentally for better understanding on impact mechanism.

This paper presents experimental measurements of COR for various sports balls, namely, golf, table tennis, hockey and cricket. The theoretical calculation for COR is presented next. The experiments and measurements using a high speed camera are explained thoroughly. Measurements of position are

Published under licence by IOP Publishing Ltd

calibrated and validated to ensure accuracy of the results. Variations of COR between various sports balls at different targets are discussed in details.

2. Coefficient of restitution (COR)

According to most of dynamics literature, coefficient of restitution specifically for one dimensional collision is defined to be the ratio of the velocity of separation (post impact velocity) to the velocity of approach (pre impact velocity). Coefficient of restitution is a common and popular parameter used in many collision models particularly for collision without friction because of its simplicity and usually applied as constant values between zeros to one [12-14]. It is usually denoted by e and also recognized as a referred parameter for energy loss due to the motion in normal direction. To comply with work-energy principle, condition $0 < e < 1$ is implemented with the end conditions corresponding to perfectly plastic ($e = 0$) and perfectly elastic impact ($e = 1$) [12, 13]. Theoretically in the latter case, no energy loss is occurred and vice versa. Furthermore, plastic deformation usually occurs in most collisions which contributed to energy dissipation, as well. Consequently, COR is neither zero nor one in actual cases [12, 13]. The coefficient of restitution depends on many elements, such as the geometry of the bodies in contact, the approach velocity, the material properties, the duration of contact and, possibly, friction [13].

In Newton's model the coefficient of restitution is defined as the ratio of final to initial velocity. This model is based on a kinematic point of view and only the initial and final values for the relative normal velocity are taken into account. Meaningly, Newton presumed that the coefficient of restitution is a material property [10-15]. In Stronge's model; based on the internal energy dissipation hypothesis, the coefficient of restitution is defined as the square root of the ratio of energy released during restitution to the energy absorbed during compression [13].

In terms of the validation of the basic theories of restitution, experimental verification is applied. As discussed by Gilardi & Sharf (2002), the final (post-impact) velocity was calculated by using initial impact velocity and the measured displacements and purposely used to compute the coefficient of restitution for a central direct impact of two spheres. The qualitative relation for both parameters is shown in figure 1, shows that the coefficient of restitution tends to level out with increase in impact velocity. Moreover, the experimental data were used to find the dependencies between different quantities, such as coefficient of restitution and the initial impact velocity, as well as to check the limits of the known impact theories. The results obtained for the coefficient of restitution clearly demonstrated the dependence of this parameter on the geometry and material of the impacting bodies, as well as the initial impact geometry and velocity. More specifically, the coefficient of restitution decreases with the increase of the initial impact velocity, and for most materials, it is significantly smaller than unity, even at very low impact speeds.

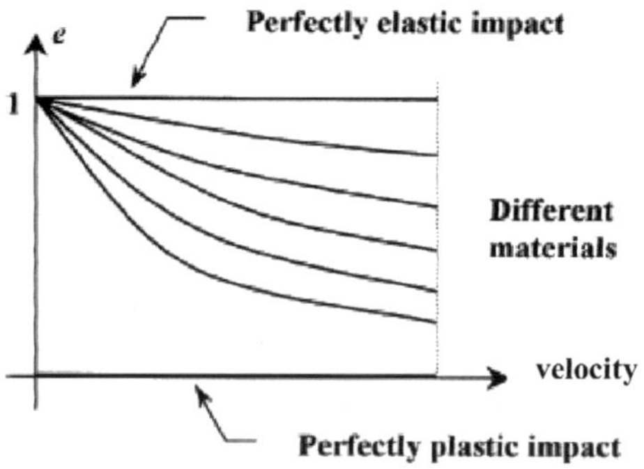

Figure 1. Coefficient of restitution as a function of approach velocity [13].

Instead of the theoretical overview regarding COR discussed above, simplification due to some approximations in calculation of COR seems to be more practically used in wide cases. For instance, the simplest experimental study on the effect of COR for impact between a spherical ball with planar surface is using the method of normal drop test [10-12]. In this case, we consider that a ball drops over a surface from an initial or inbound height of h_1 and it rebounds to a height of h_2. Moreover, as referred to previous literatures in which collision between two bodies have been investigated; air drag can be neglected during drop of bodies from small height [10-12].

According to the impact condition given, the COR is defined as:

$$e = \frac{v_2}{v_1} \tag{1}$$

where v_1 and v_2 are the inbound and rebound speed respectively [4, 6]. Thus, if the ball drops vertically with speed of v_1 impacting to a resting planar surface in normal impact configuration, it rebounds with speed of v_2. Using the energy principles, the inbound and rebound speed relate to the drop height and can be defined as follows;

$$v_1 = (2gh_1)^{1/2} \tag{2}$$

$$v_2 = (2gh_2)^{1/2} \tag{3}$$

While the change in potential energy, ΔE, is given as:

$$\Delta E = mg\,(h_2\text{-}h_1) \tag{4}$$

Using the Equations (1), (2) and (3), the COR can be related to height as:

$$e = \frac{v_2}{v_1} = \left(\frac{h_2}{h_1}\right)^{\frac{1}{2}} \tag{5}$$

3. Normal Drop Test of Sports Balls

Basically, the experimental setup consists of a drop tower with a ball releasing system, a steel table with changeable target surface that can be inclined from 0° to 45°; as an optional feature for oblique impact experiment in the future.

Furthermore, the drop tower is equipped with adjustable height mechanism up to 3 m drop height and also equipped with ball releasing system. The vacuum nozzle is attached at the top of drop tower. The sports ball is held to the nozzle by vacuum created using vacuum pump. To release the ball, the negative pressure in the hose is removed by switching off the vacuum pump. The table is made of steel and designed to be equipped with a mechanical inclination system and structurally made for changeable target surface which is embedded on top of the table. In order to minimize the effect of vibration, the impact plate compartment's is necessarily to be lined with a layer of rubber and possibly to be clamped to reduce vibration. Steel and wood plates have been chosen to represent the hard and soft impact surfaces respectively.

A high speed camera complete with accessories and an image processing system is used to capture the impact. The Fastec Inline high speed camera is equipped with 1 kW video light and 17mm high speed lens, F0.95. It is a monochrome type high speed camera with speed between 60 to 1000 frames per second(fps). Moreover, necessarily for images analyses is an image processing system. It consists of Fastec InLine Monitoring System (FIMS) software for video recording purpose, PAC Player software for playback and editing a frame subset of a video event and MIDAS Player software to convert the video to images files. At the final stage, selected images were analysed using Adobe Illustrator software to obtain the data on kinematics measurement. A schematic set-up of all the equipment used in the experiment is illustrated in figure 2 and 3.

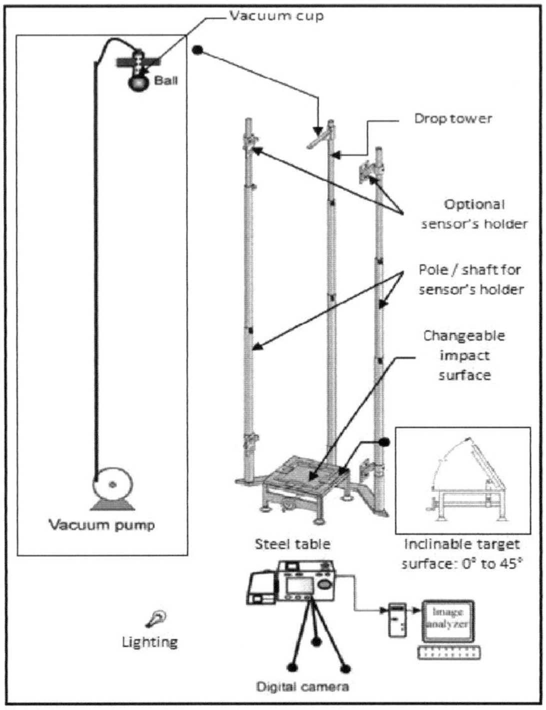

Figure 2. Schematic diagram of experimental set-up(not to scale)

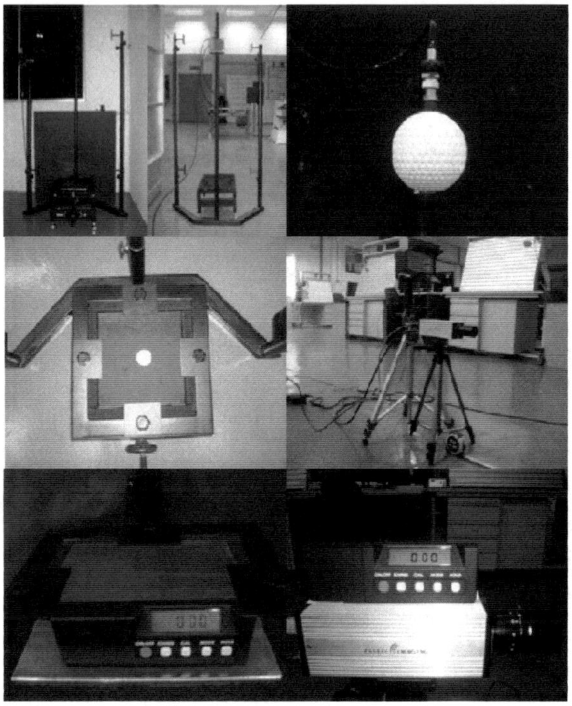

Figure 3. Apparatus used in the drop test

3.1 Technical set-up and procedures

The normal drop test of sports balls were done using test apparatus as shown in figure 2 and 3 above. Several of sports balls, namely, golf, table tennis, hockey and cricket were used in this study. The properties of the sports balls are given in table 1. The initial heights of drop for each ball were set from 1.3 m to 1.7 m with increment of 0.1 m between each drop, specifically. The hard and soft impact target of steel and wood plates with 260mm (L) x 250mm (W) x 25mm (H) in dimensions were used interchangeably during the experiments. The impact target surfaces and high speed camera used are measured to be horizontally, 0°, as required to be parallel in plane during capturing the images by using the measurement tool which is a digital inclinometer. The sports balls are dropped vertically without rotation. Purposely, the impact configuration is actually collinear and its impact property due to impact occurrence is described by the only parameter which is coefficient of restitution.

Table 1. Properties of sports balls

Type of sports ball	Mass (g)	Diameter (mm)
Golf Ball	46	42.80
Table Tennis Ball	2	39.25
Hockey Ball	162	71.00
Cricket Ball	170	71.15

Additionally, the Fastec Inline high speed camera is set to 250fps, 640 x 480 pixels; trigger position is set to END and with the duration of recording is up 17.5 seconds. Furthermore, the interfacing of this high speed camera is by Fastec InLine Monitoring System (FIMS) software with connection to computer using common computer interface cable. Network connection with the host computer should be set-up in order to operate the camera and Fastec InLine Monitoring System (FIMS) software. Briefly, InLine cameras must be connected to a host computer to operate. Two application software's are used with each InLine camera which is Fastec InLine Monitoring System (FIMS) for camera control and PAC Player for image saving and playback. FIMS will playback image clips from the on-board camera memory and download RAW image files to the host computer. Images downloaded and saved in RAW format can be played, cut and clipped and converted to the AVI file format with PAC Player. This was continued by image processing system which is involved the process of converting video file; AVI file to image file; bitmap file or jpeg file, by using MIDAS player software. At the final stage, selected images were analyzed using Adobe Illustrator software to obtain the data on kinematics measurement which is displacement of the bodies.

3.2 Experimental measurements

Regarding the experimental set-up and procedures discussed above, the measurements of displacements for each sports ball noted were calculated thoroughly. For instance, figure 4 shows the method for measuring displacement; rebound height, h_2, of a hockey ball dropped from initial height, h_1, of 1.3 m is calculated using Adobe Illustrator software. To ensure reliability and minimal errors of data, each ball tested was dropped 3 times for all set-up planned. Consequently, the displacements data's were tabulated and calculated results of v_1, v_2 and COR were incorporated in the same table; table 2, in accordance with the equations discussed earlier. The speeds, v_1 and v_2, are obtained from curve fitting on the displacement data for 5 frames each, before and after the impact respectively.

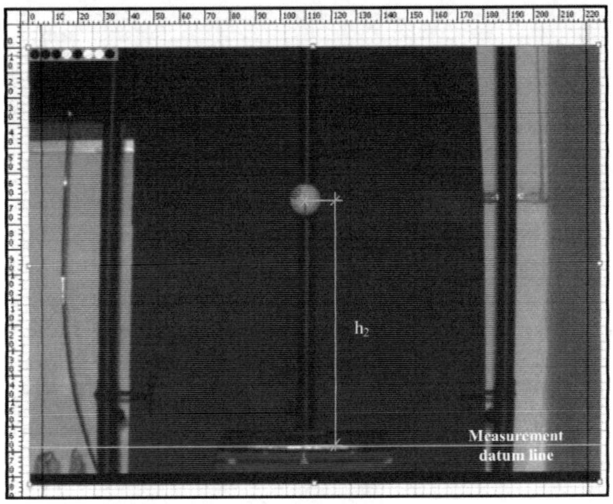

Figure 4. Captured image after the impact. The image is calibrated against known dimensions to obtain displacement of the ball

Table 2. Example of measured displacement and calculated COR for drop from 1.3 m

No. of drop	h_1 (m)	h_2 (m)	COR	v_1 (m/s)	v_2 (m/s)	COR
1	1.30	0.51	0.627	5.050	3.168	0.627
2	1.30	0.57	0.665	5.050	3.358	0.665
3	1.30	0.55	0.648	5.050	3.270	0.648
		Average	**0.647**			**0.647**

4. Results and discussions

Drop test were conducted to study variations of COR between different sports balls and the effect of target surfaces on the COR values. Table 3 and 4 below show COR of the sports balls at different target surfaces and drop heights.

Table 3. COR of sports balls for impact with **steel** target surface

Type of sports ball	Drop height, h_1 (m)				
	1.3	1.4	1.5	1.6	1.7
Golf Ball	0.897	0.896	0.894	0.892	0.889
Table Tennis Ball	0.809	0.800	0.791	0.783	0.779
Hockey Ball	0.647	0.658	0.653	0.650	0.628
Cricket Ball	0.615	0.610	0.609	0.606	0.606

Table 4. COR of sports balls for impact with **wood** target surface

Type of sports ball	Drop height, h_1 (m)				
	1.3	1.4	1.5	1.6	1.7
Table Tennis Ball	0.786	0.769	0.746	0.748	0.743
Golf Ball	0.681	0.677	0.676	0.675	0.666
Hockey Ball	0.398	0.391	0.381	0.380	0.378
Cricket Ball	0.356	0.351	0.351	0.343	0.350

In addition, COR in both tables are plotted on several graphs in order further analyse the COR regarding the energy loss and effects of target plates in relation to the material and construction of those sports balls. Figures 5 and 6 show COR vs. drop height for steel and wood target surfaces respectively.

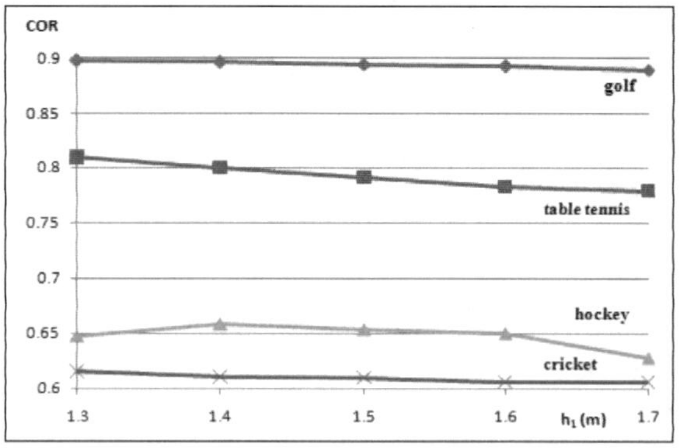

Figure 5. COR vs. drop height of sports balls with steel as impact target surface

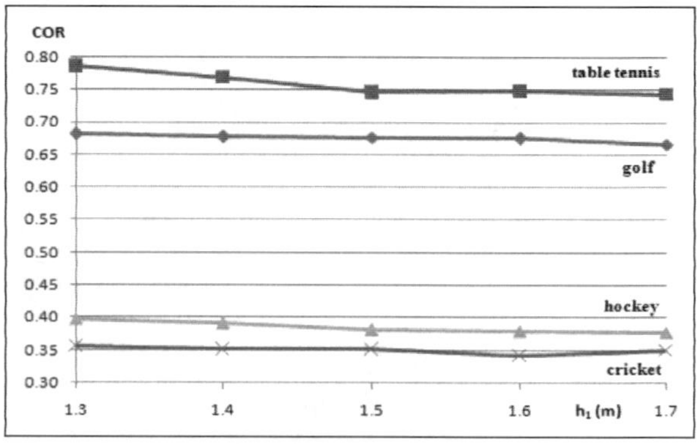

Figure 6. COR vs. drop height of sports balls with wood as impact target surface

The COR values reduce as the drop heights increase i.e. as the velocity increase. For the steel target surface, golf ball has the highest COR value at every drop height (Figure 5). This follows by table tennis, hockey and cricket balls. Golf ball is normally made from different layer of materials. Elastomer is one of its major constituents; thus it is bound to be elastic which can be seen by having the highest COR compared to the other sports balls. On the other hand, cricket ball is normally made from a core of cork, and is layered with tightly wound string and covered by leather. These materials are known to be dissipiative in nature and thus resulting the lowest COR values in all experiments.

However, when the target surface is changed to a softer target, that is wood, table tennis now has the highest COR (Figure 6). The COR values of every balls drop significantly when the target was changed from a hard to a soft surface as shown in figure 7. In details, COR of a table tennis ball drops by 4.3% in average when the target surface is changed. This follows by the golf ball with 24.46%, hockey ball by 40.41% and cricket ball by 42.51%. It seems that the percentage drop in COR is governed by the mass of the ball. Table tennis is the lightest and cricket is the heaviest among the sports balls tested; their percentage drop in COR is the smallest and largest respectively. The measured COR is also related to energy losses during the impact. When the target is changed to the soft surface, more energy is loss during the impact since the target is more deformable compared to the hard surface.

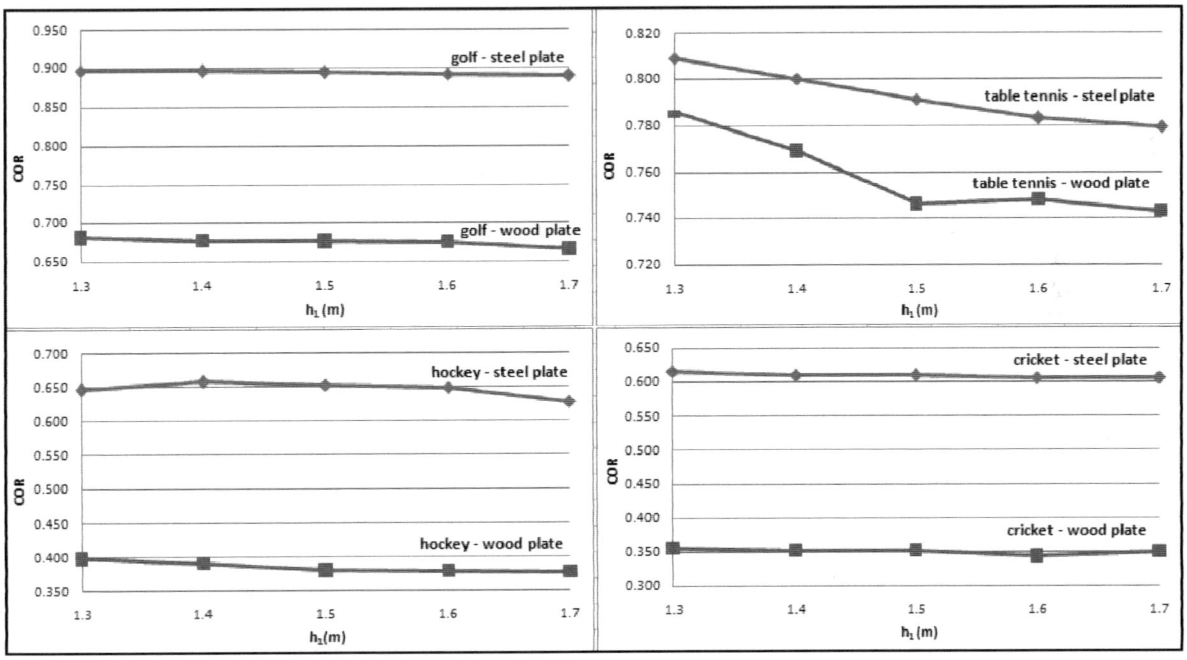

Figure 7. Comparison of COR for sports balls for impact with hard (steel) and soft (wood) impact target surfaces

5. Conclusion

Normal drop test were conducted to study variations of COR between different sports balls. Moreover, comparison on the different impact surfaces, COR for various sports balls, namely golf, table tennis, hockey and cricket due to hard and soft target surfaces were carried out from the results obtained. The energy loss in term of measured COR and effects of target plate are discussed in relation to the material and construction of these sports balls. The main conclusions from the experimental study are summarised below.

- The COR values reduce as the drop height is increased, i.e. as the impact speed increases.
- The COR values varies between different sports balls due to the material and construction of these sports balls. The measured COR is related to energy losses during the impact.
- As the impact surface changes from a hard to a soft target, the energy losses during impact increases (indicated by decreasing COR)

REFERENCES

[1] Arakawa K, Mada T, Komatsu H, Shimizu T, Satou M, Takehara K and Etoh G 2009 Dynamic Deformation Behavior of a Golf Ball during Normal Impact *Experimental Mechanics* **49** 471-7
[2] Arakawa K, Mada T, Komatsu H, Shimizu T, Satou M, Takehara K and Etoh G 2006 Dynamic Contact Behavior of a Golf Ball during an Oblique Impact *Experimental Mechanics* **46** 691- 7
[3] Arakawa K, Mada T, Komatsu H, Shimizu T, Satou M, Takehara K and Etoh G 2007 Dynamic Contact Behavior of a Golf Ball during Oblique Impact: Effect of Friction between the Ball and Target *Experimental Mechanics* **47** 277- 82
[4] Cross R 2003 Measurements of the horizontal and vertical speeds of tennis courts *Sports Engineering* **6** 95-111.
[5] Dong H and Moys M H 2006 Experimental study of oblique impacts with initial spin *Powder Technology* **161** 22-31
[6] Estivalet M, Brisson P, Cornish J, Otto S and Strangwood M 2008 Modelling The Oblique Impact Of Golf Balls *The Engineering of Sport 7*: Springer Paris pp. 669-75
[7] Carré, M, James D and Haake S 2004 Impact on a non-homogeneous sphere on a rigid surface *Proceedings of the Institution of Mechanical Engineers, Part C: Journal of Mechanical Engineering Science* **218** 273- 81.
[8] Carré, M, James D and Haake S 2006 Hybrid Method for Assessing the Performance of Sports Surfaces During Ball Impacts *Proceedings of the Institution of Mechanical Engineers, Part L: Journal of Materials: Design and Applications* **220** 31-9
[9] Goodwill S and Haake S 2001 Spring damper model of an impact between a tennis ball and racket *Proceedings of the Institution of Mechanical Engineers, Part C: Journal of Mechanical Engineering Science* **215** 133 - 41
[10] Cross R 1999 The bounce of a ball *American Journal of Physics* **67** 222-7
[11] Dong H and Moys M H 2003 Measurement of impact behaviour between balls and walls in grinding mills *Minerals Engineering* **16** 543-550
[12] Aryaei A, Hashemnia K and Jafarpur K 2010 Experimental and numerical study of ball size effect on restitution coefficient in low velocity impacts *International Journal of Impact Engineering* **37** 1037-44
[13] Gilardi G and Sharf I 2002 Literature survey of contact dynamics modelling *Mechanism and Machine Theory* **37** 1213-39
[15] Chatterjee A 1997 Rigid body collisions: some general considerations, new collision laws, and some experimental data: Cornell University, January
[15] Stronge W J 2004 *Impact Mechanics* (Cambridge: Cambridge University Press) pp 2-13

Efficient Finite Element and Differential Quadrature Methods for Heat Distribution in One-Dimensional Insulated-Tip Rectangular Fin

M. M. Fakir[1], S. Basri[2], M. M. Rahman[1] and Rosli A. Bakar[1]
[1]Faculty of Mechanical Engineering
Universiti Malaysia Pahang
26600 Pekan, Pahang, Malaysia
Phone: +6094242337; Fax: +6094242202
[2]Faculty of manufacturing Engineering
Universiti Malaysia Pahang
26600 Pekan, Pahang, Malaysia

E-mail: moslem@ump.edu.my

Abstract. There are many numerical solution techniques acquainted by the computational mechanics community including the finite element method (FEM) and differential quadrature method (DQM). Usually elements are sub-divided uniformly in FEM (conventional FEM, CFEM) to obtain temperature distribution behavior in a fin. In CFEM extra computational complexity is needed to obtain a better solution with required accuracy. In this paper, non-uniform sub-elements techniques are considered for the FEM (efficient FEM, EFEM) solution to reduce the computational complexity. Then EFEM is applied for the solution of the one-dimensional heat transfer problem in an insulated-tip thin rectangular fin. The results are compared with CFEM and efficient DQM (EDQM, with non-uniform mesh generation). It is observed that EFEM exhibits more accurate results compared to CFEM and EDQM. The proposed techniques are showing the potentiality of the heat transfer related problem.

***Keywords*:** Heat transfer, Finite element method, Differential quadrature method, Rectangular fin.

1. Introduction

There are many numerical solution techniques well familiar by the computational mechanics community. FEM is one of those numerical solution techniques to solve structural, mechanical, heat transfer, and fluid dynamics which arise in problems of engineering and physical sciences [1-4]. The conventional FEM (CFEM) means the elements that we use are of same size and uniformly distributed. In its application to the solution of engineering problems, the finite element discretization has been implemented almost to the spatial problems. For dynamic or time dependent problems whose solutions as functions of time are of interest, a step by step procedure of finite difference is usually employed with computational complexity.

For heat transfer problems, rapid changes of heat/temperature distributions take place near the element boundary (and at the boundary). It is very important to know these temperature change behavior of an element prior to its use. Hence, to get an actual picture using FEM, the element is usually subdivided into very small sub-elements uniformly (conventional FEM, CFEM), which leads

Published under licence by IOP Publishing Ltd

to huge amount of complexity, memory consumption and computational time (Park, 1996). Otherwise, error flow occurs with unreliable results [1-2,5].

On the other hand, to get a clear picture about the temperature changes near (and at) the element boundary, better to subdivide the elements into very small sub-elements at the boundary only, followed by relatively bigger elements gradually towards the mid-point of the element non-uniformly (efficient FEM, EFEM). This may serve the intended purpose without any additional burden and this is highlighted in this paper with improved accuracy (approximately 65%) compared to CFEM. Hence, here, focus is given to develop and apply efficient (non-uniform mesh density) nodal points distribution algorithm for automatic mesh (elements) generation to optimize CFEM solution.

DQM is another numerical solution technique to solve above mentioned problems efficiently [6-12]. The essence of the DQM is that the partial derivative of a function is approximated by a weighted linear sum of the function values at given discrete points. Bellman and Casti [6-7] developed this numerical solution technique in the early 1970s and since then, the technique has been successfully employed in a variety of problems in engineering and physical sciences. To make the DQM more efficient with less computational complexity, efficient DQM (EDQM) was proposed in[10-12] with non-uniformly distributed mesh points.

Hence, in this paper, one-dimensional (1-D) heat conduction problems in a thin insulated-tip rectangular fin is solved using EFEM by means of the accurate discretization and solver (code) and then the results are compared with CFEM and EDQM to verify EFEM efficiency. The paper is organized as follows. Section II presents the governing equation with EFEM rules, followed by simulation set-up and assumptions, results and discussions, and finally conclusion of the paper.

2. One-Dimensional Efficient Finite Element Method

One dimensional (1-D) heat conduction equation is shown in Eq. (1) [2,13-17].

$$\frac{d}{dx}\left(k\frac{dT}{dx}\right)+Q=0 \tag{1}$$

with the boundary conditions $T|_{x=0}=T_0$ and $q|_{x=L}=h(T_L-T_\infty)$ as shown in figure 1. Here, the heat flux $q=-k\dfrac{dt}{dx}$.

Figure 1 shows the 1-D element discretization in the x-direction. The temperature T at various nodal points are the unknowns except at node 1, where, $T_1=T_0$ with initial temperature T_0. Within a typical element 'ei or e' the local node numbers are i and $i+1$ with coordinates x_i and x_{i+1} and element length, $l_{ei}=x_{i+1}-x_i$. For example, $e1$ whose local node numbers are 1 and 2 with coordinates x_1 and x_2, and element length $l_{e1}=x_2-x_1$ respectively.

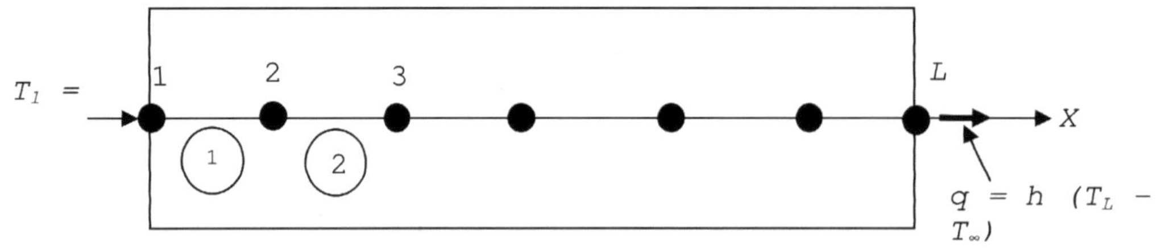

Figure 1. Boundary conditions for 1-D heat conduction

One-dimensional thin rectangular fin is shown in figure 2. Heat is transmitted along its length by conduction and the tip of the fin is insulated. The governing equation for the temperature in the fin is given in Eq. (1).

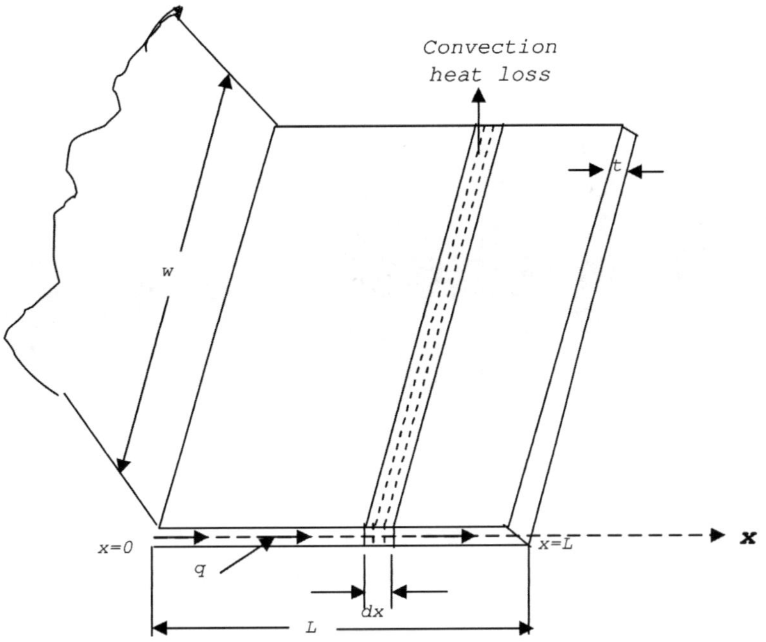

Figure 2. Thin rectangular fin

The parameter, M is given by $M^2 = \dfrac{hp}{kA_C}$,

where, p is the fin perimeter (m) and A_c is the cross sectional area of the fin [m^2]. Fin length, width and thickness are L, w and t respectively.

In this case, The heat flux is $q = h(T - T_\infty) = -k\dfrac{dT}{dx}$, perimeter, $p = 2(w+t)$, cross-section area, $A_C = w \times t$ and $\dfrac{p}{A_C} = \dfrac{2(w+t)}{w \times t} \approx \dfrac{2}{t}$.

The convection heat loss in the fin is equivalent to negative heat source and can be expressed as follows: $Q = -\dfrac{(p\,dx)h(T - T_\infty)}{A_C dx} = -\dfrac{ph}{A_C}(T - T_\infty)$.

After manipulating, Eq. (1) can be expressed in the form of Eq. (2).

$$\frac{d}{dx}\left(k\frac{dT}{dx}\right) - \frac{ph}{A_C}(T - T_\infty) = 0 \tag{2}$$

To calculate the approximate numerical solution *T(x)*, the mathematical formulation using Galerkin's approach [2] is written in Eq. (3).

$$\int_0^L \phi\left[\frac{d}{dx}\left(k\frac{dT}{dx}-\frac{ph}{A_C}(T-T_\infty)\right)\right]dx=0 \tag{3}$$

where ϕ is a test function constructed from the same basis functions as those of *T*, with $\phi(0)=0$. Integrating by parts Eq. (3) becomes,

$$\phi k\frac{dT}{dx}\bigg|_0^L-\int_0^L k\frac{d\phi}{dx}\frac{dT}{dx}dx-\int_0^L \phi\frac{ph}{A_C}(T-T_\infty)dx=0 \tag{4}$$

Since $\phi(0)=0$ and $q=h(T_L-T_\infty)$, Eq. (4) is expressed as in Eq. (5),

$$-\phi(L)h(T_L-T_\infty)-\int_0^L k\frac{d\phi}{dx}\frac{dT}{dx}dx-\int_0^L \phi\frac{ph}{A_C}(T-T_\infty)dx=0 \tag{5}$$

A global virtual temperature vector is defined as $\psi=[\psi_1,\psi_2,\psi_3,\ldots,\psi_L]^T$ then within each element, the test function becomes, $\phi(i)=N_i\psi_i$. Here, N is the element shape function and $N_L=1$ at the element boundary (figure 1). Therefore we can write as Eq. (6).

$$\phi(L)=[N\psi]_L=\psi_L \tag{6}$$

As, $\frac{dT}{dx}=\mathbf{B}_T\mathbf{T}^e$, from Eq. (6), $\frac{d\phi}{dx}=\mathbf{B}_T\psi$, then, $\left(\frac{d\phi}{dx}\right)^T\times\left(\frac{dT}{dx}\right)=\left(\mathbf{B}_T^T\boldsymbol{\psi}^T\right)\left(\mathbf{B}_\mathbf{T}\mathbf{T}^{ei}\right)$ and $\mathbf{B}_T^\mathbf{T}\mathbf{B}_T=\frac{1}{(l_{ei})^2}\begin{bmatrix}1 & -1\\ -1 & 1\end{bmatrix}$.

The element conductivity matrix is expressed in Eq. (7).

$$k_T=\frac{k_{ei}l_{ei}}{2}\int_{-1}^{1}\mathbf{B}_T^\mathbf{T}\mathbf{B}_T d\xi=\frac{k_{ei}}{l_{ei}}\begin{bmatrix}1 & -1\\ -1 & 1\end{bmatrix} \tag{7}$$

where, ξ varies from -1 to $+1$ and $\xi=\frac{2}{x_{i+1}-x_i}(x-x_i)-1$ *with* $d\xi=\frac{2}{x_{i+1}-x_i}dx$.

The element heat rate vector due to the heat source is written by Eq. (8).

$$\mathbf{R}=\mathbf{r}_Q=\frac{Q_{ei}l_{ei}}{2}\int_{-1}^{1}\mathbf{N}^\mathbf{T}d\xi=\frac{Q_{ei}l_{ei}}{2}\begin{bmatrix}1\\1\end{bmatrix} \tag{8}$$

With the help of Eqs. (3-8), Eq. (2) can be transformed into either Eq. (9) or Eq. (10)

$$-\psi_l h(T_L - T_\infty) - \sum_{ei} \boldsymbol{\psi}^{\mathrm{T}} \left(\frac{k_{ei}\, l_{ei}}{2} \int_{-1}^{1} \mathbf{B}_T^{\mathrm{T}} \mathbf{B}_T d\xi \right) \mathbf{T}^e + \sum_{ei} \boldsymbol{\psi}^{\mathrm{T}} \frac{Q_{ei}\, l_{ei}}{2} \int_{-1}^{1} \mathbf{N}^{\mathrm{T}} d\xi = 0 \tag{9}$$

Or,

$$\boldsymbol{\psi}^{\mathrm{T}} \mathbf{K}_T \mathbf{T} + \psi_L h T_L = \boldsymbol{\psi}^{\mathrm{T}} \mathbf{R} + \psi_L h T_\infty \tag{10}$$

For insulated-tip fin, the base of the fin is held at a constant temperature, T_0 and the tip of the fin is insulated, and the final global matrix shown in Eq. (10) can be written as Eq. (11).

$$\begin{pmatrix} A_{22} & A_{23} & \dots A_{2L} \\ A_{32} & A_{33} & \dots A_{3L} \\ \dots & \dots & \dots \; \dots \\ A_{L2} & A_{L3} & \dots A_{LL} \end{pmatrix} \begin{pmatrix} T_2 \\ T_3 \\ \dots \\ T_L \end{pmatrix} = \begin{pmatrix} R_{2\infty} \\ R_{3\infty} \\ \dots \\ R_{L\infty} \end{pmatrix} - \begin{pmatrix} A_{21}T_0 \\ A_{31}T_0 \\ \dots \\ A_{L1}T_0 \end{pmatrix} \tag{11}$$

Using Eq. (11) and the efficient FEM (EFEM) algorithm, the approximate solution $T(x)$ has been obtained. The 1-D EFEM algorithm (rule) is depicted in terms of self-explanatory flow chart in figure 3. Example of non-uniform and uniform sub-element distributions and their length calculations are depicted in figures 4 and 5 respectively.

Simulation Set-Up and Assumptions

Table 1 shows the parameters we consider and their corresponding values that are used to obtain simulation results using FORTRAN 90 software. We used these values to obtain the temperature distribution for EFEM, CFEM, EDQM and exact solution. We assumed $M^2 = hP/kA = 1$ and the associated assumptions (in Table I) to compare the obtained FEM results with DQM [11] and exact solution [17]. We need to remark that, to obtain 1-D DQM solutions, element material properties, fin-width and fin-thickness are not required (which is the shortcoming of the method). The errors in FEM and DQM solutions are computed and compared to the exact solutions.

Table 1. Input Parameters and Assumptions for 1-D Rectangular Fin

Input Parameters	**Assumed value for *Insulated-Tip* Fin**
Boundary and other values:	
Initial temperature (T0)	1 ^{o}C
Ambient temperature (T∞)	0 ^{o}C
Heat flux (q)	0 at x = 1
% Error threshold (eh)	0 - 0.1
Element Type (NNODE):	
Linear (for 1-D)	2
Element material properties:	
Thermal conductivity ($k_e = k$)	7.03125 $W/(m\,^0C)$
Convective heat transfer coefficient (h)	9 $W/m^2\,^0C$
Heat source (Q)	0 $W/m^3\,^0C$
Element (Fin) dimension:	
length (L) along x-axis	1 m
width (w)	Variable to make M = 1
thickness (t)	Variable to make M = 1
Number of elements (N)	11 - 104

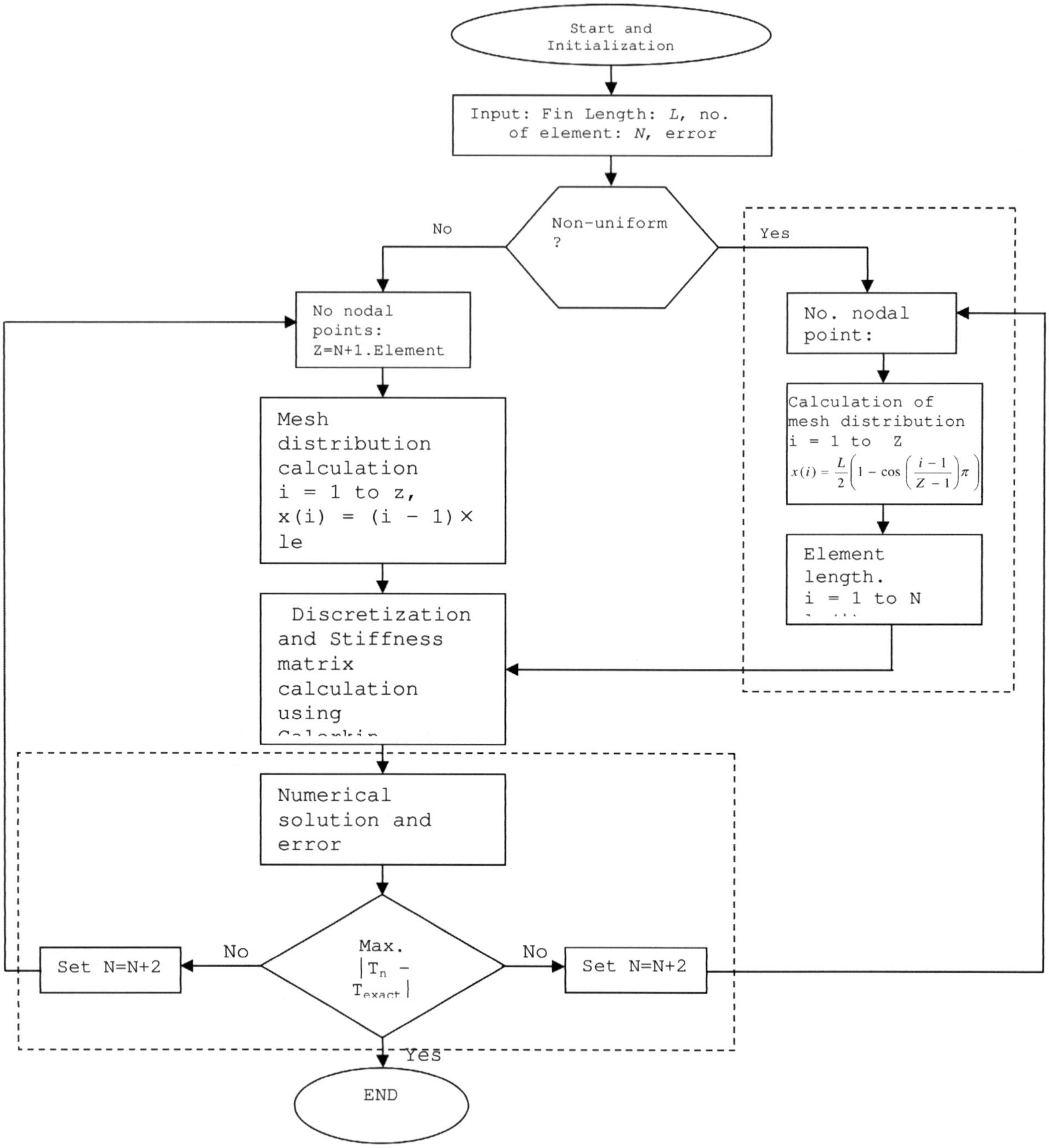

Figure 3. Efficient discretization and solution rule for 1-D FEM.

Figure 4. Example 1-D efficient (non-uniform) sub-element lengths distribution.

Figure 5. Example 1-D conventional (uniform) sub-element lengths distribution.

4. Results and Discussion

The convergence of the FEM solution for conventional and optimum mesh distribution have been compared and shown in figures 6 and 7 (magnified version). Both uniform FEM and non-uniform FEM solutions in terms of maximum % errors show a monotonic convergence with the increasing number of mesh points (shown up to $N = 104$). It is apparent that non-uniform FEM results show similar (or sometimes less) accuracy for $N \leq 11$, but yields result with higher accuracy, of one order of magnitude or more with increasing N (for $N \geq 30$) compared to that with uniform FEM. The results converge rapidly up to N = 101 and at this point the best results are obtained, whereas uniform FEM results converge slowly throughout the solution domain and then diverge for $N \geq 900$ (as tested) without showing the best results like non-uniform FEM. It happens due to the mesh point distribution strategy of equally spaced and unequally spaced nodal points in the computational domain and the inherited complexity to compute the stiffness matrix for equally spaced nodal points. This shows that for FEM, the solution for non-uniform mesh FEM is better than those of uniform mesh FEM.

Comparisons of convergence of fin-temperature in terms of maximum % error for FEM versus DQM solution are shown in figures 8 to 10. It is apparent that the uniform DQM solutions converge up to N = 41, oscillates within the range $31<N<47$, and then the solution starts deteriorating rapidly for $N \geq 47$ (shown up to N = 55, figures 8 and 9). On the contrary, non-uniform DQM converge up to N = 100 and then diverge gradually, whereas the FEM (all cases) solutions converge smoothly for all N within the solution domain (shown up to N = 104, figures 8 and 10), showing best result at N = 101. Hence, the efficiency of our (conventional and optimum FEM) results are apparent.

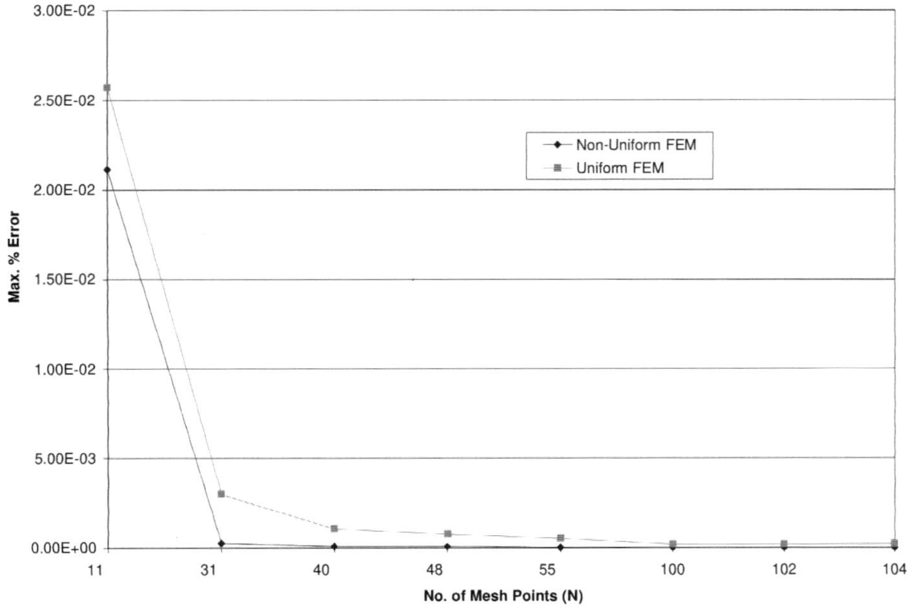

Figure 6. Comparison of convergence of fin-temperature in terms of maximum % error for conventional versus optimum FEM solution (N = 11-104)

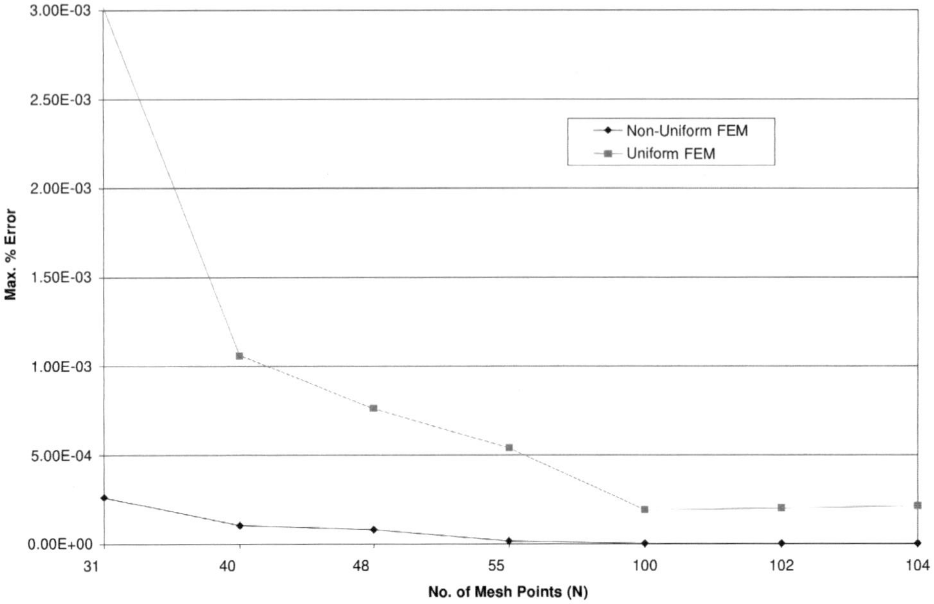

Figure 7. Comparison of convergence of fin-temperature in terms of maximum % error for conventional versus optimum FEM solution (N = 31-104)

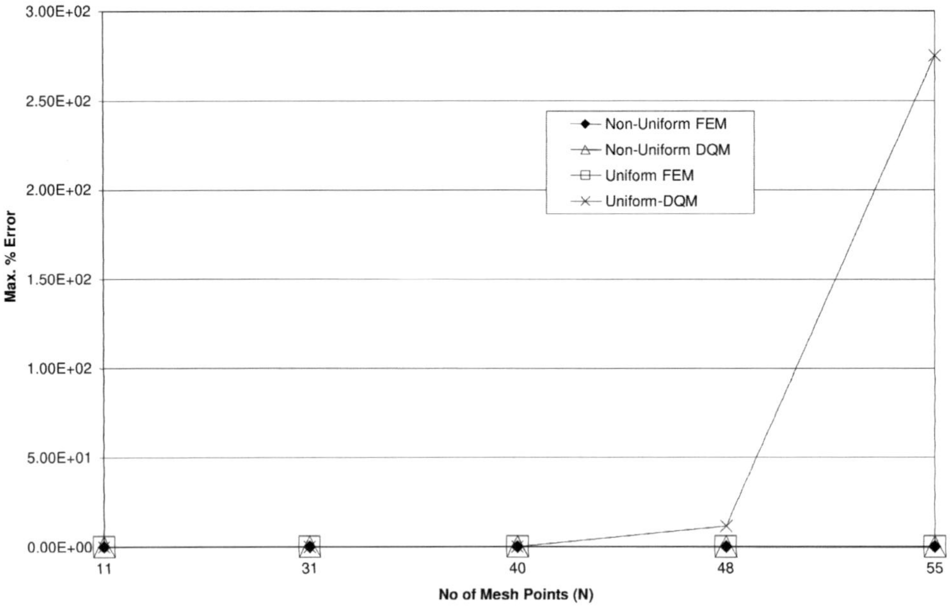

Figure 8. Comparison of convergence of fin-temperature in terms of maximum % error for all cases of FEM versus DQM solution (N = 11-55)

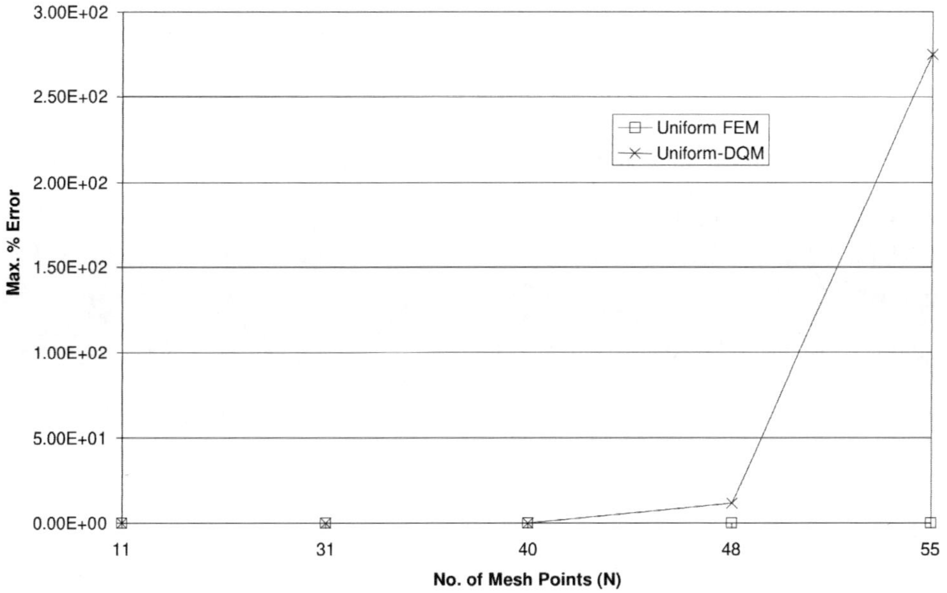

Figure 9. Comparison of convergence of fin-temperature in terms of maximum % error for conventional FEM versus DQM solution (N = 11-55)

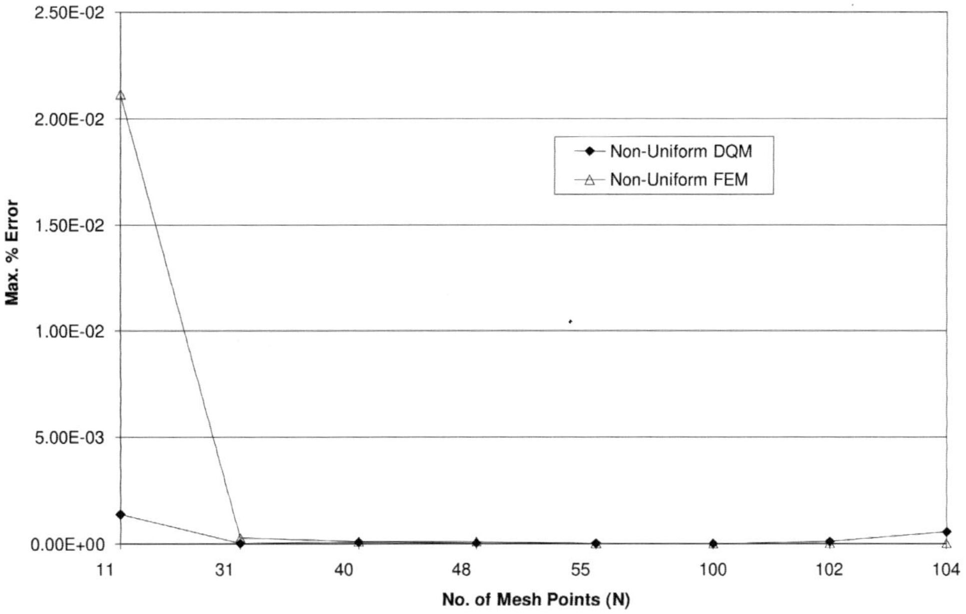

Figure 10. Comparison of convergence of fin-temperature in terms of maximum % error for optimum FEM versus DQM solution (N = 11-104).

5. Conclusions

The solutions of the temperature distribution in insulated-tip 1-D rectangular thin fin are computed numerically using FEM. The results are found to be in good agreement with the exact solution. It is found that the unequally spaced element distribution yield more accurate results than equally spaced for FEM solution. The solution converges smoothly as the number of elements reach to the optimum value. The results of EFEM shows outstanding improvement compared to CFEM and agree very well with EDQM with very minor variance showing its potentiality. Hence EFEM is suitable to test the temperature distribution scenario in any thin metal fin/plate prior to its design and practical implementation.

6. Acknowledgments

The authors would like to thank Universiti Malaysia Pahang and financial support under Universiti Research Grant (RDU110332).

7. References

[1] Strang G and Fix G J 1973 An Analysis of the Finite Element Method (New Jersey: Prentice-Hall, Inc).
[2] Tirupathi R C and Ashok D B 1997 Introduction to Finite Elements in Engineering (New Jersey: Prentice-Hall International, Inc).
[3] Li B 2004 *Numerical method for a parabolic stochastic partial differential equation* (Master's thesis, Chalmers University of Technology).
[4] Fairag F 2001 Technical Report of ArXiv: Math, NA/0109109.
[5] Park S 1996 Development and Applications of Finite Elements in Time Domain. (PhD Thesis, Virginia Polytechnic Institute and State University, USA).
[6] Bellman R E and Casti J 1971 J Math and Appl. **34** 235.
[7] Bellman R E, Kashef B G and Casti J 1972 J Comput Phys, **10** 40.
[8] Bert C W and Malik M 1996 Appl. Mech. Rev, **49** 1.
[9] Bert C W, Striz A G and Jang S K 1989 *J. Comput. Mech.* **5** 217.
[10] Shu C, Chen W, Xue H and Du H 2001 *Int. J. Numer. Methods Eng.* **51**(2) 159.
[11] Fakir M M, Mawlood M K, Asrar W, Basri S and Omar A A 2002 J. IEM **63**(4) 41.
[12] Fakir M M, Mawlood MK, Asrar W, Basri S and Omar A A 2003 J. Mech. Malaysia **15** 20.
[13] Hinton E and Owen D R J 1985 *An introduction to finite element computations* (UK: Pineridge Press Limited).
[14] Tiwari S, Biswas G, Prasad P L N and Basu S 2003 ASME J. Heat Transfer **125**(3) 413.
[15] Wang B L and Tian Z H 2005 J. Fin. Elem. Anal. Des. **41** 335.
[16] Lo S H and Wang W X 2005 J. Fin. Elem. Anal. Des. 41: 351-370.
[17] Ozisik M N 1985 *Heat Transfer: A Basic Approach* (New York: McGraw-Hill).

Design and Performance of 20 Watts Portable Solar Generator

Z A Abdul Majid[1], N Hazali[1], M A K M Hanafiah[2], A A Abdullah[3], A F Ismail[4], M H Ruslan[5], K Sopian[5] and M S Mohd Azmi[5]

[1] Kulliyyah of Allied Health Sciences, International Islamic University of Malaysia, Bandar Indera Mahkota. 25200 Kuantan, Pahang, MALAYSIA
[2] Faculty of Applied Sciences, Universiti Teknologi MARA (UiTM) Pahang, 26400 Bandar Jengka,Pahang, MALAYSIA.
[3]Faculty of Mechanical Engineering, Universiti Malaysia Pahang, 26600 Pekan, Pahang, MALAYSIA
[4]Kulliyyah of Enginering. International Islamic University of Malaysia, Jalan Gombak, 53100 Kuala Lumpur, MALAYSIA
[5]Solar Energy Research Institute, Universiti Kebangsaan Malaysia, 43600 Bangi, Selangor, MALAYSIA

E-mail: amzafri@iium.edu.my

Abstract. A new portable solar generator has been developed to generate electricity. It has the potential to replace petrol generator, widely used by peddlers at night markets *(pasar malam)*. Conventional generators are heavy, oily, have high maintenance and use fossil fuel to generate electricity. The solar generator can generate 20 Watts of electricity. This amount of power can supply up to 96 hours of electricity for the purpose of lighting and running small electrical appliances. The power output is (alternating current) AC current using 150 Watts inverter with 200 Watts surge, suitable for all commercial single phase electric appliances. Solar charge controller is used to maximize the charging rate and to protect the battery. The system has low maintenance whereby the batteries need to be changed every three to four years, depending on the usage. The main concepts of portable solar generator are to reduce installation cost and to introduce a compact design of an optimal energy sizing system. The materials used to develop the solar generator can be easily obtained from local markets, thus reducing the cost of developing the system and making it suitable for commercialization.

Keywords: photovoltaic, solar charge controller, inverter, battery

1. Introduction

Energy has become important in life nowadays. The fluctuation of energy price can affect daily human activities which contribute to political and economic stability [1]. This is due to the decline in energy sources from fossil fuel [2]. Fossil fuel pollutes the environment and increases global temperature [3][4]. Solar energy however is not popular in Malaysia. About 99 % of energy consumed in 2008 was produced from fossil fuel and hydro power [5]. The main problem with the solar system is that the price is still high in comparison to conventional energy source [6].

Generators are widely used if electricity from the grid is not available. Generators are used by hawkers such as at night markets *(pasar malam)* in Malaysia. Business at night markets in Malaysia usually starts at 5 pm until 10 pm in the evening. The main problem with generator is it needs petrol or

Published under licence by IOP Publishing Ltd

diesel to generate electricity. Furthermore, the generator requires maintenance and periodical servicing. Besides releasing gaseous pollutants, generators are also noisy, oily, heavy and consume space during its operation. In Malaysia, the demand for solar technology increasing due to the government support toward green technology. The cost of installing the grid connected photovoltaic system has increased to 29% in Malaysia [7], and for the stand alone system the cost is much higher due to extra wiring and energy storage. The objective of this research is to develop a portable solar generator using 20 Watts photovoltaic panel by combining two units of 10 Watts solar panel via a compact design. Solar generator is categorized as a stand alone system in which only the photovoltaic panel is used to provide energy to the battery.

2. Procedure

Selection of component is important in this research as it would determine the size of the system. The portable solar generator system is illustrated in Figure 1. The photovoltaic panel power is 20 Watts. Two sets of photovoltaic panels are used; each panel contributes 10 Watts power and they can be flipped or opened during charging. The photovoltaic power can be increased by using a photovoltaic extension port. It is designed in such a way to leave more space for storage and also to make it portable. Holders are placed on both sides of the system. The total weight of the system is 14.3 kg and the design is cubical with a dimension of the container are 26.0 cm (L) × 31.0 cm (W) × 22.0 cm (H), which is more compact, compared to a 650 Watts petrol generator which size approximately 36.6 cm (L) × 30.8 cm (W) × 37.6 cm (H). Photovoltaic panel is used to convert solar radiation into electric power. The type of solar panel used was polycrystalline due to its low cost and suitable for Malaysia climate. The battery used is lead acid 12 V (40 Ah) for 4 days autonomy. The maintenance cost of the solar generator is low, where only the battery needs to be changed once every three to four years. The solar charge controller used in the study is 12 V (5 A) to charge the battery and to protect the inverter.

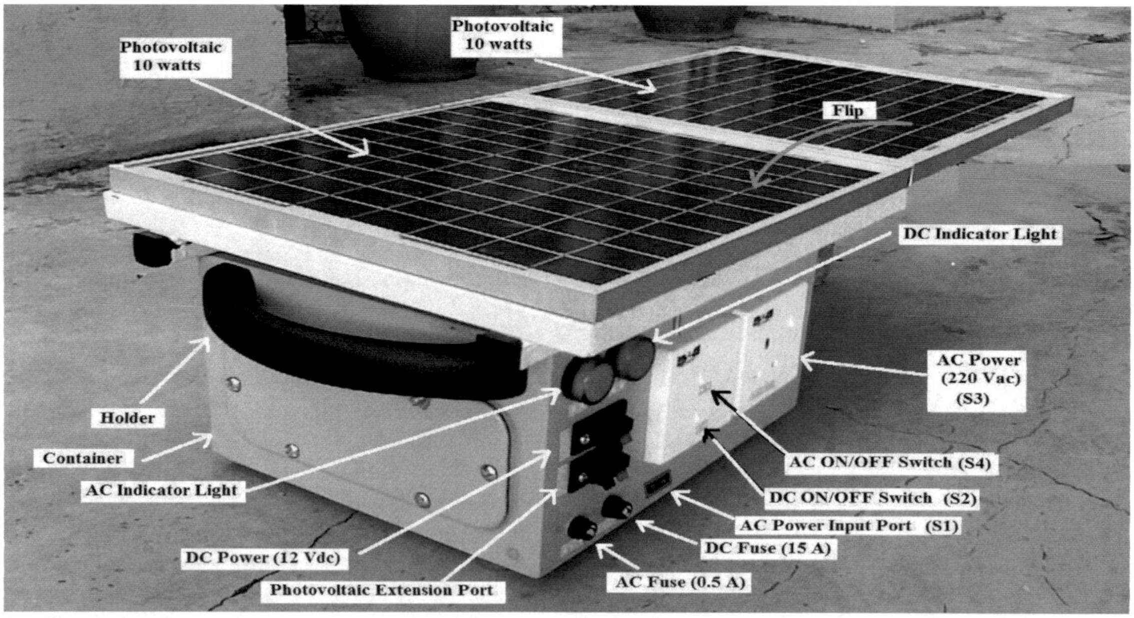

Figure 1. A 20 Watts portable solar generator

The schematic diagram of the system is shown in Figure 2. The power output of the system can be used to power the (direct current) DC and AC load simultaneously. The DC output is 12 Vdc and protected by a 15 A fuse. The AC output is a single phase 220 Vac modified sine wave using a 150 Watts peak powerinverter with a 200 Watts surge and is protected by a 0.5 A fuse. The main fuse current is 20A and is used to protect the battery and the inverter. If the solar radiation is too low because of bad weather, the system can be charged using electric power from grid by changing switch S1 at AC power input port. Switch S2 is used to switch the DC power and switch S3 is used to switch the AC power. The inverter needs to be disconnected using switch S4 from solar charge controller for the purpose of energy saving. In this research, the performances of the solar generator as well as the optimum operations were evaluated. The experiment was divided into two parts. The first experiment was conducted to obtain the maximum daily power output by the photovoltaic system while the second experiment was carried out to determine the system performance using variable loads. The load used was between 5 to 150 Watts. Temperature can affect the photovoltaic panel efficiency which can decrease the power gain [8]. The automatic fan was used to remove the heat generated by the photovoltaic system. The battery temperature was maintained at a temperaturebelow 35 °C during operation. Temperature higher than 35 °C could damage the battery or decrease the system's efficiency.

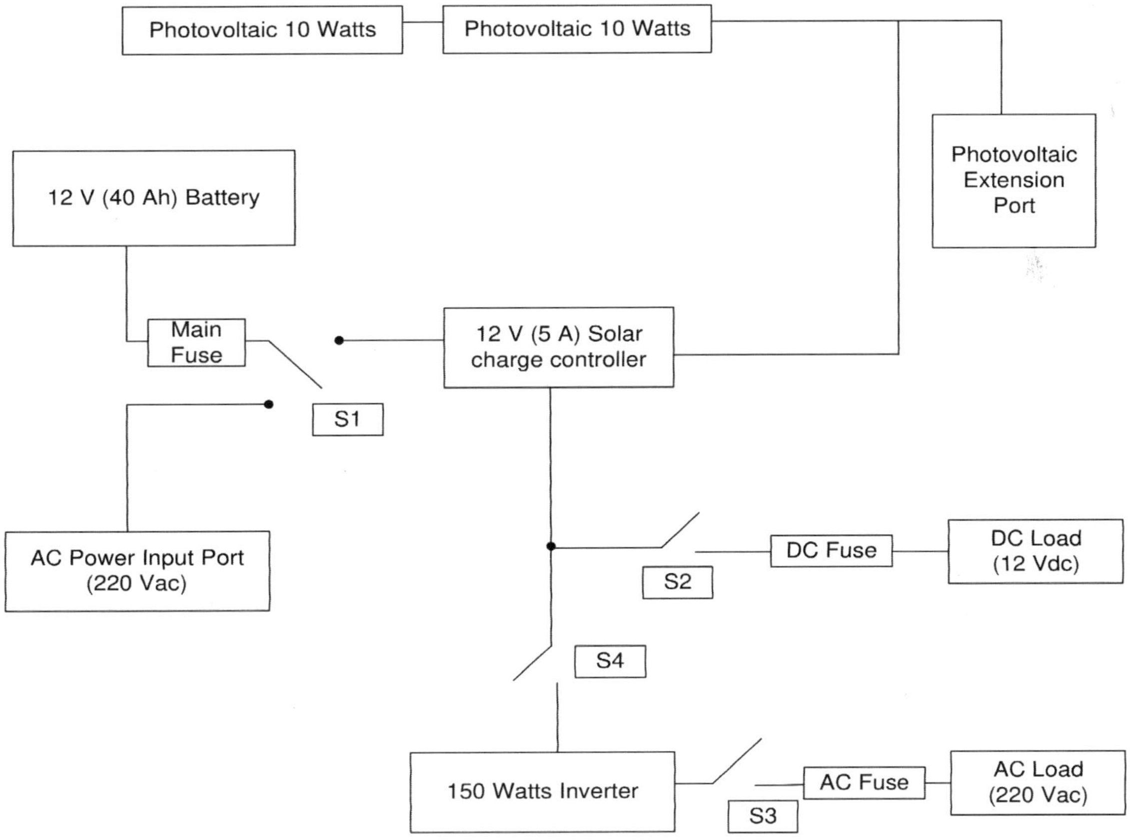

Figure 2. A 20 Watts portable solar generator schematic diagram

3. Analysis

Experimental results were obtained by using ADAM VIEW Programme. The maximum daily performance of the photovoltaic panel is shown in Figure 3. The maximum energy gain by 20 Watts photovoltaic panel was 153.1 Watts-hour. The maximum power of photovoltaic panel that could be added to the system is 60 Watts in total. This would contribute to the maximum daily energy gain of up to 459.3 Watts-hour. Since the daily average solar radiation in Malaysia is between 4 to 6 hours, the power gained by the system can be affected. Therefore, in order to get sufficient energy, the photovoltaic size should be increased. As the photovoltaic panels are almost flat, it issuggested that the collectors should be facing south in the northern hemisphere and facing north in the southern hemisphere [9]. This situation however will not affect the power gain in Malaysia because Malaysia is located near the equator. The total power gain throughout the year will not give any big difference [1].

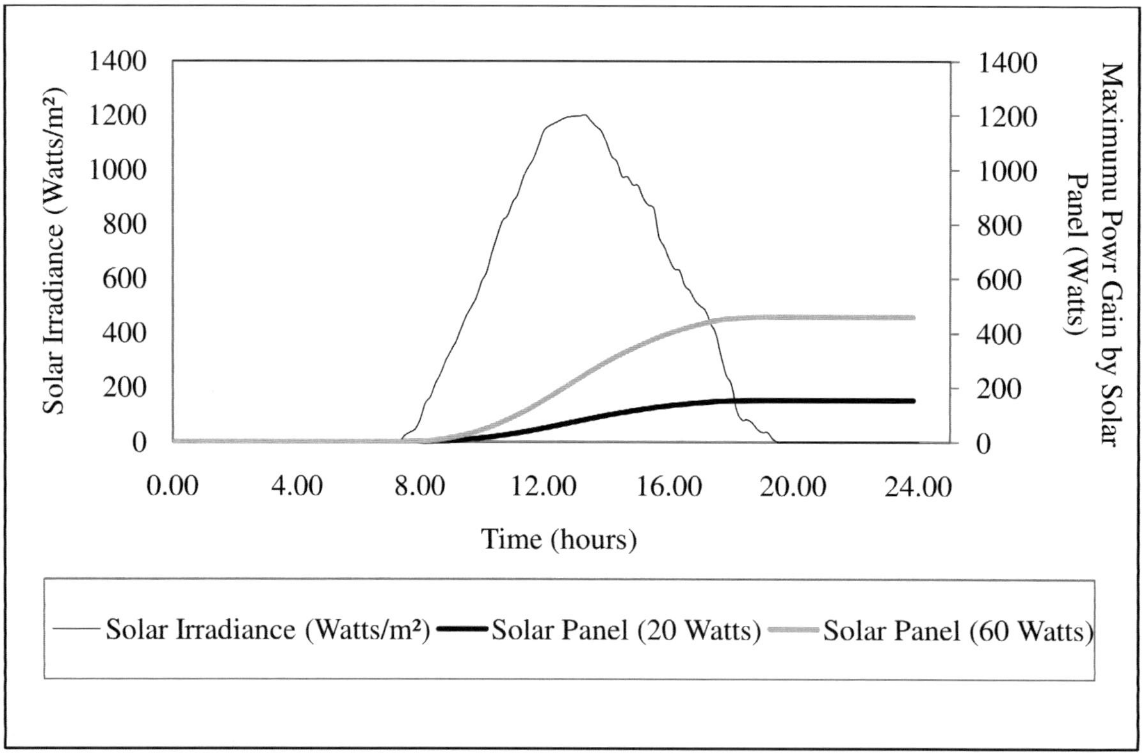

Figure 3. Maximum power gain by photovoltaic at maximum daily solar irradiance

Figure 4 shows lower load would increase the duration of the solar generator operation time. This can be explained by the gradient of the graph which increases by increasing the load. For 5 Watts load, the solar generator can be operated up to 96 hours, but reduces to 2 hours for load of 150 Watts. The energy generated by the battery cannot be completely discharged. It can reduce the battery life, and normally the battery should remain for at least 30 % of energy. The battery can be protected by using an inverter because the threshold voltage is set at 10.5 V, which will stop operating if the voltage of the battery drops lower than the set value. For DC power, the battery is protected by the solar charge controller which will stop operating if the voltage drops to 8 V. The solar generator is suitable for lighting at night market using a 40 Watts energy which can last for 3 to 4 hours.

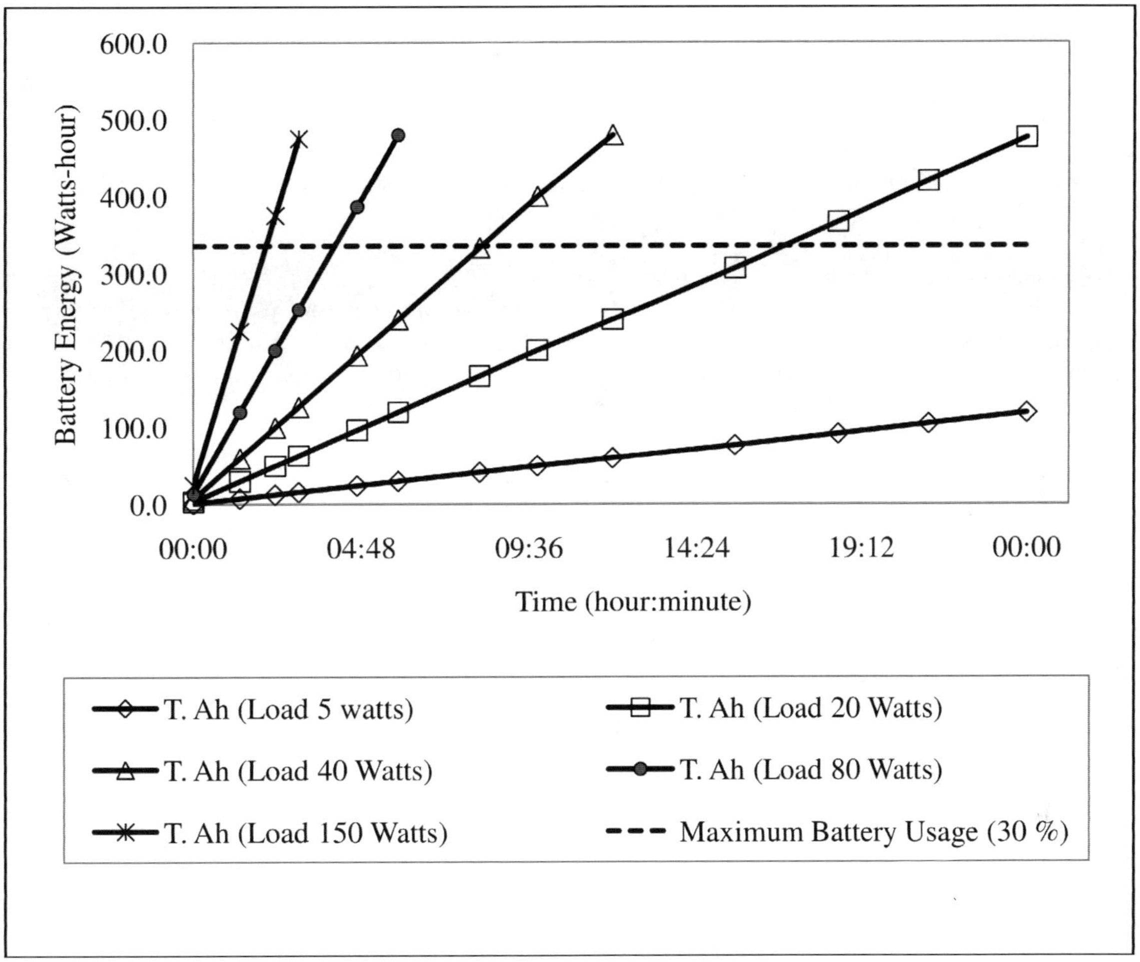

Figure 4: Load and time of operation of the portable solar generator

The main concept of solar generator is to reduce the installation cost. The installation can be done at a factory and the materials used to develop the system can be obtained from local market, thus reduces the cost of developing the system. This would make the solar generator very marketable. Besides night market, the product can be used at home or for outdoor activities such as camping or picnic. It also can be used by army, fire engine, police car and ambulance as backup power supply. However, the solar generator cannot compete with petrol generator in terms of energy output based on size. For instance, petrol generator can produce a maximum power of 650 Watts, which is the power commonly used for lighting and running fans at night market. The major drawback of petrol generators is the need for fuel for operation. A petrol generator consumes an average of 4.2 liter fuel for 8 hours operation. At night market, petrol generators are run for at least 3 hours and the cost of the petrol is RM 3/day or about RM 1100/year. As a comparison, the battery of portable solar generator costs around RM 340 and can be used up to four years. The payback value for portable solar generator is less than a year compared to petrol generator; therefore portable solar generator is more economical. Furthermore a portable solar generator can permanently store its energy without the need to warm up.

Solar generator is categorized as stand alone system and is more realistic compared to grid connected photovoltaic system in generating power from the sun and converting it into electricity. The energy comes from the photovoltaic system and stored into the battery, thus creates a greater awareness concept especially on the energy usage. Grid connected photovoltaic system however presents a concept of selling and buying energy, and at the same time being used to support the grid system especially during peak hours. This kind of practice does little to energy saving concept. Furthermore, the power generated by the grid system is always higher than the demand [10].

4. Conclusion

The portable solar generator has the potential to replace petrol generator in future especially for lighting and running small electrical appliances. The maintenance cost is low with only the battery needs to be changed once every three to four years. It is small, lighter and portable making it easy to store and suitable for use at home or night market and as backup energy supply.

5. Acknowledgment

The authors would like to thank International Islamic University of Malaysia for financial assistance in carrying this research project.

6. References

[1] Abdul Majid Z A 2011 *Sistem pengering pam haba terbantu suria dengan pengumpul suria multifungsi.* PhD Thesis, Universiti Kebangsaan Malaysia.

[2] Sopian K 2006 Bringing Renewables into the Mainstream *Proc. of the Malaysia Chemical Congress (MCC) Int. Conf. on Green Chemistry* (*Kuala Lumpur, 19-21 September 2006*) pp 234-241.

[3] Yatim B 1995 *Inovasi Menghadapi Cabaran Bekalan Tenaga* (Bangi: Penerbit Universiti Kebangsaan Malaysia) pp 31-32

[4] Sopian K 2008 *Tenaga Keterbaharuan – Laluan Masa Depan Bebas Karbon* (Bangi: Penerbit Universiti Kebangsaan Malaysia) pp 14-15

[5] Suruhanjaya Tenaga 2008 *Industri Pembekalan Elektrik di Malaysia – Maklumat Prestasi & Statistik 2008.* Kuala Lumpur Suruhanjaya Tenaga p 151

[6] Henden L, Rekstad J and Meir M 2002 *Thermal performance of combined solar systems with different collector efficiencies* Solar Energy 72(4) pp 299–305.

[7] Abdul Majid Z A, Ruslan M F, Othman M Y, and Sopian K 2008 Development of Floating Photovoltaic (FPV) Power Plant In Semenyih Dam Malaysia *Proc. 9th SENVER + 2nd ISESEE, (Shah Alam1–3 Disember 2008)* pp 513-517.

[8] Ruslan M F, Othman M Y, Sopian K, Abdul Majid Z A and Ai Him W 2008 Development of Photovoltaic-Thermal (PV/T) Water Based System (*Proc. of the World Renewable Energy Congress (WREC X), Glasgow 19-25 Julai 2008*) pp 1853-1857

[9] Kalogirou S A 2004 Solar thermal collectors and applications *Progress in Energy and Combustion Science* 30 pp 231-295

[10] Suruhanjaya Tenaga 2008 *Industri Pembekalan Elektrik di Malaysia – Maklumat Prestasi & Statistik 2008.* Kuala Lumpur Suruhanjaya Tenaga pp 12-14

Cycle Engine Modelling Of Spark Ignition Engine Processes during Wide-Open Throttle (WOT) Engine Operation Running By Gasoline Fuel

M.F. Abdul Rahim[1], M.M. Rahman and R.A. Bakar
Future Engine and Fuel Focus Group, Faculty of Mechanical Engineering, Universiti Malaysia Pahang, 26600, Pekan, Pahang. Tel: +609-4242255 Fax: +609-4242202

Email: mfadzil@ump.edu.my

Abstract: One-dimensional engine model is developed to simulate spark ignition engine processes in a 4-stroke, 4 cylinders gasoline engine. Physically, the baseline engine is inline cylinder engine with 3-valves per cylinder. Currently, the engine's mixture is formed by external mixture formation using piston-type carburettor. The model of the engine is based on one-dimensional equation of the gas exchange process, isentropic compression and expansion, progressive engine combustion process, and accounting for the heat transfer and frictional losses as well as the effect of valves overlapping. The model is tested for 2000, 3000 and 4000 rpm of engine speed and validated using experimental engine data. Results showed that the engine is able to simulate engine's combustion process and produce reasonable prediction. However, by comparing with experimental data, major discrepancy is noticeable especially on the 2000 and 4000 rpm prediction. At low and high engine speed, simulated cylinder pressures tend to under predict the measured data. Whereas the cylinder temperatures always tend to over predict the measured data at all engine speed. The most accurate prediction is obtained at medium engine speed of 3000 rpm. Appropriate wall heat transfer setup is vital for more precise calculation of cylinder pressure and temperature. More heat loss to the wall can lower cylinder temperature. On the hand, more heat converted to the useful work mean an increase in cylinder pressure. Thus, instead of wall heat transfer setup, the Wiebe combustion parameters are needed to be carefully evaluated for better results.

1. Introduction

Internal combustion engine has become major powertrain to drive modern transportation. The engine design and development process has emerged progressively with the finding of newer experimental method and instrumentation to gain insight information of the in-cylinder processes. Most of the experimental approach required huge expenses with considerably high lead time. The advent of digital computers allowed engine's performance prediction prior to the hardware development through engine modelling. Three dimensional engine modelling allowed detail spatial and temporal characterization of flow and combustion with high computing time. A cheaper means is to use zero or one-dimensional engine models which solve basic thermodynamic and gas dynamic equation. This type of engine model is also called cycle engine model which permit prediction of numerous engine performance parameter such as engine torque, brake power, volumetric efficiency and thermal efficiency. But the governing equation of such model is the combustion or energy equation where its solution will determine the rate of heat release, subsequently the rate of temperature and pressure rise. These parameters then will determine the engine indicated work, power and efficiency. One dimensional engine model has been used in current study to simulate the 4 stroke cycle engine operation. The model is developed using *GT-SUITE*™ software. The model fundamental equations comprise of one-

[1] Corresponding Author

Published under licence by IOP Publishing Ltd

dimensional equation of the gas exchange process, isentropic compression and expansion model and progressive engine combustion model based on Wiebe approach. This engine model also accounts for the heat transfer and frictional losses as well as the effect of valves overlapping.

2.0 Objective of the Study

The model is developed as a means to acquire engine performance prediction solely for spark ignition engine. The parameters of interest are engine cylinder pressure and temperature. The engine performance predictions need to be validated with experimental engine data in order to pledge trustworthiness and reliability. The engine which is currently running on gasoline is expected to be used for alternative fuel study namely compressed natural gas and hydrogen. The introduction of gaseous fuels such as hydrogen into existing engine will require the addition of gaseous fuel injector which is expected to be completed in the latter stage of the study.

3. Experimental Engine Setup

The baseline engine is a 4 cylinder, 12 valves engine which is currently running on gasoline fuel. The engine is naturally aspirated and the fresh mixture is carburetted in design. Each cylinder equipped with two intake valves of different valve diameter and one exhaust valve which has larger valve diameter. The ignition is control by mechanical distributor. For the purpose of static dynamometer experimentation, the original cooling loop has been made external and increased in terms of water capacity to improve the cooling capability during wide open throttle condition. Other modified system is the intake and exhaust system. The intake air filter is removed while the other parts are retained and attached to the surge tank with orifice airflow measurement. The exhaust pipe is made longer to connect with an exhaust gas calorimeter. The baseline engine has been tested for experimental data on a stationary test-bed coupled with hydraulic dynamometer. All required engine geometrical and operational data have been measured and embedded into this model. The following table provides detail engine specifications.

Table 1: Baseline engine specifications

Parameter	Size and Feature
Valve train type	In-line OHV, SOHC
Number of cylinders	4 cylinders
valves per cylinder	3 valves per cylinder
Combustion chamber type	Pent-Roof type
Total displacement (cm^3)	1,488
Cylinder bore (mm)	75.5
Piston stroke (mm)	82.0
Compression ratio	9.2
Intake valves open/close	15° BTDC/63° ABDC
Exhaust valve open/close	57° BBDC/ 13° ATDC
Lubrication system	Pressure load, full – flow filtration
Oil pump type	Trochoid type
Cooling system	Water – cooled forced circulation
Water pump type	Centrifugal impeller type
Maximum output	66 kW@6000 rpm
Maximum Torque	124 Nm@3000 rpm

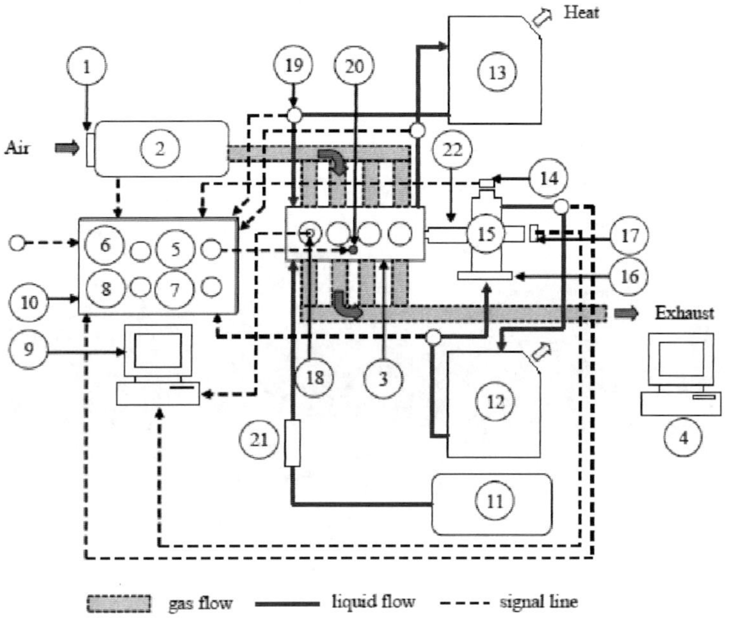

1. Orifice plate meter
2. Surge tank
3. Tested engine
4. Emission analyzer
5. Throttle controller
6. Pressure manometer
7. Load indicator
8. Temperature scanner
9. Combustion analyzer
10. Instrumentation panel
11. Fuel supply tank
12. Dynamometer water supply
13. Engine cooling system
14. Load sensor
15. Hydraulic dynamometer
16. Dynamometer controller
17. Crank angle encoder
18. Pressure Transducer
19. Temperature sensor
20. Throttle body
21. Fuel flow meter
22. Engine-dynamometer coupling

Figure 1: Schematic of the hydraulic dynamometer engine test rig

4. Engine Modelling

The engine model is developed using $GT\text{-}SUITE^{TM}$ package which has special capability to predict engine cycle performance. Based on the physical appearance of an actual engine, the engine model can be also divided as follow (1) the intake manifold system (2) the engine combustion chamber and cylinder (3) the engine crankcase, and (4) the exhaust manifold system. The quasi one-dimensional approach is mostly adopted for the solution of the gas dynamic flow within the intake and exhaust manifold whereas the engine cylinder represent the zero dimensional part of the model. Basically, the engine model is governed by the equation of mass continuity, momentum and energy equation. The combustion or rate of heat release model is commonly the mathematical equation that determines the rate of temperature and pressure rise which will further affect the engine performance parameter. Wiebe model of spark ignition engine combustion is among the simplest. The mass fraction burn rate (MFB) using Wiebe function can be expressed as Eq. (1):

$$MFB(\theta) = 1 - \exp\left[-a\left(\frac{\theta-\theta_0}{\Delta\theta}\right)^{m+1}\right] \tag{1}$$

where θ is crank angle degrees with θ_0 corresponding to the initialization of heat release and $\Delta\theta$ corresponding to the duration of burn. The equation is also defined by two constants a and m which have typical values of 5 and 3 respectively. The $\boldsymbol{a}$ and $\boldsymbol{m}$ are also adjustable parameters that can be fine-tune to fit the model with experimental data. Hence, it can be said that the Wiebe function is useful in a state where the cylinder data properties are available and it is not predictive in nature.

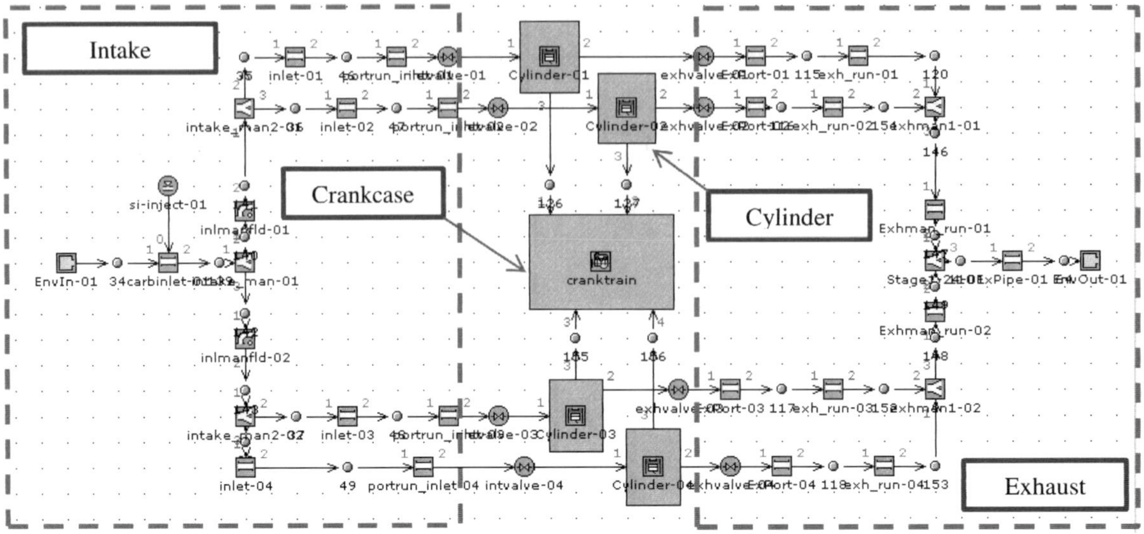

Figure 2: The layout of the engine model

5. Results and Discussion

5.1 Effect of Different Engine Speed

Figure 3 (a) and (b) present the simulated pressure-crank angle and temperature crank-angle diagram for engine rotational speed of 2000, 3000 and 4000 rpm. The engine model simulates the wide open throttle operation of the engine. Important inputs for the model are presented by Table 2. The air to fuel ratio, the burning duration (10% to 90% burning), the 50% burning crank angle values are all based on the experimental data of the engine tested at wide open throttle condition. The value of ***a*** and ***m*** (the adjustable parameters) are retained as default since the curve-fitting process required heuristic approach that will increased the time allocation for this stage of the study.

Table 2: Input for the simulated cases

Engine speed (rpm)	Air fuel ratio	Burning duration, $\Delta\theta$, (*CA unit*)	50% burning crank angle (*CA*)	a	m
2000	12.07	13.99	-1.19		
3000	13.99	14.39	-0.40		
4000	12.06	15.99	2.40		

Based on the pressure-crank angle diagram, the simulated results present an increment trend of cylinder peak pressure. The maximum peak for 2000, 3000 and 4000 rpm are 56.8 bar, 61.7 bar and 67.4 bar. All peak pressure timing is obtained at about 10° CA after top dead centre which corresponds to the optimum peak pressure timing of a natural aspirated spark ignition engine. These are also reflected the measured engine data that produce the same results. Based on the data specified by Table 2, the decisive factor that may differentiate the maximum peak pressure of each case is the air to fuel ratio, the burning duration, $\Delta\theta$ and the heat release curve anchor angle, the 50% burning crank angle. The richest air to fuel ratio input for the model are 12.07 and 12.06 for 2000 rpm and 4000 rpm respectively whereas the air fuel ratio input at medium engine speed is about 13.99, approaching the stoichiometric value for gasoline fuels. However, the air fuel ratio is not fully reflecting the trend of peak pressure. Examination on the burning duration provides a more justified reason for the increment of peak pressure. As the engine speed increase, the burning duration is also increase which initially showed an inverse proportional. However, if the unit of crank angle is converted to the time scale, the corresponding value of burning duration for 2000, 3000 and 4000 rpm are 1.16 ms, 0.8 ms and 0.67 ms respectively. Theoretically, the faster burning duration will enhance higher cylinder pressure and these results support the trend of peak cylinder pressure.

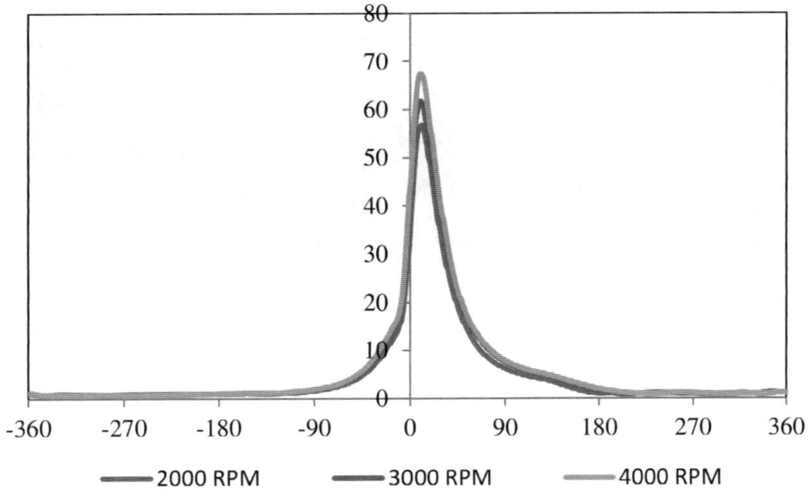

(a)

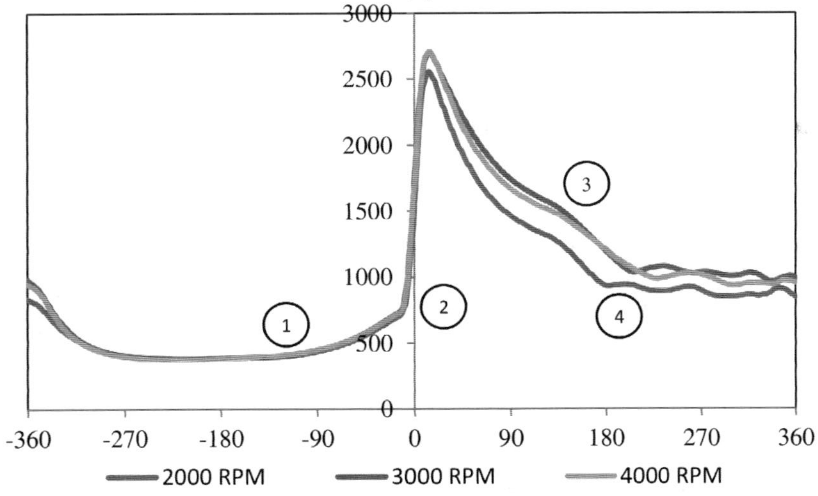

(b)

Figure 3(a): pressure-crank angle diagram for cylinder no. 1**(b):** temperature-crank angle diagram for cylinder no.1

The temperature-crank angle diagram presents a normal trend of temperature variation in engine cylinder. By overall, the trend is in agreement with the cylinder pressure data. As the engine speed increase, the cylinder pressure increase as well as the cylinder temperature. However, it is noticeable that the cylinder temperature at 4000 rpm is just slightly higher than the data at 3000 rpm. Point 1 showing the start of temperature increased corresponding to the inlet valve closing at -117° before top dead centre. Point 2 corresponding to the ignition point and the start of major heat release event. Point

3 corresponding to the start of exhaust blow down process during the start of exhaust valve opening at 123° CA after top dead centre. While point 4 corresponding to the start of exhaust stroke where the piston moved upwards expelling the burned gas within the engine cylinder.

5.2 Validation Based on Cylinder Pressure

The model is first validated by using the cylinder pressure data. Figure 4 (a) - (c) present the comparison of measured cylinder pressure (measured at cylinder no.1) with the simulated cylinder pressure. In general, the simulated cylinder pressures replicate the trend of measured data. The peak pressure timing is coincident in most cases. The simulated rate of pressure rise and it's reduce follow the trend of experimental data. The discrepancy is noticeable in terms of the cylinder peak pressure. Table 3 presents the deviation comparison between the simulated and measured cylinder peak pressure.

Table 3: Deviation of simulated cylinder peak pressure against measured data

Engine speed (rpm)	Simulated peak pressure (bar)	Measured peak pressure (bar)	Deviation (%)
2000	56.8	61.19	-7.17
3000	61.7	62.27	-0.92
4000	67.4	67.05	0.52

The largest deviation occurred at low engine speed where the model under predicted the cylinder peak pressure. In other two cases, the deviation is less than 1%. Hence, it can be concluded that the current model accuracy is adequate for the prediction of cylinder peak pressure. This means that all other related performance parameter based on cylinder pressure such as the indicated work indicated power are accurate enough for further evaluation.

5.2 Validation Based on Cylinder Temperature

The second validation is based on the cylinder temperature data. Figure 5 (a) - (c) present the comparison of measured cylinder temperature (measured at cylinder no. 1) with the simulated cylinder temperature. For the purpose of comparison, the measured cylinder temperature is not directly measured from the engine cylinder, however the values is derived based on the cylinder pressure using the ideal gas relationship within a closed volume at the instance of intake valve closing. Hence only the temperatures values during all valves closed are considered. The expression of the ideal gas is given by Eq. 2:

$$T_{cyl} = \frac{P_{cyl} \times V_{cyl} \times T_{amb}}{P_{amb} \times V_{amb}} \qquad \text{(Eq.2)}$$

In general, the simulated cylinder temperatures also replicate the trend of measured data. From the instance of intake valve closing, followed by the compression and ignition process, the simulated temperature profile fitted the measured data. The discrepancy is then obvious in terms of cylinder peak temperature and its subsequent instantaneous values. Table 4 presents the deviation comparison between the simulated and measured cylinder peak temperature. Based on the comparison, the deviation of cylinder peak cylinder temperatures are about 10%. Due to the difference in cylinder peak temperature, the subsequent instantaneous temperature values also over predicted the measured data. However, the rate of temperature reduction are seems to be comparable with the measured data except for the case of 2000 rpm where the measured data showing a moderating rate of temperature reduction. This clearly cannot be captured by the model.

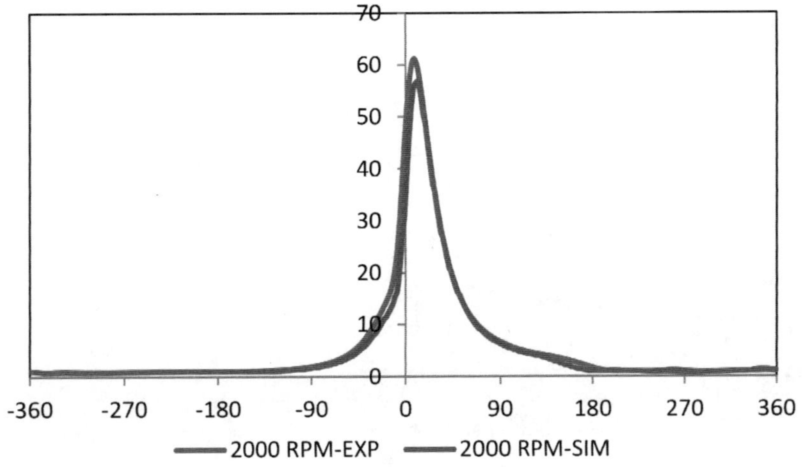

(a)

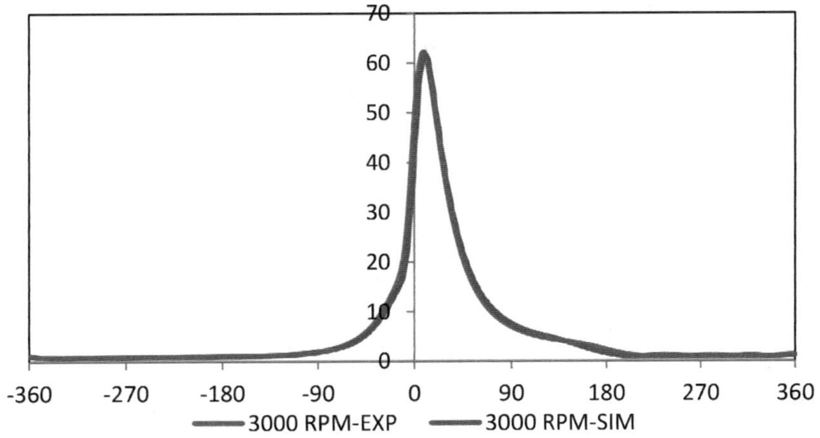

(b)

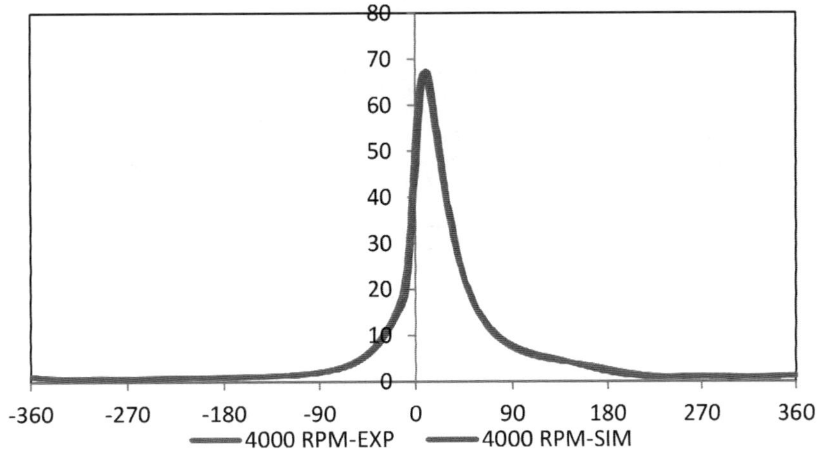

(c)

Figure 4(a)-(c): Comparison between measured and simulated cylinder pressure for engine speed of (a) 2000 rpm (b) 3000 rpm and (c) 4000 rpm.

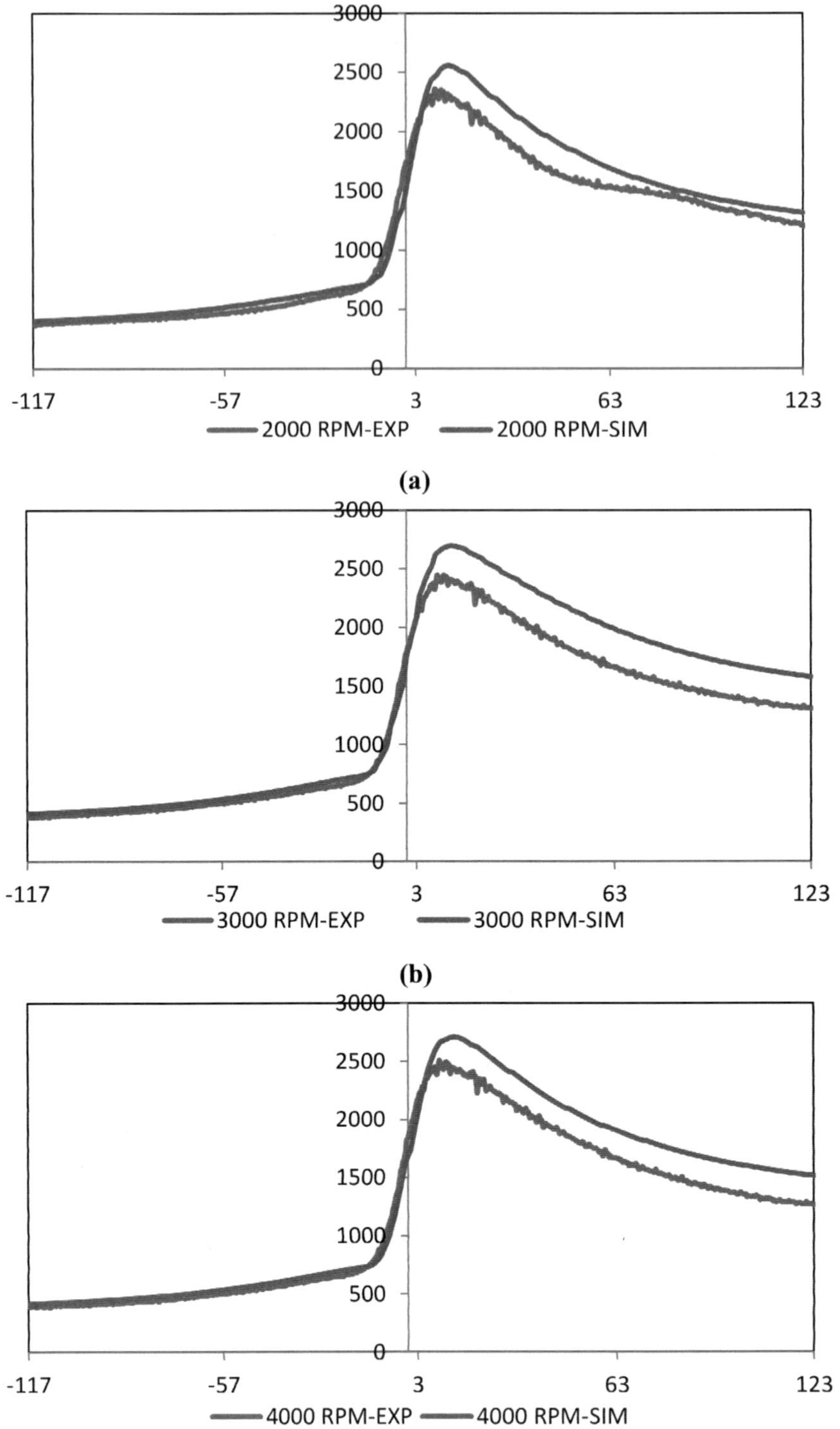

(c)

Figure 5(a)-(c): Comparison between measured and simulated cylinder temperature for engine speed of (a) 2000 rpm (b) 3000 rpm and (c) 4000 rpm.

It was believed that, the moderating rate of temperature reduction is due to the burning of end gas in actual engine combustion. Additional heat was deliberated and causing the rate of temperature reduction becoming lessened. All simulated cases showed an over prediction of cylinder peak temperature.

Table 3: Deviation of simulated cylinder peak temperature against measured data

Engine speed (rpm)	Simulated peak temperature (°C)	Measured peak temperature (°C)	Deviation (%)
2000	2556	2362	8.21
3000	2696	2447	10.17
4000	2711	2508	8.09

The difference between the simulated and measured cylinder peak temperatures are expected as the model assumed a complete combustion whereas in actual condition, the dissociation process imposed significant impact by forming intermediate combustion products. Current model also neglect the formation of nitrous oxide, carbon monoxide and other rich combustion products that will absolutely alter the maximum heat deliberated during combustion.

6. CONCLUSION AND RECOMMENDATION

Cycle engine modelling has been carried out and it was proven that this type of model capable of delivering adequate performance prediction for spark ignition engine provided that realistic input and detail geometrical configurations are available. But noticeable discrepancies are still occurred and that required further improvement on the model. Suggestion for further improvement includes the following work:

1. Optimisation of the adjustable Wiebe parameter ***a*** and ***m*** that may be affected by the type of fuel, engine geometrical details and engine operating condition.
2. Inclusion of knock and NO_x model and other rich combustion products that maybe alter the cylinder peak temperature.
3. Investigation of other engine performance parameter such as the brake quantities and engine efficiencies.
4. Switching the combustion model that is more predictive in nature which is not totally rely on existing cylinder data and includes the effect of ignition or injection timing.

REFERENCES

1. Rahim, M.F.A. (2009). Numerical Study of Combustion for a 4-Stroke Gasoline Engine Using Turbulent Flame Speed Closure Model, Master Thesis, Universiti Malaysia Pahang.
2. Rahman, M. M., Mohammed K.M. and Bakar, R.A. (2009). Effects of Air Fuel Ratio and Engine Speed on Engine Performance of Hydrogen Fueled Port Injection Engine, American Journal of Scientific Research, ISSN 1450-223X Issue 1(2009), pp. 23-33, EuroJournals Publishing, Inc.
3. Meyer, J. (2007). Engine Modeling of an Internal Combustion Engine With Twin Independent Cam Phasing, Under Graduate Thesis, Ohio State University.
4. Meyer, J. (2008). Innovations in Representation and Calibration of Residual Gas Fraction and Volumetric Efficiency in a Spark Ignited, Internal Combustion Engine, Master Degree Thesis, Ohio State University.

IOP Conf. Series: Materials Science and Engineering **36** (2012) 012042 doi:10.1088/1757-899X/36/1/012042

Evaluating tractor performance and exhaust gas emissions using biodiesel from cotton seed oil

Saddam H. Al-lwayzy[1], Talal Yusaf and Troy Jensen

National Centre for Engineering in Agriculture (NCEA), Faculty of Engineering and Surveying, University of Southern Queensland, Toowoomba, 4350 QLD Australia.

E-mail[1]: SaddamHussen.Allwayzy@usq.edu.au, yusaft@usq.edu.au, Troy.Jensen@usq.edu.au

Abstract Alternative fuels for diesel engines, such as biodiesel, have attracted much attention recently due to increasing fuel prices and the imperative to reduce emissions. The exhaust gas emissions from tractors and other agricultural machinery make a significant contribution to these emissions. The use of biodiesel in internal combustion engines (ICE) has been reported to give comparable performance to conventional diesel (CD), but with generally lower emissions. There is however, contradictory evidence of NO emissions being both higher and lower from the use of biodiesel. In this work, agriculture tractor engine performance and its emission using both CD and biodiesel from cotton seed oil (CSO-B20) mixed at a 20% blend ration has been evaluated and compared. The PTO test results showed comparable exhaust emissions between CD and CSO-B20. However, the use of CSO-B20 led to reductions in the thermal efficiency and exhaust temperature and an increase in the brake specific fuel consumption (BSFC), when compared to CD.

1. Introduction

Interest in alternative fuels has increased substantially over the last decade as the demand for energy has intensified, whilst the reserves of fossil fuels are depleting. This has contributed to higher fuel prices making alternative fuels a viable commercial alternative. Enhancing the energy security is another factor for pushing towards alternative fuels to vary the sources of energy and also reduce the dependence on petroleum oil.

The diesel engine is widely used in most industries because of eht low fuel consumption and adequate efficiency [1]. Biodiesel is a viable alternative to conventional diesel as it is renewable, environmentally friendly and does not require significant modification to the existing diesel engine technology. For more than two decades, biodiesel has been used successfully in Europe and its use in the United State has increased considerably over recent years [1]. Raheman & Phadatare [2] concluded that Karanja biodiesel (with blend rates of up to 40% with conventional diesel) can be an alternative fuel for diesel engines. Recently, cotton seed oil and its methyl ester, are increasingly used in IC engines and it is expected to be one of the alternative fuels in the future [3].

Producing biodiesel from crops involves a series of steps in which diesel fuel is used. Using alternative fuel in agriculture is the first step for clean energy. As tractors are the major power source for most agricultural power consuming processes (e.g. tillage, loading, mower and tracking), investigating the potential of using biodiesel as the fuel source is a worthwhile exercise.

Sahoo et al. [4] tested the effect of different blend ratios of jatropha, karanja and polanga biodiesel when used in a three cylinder diesel engine. They found that there was raising corresponding increase in the Brake Specific Fuel Consumptions (BSFC) with the increase in the blend ratios. Industrially

Published under licence by IOP Publishing Ltd

produced biodiesel and homemade biodiesel were tested in agricultural tractor diesel engine by Gravalos et al. [5]. The homemade biodiesel resulted same tractor performance as diesel fuel and give better performance than the commercial biodiesel. Therefore evaluating tractor engine performance and emissions is essential. Tractor power Take Off (PTO) test was conducted using conventional diesel (CD), Biodiesel from Cotton Seed Oil (CSO) 20% blended with 80% CD by volume, and microalgae Biodiesel (MB) 20% blended with CD under different PTO operation condition. In this work, the data of comparing the performance and emissions of CD and CSO-B20 at mid load was represented.

2. Materials and methods

2.1 Tractor test and experimental apparatus

The PTO tractor dynamiter test has been conducted using John Deere 4410 tractor ehydro (see Figure1). The engine in the JD4410 was a Yanmar, 3TNE88, three cylinders water cooling diesel engine with compression ratio of 18.8:1. The engine gross power is 25.8kW and the manufacture's estimated PTO power is 21.3 kW. The engine torque at rated speed is 87.9 Nm.

Figure 1. The experimental setup

2.2 Fuel consumption measurement

The fuel consumption rate was measured using the measuring cylinder shown in Figure 2. The measuring cylinder was fitted with a two-way valve at both the top and the bottom. The top two-way value allowed for the fuel returning from the injectors to either enter the cylinder or return to tank. The bottom two-way valve allowed for the fuel to either be drawn from the cylinder or to be supplied form the tank. In order to make a fuel consumption reading, the top and bottom value were open simultaneously allowing return fuel to enter the cylinder whilst fuel was being drawn from the bottom. A separate tank was connected when biodiesel was being used. The time was recorded using stop watch for fixed volume of fuel consumed. At the end of the test using CD, the system was cleaned and the fuel filter was changed to prevent any contamination from the previous test.

2.3 Dynamometer

PTO dynamometer was used on apply load to the tractor engine. The dynamiter was instrumented to measure PTO speed (rpm) and torque (Nm) and to calculate the power (kW). The instrumentation required that the dynamometer was calibrated after each series of tests. This calibration was conducted using five standard weights of 20 kg each. The calibration started by running the monitor of the dynamometer on the calibration mode. Then the zero torque was set. Then standard weights of 100 kg were suspended on an arm of one meter length to apply load on the load cell. At that time the torque of 981 Nm was set on the screen. The ratio of the engine to the PTO speeds was found to be 4.815.

2.4 Emission Analyser

A BEA 460 Bosch gas analyser was connected to the tractor exhaust to monitor the O_2, CO, CO_2, NO, and Lambda with the operating ranges and resolutions being given in Table 1. The gas analyser was calibrated by the manufacturer prior to the test.

Table 1. The gas analyser ranges, accuracy and resolution

Component	**Min.**	**Max.**	**Resolution**
CO (%)vol	0	10.00	0.001
CO_2(%)vol	0	18.00	0.01
HC (PPM)	0	9999	1
Lambda	0.5	22.00	0.001
NO (PPM)	0	5000	1

2.5 Thermocouple

A Digi-Sense thermocouple K type with four digit digital screen was used to monitor the exhaust gases temperatures with accuracy of 0.1%.

2.6 Fuel preparation

Pure CSO biodiesel (100%) was obtained from local supplier which has been converted to biodiesel though transesterification. The specifications of CSO biodiesel are given in Table 2. To prepare the CSO-B20 blend, 20% by volume CSO was mixed with 80% by volume of CD. The specification of CSO-B20 was calculated from the CD and CSO blend and is shown in Table 2.

Table 2. Properties of CD, CSO [1] and CSO-B20

Properties	**Diesel**	**CSO**	**ASTM method**	**CSO-B20**
Kinematic viscosity mm^2/s at 40°C	3.9	6.0	D445	4.3
Density kg/m^3 at 15°C	838	850	D1298	840.4
Higher calorific value kJ/kg	43.92	41.68	D5865	43.47
Cetane number	50	52	D613	50.4

Figure 2. The fuel measurement system

2.7 Engine test procedure

The test was conducted using different fuels of CD and CSO-B20 and MB-20 at different operating conditions of no load, medium load and maximum load. This work presents only the tractor performance and exhaust gas emissions results using CD and CSO-B20 for medium load operating conditions. In order to have consistence readings, the tests were conducted at same atmospheric conditions in which humidity, atmospheric pressure and temperatures are relatively constant. The tractor engine was warmed up for one hour before conducting the test using CD. The dynamometer was carefully calibrated using a standard calibration arms with different load and speed tachometer as mentioned in subsection 1.2 , this procedure was repeated three times to make sure PTO torques and speed readings are accurate. The test was performed by setting the tractor engine speed to 2000 (RPM) (without load) using the manual speed controller. This resulted in a PTO speed of approximately 415 rpm. Tests were conducted by increasing the load on the PTO, whilst engine speeds, PTO speeds, PTO torque and power, fuel consumption, exhaust temperature, and exhaust gas emission components was recorded simultaneously.

3. Results and disscussions

3.1 Tractor Engine Performance

The tractor PTO test was conducted to evaluate and compare the tractor performance using CD and CSO. The gross energy, brake power, brake specific fuel consumption, engine efficiency and exhaust temperature are presented below.

3.1.1 Gross energy and Brake Specific Fuel Consumption (BSFC)

The relationship between the gross energy and the PTO torque from CD and CSO-B20 is presented in Figure 3. The gross energy is the result of lower calorific value multiplied by the mass flow rate of fuel. Figure 3 shows that the gross energy increased over the PTO torque range between 100–400Nm. This energy rise was due to the high torque values associated with high engine speeds (1910-1710 RPM for CD and 1916-1705 RPM for CSO-B20) at this range of 100–400 Nm respectively. Figure 3 also shows that the gross energy started to decline after 400 Nm torque, caused by the engine speed reduction when higher torques were applied. The engine speed dropped to 1300 RPM for CD at 440 Nm and 1160 RPM for CSO-B20 at 430N. At PTO torques below 400 Nm, CSO-B20 shows a slightly higher gross energy than CD due to the higher fuel consumption rate of CSO-B20 compare to CD.

The BSFC is an indicator of fuel efficiency and is calculated by dividing the fuel consumption rate by the engine brake power. Figure 4 represents the relationship between the PTO torque and the BSFC for CD and CSO-B20. This figure shows a decrease in BSFC with PTO torque increases over the torque range from 100 Nm to 450 Nm. This result agrees with Forson, Oduro et al. [6] who found that increasing the engine torques will decrease the BSFC for all fuel blends of diesel and Jatropha biodiesel. The results presented in Figure 4 showed that the BSFC value of CSO-B20 is higher than the CD at all loads. The maximum BSFC for CSO is 644 g/kwh at 100 Nm, while the maximum BSFC for CD is 574 g/kwh at around 100 Nm PTO torque. The reason of this reduction is due to the lower calorific value of the CSO which means the engine started to consume more CSO-B20 to produce same power.

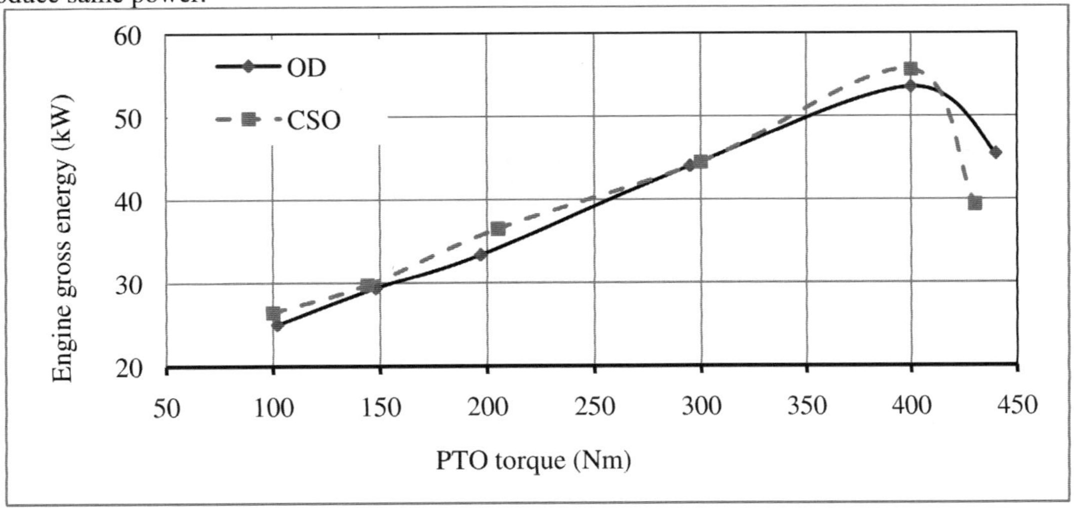

Figure 3. The relationship between PTO torques (Nm) and engine gross power (kW) for CD and CSO-B20

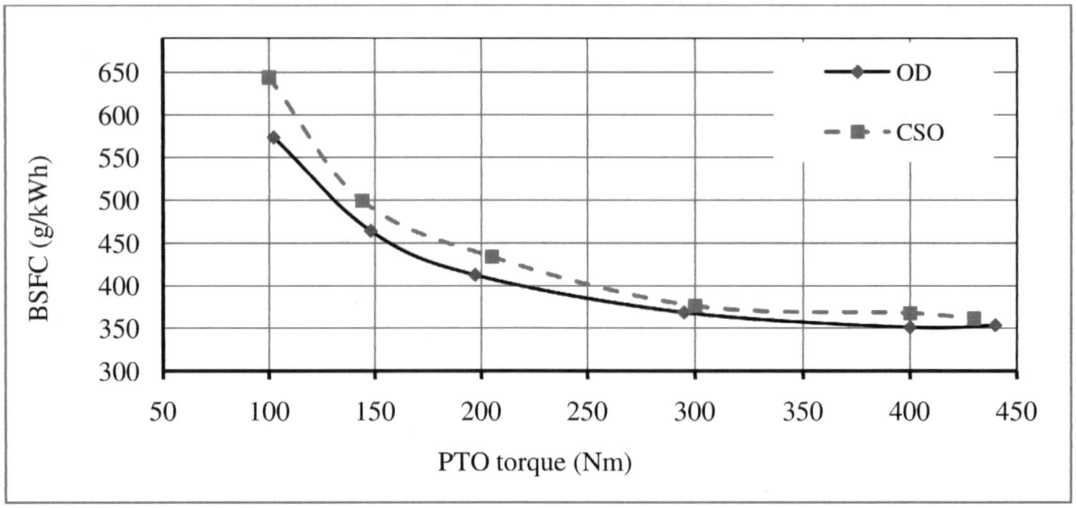

Figure 4. The relationship between PTO torques (Nm) and BCFC (g/kWh) for CD and CSO-B20

3.1.2 Engine Brake Thermal Efficiency (ηth)

Figure 5 represents the relationship between the PTO torques and the engine brake thermal efficiency (η_{th}) at medium loads. Figure 5 shows that that the brake thermal efficiency of both fuels was noticed to be increasing with the increase of PTO torque. Same trend was found by [2] in which, the thermal efficiency increased with torque increased, the reason for that was due to the power gain and the reduction in heat losses as torque increase. As discussed previously, the gross energy of CSO is higher than CD at same brake power, therefore a reduction in thermal efficiency of CSO-B20 occurred at all PTO torques. Nabi et al. [1] reported that the reduction in the engine thermal efficiency using biodiesel is due to lower heating value of the biodiesel (cotton seed blends). Aydin & Bayindir [3] stated that the higher viscosity and lower heating value of biodiesel of cotton seed oil cause a reduction on the engine efficiency in comparison with diesel. Those outcomes show an agreement with the outcome of the thermal efficiency of the present work.

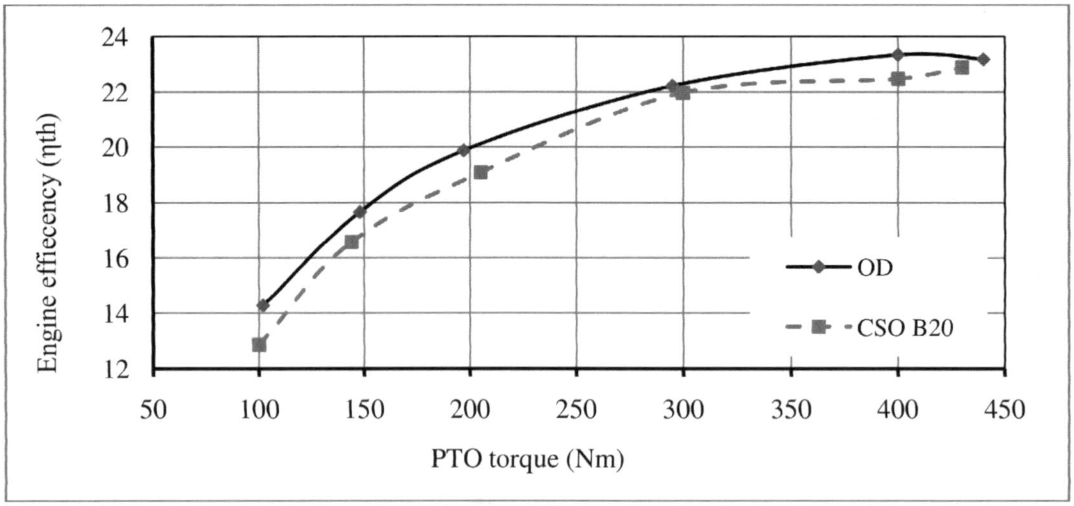

Figure 5. The relationship between PTO torques (Nm) and engine brake thermal efficiency (η_{th}) for CD and CSO-B20

3.1.3 Exhaust Temperature

Figure 6 draw the relationship between the PTO torques and the exhaust gas temperature for CD and CSO-B20. This figure shows an increase in the exhaust temperature with the increasing of PTO torques for both fuels, that is until the maximum temperature is reached at 400 Nm, and then the exhaust temperature goes down at the maximum studded torque. The maximum exhaust temperatures are 450 °C and 439 °C for CD and CSO-B20 respectively. The general trend of exhaust temperature for the engine fuelled by CSO-B20 is found to be lower than the exhaust temperature when the engine fuelled by CD. The possible reason for that difference in the exhaust temperature is due to the lower adiabatic flame temperature for the CSO-B20 than diesel.

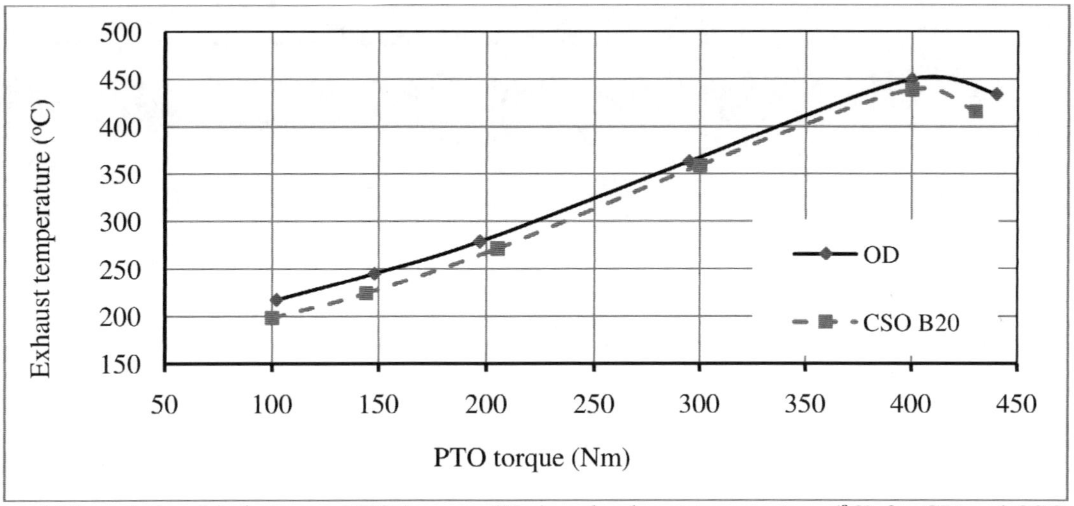

Figure 6. The relationship between PTO torques (Nm) and exhaust temperature (°C) for CD and CSO-B20

3.2 Engine Gases Emissions

3.2.1 Carbon Monoxide (CO) and Carbon Dioxide (CO_2)

Excess CO concentration in the exhaust gas means the reactions in the combustion chamber is incomplete. The CO emission measured in percentage versus the PTO torques using CD and CSO-B20 is presented in Figure 7. This figure shows that the CO percentage at the toques below 300 Nm is almost constant, while the CO percentage at the torques above 300 Nm increased sharply for both fuels. Both fuels give comparable results at the PTO torques below 400 Nm because of the lean combustion indicated by the Lambda values in Figure 11. This Lambda results show that and amount of oxygen is higher than the stoichiometric for both fuels. On the other hand, CSO-B20 shows higher CO level than CD at the torque up to 400 Nm.

Figure 8 displays the CO_2 percentage against the PTO torques for CD and CSO-B20. It can be seen that the CO_2 level increase with the PTO torque increase for both fuels. CSO-B20 produces CO_2 level slightly higher than CD for the torques below 300 Nm. This results agree with Murugesan et al. [7] outcome in which, there were no significant effect of vegetable biodiesel B20 on CO_2 concentration in the exhaust gas emission.

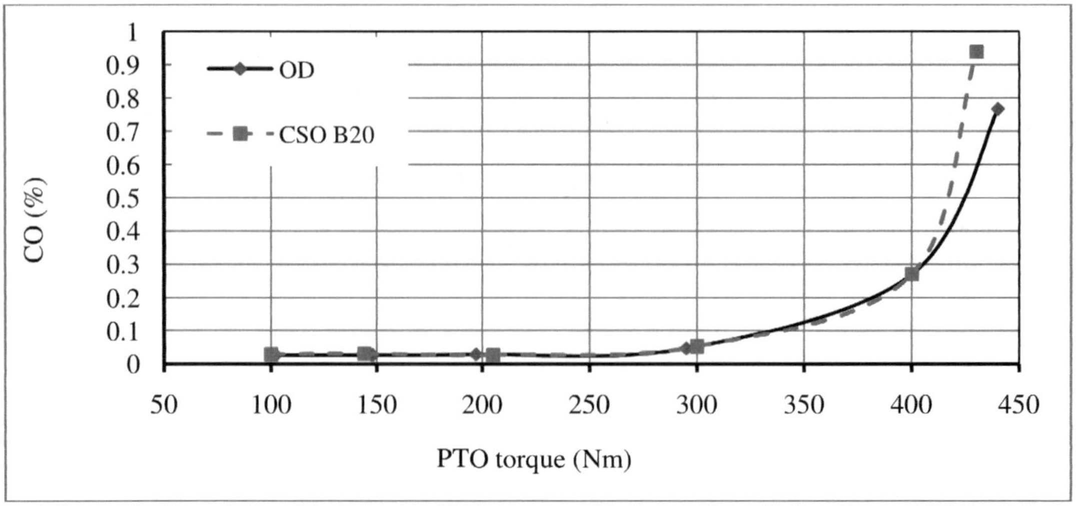

Figure 7. The relationship between PTO torques (Nm) and CO (%) for CD and CSO-B20

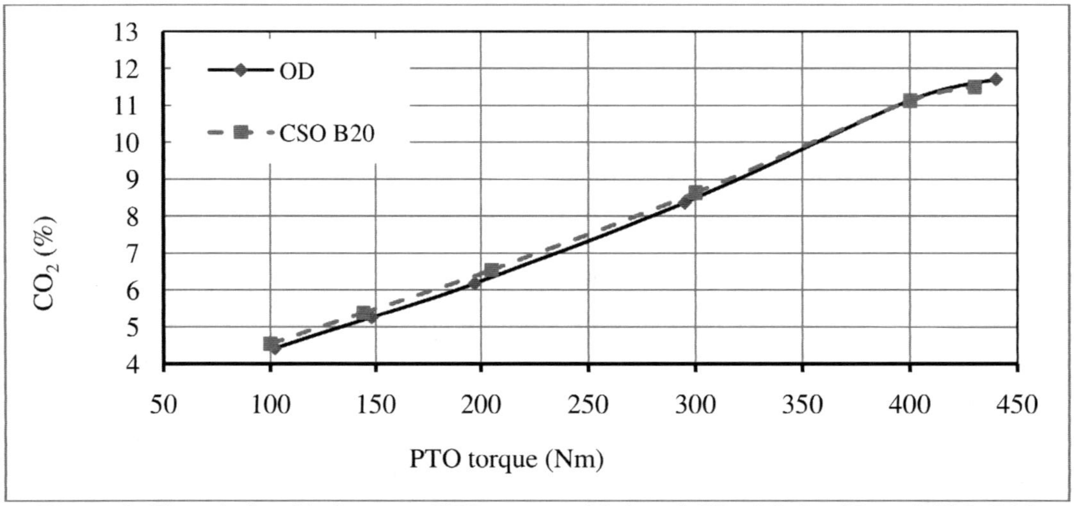

Figure 8. The relationship between PTO torques (Nm) and CO_2 (%) for CD and CSO-B20

3.2.2 Nitrogen Monoxide (NO)

The relationship between the level of NO (PPM) and the PTO torque is given in Figure 9 for CD and CSO-B20. The general trend of both fuels shows that increasing the PTO torque led to an increase in NO concentration (measured in PPM). The NO rise can be due to higher loads and exhaust temperature. Murugesan et al. [7] results showed a reduction in NO with the percentage of biodiesel increased for different engine load. This results show that there was no considerable difference between diesel and B20 of vegetable oil.

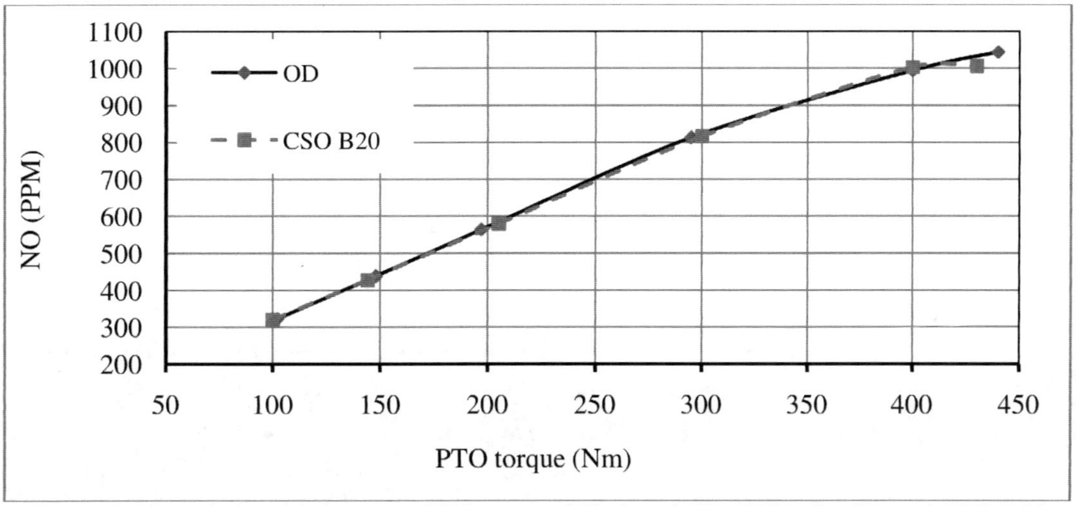

Figure 9. The relationship between PTO torques (Nm) and NO (PPM) for CD and CSO-B20

3.2.3 Oxygen (O_2) and lambda (λ)

Figure 10 represent the relationship between the PTO torque and the O_2 level for CSO-B20 and CD. This figure shows a decline in O_2 percentage from 14.7% and 14.4% to 3.6% and 4.5% for CD and CSO-B20 respectively at different PTO torque values. The O_2 percentage was found to be comparable between CSO-B20 and CD at all PTO torques.

The relationship between lambda and the PTO torques for CD and CSO-B20 is shown in Figure 11. Lambda is the actual air-fuel ratio divided by the stoichiometric air-fuel ratio. When Lambda is higher than one, the engine runs lean, which is a normal case in diesel engines [8]. Figure 11 shows that both fuels are running lean over different PTO torques. The lambda's values were noticed to be decreased with decreasing the load applied on the engine for both fuels. This reduction is a result of high fuel requirement to produce more power. Andersson [9] reported that diesel engines run lean at ideal and low load because of the air flow is constant while the fuel can be controlled through acceleration. The results presented in this work showed good agreement with Andersson [9] results.

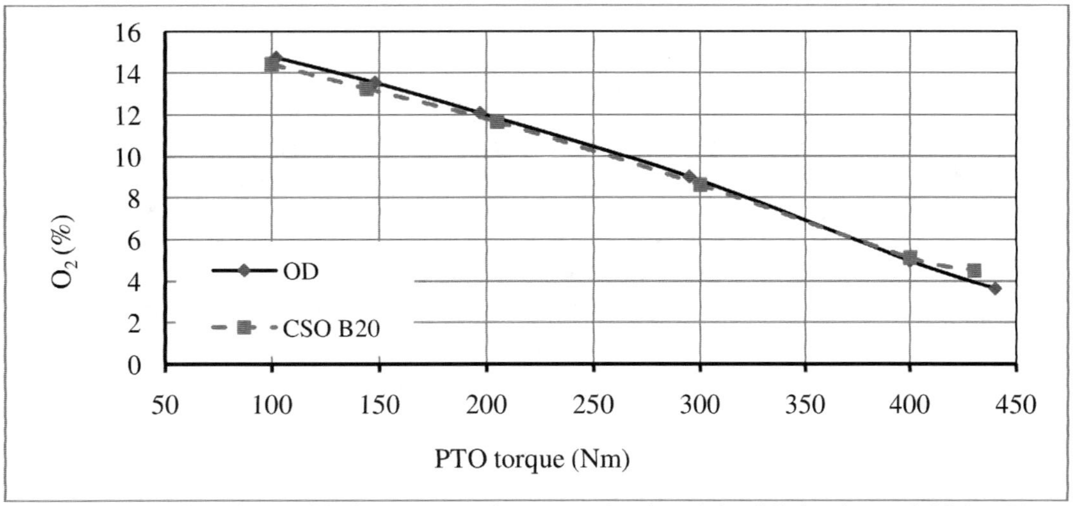

Figure 10. The relationship between PTO torques (Nm) and O_2 (%) for CD and CSO-B20

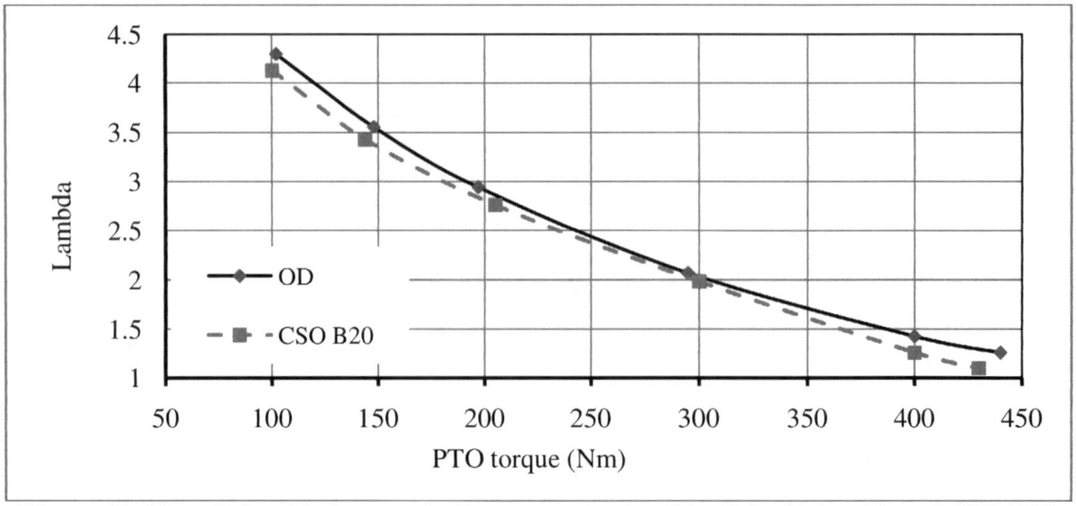

Figure 11. The relationship between PTO torques (Nm) and lambda (λ) for CD and CSO-B20

4. Conclusion

The medium load engine test results demonstrated that biodiesel from cotton seed oil can be a good alternative fuel for diesel engine at B20 blend without engine modification. It was found that CSO-B20 can produce very comparable engine performance and exhaust gas emission results at medium load. The experimental results showed CSO-B20 produce higher BSFC and lower engine thermal efficiency in comparison with CD across a wide range of PTO torques. Further experimental work is needed to investigate the affect of different blend ratio on engine performance and exhaust gas percentage.

References

[1] Nabi, M.N., M.M. Rahman, and M.S. Akhter, *Biodiesel from cotton seed oil and its effect on engine performance and exhaust emissions.* Applied Thermal Engineering, 2009. **29**(11-12): p. 2265-2270.

[2] Raheman, H. and A.G. Phadatare, *Diesel engine emissions and performance from blends of karanja methyl ester and diesel.* Biomass and Bioenergy, 2004. **27**(4): p. 393-397.

[3] Aydin, H. and H. Bayindir, *Performance and emission analysis of cottonseed oil methyl ester in a diesel engine.* Renewable Energy, 2010. **35**(3): p. 588-592.

[4] Sahoo, P.K., et al., *Comparative evaluation of performance and emission characteristics of jatropha, karanja and polanga based biodiesel as fuel in a tractor engine.* Fuel, 2009. **88**(9): p. 1698-1707.

[5] Gravalos, I., et al., *Comparison of performance characteristics of agricultural tractor diesel engine operating on home and industrially produced biodiesel.* International Journal of Energy Research, 2009. **33**(12): p. 1048-1058.

[6] Forson, F.K., E.K. Oduro, and E. Hammond-Donkoh, *Performance of jatropha oil blends in a diesel engine.* Renewable Energy, 2004. **29**(7): p. 1135-1145.

[7] Murugesan, A., et al., *Bio-diesel as an alternative fuel for diesel engines--A review.* Renewable and Sustainable Energy Reviews, 2009. **13**(3): p. 653-662.

[8] Fernando, S., C. Hall, and S. Jha, *NOx Reduction from Biodiesel Fuels.* Energy & Fuels, 2005. **20**(1): p. 376-382.

[9] Andersson, C., *Observations on Electric Hybrid Bus Design*, in *department of Industrial Electical Engineering and Automation.* 2001, Lund Institute of Technology: Lund. p. 117.

IOP Conf. Series: Materials Science and Engineering **36** (2012) 012043 doi:10.1088/1757-899X/36/1/012043

Study of Henna (*Lawsonia inermis*) as Natural Corrosion Inhibitor for Aluminum Alloy in Seawater

W B Wan Nik[1], F Zulkifli[1], O Sulaiman[1], K B Samo[1] & R Rosliza[2]

[1]Department of Maritime Technology, Faculty of Maritime Studies and Marine Science, Universiti Malaysia Terengganu 21030 Kuala Terengganu, Terengganu, Malaysia.
[2]TATi University College, Teluk Kalong, 24000 Kemaman, Terengganu, Malaysia.
E-mail[1]: niksani@umt.edu.my

Abstract. Commercial henna (*Lawsonia inermis*) was investigated to inhibit the corrosion of aluminum alloy through immersion in seawater. The aluminum alloy (5083) was prepared in size of 25mm x 25mm x 3mm. The immersion test was conducted in seawater with different concentration of henna, 100ppm, 300ppm, 500ppm for duration of 60 days. Four characterizations were performed in this study which was weight loss study, Fourier Transform Infrared (FTIR), Electrochemical Impedance Spectroscopy (EIS) and adsorption isotherm. The results indicated that henna has major constituents of lawsone which contributed to the chemisorptions or adsorption process by forming an isolation layers on the aluminum alloy surface which follows the Langmuir adsorption isotherm. It was found that the protection layer attached on metal was not permanent and precipitation occurred as the time increases. The highest inhibition efficiency was found at 88% (500ppm). This research found that henna is an excellent natural inhibitor for aluminum alloy in seawater.

1. Introduction

At present, aluminum alloys are widely used in shipbuilding because they are strong and light in weight. These enable the aluminum ships to move faster, load more cargoes and travel further while maintaining stability and fuel consumed efficiency. However, aluminum is typically a reactive metal in the galvanic series. When it is in contact with other metals, it will act as the anode and may begin to corrode [1].

Although aluminum alloys have high corrosion resistance but it will react very actively once exposed to the saline environment especially seawater which contained mostly sodium chloride that destabilizes the protective oxide film, leading the localized attack, or pitting. Hence, alternative way is to be discovered in order to reduce the corrosion especially on aluminum alloy. Nowadays, anti-corrosion paint is introduced for the ship to reduce the rate of corrosion. However, most of the anti-corrosion paints are found to consist of high amount of chromate which is extremely toxic and harmful to environment [2].

The increasing concern about the toxicity of corrosion inhibitors in the industries has replaced with more environmental friendly organic inhibitors. Thus, plant extracts are becoming important as an environmentally acceptable, readily available and renewable source for wide range of inhibitors. They are the rich sources of ingredients which have very high inhibition efficiency. The corrosion inhibition activity in many plant extracts could be due to the presence of heterocyclic constituents like alkaloid

Published under licence by IOP Publishing Ltd

IOP Conf. Series: Materials Science and Engineering **36** (2012) 012043 doi:10.1088/1757-899X/36/1/012043

and flavonoid. The presence of tannin, cellulose and polycyclic compound could enhance the film formation over the metal surface, thus aiding corrosion [3].

The previous study had found that henna has four main constituents which are lawsone, gallic acid, a-D-Glucose and tannic acid. These constituents have great inhibition ability which could combined with the metal cations and adsorbed on the metal surface as a formation of insoluble complex compounds to protect the metal from corrosion attack [4].

Therefore, the aim of this research was to investigate the feasibility use of henna (*Lawsonia inermis*) as natural corrosion inhibitor for aluminum alloy in seawater. The components of henna were determined by using Fourier Transfrom Infrared Spectroscopy while its corrosion inhibition ability was tested through immersion method in seawater and the corrosion inhibition efficiency was analyzed by electrochemical measurement.

2. Experiment Procedures and Method

2.1 Sample Preparation

Aluminum alloy (5083) were purchased in standard size of 25mm x 25mm x 3mm as test coupons. It was alloyed by 0.40% Si, 0.40% Fe, 0.10% Cu, 0.40-1.0% Mn, 4.0-4.9% Mg, 0.05-0.25% Cr, 0.25% Zn, 0.15% Ti, and the remainder are aluminum. The coupons were cleaned and polished by sandpaper following grit of 400, 600, 800, 1000 and 1200 grit. Then, they were rinsed and cleaned with distilled water, degreased with acetone and dried in the desiccators before the experiment started. The commercial henna powder made in India was prepared and the components were analyzed through Fourier Transfrom Infrared Spectroscopy (FTIR). Then it was dissolved in seawater with various concentration of 100ppm, 300ppm, and 500ppm.

2.2 Weight Loss Measurement

Before the coupons were cleaned with acetone, the coupons were weighed for the original weight by using Sartorius Cole-Polmer analytical balance (readability: 0.0001g) and hung in the aquarium containing seawater for 60 days. One sample was taken out from the aquarium in 10 days interval for 6 weeks. Before weighing, the sample was cleaned with distilled water and dried. Then the sample was immersed in nitric acid (HNO_3) to remove the corrosion products. Finally, the sample was washed with distilled water, dried and weighed in order to obtain the final weight.

2.3 Fourier Transform Infrared Spectroscopy (FTIR)

The commercial henna powder was placed and exposed under range of infrared ray beams. The transmittance and reflectance of the infrared rays at different frequencies was translated into an IR absorption plot consisting of reverse peaks. The spectral pattern was analyzed and matched according to IR absorption table to identify the functional group contained in the henna powder.

2.4 Immersion Test

Aquariums were prepared with specific amount seawater where one as control medium while the rest of the solutions were dissolved with commercial henna powder in different concentration such as 100ppm, 300ppm and 500ppm. It was carried out for duration of 60 days in enclosed condition without any disturbance. The coupons were taken out for each 10 days for further weight loss measurement, polarization test, corrosion rate analysis and inhibition efficiency calculation.

2.5 Electrochemical Impedance Spectroscopy (EIS)

For EIS measurements, the test was conducted by using AC signal of impedance measurements by the use of Autolab PGSTAT302N and run at the corrosion potential. All the potentials referred were relative to SCE. The impedance measurements were conducted over a frequency range of 1 MHz down to 10 mHz. The results were analyzed using the fit program FRA. The inhibition efficiency, *IE* (%) was calculated by using the Equation 1 where R_2 and R_1 are the polarization resistance in the presence and absence of inhibitors [4]:

$$IE(\%) = \frac{(R_2 - R_1)}{R_2} \times 100 \qquad (1)$$

3. Result and Discussion

3.1 Fourier Transform Infrared Spectroscopy (FTIR)

Figure 1 shows the functional group of henna. The presence of hydroxyl group appeared at 3299cm^{-1}. There are two α,β-unsaturated carbonyl bands at 1735.2 cm^{-1} and 1625.1cm^{-1} (C=O chelated with an α-hydroxyl) signify the presence of two carbonyl groups conjugated with a double bond (α,β - unsaturated carbonyls) and thus endorsing the possibility of having a 1,4-naphthoquinone [5-6]. It was found that henna consists of a major component which is Lawsone. This compound has molecular formula of $C_{10}H_6O_3$ with molecular mass of 174.153. The phenol group of lawsone would donate electron to the metal to achieve its noble state of orbit, while the metal would receive the electron to become more stable. This indirectly retard further redox reaction and could resist metal from corrosion attack [7].

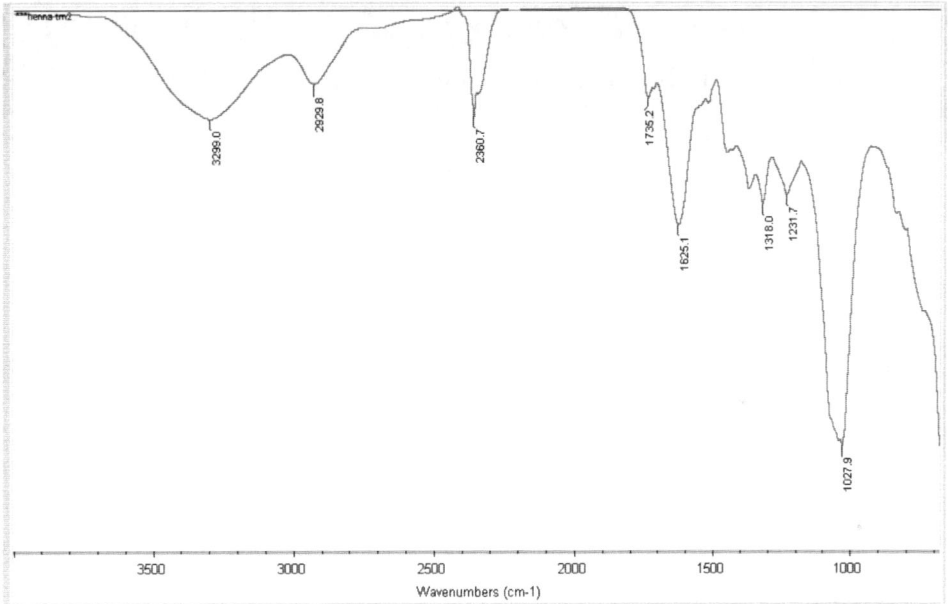

Figure 1. Functional group in commercial henna powder analyzed by FTIR

Henna constituents have antioxidant characteristic especially by the naphthoquinones. The possible mechanism for chemisorptions can be attributed to the complex formation between lawsone and metal cations. The main cathodic reaction in the surface is hydrogen evolution. In this case, oxygen reduction does not play a significant role in total cathodic reaction. Therefore, the oxygen scavenging characteristic of henna is not the main responsible mechanism for the observed inhibition of henna extract yet is the chemisorption of inhibitor molecules on the surface. So, it is concluded that the chemisorption of inhibitor molecules on steel surface is the main inhibition mechanism while the inhibitive action is slightly enhanced via oxygen scavenging [7].

3.2 Electrochemical Impedance Spectroscopy (EIS)

EIS test was conducted to obtain interfacial information or working electrodes in electrolyte [8]. The Nyquist plot for AA5083 immersed in seawater with henna as inhibitor is presented in Figure 2. As seen from the figure, the plot of impedance shows single capacitive semicircle which indicates the process was controlled by charge transfer and could be assigned to the oxide layer of AA5083 [9]. It was found that the addition of the inhibitor could increase the corrosion resistance of the coating formed on metal surface. Upon incresasing of inhibitor concentration, the value of polarization resistace was also increased indicating that there are gradual formation of passive film on the sample surface.

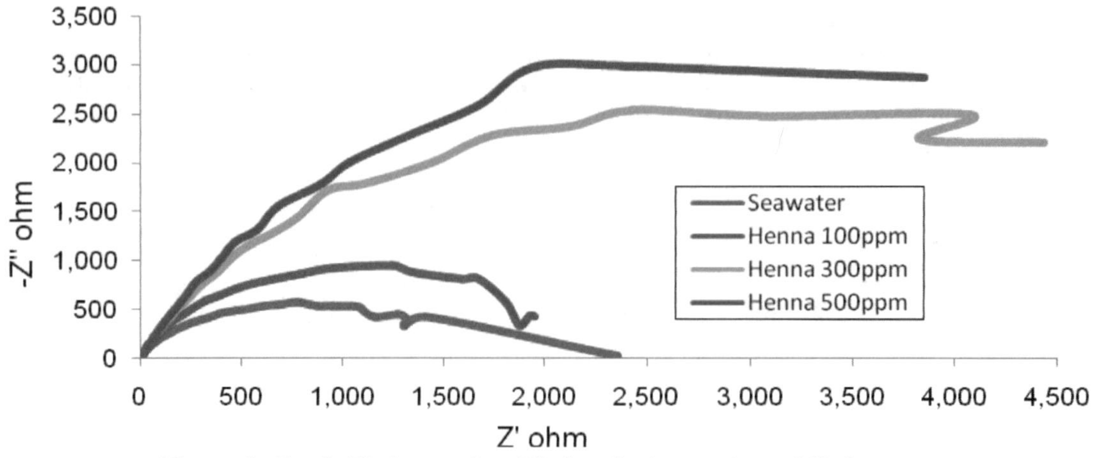

Figure 2. Graph Z' ohm against Z" ohm for immersion of 30 days

The semi-circle curves represented that the henna had absorbed on the aluminum surface and produced a coating to separate the metal from the seawater. In this case, there were no direct exposure of the metal to the saline environment. On the other hand, the aluminum alloy without inhibitor had indicated that the aluminum alloy had reacted with oxygen to form aluminum oxide, Al_2O_3 and these become the film to protect the metal from further redox reaction or corrosion attack. The diameter of semicircle represented that the corrosion resistance of aluminum oxide layer formed was not good enough compared to inhibitor coating by henna [7]. AC impedance results showed that the decrease in C_{dl} value, as shown in Table 1, was a result of the formation of a protective layer by the adsorption of inhibitor molecules on the AA5083 surface. The adsorption is influenced from the molecule's chemical structure, the solution's chemical composition, the nature of the metal surface, the temperature and the electrochemical potential at the metal/solution interface. The changes in C_{dl} phenomenon is influenced from the molecule's structure, the solution's chemical composition, the nature of the metal surface, the temperature and the electrochemical potential at the metal/solution interface [10].

Table 1. Impedance parameter for AA5083 immersed in seawater

Concentration	R_s (Ω)	R_p (Ω)	C_{dl} (F)
0 ppm	4.5374	1233.99	2.50E-05
100 ppm	2.3220	2522.04	4.46E-05
300 ppm	2.3025	6762.92	3.50E-05
500 ppm	2.1887	9883.31	5.07E-05

Table 2. Inhibition efficiency gained from impedance parameter

Inhibitor	Inhibition Efficiency (%)
0 ppm	-
100 ppm	51
300 ppm	82
500 ppm	88

3.3 Weight Loss Measurement

The weight loss measurement is a classical way to determine the corrosion rate of aluminium alloy. The result for the weight loss measurement was summarized in Figure 3. The graph shown in Figure 3 reveals that there was an increasing in weight loss with respect to immersion time. The weight loss of coupons immersed in seawater in the presence of henna was lower compared in absence of henna which indicates that henna has retarded the corrosion process of AA5083. As the concentration increases, the weight loss decreases due to rises in surface area covered by adsorbed molecules of henna during the increment of concentration [11]. The existences of various compounds like oxygen (polyphenols) as well as nitrogen and sulphur (glucosinate), it is hard to determine which components inhibit the metal from corrosion [12]. It may be one or all of them involve in synergism action [13]. There was considerable reduction in weight loss and hence reduces the corrosion rate with the introduction of inhibitor in corrosive media [14]. Higher concentration of henna was effectively inhibiting the corrosion compares to lower concentration of henna.

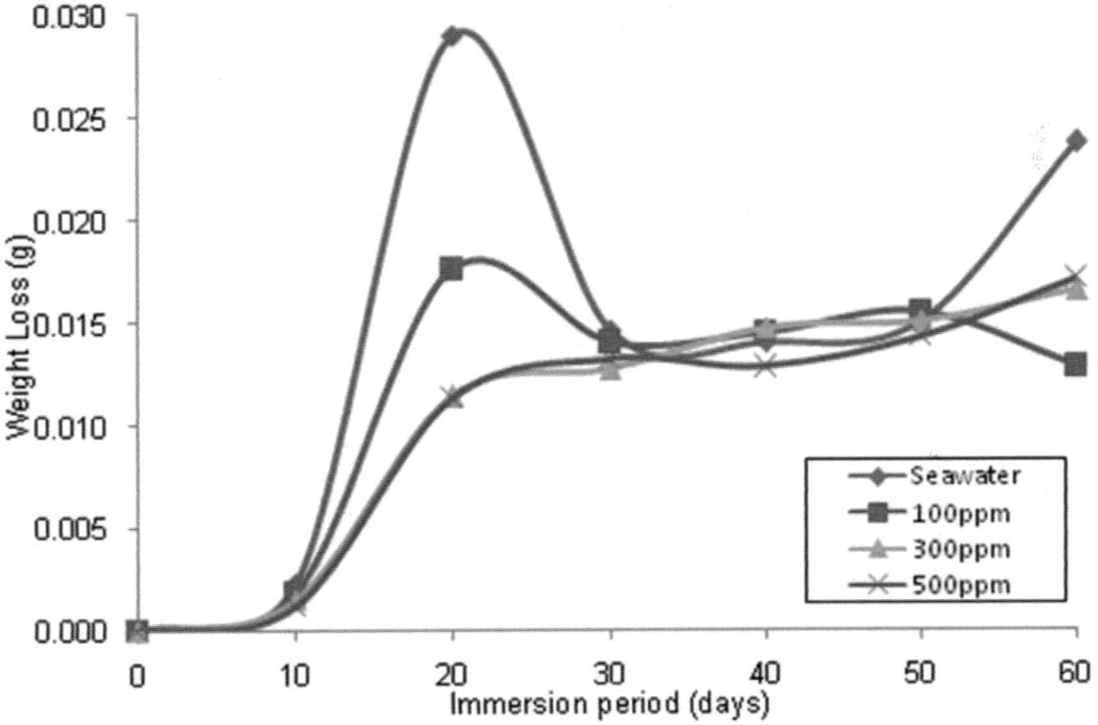

Figure 3. Graph weight loss (g) against immersion period (days)

3.4 Adsorption Isotherm

Graph inhibitor concentration (*C*) against C/θ plotted in Figure 4 was illustrated by mere straight lines which indicated that the adsorption of henna on aluminum alloy 5083 follows the Langmuir adsorption isotherm. The observed inhibitive action of henna could be due to the adsorption of its molecules on the aluminum alloy surface which produce a layer or a barrier for the charge and mass transfer between the metal and the surrounding. As the concentration increased the fraction of metal surface covered by the adsorbed molecules (θ) increased and lead to higher inhibition efficiency [7].

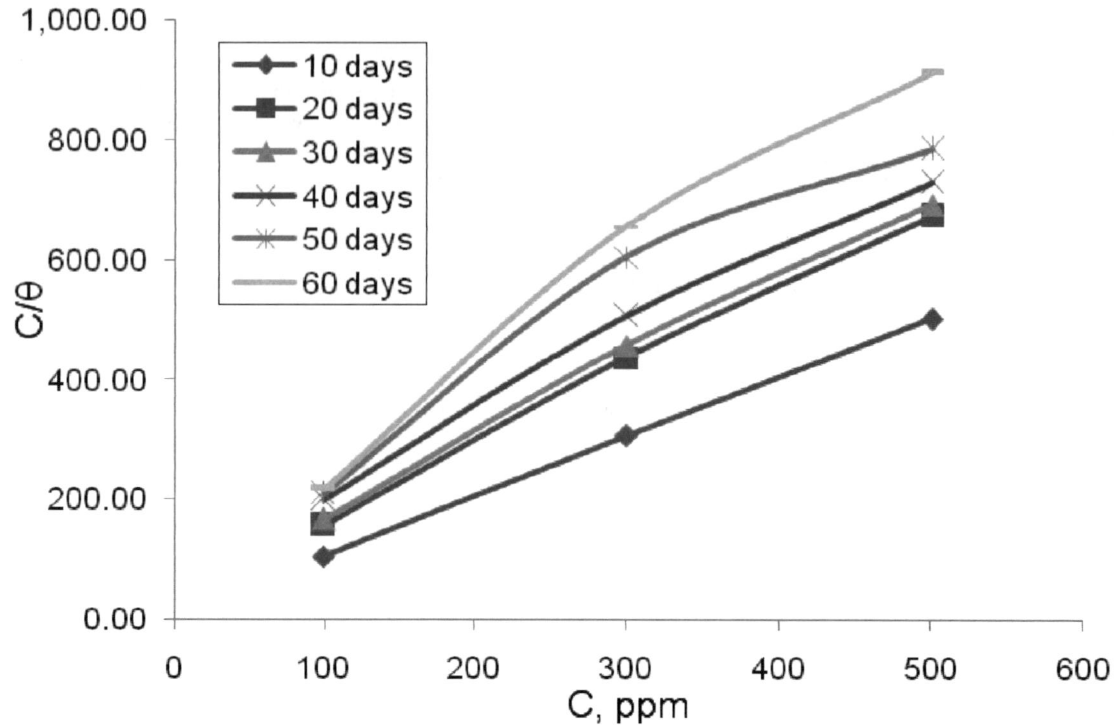

Figure 4. Adsorption isotherm of various inhibitor concentration on aluminum alloy surface for immersion duration of 60 days

4. Conclusion

Henna (*Lawsonia inermis*) contain major constituents of lawsone which contributed to the chemisorptions or adsorption process by forming an isolation layers on the aluminum alloy surface which obey the Langmuir adsorption isotherm. However, the protection layer attached on metal was not permenant when exposed to saline environment and precipitation occurred as the time increased. The inhibition efficiency is proportional to its concentration dissolved. It reached the highest 88% at concentration of 500ppm. The inhibition efficiency gained from EIS measurement is still in acceptable aggrement. This research found that henna is an excellent natural inhibitor for aluminum alloy in seawater.

References

[1] Perryman J 2007 Corrosion Resistance of Aluminum. Waterfront Solution. *Crane Materials International.*

[2] El-Meligi A A 2010 Corrosion preventive strategies as a crucial need for decreasing environmental pollution and saving economics. *Recent Patents on Corrosion Science* **2** 22-33

[3] Raja P B and Sethuraman M G 2008 Natural products as corrosion inhibitor for metals in corrosive media. *Materials Letters* **62** 113–116

[4] Ostovari A, Hoseinieh S M, Peikari M, Shadizadeh S R and Hashemi S J 2009 Corrosion inhibition of mild steel in 1 M HCl solution by henna extract: A comparative study of the inhibition by henna and its constituents (Lawsone, Gallic acid, a-D-Glucose and Tannic acid) *Corrosion Science* **51** 1935–19

[5] Harborne J B 1984 Phytochemical Methods;A Guide to Modern Techniques of Plant Analysis 2nd ed Chapman and Hall London 22-2

[6] Botros R M, Farid A B, Galal T M and Mohamed M A 2004 Antioxidant and immunomodulatory constituents of henna leaves *Z. Naturforsch* **59c** 468-476

[7] El-Etre A Y, Abdallah M and El-Tantawy Z E 2005 Corrosion inhibition of some metals using lawsonia extract *Corrosion Science* **47** 385–395

[8] Shi H, Han E and Liu F 2011 Corrosion protection of aluminium alloy 2024-T3 in 0.05 NaCl by cerium cinnamate *Corrosion Science* **53** 2374-2384
[9] Rosliza R, Wan Nik W B, Izman S and Prawoto Y 2010 Anti corrosive of natural honey on Al-Mg-Si alloy in seawater *Current Applied Physics* **10** 923-929
[10] Muzaffer ¨ Ozcana, Ramazan Solmazb G, Ulfeza Kardas and ˙Ilyas Dehrib 2008 Adsorption properties of barbiturates as green corrosion inhibitors on mild steel in phosphoric acid *Colloids and Surfaces A: Physicochem. Eng. Aspects* **325** 57–63
[11] Hamdy A, Farag A B, EL-Bassoussi A A, Salah B A and Ibrahim O M 2010 Electrochemical behavior of brass alloy in different saline media: effect of benzotriazole *World Appl. Sci. J* **8(5)** 565-571
[12] Radojcˇic´ I, Berkovic´ K, Kovacˇ S and Vorkapic´-Furacˇ J 2009 Natural henna and black radish juice as tin corrosion inhibitors *Corrosion Science* **50** 1498–1504
[13] El-Etre A Y and Abdallah M 2000 Natural henna as corrosion inhibitor for metals and alloys. II. C-steel in high saline water *Corrosion Science* **42** 731-738
[14] Rosliza R, Wan Nik W B, Senin H B 2008 The effect of inhibitor on the corrosion of aluminum alloys in acidic solutions *Materials Chemistry and Physics* **107** 281–288

Optimization of control parameters for SR in EDM injection flushing type on stainless steel 304 workpiece

M.S. Reza*, A. R. Yusoff†, M.A. Shaharun†

Faculty of Mechanical Engineering, Universiti Malaysia Pahang, 26600 Pekan, Pahang, Malaysia

*Corresponding author

Abstract. The operating control parameters of injection flushing type of electrical discharge machining process on stainless steel 304 workpiece with copper tools are being optimized according to its individual machining characteristic i.e. surface roughness (SR). Higher SR during EDM machining process results for poor surface integrity of the workpiece. Hence, the quality characteristic for SR is set to lower-the-better to achieve the optimum surface integrity. Taguchi method has been used for the construction, layout and analysis of the experiment for each of the machining characteristic for the SR. The use of Taguchi method in the experiment saves a lot of time and cost of machining the experiment samples. Therefore, an L18 Orthogonal array which was the fundamental component in the statistical design of experiments has been used to plan the experiments and Analysis of Variance (ANOVA) is used to determine the optimum machining parameters for this machining characteristic. The control parameters selected for this optimization experiments are polarity, pulse on duration, discharge current, discharge voltage, machining depth, machining diameter and dielectric liquid pressure. The result had shown that the lower the machining diameter, the lower will be the SR.

Keyword—ANOVA, EDM, Injection Flushing, L18 Orthogonal Array, SR, Stainless Steel 304

1. Introduction

Electrical discharge machining, commonly known as EDM is a non-traditional machining technique that is used to remove metal through the erosive action of an electrical discharge of short duration and high current density between the tool and the workpiece in an immersed dielectric fluid. This dielectric fluid is used to continuously flush the eroded particles (mainly in the form of hollow spheres) from the machining gap besides performs as a coolant and reduces the extremely high temperatures in the arc gap. Improper flushing such as wrong selection of flushing method during EDM machining, insufficient flushing pressure during machining would contribute to erratic cutting, poor machining rate [1] and occurrence of short-circuits [2]. Hence, a proper flushing method is critical to achieve good machining condition such as low surface roughness (SR) of the machined workpiece.

Published under licence by IOP Publishing Ltd

There a many types of flushing method exist in EDM machining and one of it is side flushing. Side flushing is the least effective or inefficient method of removing suspended particles from the spark gap. This is because the pressurized dielectric would not effectively being flushed away especially when the tool electrode is cycled down in negative Z-axis; it only pushes out the contaminated oil without the pressurized side flushing goes between the machining gap. Hence, injection flushing is the proper method to be used to help the dielectric being flushed away effectively [3]. Injection flushing is the method where dielectric fluid is forced down through a flushing hole in the electrode. Great advantage of injection flushing is the operator can visually see the amount of oil that is being used for flushing and with pressure gauge this method is ease and simple to used [2, 4].

In electrical discharge machining (EDM), it is important to select machining parameters for achieving optimal machining performance with the proper flushing method used. Usually, the desired machining parameters are determined based on experience or on handbook values. However, this does not ensure that the selected machining parameters may results in optimal or near optimal machining performance such as surface roughness (SR) for that particular electrical discharge machine and environment. Minimizing the surface roughness to obtained a good surface finish of the workpiece with optimizing the input parameters will affect the machining performance [5]. The Taguchi Method using L18 orthogonal array is used in carrying out experiments for solving the optimization process. [6]. Before the use of Taguchi method in the Design of Experiment, experiments were selected with a full combination of factors that the total number of factors is so large. The effect of this, the number of experiments increases rapidly. Hence, by using Taguchi Method fewer experiments is done compared to the use of full combinations of factors [7].

2. Experimental Setup

Design of Experiments Using L18 Orthogonal Array

By using Taguchi method, the control factors table is used as a reference guide to start and execute the experiment. This table will be used with L18 Orthogonal Array table where different levels of factors will be selected for each experiment. L18 Orthogonal Array is used because among of all the control factors, one of it has one-2 level control factors and others has six-3-level of control factors [8]. Table 1 shows the control factors table.

Table 1: Control Factors

Factor	**Description**	**L1**	**L 2**	**L 3**	**Units**
A	Polarity	W/P(+) Tool (-)	W/P (-) Tool (+)	-	Positive (+) Negative (-)
B	Pulse-on time	4	6	8	Microsecond
C	Discharge Current	57	66	75	Ampere
D	Discharge Voltage	60	90	120	Voltage
E	Machining Depth	1.5	2.0	2.5	mm
F	Machining Diameter	9.5	11.0	12.5	mm
G	Dielectric Liquid Pressure	1.0	1.5	2.0	bar

Description of Factors

i) Polarity – refers to the direction of current flow in relation to the electrode. The polarity can be either positive or negative. Usually positive terminal of workpiece determines positive polarity and negative terminal of workpiece determines negative polarity.
ii) Pulse on duration – The amount of time current runs into the gap before it is turned off.
iii) Discharge current – Electrical discharge known as plasma sustained by a direct current (DC) through an ionized medium.
iv) Discharge voltage – Discharge voltage is referred as electrical pressure, force or electromotive force (EMF). It is the pressure that makes amperes flow in the form of a spark. Besides that, it is normally known as the potential difference between the electrode and the workpiece.
v) Machining Depth – The hole depth of cut in EDM machining
vi) Machining Diameter – The hole diameter of cut in EDM machining
vii) Dielectric Liquid Pressure – The pressurized dielectric liquid flow through the electrode.

Workpiece Material

Stainless Steel 304 was used as the workpiece in the experiment. It is being selected as it is often used in producing springs, nuts, bolts and screws. Table 2 shows the physical properties of stainless steel 304.

Table 2: Physical Properties of Stainless Steel 304

Physical properties	**Stainless Steel 304**
Density [g/cm^3]	8.03
Electrical conductivity [x 10^5 / Ω cm]	11.6
Thermal conductivity [W/(cm K)]	0.162
Melting point [K]	1644
Boiling point [K]	1672

Tool Material and Shape

A pipe copper tool is being selected for the machining because it is a highly conductive tool, low cost, low wear ratio, good machinability and finishing. The initial cubic cylindrical copper tool is being drilled to produce a pipe copper tool. It is mainly for the purpose of the dielectric could be flow through and being injected to the workpiece during the machining and it is called injection flushing machining. Table 3 shows the copper pipe tool electrode properties.

Table 3: Properties of copper pipe tool electrode

Properties	**Copper**
Density [g/cm^3]	8.96
Electrical conductivity [x 10^5 / Ω cm]	5.88
Thermal conductivity [W/(cm K)]	3.98
Melting point [K]	1356

Machining Characteristic Calculation

A profilometer measured the machined surface roughness. The centre-line average surface roughness R_a is measured to quantitatively evaluate how EDM parameters affect the surface finish. 2 measurements were taken per workpiece after experiment. Figure 1 shows the respective machine used for this purpose.

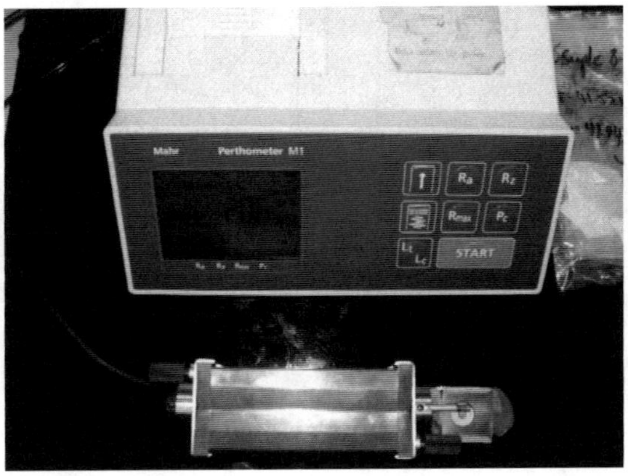

Figure 1: Surface Roughness Measuring Machine

Optimum Condition and Verification

The optimum condition for surface roughness is determined using response graph where the quality characteristics should be known to select which point exhibits the relation with the quality characteristics. In this case, for SR quality characteristic is lower-the-better, where minimum points are chosen from the response graph.
Verification of the optimum condition is done where confirmation result for machining characteristics (after determine the optimum condition) is not less than 90% difference with the predicted values to proof that the optimum condition obtained is acceptable.

Significant Factor Analysis using Analysis of Variance (ANOVA)

The significant factors that contributed to the optimized machining characteristic (SR) is determined by using Analysis of Variance (ANOVA). ANOVA is a method of calculating the statistical significant of factors in the experimentation and it represents the relationship between the machining parameters with overall process performance.

Results and Discussions

Figure 2 shows the response graph for SR machining characteristic.

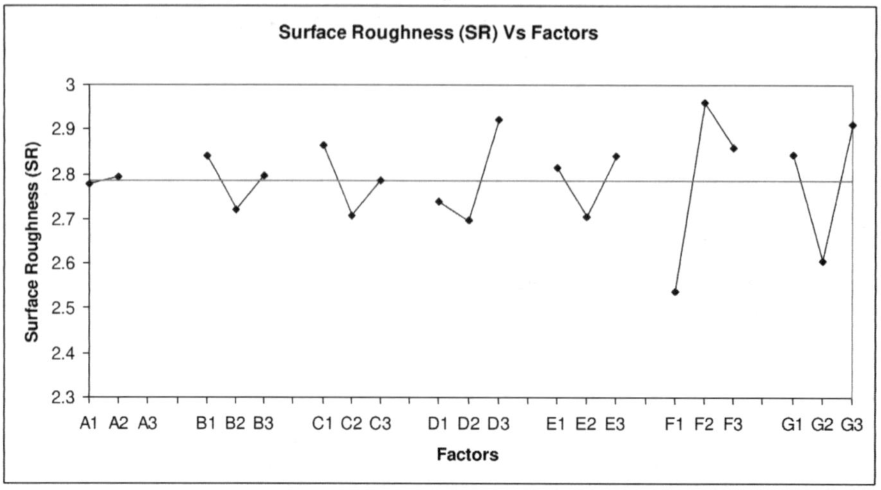

Figure 2: Response Graph for SR

According to lower-the-better quality characteristic for SR, the optimum condition for each factor indicated is A1 (positive polarity), B2 (6μs), C2 (66A), D2 (90V), E2 (2.0 mm), F1 (9.5 mm), G2 (1.5 bar). Consequently, based from confirmation experiment, only 8.08% difference from predicted value, PV that still below the maximum allowance percentage for quality which is 10%.

The significant factors from the above graph were calculated using ANOVA equation and the result is shown on the following table 4.

Table 4: ANOVA for SR

Factor	Description	F-ratio	% contribution ratio, ρ
A	Polarity, P	0.04650	1.125
B	Pulse-on-duration, τ_{on}	0.88890	0.261
C	Discharge Current, I_d	1.4423	1.047
D	Discharge Voltage, V_d	3.4219	5.718
E	Machining Depth, l_i	1.2301	0.543
F	Machining Diameter, d	11.718**	25.30
G	Dielectric Liquid Pressure, P_l	6.131*	12.11

The highest F-ratio contributes to be the most significant factor which is machining diameter, d followed with 2nd highest F-ratio to be the significant factor which is dielectric liquid pressure, P_l.

IOP Conf. Series: Materials Science and Engineering **36** (2012) 012044 doi:10.1088/1757-899X/36/1/012044

SR Most Significant Factor – Machining Diameter

Machining diameter is the most significant factor that affects surface roughness for this research. From the review of the previous research papers, there was no direct research made due to effects of machining diameter on surface roughness. However, there was one paper reported by Lee H.T. [5], he reported that surface roughness is caused by surface cracks. In their reports said reducing the electrode diameter can prevent the formation of surface cracks as well as surface roughness. Electrical discharge machining uses erosion process by means of recurring electrical discharge to remove material. Effect to this process, it will produce heat energy on a unit surface area affected that the temperature on the surface and the body is different. Surface cracks may happen due to the internal stress occur between this two body differences in temperature phenomenon. According to the basic heat conduction formula which local heat flux, q is equal to the product of thermal conductivity, k, and the negative local temperature gradient, ΔT. The heat flux density is the amount of energy that flows through a unit area per unit time. By relating with this formula, the surface crack may be eliminated or reduced if the workpiece heat flux density is increased by means of increasing the temperature gradient, ΔT with fixed unit area for energy to flow. Hence, this is why the electrical discharge should be set higher than the minimum setting (referred to the response graph for current) to increase the workpiece heat flux. This agrees the paper reported by Lee H. T. [5].

SR Significant Factor – Dielectric Liquid Pressure

Dielectric liquid pressure is the significant factor that affects surface roughness for this research. As agreed by Wang C.C. [9], surface roughness will decrease slightly if flushing pressure is not applied. Flushing is the removal of the metal particles generated in the spark gap. If there is no flushing pressure to flow through the machining gap, then it would increase dielectric liquid temperature to a very high temperature. This phenomenon will affect more crack formation especially on the machined surface that soon will increase the surface roughness. As previously explained, surface crack formation is related to the surface roughness and it was being affected by the heat energy supplied during EDM machining. Increasing of temperature gradient, ΔT may contributes to the decrease of machined workpiece surface roughness where dielectric liquid may decrease the workpiece temperature during machining along with the initial high temperature obtained from the electrical discharge phenomenon . Hence, increasing the heat flux, q may improve the surface roughness and this was proved by the confirmation experiment made.

3. Conclusions

This research has investigated the optimization of control parameters in injection flushing type of electrical discharge machining on stainless workpiece with copper tools. The main conclusions of this research are as follows:-

1. Machining performance in the EDM process can be improved effectively by using optimum factors that had been determined for SR machining characteristic and it has been proved by the confirmation experiment made.
2. Machining diameter and flushing pressure are most significant and significant factors that affect SR.
3. Optimum condition for SR are being set at positive polarity (workpiece positive and tool negative), 6 microseconds of pulse on duration, 66 Ampere of discharge current, 90 volts of discharge voltage, 2.0 mm of machining depth, 9.5 mm of machining diameter and 1.5 bar of dielectric liquid pressure.
4. For future research work, other single objective performance characteristic material removal rate (MRR), and electrode wear ratio (EWR) could be made using the same method.

References

[1] X.P. Li (2001). Study of the effect of machining parameters on the machining characteristics in electrical discharge machining of tungsten carbide, *J. of Materials Processing Technology* **115.** 344-358.

[2] Can Cogun (2006). An experimental investigation of tool wear in electric discharge machining, *Int J Adv Manuf Technol* **27** 488-500.

[3] Lin C.L. (2002). Optimisation of the EDM Process Based on the Orthogonal Array with Fuzzy Logic and Grey Relational Analysis Method", *Int J. Adv Manuf Technol* **19**. 271-277.

[4] Yan B.H. (2000). Feasibility study of rotary electrical discharge machining with ball burnishing for Al_2O_3/6061Al composite, *International Journal of Machine Tools & Manufacture* **40**. 1403-1421.

[5] Lee H.T. (2003). Relationship between EDM parameters and surface crack formation, *Journal of Materials Processing Technology* **142**. 676-683.

[6] Nicolo Belavendram (2005), Student Notes, Robust Design, KCEC 4301 Semester 1 *Department of Design and Manufacturing, Faculty of Engineering, University of Malaya.*

[7] Lee C.M. (2006). A study on the optimal cutting condition of a high speed feeding type laser cutting machine by using Taguchi method, *International Journal of Precision Engineering and Manufacturing* **Vol. 7.** No.1.

[8] Jiju Antony (2001). Teaching the Taguchi method to industrial engineers, *Work Study* **Volume 50 Number 4**. pp. 141-149, MCB University Press ISSN 0043-8022.

[9] Wang C.C. (1999). The machining characteristics of Al_2O_3/6061 Al composite using rotary electro-discharge machining with a tube electrode, *J. of Materials Processing Technology* **95**. 222-231.

AUTHOR INDEX

AUTHOR INDEX

Institute of Physics
Dirac House, Temple Back
Bristol BS1 6BE UK

CURRAN ASSOCIATES INC
proceedings
.com